飞行计划

谢春生　王玉　刘薇　闫凤良　王晓敏　编著

大连海事大学出版社

DALIAN MARITIME UNIVERSITY PRESS

ⓒ谢春生　王玉　刘薇　闫凤良　王晓敏　**2024**

图书在版编目(CIP)数据

飞行计划／谢春生等编著. — 大连：大连海事大
学出版社，2024. 12. — ISBN 978-7-5632-4595-6

Ⅰ. V323.1

中国国家版本馆 CIP 数据核字第 2024V3Z823 号

大连海事大学出版社出版

地址:大连市黄浦路523号　邮编:116026　电话:0411-84729665(营销部)　84729480(总编室)
http://press.dlmu.edu.cn　E-mail:dmupress@ dlmu.edu.cn

大连天骄彩色印刷有限公司印装　　　　大连海事大学出版社发行

2024 年 12 月第 1 版　　　　　　　　2024 年 12 月第 1 次印刷
幅面尺寸:184 mm×260 mm　　　　　　　　印张:22
字数:556 千　　　　　　　　　　　　　印数:1～2000 册

出版人:刘明凯

责任编辑:董洪英　　　　　　　　　　　责任校对:王　琴
封面设计:张爱妮　　　　　　　　　　　版式设计:张爱妮

ISBN 978-7-5632-4595-6　　　定价:62.00 元

前　言

现代民用飞机是高技术产物，是先进的运输工具，价格高昂，燃油消耗量很大，燃油成本在使用成本中占有相当大的比重。为了安全有效地使用飞机，获得高的经济效益，应该提前为每个航班制作飞行计划，算出最大允许业载、应加的总油量、消耗的油量、飞行时间、起飞重量等数据。航空公司使用的系统运行控制中心(SOC)或飞行运行控制中心(FOC)的计算机飞行计划(CFP)，还可以给出预计到达各航路点的时间、剩余油量、真空速、航向等参数供飞行员参考。制作飞行计划的航务工程人员应该掌握制作飞行计划的原理和方法。飞行员是飞行计划的使用者，了解制作飞行计划的原理和方法也是有益的。

本书在编者以往教材的基础上，结合了编者多年的教学实践和飞行计划课题的研究，并采纳了专业课老师及航空公司行业专家提出的宝贵意见，参考了相关专业书籍、资料和飞机手册等，经多次修改而编写完成。本书是为将要从事民航航务工作的大学本科高年级学生编写的，读者需具备一定的空气动力学和飞机性能等相关的基础知识。

本书主要包括飞行计划制作、装载与配平以及供氧分析三部分。全书共分为六章，第1章是飞行计划基础，主要介绍了《大型飞机公共航空运输承运人运行合格审定规则》(CCAR-121-R8)对制作飞行计划时与天气、备降场和燃油有关的规定，以及飞行计划制作中常用的术语和相关数据的计算；第2章是常规飞行计划制作，主要介绍了常用的飞行计划制作方法，飞机使用手册中飞行计划用的图表，以及飞行计划中相关数据的计算；第3章是特殊飞行计划制作，主要介绍了各种类型的特殊飞行计划；第4章是延程运行，主要介绍了延程运行的历史发展和运行条件，以及延程运行飞行计划的制作方法；第5章是装载与配平，主要介绍了飞机的载重与平衡问题，包括飞机最大业务载重量的计算、飞机重心位置的确定、客货飞机装载与配平的计算方法和载重平衡图的填制等；第6章是供氧分析，主要介绍了民航规章对飞行过程中有关旅客供氧的规定，飞机上的氧气系统以及座舱失压后的供氧越障分析。

本书第1章和附录1由于晓敏负责编写，第2章、第6章和附录2由王玉负责编写，第3章和附录4、附录5由刘薇和谢春生负责编写，第4章、第5章和附录3由闫凤良负责编写。本书在编写过程中得到了中国民航大学空中交通管理学院的大力支持，在此深表谢意。

由于编者水平有限，书中疏漏和错误在所难免，敬请读者批评指正。

编　者

2023 年 11 月于中国民航大学

目 录

第1章

飞行计划基础

飞行计划最基本的内容是针对每一航班算出最大允许业载、轮挡油量、备份燃油、起飞总油量、轮挡时间等各项数据,详细的飞行计划还应算出到达各航路点的时间,所消耗的油量(或剩余油量),在各航路点的速度、航向等。为了安全有效地使用飞机,得到更高的经济效益,应该提前制作飞行计划,这样可以避免多加油和因此造成的减载。多加的油本身也要消耗燃油,例如:对于 1500~2000 km 航程的航线,每多加 1 t 油将多消耗燃油 7%~9%,即 70~90 kg,航程越长或顶风越大,多加的这部分油消耗的燃油越多,这会造成资源的浪费,也降低了公司的经济效益。多加油还会使起飞重量增加,虽然只要实际起飞重量不超过最大允许起飞重量,就可以保证起飞安全,但重量增加将使安全余度减小,此外,还将使飞机的高度能力、机动能力下降,使飞机承受过载的能力下降、调速范围减小。因此,应该避免多加油。

在空中交通管制允许的情况下,制作飞行计划时应选择合适的巡航高度,对较长航线可以采用阶梯巡航来充分发挥飞机的性能,减少燃油消耗,提高经济效益。在各地油价不同时,制作飞行计划时可以考虑如何利用这个差价来节省燃油费用。在知道成本指数的情况下,可以制作最小成本飞行计划,有效地降低航班成本。

现代飞机上有飞行管理系统,可以对飞行性能进行优化,以保证飞行安全,提高经济效益,但仍然有必要制作飞行计划,因为起飞前飞行员要向飞行管理系统输入飞机重量(或无油重量)、总油量和备份燃油(总油量可不输入,油箱中的油量可以自动测出,但必须知道应加油量),飞行管理系统是根据这些输入的数据进行优化管理的,而这些数据是通过制作飞行计划得到的。如果不制作飞行计划,油加多了,也会多耗油量,加多了油也会减少业载。

在制作飞行计划之前应先做机场分析,即算出起飞机场的最大允许起飞重量、目的地机场和备降机场的最大允许着陆重量,还要知道航路(航线)的情况(比如航程、各航路点的位置、各航段的距离和航向等)和气象信息(航路上的风向、风速及气温等)。制作飞行计划时计算燃油量的方法应当符合民航规章中有关燃油的规定及航空公司的燃油政策。

1.1 飞行剖面

飞行剖面是为完成某一特定飞行任务而绘制的飞机航迹图形,能形象地表达飞行任务。飞行剖面按航迹所在平面分为垂直飞行剖面和水平飞行剖面。经常使用的是垂直飞行剖面。制作飞行计划所使用的飞行剖面主要是垂直飞行剖面(后文简称飞行剖面),以简明示意图的形式说明计算步骤,方便计算,如图 1-1-1 和图 1-1-2 所示。飞行剖面以起飞机场为起点,由滑行(包括开车及滑出)、起飞、爬升、巡航、下降、机动进近、着陆和复飞等若干个飞行阶段组成。

1.1.1 常规航线的飞行剖面(以涡轮动力飞机为例)

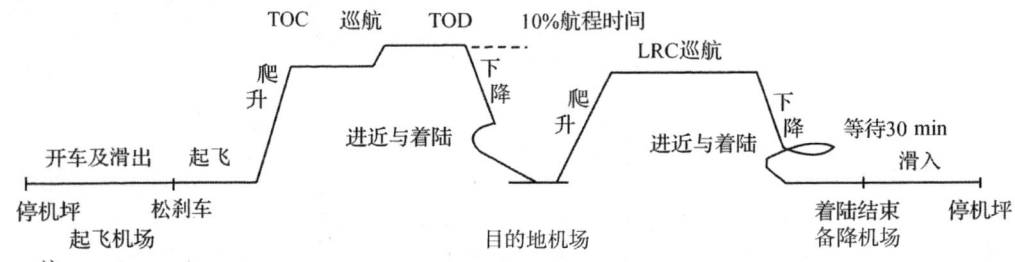

注:
不可预期燃油量取①和②两者中较大值:
①为计划航程燃油10%的燃油量;
②为以等待速度在目的地机场上空450 m(1500 ft)高度上在标准条件下飞行15 min所需的燃油量。

图 1-1-1　常规航线的飞行剖面(涡轮动力飞机)

涡桨动力飞机和非涡轮动力飞机常规航线的飞行剖面与上述基本相同,只是对等待时间的规定有所不同,制作飞行计划的方法是类似的。

1.1.2 特殊航线飞行剖面

(1)在目的地机场可以加油的飞行剖面

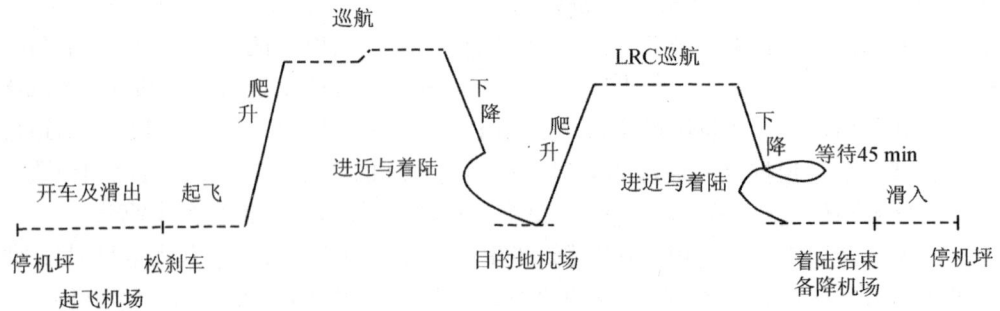

图 1-1-2　在目的地机场可以加油的飞行剖面

（2）在目的地机场不能加油的飞行剖面

假设从某机场（DEPART）起飞飞往不能加油的机场 DEST1，它的备降机场为 ALT1，由 DEST1 起飞执行飞往 DEST2 的航班，它的备降机场为 ALT2。为计算从 DEPART 起飞的总油量等数据，应取如图 1-1-3 所示的飞行剖面。

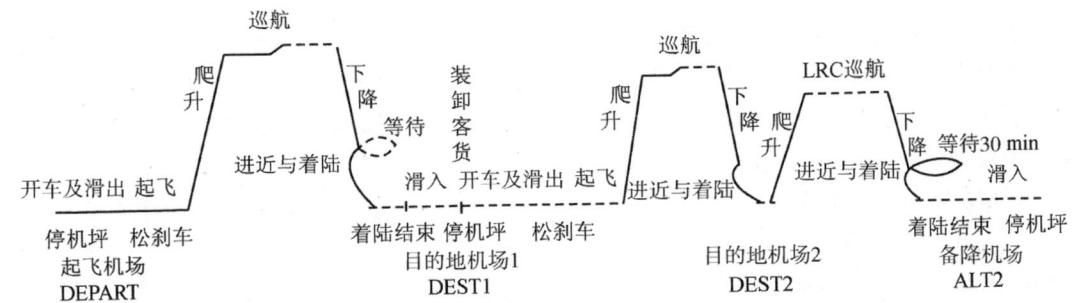

图 1-1-3　在目的地机场 DEST1 不能加油的飞行剖面

注：在目的地机场 DEST1 不能加油的飞行剖面同图 1-1-3，对于由 DEPART 到 DEST1 的航线，如公司有特殊要求，可额外增加一定时间的等待燃油。由 DEST1 到 DEST2 的航线，在 ALT2 上空的等待时间为 30 min，并加上由 DEST1 到 DEST2 航程的不可预期燃油。

1.2　备降机场和燃油量的相关规定

《大型飞机公共航空运输承运人运行合格审定规则》（CCAR-121）于 1999 年 5 月 5 日公布，并于 2024 年 4 月 23 日第八次修订（CCAR-121-R8）。在 CCAR-121 的 U 章"签派和飞行放行"的第 121.637、121.639、121.641、121.642、121.643、121.657、121.659、121.663 等条款中做了有关备降机场和燃油量的规定；《小型航空器商业运输运营人运行合格审定规则》（CCAR-135）于 2005 年 9 月 20 日发布，并于 2018 年 11 月 6 日进行了第二次修订（CCAR-135-R2），后于 2022 年 1 月 4 日发布了《小型商业运输和空中游览运营人运行合格审定规则》（CCAR-135-R3），在第 135.133、135.135、135.137、135.145、135.183、135.185、135.187、135.189、135.191 条也对有关备降机场、燃油量和运行的天气标准进行了规定。

1.2.1　CCAR-121 关于备降机场的规定

第 121.637 条　起飞备降机场

（a）如果起飞机场的气象条件低于合格证持有人运行规范中为该机场规定的着陆最低标准，在签派或者放行飞机前应当按照下述规定选择起飞备降机场：

（1）对于双发动机飞机，备降机场与起飞机场的距离不大于飞机使用一发失效的巡航速度在静风条件下飞行 1 小时的距离；

（2）对于装有三台或者三台以上发动机的飞机，备降机场与起飞机场的距离不大于飞机使用一发失效时的巡航速度在静风条件下飞行 2 小时的距离。

（b）对于本条（a）款，备降机场的天气条件应当满足本规则第 121.643 条的要求。

(c)在签派或者放行飞机前,签派或者飞行放行单中应当列出每个必需的起飞备降机场。

第121.639条 仪表飞行规则国内定期载客运行的目的地备降机场

(a)按照仪表飞行规则签派飞机飞行前,应当在签派单上至少为每个目的地机场列出一个备降机场。当目的地机场和第一备降机场的天气条件预报都处于边缘状态时,应当再指定至少一个备降机场且在签派单上列出。

(b)如果目的地机场符合下列条件并且合格证持有人在飞机与运行控制中心之间建立了独立可靠的通信系统进行全程监控,则可以不选择目的地备降机场:

(1)如果天气实况报告、预报或者两者的组合表明,在飞机预计到达目的地机场时刻前后至少1小时的时间段内:

(i)机场云底高度至少在公布的最低的仪表进近最低标准之上450米(1500英尺),或者在机场标高之上600米(2000英尺),取其中较高值;

(ii)机场能见度至少为5000米。

(2)有独立可用的多条跑道,且其中一条跑道的仪表进近程序处在可用状态。

(c)按照本条规定选择的目的地备降机场的天气条件应当满足本规则第121.643条的要求。

第121.641条 国际定期载客运行的目的地备降机场

(a)按照仪表飞行规则签派飞机飞行前,应当在签派单上为每个目的地机场至少列出一个备降机场。但在下列情形下,如果在每架飞机与运行控制中心之间建立了独立可靠的通信系统进行全程监控,则可以不选择目的地备降机场:

(1)当预定的飞行不超过6小时,且相应的天气实况报告、预报或者两者的组合表明,在预计到达目的地机场时刻前后至少1小时的时间内,目的地机场的天气条件符合下列规定:

(i)机场云底高度符合下列两者之一:

(A)如果该机场需要并准许盘旋进近,至少在最低的盘旋进近最低下降高度(MDA)之上450米(1500英尺);

(B)至少在公布的最低的仪表进近最低标准之上450米(1500英尺),或者机场标高之上600米(2000英尺),取其中较高者。

(ii)机场能见度(VIS)至少为5000米(或者3英里),或者高于计划使用的目的地机场仪表进近程序能见度最低标准3200米(或者2英里)以上,取其中较大者。

(2)有独立可用的多条跑道,且其中一条跑道的仪表进近程序处在可用状态。

(3)该次飞行是在前往无可用备降机场的特定目的地机场的航路上进行的,而且飞机有足够的燃油来满足本规则第121.659条的要求。

(b)按照本条规定选择的目的地备降机场的天气条件应当满足本规则第121.643条的要求。

第121.642条 仪表飞行规则补充运行的目的地备降机场

(a)除本条(b)款规定外,当放行飞机按照仪表飞行规则进行补充运行时,应当在飞行放行单中至少为每个目的地机场列出一个备降机场。

(b)对于在国外飞行的航路上,当特定目的地机场无可用备降机场时,如果飞机装载了本规则第121.659条规定的燃油,在仪表飞行规则下可以不指定备降机场。

(c)根据本条(a)款的规定,备降机场天气条件应当符合第121.643条规定的标准。

(d)除非放行单上列出了每个必需的备降机场,否则不得放行飞机。

第 121.643 条　备降机场最低天气标准

（a）对于签派或者飞行放行单上所列的备降机场，应当有相应的天气实况报告、预报或者两者的组合表明，当飞机到达该机场时，该机场的天气条件等于或者高于合格证持有人运行规范规定的备降机场最低天气标准。

（b）在确定备降机场天气标准时，合格证持有人不得使用标注有"未批准备降机场天气标准"的仪表进近程序。

（c）在确定备降机场天气标准时，应当考虑风、条件性预报、最低设备清单条款限制等影响因素。

（d）在合格证持有人运行规范中，签派或者放行的标准应当在经批准的该机场的最低运行标准上至少增加下列数值，作为该机场用作备降机场时的最低天气标准：

（1）对于至少有一套可用进近设施的机场，其进近设施能提供直线非精密进近程序（NPA）或者直线Ⅰ类精密进近程序（PA），或者在适用时可以从仪表进近程序改为盘旋机动，最低下降高（MDH）或者决断高（DH）增加 120 米（400 英尺），能见度（VIS）增加 1600 米（1 英里）；

（2）对于至少有两套能够提供不同跑道直线进近的可用进近设施的机场，其进近设施能提供直线非精密进近程序（NPA）、直线有垂直引导的进近程序（APV）或者直线Ⅰ类精密进近程序（PA），应当选择两个服务于不同适用跑道的进近设施，在相应直线进近程序的决断高（DH）或者最低下降高（MDH）较高值上增加 60 米（200 英尺），在能见度（VIS）较高值上增加 800 米（1/2 英里）。

（e）如选择具备Ⅱ类或者Ⅲ类精密进近程序（PA）的机场作为备降机场计算备降机场天气标准，合格证持有人必须确保机组和飞机具备执行相应进近程序的资格，且飞机还应当具备Ⅲ类一发失效进近能力。此时，签派或者放行标准应当按下列数值确定：

（1）对于至少一套Ⅱ类精密进近程序（PA）的机场，云高不得低于 90 米，能见度（VIS）或者跑道视程（RVR）不得低于 1200 米；

（2）对于至少一套Ⅲ类精密进近程序（PA）的机场，云高不得低于 60 米，能见度不得低于 800 米，或者云高不得低于 60 米，跑道视程（RVR）不得低于 550 米。

（f）如选择具备基于 GNSS 导航源的有垂直引导的进近程序（APV）的机场作为备降机场计算备降机场天气标准时，合格证持有人应当经过局方批准并确保：

（1）机组和飞机具备执行相应进近程序的资格；

（2）在签派或放行时，不得在目的地机场和备降机场同时计划使用类精密进近程序；

（3）对使用基于 GNSS 导航源的有垂直引导的进近程序（APV）的机场，应当检查航行资料或者航行通告并进行飞行前接收机自主完好性（RAIM）预测；

（4）对于使用 RNP AR 程序的备降机场，计算备降机场天气标准所基于的 RNP 值不得低于 RNP0.3；

（5）在目的地机场有传统进近程序可用；

（6）在确定本条（d）款中的进近导航设施构型时，应当将基于同一 GNSS 星座的仪表进近程序当作一套进近导航设施。

1.2.2 CCAR-121 关于燃油量的规定

第 121.657 条　燃油量要求

(a)飞机必须携带足够的可用燃油以安全地完成计划的飞行并从计划的飞行中备降。

(b)飞行前对所需可用燃油的计算必须包括:

(1)滑行燃油:考虑到起飞机场的当地条件和辅助动力装置(APU)的燃油消耗,起飞前预计消耗的燃油量。

(2)航程燃油:考虑到121.663条的运行条件,允许飞机从起飞机场或从重新签派或放行点飞到目的地机场着陆所需的燃油量。

(3)不可预期燃油:为补偿不可预见因素所需的燃油量。根据航程燃油方案使用的燃油消耗率计算,它占计划航程燃油10%的所需燃油,但在任何情况下不得低于以等待速度在目的地机场上空450米(1500英尺)高度上在标准条件下飞行15分钟所需的燃油量。

(4)备降燃油:飞机有所需的燃油以便能够:

(i)在目的地机场复飞;

(ii)爬升到预定的巡航高度;

(iii)沿预定航路飞行;

(iv)下降到开始预期进近的一个点;

(v)在放行单列出的目的地的最远备降机场进近并着陆。

(5)最后储备燃油:使用到达目的地备降机场,或者不需要目的地备降机场时,到达目的地机场的预计着陆重量计算得出的燃油量;对于涡轮发动机飞机,以等待速度在机场上空450米(1500英尺)高度上在标准条件下飞行30分钟所需的油量。

(6)酌情携带的燃油:合格证持有人决定携带的附加燃油。

(c)合格证持有人应按照四舍五入方式为其机队每种型别飞机和衍生型确定一个最后储备燃油值。

(d)除非机上可使用的燃油按照要求符合本条(b)款的要求,否则不得开始飞行;除非机上可使用的燃油按照要求符合本条(b)款除滑行燃油以外的要求,否则不得从飞行中重新签派点继续飞往目的地机场。

第 121.659 条　特定情况燃油要求

(a)特定情况下目的地备降机场燃油的计算:

(1)当不需要有目的地备降机场时,所需油量能够使飞机在目的地机场上空450米(1500英尺)高度上在标准条件下飞行15分钟。

(2)预定着陆机场是一个孤立机场(无可用备降机场的特定目的地机场):

(i)能够以正常燃油消耗率在目的地机场上空飞行2小时的所需油量,包括最后储备燃油。

(ii)当按照本规则第641条(a)款第(2)项或第642条(b)款放行飞机前往孤立机场(无可用备降机场的特定目的地机场)时,需满足以下条件:

(1)在飞机与运行控制中心之间建立了独立可靠的语音通信系统进行全程监控;

(2)必须为每次飞行至少确定一个航路备降机场和与之对应的航线临界点;

(3)除非气象条件、交通和其他运行条件表明在预计使用时间内可以安全着陆,否则飞往无可用备降机场的特定目的地机场的飞行不得继续飞过航线临界点。

（b）对于涡轮螺旋桨发动机飞机的国际定期载客运行或者包括有至少一个国外机场的补充运行，不可预期燃油不得低于以正常巡航消耗率飞往本规则第 657 条（b）款第（2）、（4）项规定的机场所需总时间的 15% 所需的油量，或者以正常巡航消耗率飞行 60 分钟油量，两者当中取其中较短的飞行时间。

（c）如果根据本规则 121.657 条计算的最低燃油不足以完成下列飞行，则应当要求额外燃油：

（1）假定在航路最困难临界点发动机发生失效或者丧失增压需要更多燃油的情况下，允许飞机在必要时下降并飞行到某一备降机场：

（i）以等待速度在该机场上空 450 米（1500 英尺）高度上在标准条件下飞行 15 分钟；

（ii）在该机场进近并着陆。

（2）延程运行（EDTO）的飞机应当遵守经批准的延程运行临界燃油方案。

（3）满足上述未包含的其他规定。

第 121.663 条　计算所需燃油应当考虑的因素

（a）携带的可用燃油量必须至少基于下列数据：

（1）如果有的话，从燃油消耗监测系统获得的特定飞机的目前数据；

（2）如果没有特定飞机的目前数据，则采用飞机制造商提供的数据。

（b）计算燃油量须考虑计划飞行的运行条件，包括：

（1）风和其他天气条件预报；

（2）飞机的预计重量；

（3）航行通告；

（4）气象实况报告或气象实况报告、预报两者的组合；

（5）空中交通服务程序、限制及预期的延误；

（6）延迟维修项目和/或构型偏离的影响；

（7）空中释压和航路上一台发动机失效的情况；

（8）可能延误飞机着陆的任何其他条件。

（c）尽管有本规则第 657 条和第 659 条的规定，若安全风险评估结果表明合格证持有人能够保持同等的安全水平，局方仍可以颁发运行规范批准合格证持有人使用不同的燃油政策。

（d）本条中的所需燃油是指不可用燃油之外的燃油。

1.2.3　CCAR-135 关于目视/仪表飞行规则的运行限制和天气要求

第 135.133 条　目视飞行规则飞行的最低高度要求

除航空器起飞和着陆外，按照目视飞行规则（VFR）运行的航空器应当满足下列最低高度要求：

（a）对于飞机：

（1）昼间飞行时，离地面、水面的高度不得低于 150 米（500 英尺），并且离障碍物的水平距离不得小于 150 米（500 英尺）。

（2）夜间飞行时，飞行高度应当高于离预定飞行航路水平距离 8 千米（5 英里）范围内的最高障碍物至少 300 米（1000 英尺）。在山区，飞行高度应当高于离预定飞行航路水平距离

8 千米(5 英里)范围内的最高障碍物至少 600 米(2000 英尺)。

（b）对于直升机，在飞越人口稠密区上空时，离地高度不得低于 90 米(300 英尺)。

第 135.135 条　目视飞行规则飞行的能见度要求

（a）在运输机场空域以外的空域按照目视飞行规则(VFR)运行飞机时，如果云底高小于 300 米(1000 英尺)，则飞行能见度不得小于 3200 米(2 英里)。

（b）在修正海平面气压高度 900 米(3000 英尺)以下或者离地高度 300 米(1000 英尺)以下(以高者为准)按照目视飞行规则(VFR)运行直升机时，飞行能见度在昼间不得小于 800 米(1/2 英里)，在夜间不得小于 1600 米(1 英里)。

第 135.137 条　直升机目视飞行规则飞行中的目视参考要求

按照目视飞行规则(VFR)运行直升机时，驾驶员应当建立足够的目视地面或者水面参考。对于夜间飞行，建立足够的目视地面或者水面灯光参考，能够保证其安全操作直升机。

第 135.145 条　目视飞行规则飞行的燃油供应要求

（a）按照目视飞行规则(VFR)运行飞机时，应当在考虑风和预报的天气条件后，有足够的燃油飞至第一个预计着陆点，并且以正常巡航燃油消耗率完成下列飞行：

（1）对于昼间运行，至少再飞行 30 分钟。

（2）对于夜间运行，至少再飞行 45 分钟。

（b）按照目视飞行规则(VFR)运行直升机时，应当在考虑风和预报的天气条件后，有足够的燃油飞至第一个预计着陆点，并且以正常巡航速度再飞行 20 分钟。

第 135.183 条　仪表飞行规则起飞限制

当天气条件不低于起飞最低标准，但低于经批准的仪表飞行规则(IFR)着陆最低标准时，仅当在距起飞机场 1 小时飞行时间(在静止空气中以正常巡航速度飞行)的距离内有一备降机场，方可按照仪表飞行规则(IFR)起飞航空器。

第 135.185 条　仪表飞行规则目的地机场最低天气标准

仅当最新的天气报告、预报或者两者的组合表明，在航空器到达预定着陆机场的预计时刻，天气条件达到或者高于经批准的仪表飞行规则(IFR)着陆最低标准，方可按照仪表飞行规则(IFR)起飞航空器或者进入仪表飞行规则(IFR)飞行或者云上运行。

第 135.187 条　仪表飞行规则备降机场最低天气标准

（a）对于仪表飞行规则(IFR)飞行中所用的备降机场，应当有相应的天气实况报告、预报或者两者的组合表明，当航空器到达该机场时，该机场的天气条件等于或者高于备降机场最低天气标准。

（b）对于按本规则运行的飞机，合格证持有人应当在经批准的机场最低运行标准上增加至少下列数值，作为该机场用作备降机场时的最低天气标准：

（1）对于只有一套进近设施与程序的机场，决断高度/高(DA/DH)或者最低下降高度/高(MDA/MDH)增加 120 米(400 英尺)，能见度增加 1600 米(1 英里)。

（2）对于具有两套(含)以上精密或者非精密进近设施与程序并且能提供不同跑道进近的机场，决断高度/高(DA/DH)或者最低下降高度/高(MDA/MDH)增加 60 米(200 英尺)，能见度增加 800 米(1/2 英里)，在两条较低标准的跑道中取较高值。

（c）对于按照本规则运行的直升机，合格证持有人应当在经批准的机场进近程序决断高度/高(DA/DH)或者最低下降高度/高(MDA/MDH)上增加 60 米，能见度至少 1600 米，但是不小于所用进近程序最低能见度标准，作为该机场用作备降机场时的最低天气标准。

第 135.189 条　仪表飞行规则燃油及备降机场要求

（a）除本条（b）款规定的情况外，仅当在考虑到天气报告、预报或者两者的组合后，航空器上携带了能完成下列飞行的燃油，方可在仪表飞行规则（IFR）条件下运行航空器：

（1）完成到达第一个预定着陆机场的飞行。

（2）从该机场飞至备降机场。

（3）此后以正常巡航速度飞行 45 分钟。对于直升机，以正常巡航速度飞行 30 分钟。

（b）如果第一个预定着陆机场具有局方公布的标准仪表进近程序，并且相应的天气报告、预报或者两者的组合表明，在预计到达时刻前后至少 1 小时的时间段内达到下列天气条件，则可以不选择备降机场，本条（a）款（2）项不适用：

（1）云高在盘旋进近的最低下降高度/高（MDA/MDH）之上至少增加 450 米（1500 英尺）。或者，如果该机场没有局方公布的仪表盘旋进近程序，云高为公布的最低标准之上至少 450 米（1500 英尺）或者机场标高之上至少 600 米（2000 英尺）（取两者中较高者）。

（2）在目的地机场实施仪表进近程序时，该机场预报的能见度至少为 4.8 千米（3 英里），或者至少比最低的适用能见度最低标准大 3.2 千米（2 英里）（取两者中较大者）。

（3）对于直升机，云高高于机场标高 300 米或者高于适用的进近最低标准之上 120 米（以高者为准），能见度 3000 米。

第 135.191 条　仪表飞行规则起飞、进近和着陆最低标准

（a）航空器在某一机场实施仪表进近程序前，应当满足下列条件：

（1）该机场具有符合局方规定的民用航空气象服务机构。

（2）该民用航空气象服务机构发布的最新气象报告表明，天气条件达到或者高于该机场经批准的仪表飞行规则（IFR）着陆最低标准。

（b）当本条（a）款（1）项规定的民用航空气象服务机构发布的最新天气报告表明天气条件达到或者高于经批准的仪表着陆最低标准时，航空器驾驶员方可进入仪表进近程序中的最后进近阶段继续实施进近。

（c）当驾驶员已经按照本条（b）款规定开始了仪表进近程序中的最后进近阶段，并在此后收到后续的气象报告表明天气条件低于着陆最低标准，驾驶员仍可以操作航空器继续进近。当航空器进近至经批准的决断高度/高（DA/DH）或者最低下降高度/高（MDA/MDH）时，如果驾驶员断定实际的天气条件不低于该机场的最低着陆天气标准，则可以继续进近并完成着陆。本款所述的最后进近阶段是指下列情况之一：

（1）航空器实施仪表着陆系统（ILS）进近或者基于性能导航（PBN）的仪表进近时，已经通过最后进近定位点。

（2）航空器实施机场监视雷达（ASR）或者精密进近雷达（PAR）进近时，已经移交至最后进近管制员。

（3）航空器使用甚高频全向信标台（VOR）、无方向性导航台（NDB）实施进近或者实施其他类似方法的进近时，该航空器已经通过相应的设施或者最后进近定位点，或者在没有规定最后进近定位点时，已经完成了程序转弯并且位于程序规定的距离内，按照最后进近航道向机场归航。

（d）对于在该型别飞机上担任机长时间未达到 100 小时的涡轮发动机飞机机长，应当在局方公布的机场运行最低标准或者合格证持有人的运行规范中规定的决断高度/高（DA/DH）或者最低下降高度/高（MDA/MDH）之上增加 30 米（100 英尺），能见度在着陆最低标准上增加 800 米（1/2 英里），但不超过合格证持有人将该机场作为备降机场时使用的着陆最低标准。

（e）驾驶员在军方或者国外机场实施仪表飞行规则（IFR）起飞、进近和着陆时，应当遵守该机场规定的仪表进近程序和适用的最低天气标准。如果该机场没有规定最低天气标准，应当遵守下列标准：

（1）按照仪表飞行规则（IFR）起飞时，能见度不得低于 1600 米（1 英里）。

（2）进行仪表进近时，能见度不得低于 800 米（1/2 英里）。

（f）当本条（a）款（1）项规定的民用航空气象服务机构所报告的天气条件低于局方公布的机场运行最低标准或者合格证持有人运行规范中规定的起飞最低标准时，航空器驾驶员不得按照仪表飞行规则（IFR）起飞航空器。

（g）除本条（h）款规定的情况外，当局方没有为该起飞机场规定起飞最低标准，本条（a）款（1）项规定的民用航空气象服务机构所报告的天气条件低于 CCAR-91 或者合格证持有人运行规范中规定的起飞最低标准时，航空器驾驶员不得按照仪表飞行规则（IFR）起飞航空器。

（h）除另有限制的机场外，在具有局方公布的直接仪表进近程序的机场，当本条（a）款（1）项规定的民用航空气象服务机构所报告的天气条件不低于直接进近着陆最低标准时，如果满足下列条件，航空器驾驶员可以按照仪表飞行规则（IFR）起飞航空器：

（1）起飞时刻所用跑道的风向和风速可以允许在该跑道上实施直接仪表进近。

（2）有关的地面设施和机载设备工作正常。

（3）合格证持有人已经被批准实施此种运行。

1.3　飞行计划中使用的术语及参数

1.3.1　术语

（1）使用空机重（OEW，Operational Empty Weight）或称运行空机重：飞机结构重量+机组、乘务组及其行李+随机工具、资料+救生、应急设备+配餐等处于可使用状态的重量。同一机型各飞机使用空机重可能不同，同一飞机运行项目不同也会导致使用空机重不同。

（2）停机坪重量（RPW，Ramp Weight）或滑行重量（Taxi Weight），即装完业载和所需油量可以开始滑行的重量=使用空机重+业载+全部油量（起飞总油量）。

（3）起飞重量（TOW，Take-off Weight）或松刹车重量（BRW，Brake Release Weight）：滑出重量-开车及滑出油量。

（4）无油重量（ZFW，Zero Fuel Weight）=使用空机重+业载，应不大于最大无油重量。

（5）航程油量（TRF，Trip Fuel）和航程时间（TRT，Trip time）：从松刹车加速起飞、爬升、巡航、下降、进近直到在目的地机场上着陆所用的油量及时间。

（6）轮挡油量（BLF，Block Fuel）和轮挡时间（BLT，Block Time）：航程油量（时间）+开车及滑出油量（时间）+滑入油量（时间）。

（7）备降燃油和备降时间：从目的地机场复飞、爬升、巡航、下降、进近直至在备降机场着陆（不含等待）所用的油量及时间。

（8）不可预期燃油：为补偿不可预见因素所需的燃油量。根据航程燃油方案使用的燃油消耗率计算,它占计划航程燃油 10% 的所需燃油,但在任何情况下不得低于以等待速度在目的地机场上空 450 m(1500 ft)高度上在标准条件下飞行 15 min 所需的燃油量。

(9)最后储备燃油:对于某次飞行,在指定目的地备降机场时,是指使用到达目的地备降机场的预计着陆重量计算得出的燃油量;或者在未指定目的地备降机场时,是指按照到达目的地机场的预计着陆重量计算得出的燃油量。

(a)对于活塞式发动机飞机,以等待速度在机场上空 450 m(1500 ft)高度上在标准条件下飞行 45 min 所需的油量;或

(b)对于涡轮发动机飞机,以等待速度在机场上空 450 m(1500 ft)高度上在标准条件下飞行 30 min 所需的油量。

(10)酌情携带的燃油:合格证持有人决定携带的附加燃油。例如,可以规定为完成上述飞行剖面后在备降机场停机坪还剩的保底油量,或规定为按照航程油量的某一百分比增加的油量等,这种规定可以写在公司的燃油政策中。

(11)备份燃油:备降燃油+不可预期燃油+最后储备燃油+酌情携带的燃油。

(12)起飞总油量(即停机坪油量):轮挡油量+备份燃油。

(13)在备降机场滑入停机坪后的飞机重量=使用空机重+业载+酌情携带的燃油。

进近油量(时间)和离场油量(时间)即出航油量(时间)和入航油量(时间)都算在航程油量(时间)中。

一架飞机可以使用 15~20 年。在使用时间长了之后,飞机及其发动机老化,性能衰退,使耗油量增加,使用手册上的数据没有计及这种衰退影响。对老龄飞机,可把算出的航程油量多加 2%~3%。多加多少取决于飞机的老化程度。为此,应对飞机进行性能监控,利用 QAR 数据和 APM 软件计算出各架飞机的阻力因子(阻力增长系数)和燃油流量因子(燃油流量增长系数),判断飞机老化程度,针对每架飞机确定应多加的油量,在制作飞行计划时输入各架飞机的燃油流量因子得到的飞行计划,即所谓的"尾号飞行计划"。此外,还应该修改飞机上 FMC 中的阻力因子和燃油流量因子,否则 FMC 预测的耗油量和剩余燃油就不准确了。在 MCDU 的 AIRLINE POLICY 页面(B757 是在 IDENT 页面)可以修改阻力因子和燃油流量因子,如图 1-3-1 所示。

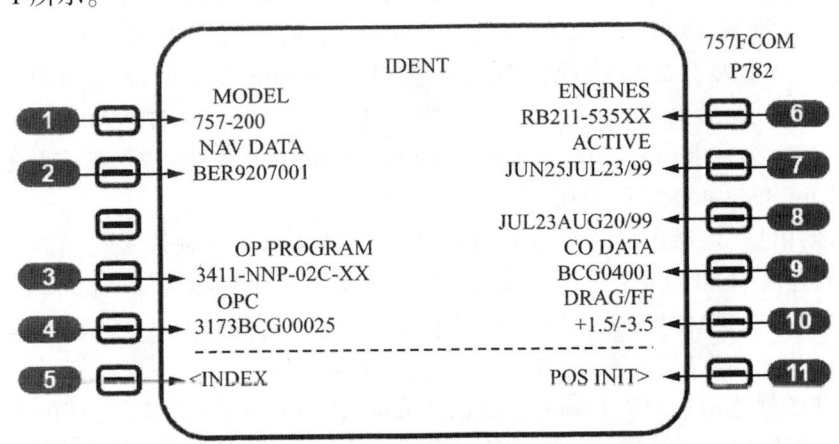

图 1-3-1　MCDU 上修改阻力因子和燃油流量因子的界面

注:"键 10"——阻力/燃油流量因子。

1.3.2 制作飞行计划所需的数据

（1）起飞机场的最大允许起飞重量（MTOW）、目的地机场及备降机场的最大允许着陆重量（MLDW）。这里所讲的最大允许起降重量是机场分析得出的结果，是考虑了机场、气象条件、飞机性能、结构限制、道面强度限制后得出的最小值。道面强度限制的起降重量用 ACN-PCN 方法得出，在本书第 1.6 节中介绍了 ACN-PCN 方法。

（2）最大无油重量（MZFW）。该重量是考虑飞机受到结构强度和适航性要求的限制，在加入燃油之前所允许的最大重量。最大无油重量对飞机载重有一定的限制作用，也就是在装载重量达到最大无油重量后剩余的载荷只能装载燃料。

（3）使用空机重（OEW）。如果需额外配氧气瓶，则氧气瓶重量也应计入使用空机重内。

（4）油箱最大油量。油箱能够装载的油量是利用油箱最大容积与燃油密度（γ）计算出来的，油箱最大油量＝燃油密度×油箱最大容积。燃油密度与燃油的牌号和温度有关，不同温度下燃油的密度不同。例如：15.6 ℃时 JP4 牌号的燃油密度为 0.775 kg/L，JP5 牌号的燃油密度为 0.828 kg/L。

（5）最大结构业载（MPLS）；最大结构业载＝最大无油重量－使用空机重。

（6）地面滑行耗油率：飞机在地面滑行时每分钟的耗油量。

（7）辅助动力装置（APU）的耗油率：飞机在地面或空中运行 APU 时，APU 每小时的耗油量。有些机场无电源车、气源车，为了进行地面勤务或空调，需要 APU 运转供气供电。大部分飞机的 APU 在空中一般不使用。如：B757-200（RB211-535E4）的 APU 耗油：地面 235 lb/h，空中 200 lb/h；B737-800（CFM56-7B26）的 APU 耗油：地面 225 lb/h，空中 200 lb/h。

（8）进近（放下襟翼的机动飞行）每分钟耗油。例如，B757-200 是 155 lb/min，B737-800 是 130 lb/min。

（9）FAR 第 91 部 70 款规定：未经批准任何飞机在 10000 ft 以下不得以大于 250 kt 的表速飞行，如飞机最小安全速度大于 250 kt，则允许以最小安全速度飞行。飞机使用手册中的爬升数据表，有的没有考虑 10000 ft 以下 250 kt 表速限制，如 B757-200 爬升数值表上注明爬升规律是 290/M0.78，即从 1500 ft 开始就以表速 290 kt 爬升，没有考虑这个限制。如 ATC 要求 10000 ft 以下表速不大于 250 kt，爬升用油将稍多些，多耗的油与机型有关，可由飞机使用手册查出。多耗的油一般在 50~600 lb，个别机型大一些，如 B737-800 多耗 50 lb，B737-300 多耗 100 lb，B757-200 多耗 100 lb，B767-200 和 B767-300 多耗 200 lb，B777-200 多耗 600 lb，B747-200 多耗 590 lb，B747-400 多耗 1800 lb。

（10）防冰用油。巡航中使用发动机或发动机和机翼防冰将额外多耗油，如 B757-200 发动机防冰耗油速度为 200 lb/h，同时用发动机和机翼防冰耗油速度为 300 lb/h。

（11）航路上的风及温度（或实际温度与标准大气温度的偏差）。

（12）允许飞的高度层，见表 1-3-1、表 1-3-2、表 1-3-3。

（13）燃油的低热值（LHV，Lower Heating Value）。热值，或称燃烧值，是指单位重量的燃料燃烧时放出的热量。航空燃油的燃烧产物中含有水，如其中的水是气态的，则燃烧时放出的热量比水是液态时放出的热量少，即热值较低。发动机中的燃烧产物是高温燃气，水是气态的，应当用低热值。波音公司的手册上给出的燃油流量等都是按 $LHV = 18580$ BTU/lb 算出的，BTU 是英国热量单位，lb 为英制重量单位，即磅；1 BTU = 252.435 cal。如所使用的燃油的

$LHV \neq 18\,580$ BTU/lb,为准确起见,应把按手册算出的油量乘以修正值($18580/LHV$)。

表 1-3-1　中国民航飞行高度层对照表

180°~359°T　向西飞行		000°~179°T　向东飞行	
←		→	
管制员发布的米制高度	机组设定的英制高度	管制员发布的米制高度	机组设定的英制高度
ETC. ↑	ETC. ↑	ETC. ↑	ETC. ↑
15500	50900	14900	48900
14300	46900	13700	44900
13100	43000		
		12500	41100
12200	40100	11900	39100
11600	38100	11300	37100
11000	36100	10700	35100
10400	34100	10100	33100
9800	32100	9500	31100
9200	30100	8900	29100
8400	27600	8100	26600
7800	25600	7500	24600
7200	23600	6900	22600
6600	21700	6300	20700
6000	19700	5700	18700
5400	17700	5100	16700
4800	15700	4500	14800
4200	13800	3900	12800
3600	11800	3300	10800
3000	9800	2700	8900
2400	7900	2100	6900
1800	5900	1500	4900
1200	3900	900	3000
600	2000	—	—
m	ft	m	ft
←		→	

备注:①高度层选择采用东单西双原则;
　　　②我国采用米制高度层,600~12500 m 为 300 m 一个高度层(8400~8900 m 为 500 m 一个高度层);
　　　12500 m 以上为 600 m 一个高度层。

表 1-3-2　国际高度层规定　　　　　　　　　　　　　　　　　　　（单位：ft）

磁航迹角 180°~359°	磁航迹角 0°~179°
43000	45000
39000	41000
35000	37000
31000	33000
28000	29000
26000	27000
24000	25000
22000	23000
20000	21000
18000	19000
16000	17000
14000	15000
12000	13000
10000	11000
8000	9000
6000	7000
4000	5000
2000	3000
不同国家高度层规定可能不同	

表 1-3-3　国际通用的缩小垂直间隔　　　　　　　　　　　　　　　（单位：ft）

磁航迹角 180°~359°	磁航迹角 0°~179°
40000	41000
38000	39000
36000	37000
34000	35000
32000	33000
30000	31000
	29000

1.4　航路数据计算

1.4.1　由经纬度计算航段距离和航向

航段距离是指过其两端点的大圆距离，航段的真航向指这种大圆航线与真北的夹角。如已知一个航段两个端点的经纬度坐标,则该航段的距离及航向均可由经纬度算出。

①航段距离——任意两点间大圆距离的计算

如图 1-4-1 和图 1-4-2 所示,设起点 P_1 的纬度为 N_1、经度为 E_1,终点 P_2 的纬度为 N_2、经度为 E_2,P_1、P_2 点的球坐标为:

$\theta_1 = 90° - N_1$(南纬代入负值)

$\varphi_1 = E_1$(东经)或 $360° + E_1$(西经,代入负值)

$\theta_2 = 90° - N_2$(南纬代入负值)

$\varphi_2 = E_2$(东经)或 $360° + E_2$(西经,代入负值)

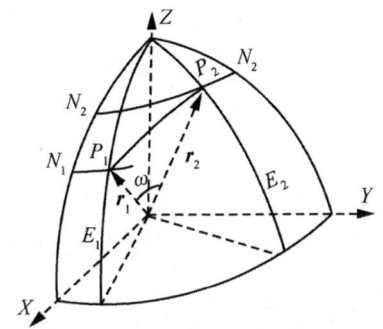

图 1-4-1　球面上的航段

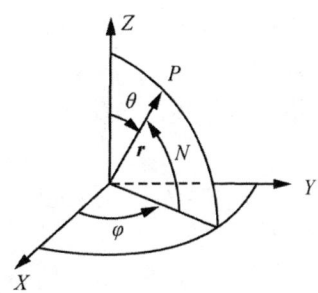

图 1-4-2　球坐标与经纬度

设地球半径为 R,则 P_1、P_2 点的矢径为:

$$r_1 = R\sin\theta_1\cos\varphi_1 i + R\sin\theta_1\sin\varphi_1 j + R\cos\theta_1 k$$

$$r_2 = R\sin\theta_2\cos\varphi_2 i + R\sin\theta_2\sin\varphi_2 j + R\cos\theta_2 k$$

式中:i、j、k 为 X、Y、Z 轴向的单位矢量。

矢量 r_1、r_2 的数量积:

$$r_1 \cdot r_2 = R^2\cos\omega$$

式中:ω 为矢量 r_1、r_2 的夹角。

$$\cos\omega = \sin\theta_1\sin\theta_2\cos(\varphi_1 - \varphi_2) + \cos\theta_1\cos\theta_2$$
$$= \cos N_1\cos N_2\cos(E_1 - E_2) + x\sin N_1\sin N_2$$

航段距离,即这两个端点之间的大圆距离:

$$D = R\omega \tag{1-4-1}$$

式中:R 为地球半径,可取为 6370 km(3440 NM)。

$$\omega = \arccos[\cos N_1\cos N_2\cos(E_1 - E_2) + \sin N_1\sin N_2] \tag{1-4-2}$$

注:N_1、N_2 为南纬,E_1、E_2 为西经时应为负值代入公式(下同)。

ω 应以弧度值代入式(1-4-1),或以度代入 $D = 60\omega$ 得到以海里(NM)为单位的距离。

②航段起点大圆航向的计算

该航段的大圆航向即起点的真航向,亦即航段大圆弧与起点的经线的夹角 β_1,如图 1-4-3 和图 1-4-4 所示。

在起点 P_1 沿大圆航线的单位切矢量：

$$t = (r_1 \times r_2) \times r_1 / \left| (r_1 \times r_2) \times r \right|$$

$$= \frac{R^2 r_2 - (r_1 \cdot r_2) r_1}{\left| R^2 r_2 - (r_1 \cdot r_2) r_1 \right|}$$

$$= (r_2 - \cos\omega r_1) / |r_2 - \cos\omega r_1|$$

$$= [(\sin\theta_2\cos\varphi_2 - \cos\omega\sin\theta_1\cos\varphi_1)i + (\sin\theta_2\sin\varphi_2 - \cos\omega\sin\theta_1\sin\varphi_1)j + (\cos\theta_2 - \cos\omega\cos\theta_1)k] / \sin\omega$$

在起点 P_1 沿经线北指的单位切矢量：

$$\tau = -\cos\theta_1\cos\varphi_1 i - \cos\theta_1\sin\varphi_1 j + \sin\theta_1 k$$

在起点 P_1 沿矢径的单位切矢量：

$$r^\circ = \sin\theta_1\cos\varphi_1 i + \sin\theta_1\sin\varphi_1 j + \cos\theta_1 k$$

在起点 P_1 沿纬线东指的单位切矢量：

$$e = \tau \times r^\circ = -\sin\varphi_1 i + \cos\varphi_1 j$$

$$\cos\beta_1 = t \cdot \tau = [\sin\theta_1\cos\theta_2 - \cos\theta_1\sin\theta_2\cos(\varphi_1 - \varphi_2)] / \sin\omega$$

即

$$\cos\beta_1 = [\cos N_1\sin N_2 - \sin N_1\cos N_2\cos(E_1 - E_2)] / \sin\omega \tag{1-4-3}$$

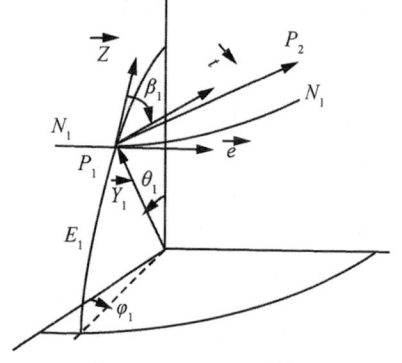

图 1-4-3　航段大圆航向

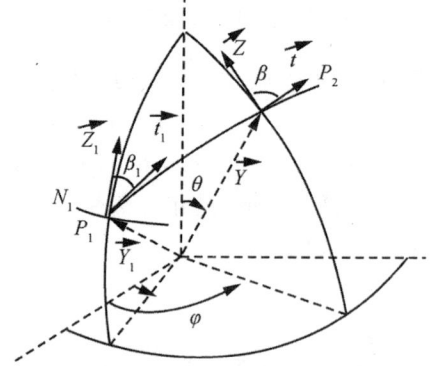

图 1-4-4　大圆航线上任一点的真航向

式(1-4-3)给出的 $\beta_1 \in [0°, 180°]$，当终点在起点西边时，大圆航向 = $360° - \beta_1$，如图 1-4-5 所示。

在编制程序计算时，可由 $\cos\gamma = t \cdot e$，即 $\cos\gamma = \sin\theta_2\sin(\varphi_2 - \varphi_1)/\sin\omega = \cos N_2\sin(E_2 - E_1)/\sin\omega$ 的正负判断 β_1 的区间。$\cos\gamma > 0$，$\beta_1 < 180°$，否则 $\beta_1 > 180°$。不难看出：

$$\sin\beta_1 = \cos\gamma = \sin\theta_2\sin(\varphi_2 - \varphi_1)/\sin\omega = \cos N_2\sin(E_2 - E_1)/\sin\omega \tag{1-4-4}$$

由 $\sin\beta_1 = \sqrt{1 - \cos^2\beta_1}$ 也可导出上式。

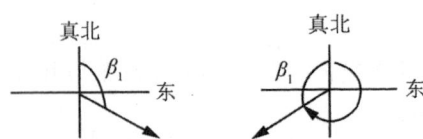

图 1-4-5　大圆航向角的确定

③大圆航线各点真航向计算

沿大圆航线飞行时各点的真航向是变化的，各点的真航向角 β 可按以下方法确定。

大圆航线所在平面的法向单位矢量：

$$n = r_1 \times r_2 / |r_1 \times r_2| = r_1 \times r_2 / (R^2 \sin\omega)$$

大圆航线上任一点的矢径 $r = R\sin\theta\cos\varphi i + R\sin\theta\sin\varphi j + R\cos\theta k$。

矢径方向的单位矢 $r° = r/R$，该点航线上的单位切矢量 $t = n \times r°$，即

$$t = \frac{(r_1 \times r_2) \times r}{R^3 \sin\omega} = \frac{(r_1 \cdot r) r_2 - (r_2 \cdot r) r_1}{R^3 \sin\omega}$$

该点沿经线北指的单位切矢量：

$$\tau = -\cos\theta\cos\varphi i - \cos\theta\sin\varphi j + \sin\theta k$$

该点真航向角 β 的余弦 $\cos\beta = \tau \cdot t$，即

$$\begin{aligned}
\cos\beta &= \left[(r_1 \cdot r)(\tau \cdot r_2) - (r_2 \cdot r)(\tau \cdot r_1) \right] / (R^3 \sin\omega) \\
&= \left[\sin\theta_1\cos\theta_2\cos(\varphi-\varphi_1) - \sin\theta_2\cos\theta_1\cos(\varphi-\varphi_2) \right] / \sin\omega \\
&= \left[\cos N_1\sin N_2\cos(E-E_1) - \cos N_2\sin N_1\cos(E-E_2) \right] / \sin\omega
\end{aligned} \tag{1-4-5}$$

当 $|\varphi_2-\varphi_1| < 180°$ 时，如 $\varphi_2 > \varphi_1$，β 取 $0° \sim 180°$；如 $\varphi_2 < \varphi_1$，β 取 $180° \sim 360°$。当 $|\varphi_2-\varphi_1| > 180°$ 时，如 $\varphi_2 > \varphi_1$，β 取 $180° \sim 360°$；如 $\varphi_2 < \varphi_1$，则 β 取 $0° \sim 180°$。

利用式(1-4-3)、式(1-4-4)及 $\cos(\varphi-\varphi_2) = \cos\left[(\varphi-\varphi_1) + (\varphi_1-\varphi_2) \right]$，式(1-4-5)可改写为：

$$\begin{aligned}
\cos\beta &= \cos\beta_1\cos(\varphi-\varphi_1) - \sin\beta_1\cos\theta_1\sin(\varphi-\varphi_1) \\
&= \cos\beta_1\cos(E-E_1) - \sin\beta_1\sin N_1\sin(E-E_1)
\end{aligned} \tag{1-4-6}$$

$$\beta = \arccos\left[\cos\beta_1\cos(E-E_1) - \sin N_1\sin\beta_1\sin(E-E_1) \right] \tag{1-4-7}$$

式中：β_1 为大圆航向，即起点的真航向；

E 为计算真航向的点的经度。

当 $\beta_1 > 180°$ 时，真航向 $= 360° - \beta$，因为式(1-4-6)给出的 $\beta \in [0°, 180°]$。各点真航向角 β 既可用式(1-4-5)计算，也可用式(1-4-6)计算。在知道两端点的坐标时可直接用式(1-4-5)计算，这样可不算起点的真航向 β_1；若给出的是起点的真航向和坐标，则只能用式(1-4-6)计算。

航段各点的真航向角 β 亦可按以下方法确定：

$$各点的真航向角 \beta = 大圆航向 \beta_1 + 该点的经线收敛角 \delta \tag{1-4-8}$$

$$经线收敛角 \delta = (E-E_1)\sin\varphi$$

式中：$\varphi = (N+N_1)/2$。

如图 1-4-6 所示，E、N 分别为计算真航向的点的经纬度，E_1、N_1 分别为航段起点的经纬度，南纬、西经代入负值。对于跨 $180°$ 经线的航段，西经应取负值后再加 $360°$。

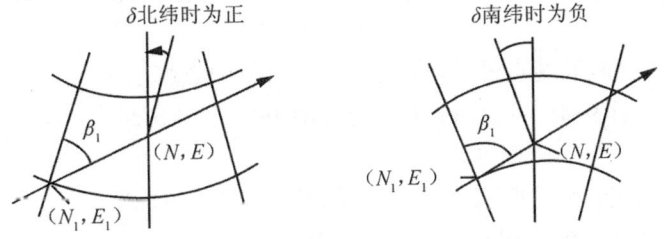

图 1-4-6　经线收敛角与真航向角

中国航图上标注的航段的磁航迹角是航段中点的磁航向，即：

$$航段的磁航迹角(磁航向) = 航段中点的真航向 - 中点磁差(西磁差为负)$$

注意：计算各点真航向角的公式即式(1-4-8)是一个近似公式，但其精度比较高，只有在很

长的大圆航线上才有较大的误差,在大圆航线上距起点 5000 km、7000 km、10000 km 处真航向的误差分别为 0.7°、1.7°、11° 左右,一般航段的长度远小于 5000 km,用式(1-4-5)~式(1-4-8)计算的结果没什么差别。

④大圆航线上各点经纬度满足的关系式

大圆航线上任一点的经纬度都应满足:$\boldsymbol{r} \cdot (\boldsymbol{r}_1 \times \boldsymbol{r}_2) = 0$,即

$$\begin{vmatrix} \sin\theta\cos\varphi & \sin\theta\sin\varphi & \cos\theta \\ \sin\theta_1\cos\varphi_1 & \sin\theta_1\sin\varphi_1 & \cos\theta_1 \\ \sin\theta_2\cos\varphi_2 & \sin\theta_2\sin\varphi_2 & \cos\theta_2 \end{vmatrix} = 0$$

化简后得:

$$\sin\theta\left[\sin\theta_1\cos\theta_2\sin(\varphi_1-\varphi)+\sin\theta_2\cos\theta_1\sin(\varphi-\varphi_2)\right]+\cos\theta\sin\theta_1\sin\theta_2\sin(\varphi_2-\varphi_1)=0 \quad (1\text{-}4\text{-}9)$$

$$\tan\theta = \frac{\sin\theta_1\sin\theta_2\sin(\varphi_2-\varphi_1)}{\sin\theta_1\cos\theta_2\sin(\varphi-\varphi_1)-\sin\theta_2\cos\theta_1\sin(\varphi-\varphi_2)}$$

$$= \frac{\cos N_1\cos N_2\sin(E_2-E_1)}{\cos N_1\sin N_2\sin(E-E_1)-\cos N_2\sin N_1\sin(E-E_2)} \quad (1\text{-}4\text{-}10)$$

利用起点大圆航向 β_1,式(1-4-10)可改写为:

$$\tan\theta = \frac{\sin\beta_1\sin\theta_1}{\cos\beta_1\sin(\varphi-\varphi_1)+\sin\beta_1\cos\theta_1\cos(\varphi-\varphi_1)}$$

$$= \frac{\sin\beta_1\cos N_1}{\cos\beta_1\sin(E-E_1)+\sin\beta_1\sin N_1\cos(E-E_1)} \quad (1\text{-}4\text{-}11)$$

若 $E_1=E_2$ 或 $\beta_1=0°$、$180°$,上述两式不能用,此时大圆航线即经线,不能计算;若 $N_1=N_2=0°$ 或 $N_1=0°$、$\beta_1=90°$,上述两式不能用,此时大圆航线即赤道,不必计算。

此外,当 $\tan\theta>0$ 时,θ 取 $0°\sim90°$,否则 θ 取 $90°\sim180°$,$N=90°-\theta$,N 为负值则表示南纬。利用这两个公式就可以把大圆航线上各点的经纬度计算出来。

注意:这里的 θ 是球坐标,是相对 Z 轴度量的角度:$\theta=90°-N$。

例 1-4-1:太原经纬度:N3745.1,E11236.9,磁差 4.25°W,大王庄经纬度:N3911.0,E11635.0,磁差 5.6°W,计算太原至大王庄航段的距离及磁航迹。

$\omega = \arccos\left[\cos37.752°\cos39.183°\cos3.9683°+\sin37.752°\sin39.183°\right]$

$\approx \arccos0.9982184 \approx 3.420597° = 0.0597$ 弧度

$D = 6370\times0.0597 \approx 380.29$ km $= 205.23$ NM

$\cos\beta_1 = \left[\cos37.7517°\sin39.183°-\sin37.7517°\cos39.183°\cos3.9683°\right]/\sin\omega = 0.4378176$

$\beta_1 \approx 64.0353°$

航段中点经纬度近似取端点经纬度的平均值,航段中点真航向:

$\beta = \arccos\left[\cos\beta_1\cos1.984°-\sin\beta_1\sin37.7512°\sin1.984°\right] \approx 65.26°$

或

$\delta = 1.984°\times\sin\left[((37.7517°+39.183°)/2+37.7517°)/2\right] \approx 1.224°$

$\beta \approx 64.0353°+1.224° = 65.2593°$

航段的磁航迹角(磁航向)$\approx 65.2593°-(-4.25°-5.6°)/2 \approx 70.185°$,航图上标注的是380 km(205 NM)和 70°,与计算结果是一致的。

注意:对于高纬度地区,航段中点经纬度与端点经纬度平均值相差很大,应先算出航段中点经纬度,再计算航段中点收敛角及真航向。

1.4.2 计算航段上到端点距离已知的点的经纬度

如图 1-4-7 所示,设航段起点 P_1 经纬度为 (N_1,E_1),终点 P_2 经纬度为 (N_2,E_2),P 点与终点的距离为 D,地球半径为 R,现在的问题是确定 P 点的经纬度。阶梯巡航时往往爬升起点在一个航段中间某处,计算爬升起点的经纬度即属于此类问题。

P_1、P_2、P 三点在球坐标下的角度为:

$\theta_1=90°-N_1,\varphi_1=E_1($东经$)$或 $360°+E_1($西经$)$

$\theta_2=90°-N_2,\varphi_2=E_2($东经$)$或 $360°+E_2($西经$)$

$\theta=90°-N,\varphi=E($东经$)$或 $360°+E($西经$)$

E 若为西经则代入负值。

P_1、P_2、P 点的矢径为:

$\boldsymbol{r}_1=R\sin\theta_1\cos\varphi_1\boldsymbol{i}+R\sin\theta_1\sin\varphi_1\boldsymbol{j}+R\cos\theta_1\boldsymbol{k}$

$\boldsymbol{r}_2=R\sin\theta_2\cos\varphi_2\boldsymbol{i}+R\sin\theta_2\sin\varphi_2\boldsymbol{j}+R\cos\theta_2\boldsymbol{k}$

$\boldsymbol{r}=R\sin\theta\cos\varphi\boldsymbol{i}+R\sin\theta\sin\varphi\boldsymbol{j}+R\cos\theta\boldsymbol{k}$

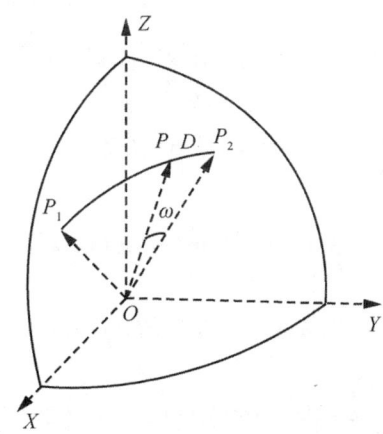

图 1-4-7 航段上任一点经纬度的确定

由三矢量共面条件: $\boldsymbol{r}\cdot\boldsymbol{r}_1\times\boldsymbol{r}_2=0$,得:

$$A\sin\theta\cos\varphi-B\sin\theta\sin\varphi+C\cos\theta=0$$

或

$$A\cos N\cos E-B\cos N\sin E+C\sin N=0 \tag{1-4-12}$$

式中:

$$\begin{aligned}A&=\sin\theta_1\sin\varphi_1\cos\theta_2-\sin\theta_2\sin\varphi_2\cos\theta_1\\&=\cos N_1\sin E_1\sin N_2-\cos N_2\sin E_2\sin N_1\\B&=\sin\theta_1\cos\varphi_1\cos\theta_2-\sin\theta_2\cos\varphi_2\cos\theta_1\\&=\cos N_1\cos E_1\sin N_2-\cos N_2\cos E_2\sin N_1\\C&=\sin\theta_1\cos\varphi_1\sin\theta_2\sin\varphi_2-\sin\theta_2\cos\varphi_2\sin\theta_1\sin\varphi_1\\&=\sin\theta_1\sin\theta_2\sin(\varphi_2-\varphi_1)=\cos N_1\cos N_2\sin(E_2-E_1)\end{aligned} \tag{1-4-13}$$

由于 $D=R\omega$,即 $\omega=D/R$,故 $\cos\omega=\cos(D/R)$,令 $S=\cos(D/R)$,由式(1-4-2)得:

$$\cos N_2\cos N\cos(E-E_2)+\sin N\sin N_2=S \tag{1-4-14}$$

P 点经纬度应同时满足式(1-4-12)和式(1-4-14),可由此两式联立解出 P 点经纬度,由式(1-4-12)得:

$$\tan N = (B\sin E - A\cos E)/C \qquad (1\text{-}4\text{-}15)$$

代入式(1-4-14)得:

$$\left[\cos N_2 \cos(E-E_2) + \tan N \sin N_2\right] = S\sqrt{1+\tan^2 N}$$

即

$$\left[\cos N_2 \cos(E-E_2) + \sin N_2(B\sin E - A\cos E)/C\right]^2 = S^2\left[1+(B\sin E - A\cos E)^2/C^2\right]$$

化简后得:

$$(p^2 - B^2 S^2)\sin^2 E + (q^2 - A^2 S^2)\cos^2 E + 2(ABS^2 - pq)\sin E\cos E - C^2 S^2 = 0$$

即

$$(p^2 - B^2 S^2 - C^2 S^2)\sin^2 E + 2(ABS^2 - pq)\sin E\cos E + (q^2 - A^2 S^2 - C^2 S^2)\cos^2 E = 0$$

式中:

$$p = B\sin N_2 + C\cos N_2 \sin E_2$$
$$q = A\sin N_2 - C\cos N_2 \cos E_2 \qquad (1\text{-}4\text{-}16)$$

再令

$$\lambda = p^2 - B^2 S^2 - C^2 S^2$$
$$\xi = ABS^2 - pq$$
$$\eta = q^2 - A^2 S^2 - C^2 S^2$$

则有

$$\lambda\sin^2 E + 2\xi\sin E\cos E + \eta\cos^2 E = 0 \qquad (1\text{-}4\text{-}17)$$

即

$$\lambda\tan^2 E + 2\xi\tan E + \eta = 0 \qquad (1\text{-}4\text{-}18)$$

如 $\lambda = 0$,由式(1-4-18)得:

$$E = 90°\text{、}-90°,\tan E = -\eta/(2\xi)$$

如 $\lambda \neq 0$,$\xi^2 - \lambda\eta = S^2\left[(Ap-Bq)^2 + C^2(p^2+q^2) - C^2 S^2\right] \geq 0$,由式(1-4-17)得:

$$\tan E = (-\xi \pm \sqrt{\xi^2 - \lambda\eta})/\lambda$$

P 点的经度 E 应取点 P_1 和 P_2 之间的值,确定了 E 值之后,再由式(1-4-15)计算 $\tan N$,求出的 N 值即 P 点的纬度。若 $\tan N > 0$,则 N 为北纬;若 $\tan N < 0$,则 N 为南纬。

对于 $E_1 = E_2$,即航段沿经线的特殊情况,$C=0$,式(1-4-15)不能用,P 点的经纬度可按以下方法计算:

由于 P 点在航段上,显然有 $E = E_1 = E_2$。此结论也可由公式导出:

由于 $C=0$,故:

$$p = B\sin N_2 = \cos E_1 \sin(N_2 - N_1)\sin N_2$$
$$q = A\sin N_2 = \sin E_1 \sin(N_2 - N_1)\sin N_2$$
$$\xi^2 - \lambda\eta = S^2(Ap - Bq)^2 = 0$$

则

$$\tan E = -\xi/\lambda = (pq - ABS^2)/(p^2 - B^2 S^2) = A/B = \tan E_1$$

所以

$$E = E_1$$

由式（1-4-14）得：

$$\cos(N - N_2) = \cos(D/R)$$

即

$$N - N_2 = \pm D/R$$

P 点纬度：

$$N = N_2 \pm D/R（弧度）$$

N 应取点 P_1 和 P_2 之间的纬度值，此式也可写为：

$$N = N_2 + 57.29578 \times (D/R) \times (N_1 - N_2)/|N_1 - N_2|（度）$$

N_1、N_2 分别为点 P_1、P_2 的纬度，如为南纬则代入负值。若 $N > 0$，则 N 为北纬；否则，为南纬。

也可以用迭代法计算 P 点经纬度，参见图 1-4-8 与图 1-4-9。

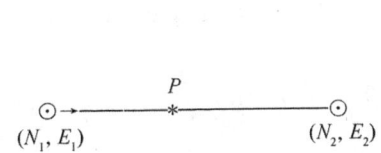

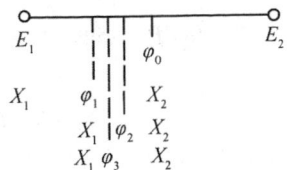

图 1-4-8　航段上任一点经纬度的确定　　图 1-4-9　迭代计算的示意图

取 $E = \varphi_0 = (E_1 + E_2)/2$，代入式（1-4-10），即：

$$\tan\theta = \cos N_1 \cos N_2 \sin(E_1 - E_2)/[\cos N_1 \sin N_2 \sin(E_1 - E) - \cos N_2 \sin N_1 \sin(E_2 - E)]$$

求出 θ（取 $[0°, 180°]$）后代入下式计算：

$$\Delta Y = \sin\theta \cos N_2 \cos(E_2 - E) + \cos\theta \sin N_2 - \cos(D/R) \tag{1-4-19}$$

注：南纬、西经以负值代入。若 $\Delta Y > 0$，表示 P 在 P_i 左面。

若 $\Delta Y > 0$，取 $E = \varphi_1 = (\varphi_0 + E_1)/2$，且令 $X_1 = E_1$，$X_2 = \varphi_0$，

若 $\Delta Y < 0$，取 $E = \varphi_1 = (\varphi_0 + E_2)/2$，且令 $X_1 = \varphi_0$，$X_2 = E_2$，

代入式（1-4-10），再次计算 θ、ΔY，若 $\Delta Y \neq 0$：

若 $\Delta Y > 0$，取 $E = \varphi_2 = (X_1 + \varphi_1)/2$，同时令 $X_2 = \varphi_1$，

若 $\Delta Y < 0$，取 $E = \varphi_2 = (\varphi_1 + X_2)/2$，同时令 $X_1 = \varphi_1$，

代入式（1-4-10），再次计算 θ、ΔY，若 $\Delta Y \neq 0$，则重复上述步骤，直到 $\Delta Y = 0$ 或 $|\varphi_n - \varphi_{n-1}|$ 小于指定误差。最后一次的 E 即为所求的经度，而纬度 $N = 90° - \theta$（负值表示南纬）。

例 1-4-2：设航段起点 P_1 的经纬度为：N20°，E085°，终点 P_2 的经纬度为：N30°，E095°，距终点 P_2 800 km 的 P 点的经纬度是：N24°44.6′，E089°27.5′。

例 1-4-3：设航段起点 P_1 的经纬度为：S10°，E005°，终点 P_2 的经纬度为：N05°，W006°，距终点 P_2 400 km 的 P 点的经纬度是：N02°05.5′，W003°52.6′。

例 1-4-4：设航段起点 P_1 的经纬度为：N20°，E008°，终点 P_2 的经纬度为：N20°，W005°，两点间大圆距离为 1357.4 km，起点大圆航向（真航向）272.23°，距终点 P_2 400 km 的 P 点的经纬度是：N20°05.9′，W001°10.2′。

1.4.3　计算距两定点大圆距离为定值的轨迹交点的经纬度

制作 ETOPS 飞行计划时可能要确定这种交点。设两定点的经纬度分别是 (N_1, E_1) 和

(N_2,E_2)，距这两点大圆距离分别是 D_1 和 D_2 的轨迹的交点为 G_1 和 G_2，交点的经纬度可按以下方法确定。

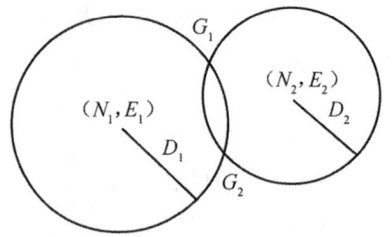

图 1-4-10　距二定点大圆距离为定值的交点

利用大圆距离公式即式(1-4-2)建立方程组：

$$D_1/R = \arccos\left[\cos N_1 \cos N\cos(E_1-E)+\sin N_1\sin N\right] \tag{1-4-20}$$

$$D_2/R = \arccos\left[\cos N_2 \cos N\cos(E_2-E)+\sin N_2\sin N\right] \tag{1-4-21}$$

令：

$$A=\cos(D_1/R),\; B=\cos(D_2/R)$$

则有：

$$\cos N_1 \cos N\cos(E_1-E)+\sin N_1\sin N = A \tag{1-4-22}$$

$$\cos N_2 \cos N\cos(E_2-E)+\sin N_2\sin N = B \tag{1-4-23}$$

由以上两式解出：

$$\cos N = \frac{A\sin N_2 - B\sin N_1}{\sin N_2\cos N_1\cos(E_1-E)-\sin N_1\cos N_2\cos(E_2-E)} \tag{1-4-24}$$

由式(1-4-22)得：

$$\left[A-\cos N_1\cos N\cos(E_1-E)\right]^2 = (\sin N_1\sin N)^2 = \sin^2 N_1 - \sin^2 N_1\cos^2 N$$

将式(1-4-24)代入上式，整理后得到关于 $\tan(E_1-E)$ 的方程，令：

$$X = \tan(E_1-E)$$

则有：

$$PX^2+QX+R=0 \tag{1-4-25}$$

式中：$P=(A\sin N_2-B\sin N_1)^2+\cos^2 N_2(A^2-\sin^2 N_1)\sin^2(E_2-E_1)$

$Q=-2\cos N_2\sin(E_2-E_1)\left[\cos N_2\cos(E_2-E_1)(A^2-\sin^2 N_1)+\cos N_1(\sin N_1\sin N_2-AB)\right]$

$R=(B^2-\sin^2 N_2)\cos^2 N_1+(A\sin N_2-B\sin N_1)^2+\cos(E_2-E_1)\cos N_2\times$

$\left[\cos(E_2-E_1)\cos N_2(A^2-\sin^2 N_1)+2\cos N_1(\sin N_1\sin N_2-AB)\right]$

当 $Q^2-4PR<0$ 时，两个圆无交点；当 $Q^2-4PR\geq0$ 时，可由式(1-4-25)解出 X，求反正切得出 $\Delta E=E_1-E$，当 $X>0$ 时，$\Delta E>0$，反之，$\Delta E<0$。$E=E_1-\Delta E$，当 E_1 为西经时，以 $360°-E_1$(E_1 本身为正)作为 E_1 代入上式计算 E。当求出 $0°<E<180°$ 时，E 为东经；当求出 $E>180°$ 时，取 $360°-E$ 为西经；当求出 $E>360°$ 时，取 $E-360°$ 为东经；当求出 $E<0°$ 时，取 $-E$ 为西经。确定 E 之后，由式(1-4-24)确定 $\cos N$(恒为正)，式(1-4-22)确定 $\sin N$，求反正弦确定 N，若 $N>0$，为北纬；否则，取 $-N$ 为南纬。

几个特例：

(1)如 $E_1=E_2$，则 $Q=0$，式(1-4-25)成为：

$$PX^2+R=0 \tag{1-4-26}$$

$$P = (A\sin N_2 - B\sin N_1)^2$$
$$R = A^2 + B^2 - 2AB\cos(N_1 - N_2) - \sin^2(N_1 - N_2)$$

（2）如 $N_1 = N_2$，则：

$$P = (A-B)^2\sin^2 N_1 + \cos^2 N_1(A^2 - \sin^2 N_1)\sin^2(E_2 - E_1)$$
$$Q = -2\cos^2 N_1\sin(E_2 - E_1)\left[\cos(E_2 - E_1)(A^2 - \sin^2 N_1) + \sin^2 N_1 - AB\right] \tag{1-4-27}$$
$$R = (B^2 - \sin^2 N_1)\cos^2 N_1 + (A-B)^2\sin^2 N_1 + \cos(E_2 - E_1)\cos^2 N_1 \times$$
$$\left[\cos(E_2 - E_1)(A^2 - \sin^2 N_1) + 2(\sin^2 N_1 - AB)\right]$$

（3）如 $N_1 = N_2$，且 $D_1 = D_2$，即 $A = B$，则：

$$P = \cos^2 N_1(A^2 - \sin^2 N_1)\sin^2(E_2 - E_1)$$
$$Q = -2\cos^2 N_1\sin(E_2 - E_1)\left[\cos(E_2 - E_1)(A^2 - \sin^2 N_1) + \sin^2 N_1 - A^2\right] \tag{1-4-28}$$
$$R = (A^2 - \sin^2 N_1)\cos^2 N_1 + \cos(E_2 - E_1)\cos^2 N_1 \times \left[\cos(E_2 - E_1)(A^2 - \sin^2 N_1) + 2(\sin^2 N_1 - A^2)\right]$$

此时 $Q^2 - 4PR = 0$，$X = -Q/(2P)$，即 $\tan(E_1 - E) = \tan\dfrac{E_1 - E_2}{2}$，$E = \dfrac{E_1 + E_2}{2}$。

式（1-4-24）成为 0/0 型，此时可由式（1-4-22）导出：

$$\sin N = \frac{A\sin N_1 \pm \cos N_1\cos(E_1 - E)\sqrt{\sin^2 N_1 - A^2 + \cos^2 N_1\cos^2(E_1 - E)}}{\sin^2 N_1 + \cos^2 N_1\cos^2(E_1 - E)} \tag{1-4-29}$$

或

$$\tan N = \frac{-\sin N_1\cos N_1\cos(E_1 - E) \pm A\sqrt{\sin^2 N_1 - A^2 + \cos^2 N_1\cos^2(E_1 - E)}}{\sin^2 N_1 - A^2} \tag{1-4-30}$$

确定 E 之后，由式（1-4-29）或式（1-4-30）确定 N。$N > 0$，则为北纬；反之，则为南纬。

确定 E 时，若 E_1、E_2 为东经，则 $E = (E_1 + E_2)/2$；若 E_1、E_2 为西经，则用 $\varphi_1 = 360° - E_1$，$\varphi_2 = 360° - E_2$（E_1、E_2 本身为正）分别作为 E_1、E_2 代入 $E = (E_1 + E_2)/2$ 确定 E，$E > 180°$ 时，取 $360° - E$ 为西经；若 φ_1、φ_2 的差值大于 $180°$，则计算出的 E 值加 $180°$ 才是经度值，如 $E' = E + 180° > 360°$，则 $E' - 360°$（即 $E - 180°$）为东经值。

如：P_1（N20°，E50°），P_2（N20°，W90°），$\varphi_1 = 50°$，$\varphi_2 = 360° - 90° = 270°$，$E = 160°$，交点的经度 $= 160° + 180°$，即 E340° 或 W20°。又如：P_1（N20°，E50°），P_2（N20°，W30°），$\varphi_1 = 50°$，$\varphi_2 = 330°$，$E = 190°$，$E' = 370°$，交点的经度 $= 370° - 360° = $ E10°，见图 1-4-11。

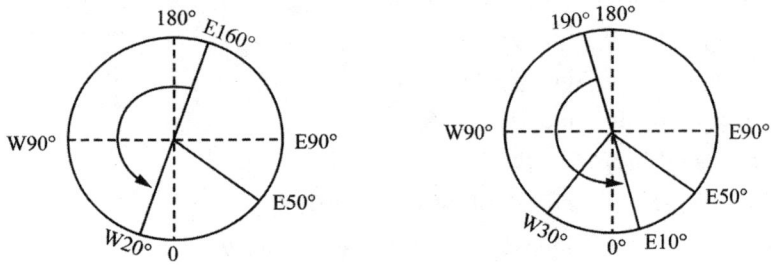

图 1-4-11 计算交点经纬度的示意图

例 1-4-5：设点 P_1 经纬度为 S05°，E130°，点 P_2 经纬度为 N05°，E145°，距 P_1 点大圆距离为 600 NM（1111.2 km）的轨迹和距 P_2 点大圆距离为 650 NM（1203.8 km）的轨迹的两个交点经纬度分别是：N04°05.2′，E134°10.4′和 S04°37.5′，E140°01.6′。

例 1-4-6：设点 P_1 经纬度为 S05°，E175°，点 P_2 经纬度为 N05°，W170°，距 P_1 点大圆距离为 600 NM（1111.2 km）的轨迹和距 P_2 点大圆距离为 650 NM（1203.8 km）的轨迹的两个交点经纬度分别是：N04°05.2′，E179°10.4′和 S04°37.5′，W174°58.4′。

1.4.4　计算两个航段交点的经纬度

如图 1-4-12 所示，在两个航段交叉或一个航段穿越飞行指挥区边界时，需要给出交点的经纬度，交点的经纬度可按以下方法确定。设航段 1 两端点的经纬度分别是 $P_{11}(N_{11}，E_{11})$ 和 $P_{12}(N_{12}，E_{12})$，航段 2（或飞行指挥区的一段边界）两端点的经纬度分别是 $P_{21}(N_{21}，E_{21})$ 和 $P_{22}(N_{22}，E_{22})$，其交点为 $P(N，E)$。

把各点的经纬度按下述公式换算为关于 θ_{11}、φ_{11}、θ_{12}、φ_{12}、θ_{21}、φ_{21}、θ_{22}、φ_{22} 和 θ、φ 的球坐标：

$\theta = 90° - N$（南纬代入负值）

$\varphi = E$（东经，E 取正值）

$\varphi = 360° + E$（西经，E 取负值）

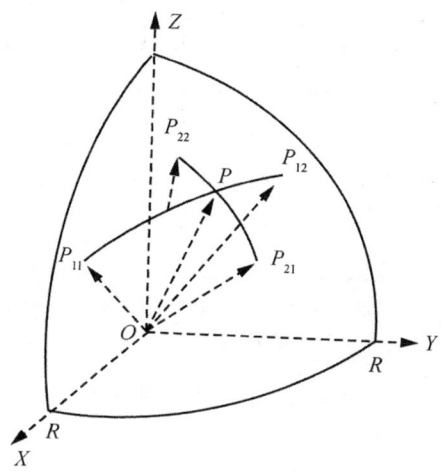

图 1-4-12　球面上两个航段的交点

设地球半径为 R，则 P_{11}、P_{12}、P_{21}、P_{22}、P 点的矢径为：

$$\boldsymbol{r}_{11} = R\sin\theta_{11}\cos\varphi_{11}\boldsymbol{i} + R\sin\theta_{11}\sin\varphi_{11}\boldsymbol{j} + R\cos\theta_{11}\boldsymbol{k}$$

$$\boldsymbol{r}_{12} = R\sin\theta_{12}\cos\varphi_{12}\boldsymbol{i} + R\sin\theta_{12}\sin\varphi_{13}\boldsymbol{j} + R\cos\theta_{12}\boldsymbol{k}$$

$$\boldsymbol{r}_{21} = R\sin\theta_{21}\cos\varphi_{21}\boldsymbol{i} + R\sin\theta_{21}\sin\varphi_{21}\boldsymbol{j} + R\cos\theta_{21}\boldsymbol{k}$$

$$\boldsymbol{r}_{22} = R\sin\theta_{22}\cos\varphi_{22}\boldsymbol{i} + R\sin\theta_{22}\sin\varphi_{23}\boldsymbol{j} + R\cos\theta_{22}\boldsymbol{k}$$

$$\boldsymbol{r} = R\sin\theta\cos\varphi\boldsymbol{i} + R\sin\theta\sin\varphi\boldsymbol{j} + R\cos\theta\boldsymbol{k}$$

由三矢量共面条件：$\boldsymbol{r} \cdot \boldsymbol{r}_{11} \times \boldsymbol{r}_{12} = 0$ 和 $\boldsymbol{r} \cdot \boldsymbol{r}_{21} \times \boldsymbol{r}_{22} = 0$，得：

$$\begin{vmatrix} \sin\theta\cos\varphi & \sin\theta\sin\varphi & \cos\theta \\ \sin\theta_{11}\cos\varphi_{11} & \sin\theta_{11}\sin\varphi_{11} & \cos\theta_{11} \\ \sin\theta_{12}\cos\varphi_{12} & \sin\theta_{12}\sin\varphi_{12} & \cos\theta_{12} \end{vmatrix} = 0$$

和

$$\begin{vmatrix} \sin\theta\cos\varphi & \sin\theta\sin\varphi & \cos\theta \\ \sin\theta_{21}\cos\varphi_{21} & \sin\theta_{21}\sin\varphi_{21} & \cos\theta_{21} \\ \sin\theta_{22}\cos\varphi_{22} & \sin\theta_{22}\sin\varphi_{22} & \cos\theta_{22} \end{vmatrix} = 0$$

即

$$A_1\sin\theta\cos\varphi - B_1\sin\theta\sin\varphi + C_1\cos\theta = 0 \tag{1-4-31}$$

$$A_2\sin\theta\cos\varphi - B_2\sin\theta\sin\varphi + C_2\cos\theta = 0 \tag{1-4-32}$$

式中：

$$\begin{aligned}
A_1 &= \sin\theta_{11}\sin\varphi_{11}\cos\theta_{12} - \sin\theta_{12}\sin\varphi_{12}\cos\theta_{11} \\
&= \cos N_{11}\sin E_{11}\sin N_{12} - \cos N_{12}\sin E_{12}\sin N_{11} \\
B_1 &= \sin\theta_{11}\cos\varphi_{11}\cos\theta_{12} - \sin\theta_{12}\cos\varphi_{12}\cos\theta_{11} \\
&= \cos N_{11}\cos E_{11}\sin N_{12} - \cos N_{12}\cos E_{12}\sin N_{11} \\
C_1 &= \sin\theta_{11}\cos\varphi_{11}\sin\theta_{12}\sin\varphi_{12} - \sin\theta_{12}\cos\varphi_{12}\sin\theta_{11}\sin\varphi_{11} \\
&= \sin\theta_{11}\sin\theta_{12}\sin(\varphi_{12}-\varphi_{11}) = \cos N_{11}\cos N_{12}\sin(E_{12}-E_{11}) \\
A_2 &= \sin\theta_{21}\sin\varphi_{21}\cos\theta_{22} - \sin\theta_{22}\sin\varphi_{22}\cos\theta_{21} \\
&= \cos N_{21}\sin E_{21}\sin N_{22} - \cos N_{22}\sin E_{22}\sin N_{21} \\
B_2 &= \sin\theta_{21}\cos\varphi_{21}\cos\theta_{22} - \sin\theta_{22}\cos\varphi_{22}\cos\theta_{21} \\
&= \cos N_{21}\cos E_{21}\sin N_{22} - \cos N_{22}\cos E_{22}\sin N_{21} \\
C_2 &= \sin\theta_{21}\cos\varphi_{21}\sin\theta_{22}\sin\varphi_{22} - \sin\theta_{22}\cos\varphi_{22}\sin\theta_{21}\sin\varphi_{21} \\
&= \sin\theta_{21}\sin\theta_{22}\sin(\varphi_{22}-\varphi_{21}) = \cos N_{21}\cos N_{22}\sin(E_{22}-E_{21})
\end{aligned} \tag{1-4-33}$$

几种特殊情况：

①若 $\varphi_{11}=\varphi_{12}$，$\varphi_{21}\neq\varphi_{22}$，即航段 1 为经线上的一段，航段 2 不沿经线，则：

$$C_1=0, A_1=\sin\varphi_{11}\sin(\theta_{11}-\theta_{12}), B_1=\cos\varphi_{11}\sin(\theta_{11}-\theta_{12}), C_2\neq0$$

代入式（1-4-31）得：

$$\sin(\varphi_{11}-\varphi)=0$$

即 $\varphi=\varphi_{11}$，这个结果是很显然的。

由式（1-4-32）得：

$$\tan\theta = C_2/(B_2\sin\varphi_{11}-A_2\cos\varphi_{11})$$

或

$$\tan N = (B_2\sin\varphi_{11}-A_2\cos\varphi_{11})/C_2 \tag{1-4-34}$$

θ 应取 $(0°,\pi)$。φ 应取 $(\varphi_{21},\varphi_{22})$，且 θ 应取 $(\theta_{11},\theta_{12})$，否则两个航段本身不相交，而是交于延长线上。

②若 $\varphi_{21}=\varphi_{22}$，$\varphi_{11}\neq\varphi_{12}$，即航段 2 为经线上的一段，航段 1 不沿经线，则：

$$C_2=0, A_2=\sin\varphi_{21}\sin(\theta_{21}-\theta_{22}), B_2=\cos\varphi_{21}\sin(\theta_{21}-\theta_{22}), C_1\neq0$$

代入式（1-4-32）得：

$$\sin(\varphi_{21}-\varphi)=0$$

即

$$\varphi=\varphi_{21}$$

由式（1-4-31）得：

$$\tan\theta = C_1/(B_1\sin\varphi_{21}-A_1\cos\varphi_{21})$$

或

$$\tan N = (B_1 \sin\varphi_{21} - A_1 \cos\varphi_{21})/C_1 \qquad (1\text{-}4\text{-}35)$$

θ 应取 $(0°, \pi)$。φ 应取 $(\varphi_{11}, \varphi_{12})$，且 θ 应取 $(\theta_{21}, \theta_{22})$，否则两个航段本身不相交，而是交于延长线上。

③若 $\varphi_{21} = \varphi_{22}$，$\varphi_{11} = \varphi_{12}$，即两个航段都为经线上的一段，则：$C_1 = 0$，$C_2 = 0$，$\theta = 0°$ 或 π，即 $N = \pm 90°$，两个航段本身或延长后交于南、北两极。

对于一般情况：

由式 $(1\text{-}4\text{-}31) \times C_2 -$ 式 $(1\text{-}4\text{-}32) \times C_1$ 得：

$$\tan\varphi = (A_2 C_1 - A_1 C_2)/(B_2 C_1 - B_1 C_2) \qquad (1\text{-}4\text{-}36)$$

当 $\tan\varphi > 0$ 时，φ 取 $(0°, 90°)$ 或 $(180°, 270°)$，否则 φ 取 $(90°, 180°)$ 或 $(270°, 360°)$。然后由式 $(1\text{-}4\text{-}31)$ 得：

$$\tan\theta(A_1 \cos\varphi - B_1 \sin\varphi) + C_1 = 0$$

若 $(A_1 \cos\varphi - B_1 \sin\varphi) \neq 0$，则：

$$\tan\theta = C_1/(B_1 \sin\varphi - A_1 \cos\varphi) \qquad (1\text{-}4\text{-}37)$$

否则：$\theta = \pi/2 = 90°$。

φ 应取 $(\varphi_{11}, \varphi_{12})$、$(\varphi_{21}, \varphi_{22})$，$\theta$ 应取 $(\theta_{11}, \theta_{12})$、$(\theta_{21}, \theta_{22})$。$\varphi$ 为 $0° \sim 180°$，则为东经；φ 为 $180° \sim 360°$，则 $360° - \varphi$ 为西经。$90° - \theta$ 为纬度，正值为北纬，负值为南纬。

例 1-4-7：航段 1 起点（朝阳）经纬度为 N41°32.3′，E120°25.9′，终点（开原）经纬度为 N42°34.2′，E124°00.4′；航段 2 起点（大虎山）经纬度为 N41°38.2′，E122°07.7′，终点（P186）经纬度为 N44°45.0′，E124°52.2′，航段 2 实际是扇区分界线。计算结果是两段交点坐标为 N42°10.5′，E122°34.8′，航图上标注的是 N42°09.7′，E122°34.9′。

例 1-4-8：航段 1 起点 P_{11} 经纬度为 N05°0′，W03°0′，终点 P_{12} 经纬度为 S08°0′，E5°0′；航段 2 起点 P_{21} 经纬度为 S07°0′，W04°0′，终点 P_{22} 经纬度为 N04°0′，E6°0′。计算结果是：两段交点坐标为 S01°31.3′，E000°59.9′。

例 1-4-9：航段 1 起点 P_{11} 经纬度为 N5°0′，E178°0′，终点 P_{12} 经纬度为 S4°0′，W177°0′；航段 2 起点 P_{21} 经纬度为 S3°0′，E177°0′，终点 P_{22} 经纬度为 N4°0′，W178°0′。计算结果是：两段交点坐标为 N01°17.5′，W179°56.2′。

1.4.5 计算航段上切点的经纬度

如图 1-4-13 所示，若 P_0 是航段 $P_1 P_2$ 之外的一点，P_0 在该航段上投影点 P，即由 P_0 向航段引垂线所得的垂足 P，称为航段与 P_0 相切之切点，简称切 P_0。设航段两端点的经纬度分别是 $P_1(N_1, E_1)$ 和 $P_2(N_2, E_2)$，P_0 点的经纬度为 (N_0, E_0)，切 P_0 点 P 的经纬度 $P(N, E)$ 可按以下方法导出。

把各点的经纬度按下述公式换算为关于 θ_1、φ_1、θ_2、φ_2、θ_0、φ_0、θ、φ 的球坐标：

$\theta = 90° - N$（南纬代入负值）

$\varphi = E$（东经，E 取正值）

$\varphi = 360° + E$（西经，E 取负值）

设地球半径为 R，则 P_1、P_2、P 点的矢径为：

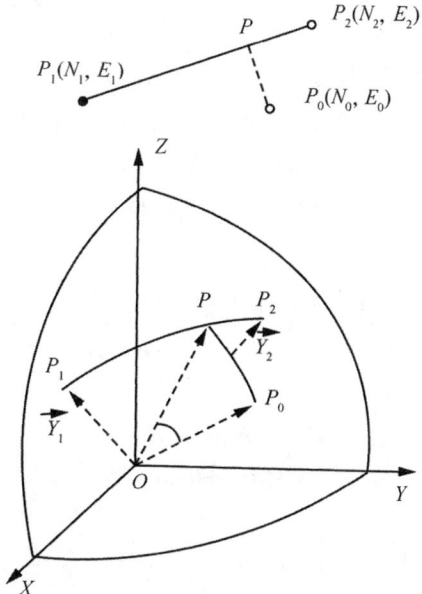

图 1-4-13　球面上航段的切点

$$r_1 = R\sin\theta_1\cos\varphi_1 i + R\sin\theta_1\sin\varphi_1 j + R\cos\theta_1 k$$
$$r_2 = R\sin\theta_2\cos\varphi_2 i + R\sin\theta_2\sin\varphi_3 j + R\cos\theta_2 k$$
$$r = R\sin\theta\cos\varphi i + R\sin\theta\sin\varphi j + R\cos\theta k$$

由三矢量共面条件：$r \cdot r_1 \times r_2 = 0$，得：

$$\begin{vmatrix} \sin\theta\cos\varphi & \sin\theta\sin\varphi & \cos\theta \\ \sin\theta_1\cos\varphi_1 & \sin\theta_1\sin\varphi_1 & \cos\theta_1 \\ \sin\theta_2\cos\varphi_2 & \sin\theta_2\sin\varphi_2 & \cos\theta_2 \end{vmatrix} = 0$$

即

$$A\sin\theta\cos\varphi - B\sin\theta\sin\varphi + C\cos\theta = 0$$

或

$$A\cos N\cos E - B\cos N\sin E + C\sin N = 0 \tag{1-4-38}$$

式中：

$$\begin{aligned} A &= \sin\theta_1\sin\varphi_1\cos\theta_2 - \sin\theta_2\sin\varphi_2\cos\theta_1 \\ &= \cos N_1\sin E_1\sin N_2 - \cos N_2\sin E_2\sin N_1 \\ B &= \sin\theta_1\cos\varphi_1\cos\theta_2 - \sin\theta_2\cos\varphi_2\cos\theta_1 \\ &= \cos N_1\cos E_1\sin N_2 - \cos N_2\cos E_2\sin N_1 \\ C &= \sin\theta_1\cos\varphi_1\sin\theta_2\sin\varphi_2 - \sin\theta_2\cos\varphi_2\sin\theta_1\sin\varphi_1 \\ &= \sin\theta_1\sin\theta_2\sin(\varphi_2-\varphi_1) = \cos N_1\cos N_2\sin(E_2-E_1) \end{aligned} \tag{1-4-39}$$

P_0 与航路上任一点 P 的矢径的数量积 y 为：

$$y = R^2\cos\omega = R^2\left[\sin\theta\sin\theta_0\cos(\varphi-\varphi_0) + \cos\theta\cos\theta_0\right]$$

或

$$y = R^2\cos\omega = R^2\left[\cos N\cos N_0\cos(E-E_0) + \sin N\sin N_0\right] \tag{1-4-40}$$

当 P 为切 P_0 点，即 P_0 的垂足时，PP_0 弧最短，ω 最小，所以切点 P 使 $d\cos\omega/d\varphi = 0$ 或使

$\mathrm{d}\cos\omega/\mathrm{d}E=0$，注意：$P$ 点的经纬度要满足式（1-4-38），故 θ 是 φ 的函数、N 是 E 的函数。由式（1-4-38）得：

$$\tan N=(B\sin E-A\cos E)/C \tag{1-4-41}$$

代入由式（1-4-40）得到的：$\cos\omega=\cos N\cos N_0\cos(E-E_0)+\sin N\sin N_0$，则有：

$$\cos\omega=\cos N\left[\cos N_0\cos(E-E_0)+\sin N_0\tan N\right]$$

$$=\left[\cos N_0\cos(E-E_0)+\sin N_0\tan N\right]/\sqrt{\tan^2 N+1}$$

$$=\left[\cos N_0\cos(E-E_0)+\sin N_0(B\sin E-A\cos E)/C\right]\times|C|/\sqrt{(B\sin E-A\cos E)^2+C^2}$$

由此式计算 $\mathrm{d}\cos\omega/\mathrm{d}E$ 并令其等于 0，经整理化简后得：

$$\left[C^2\eta+A(\xi B+\eta A)\right]\cos E-\left[C^2\xi+B(\xi B+\eta A)\right]\sin E=0$$

式中：

$$\xi=\cos N_0\cos E_0-A\sin N_0/C \tag{1-4-42}$$

$$\eta=\cos N_0\sin E_0+B\sin N_0/C$$

由此解出：

$$\tan E=\left[C^2\eta+A(\xi B+\eta A)\right]/\left[C^2\xi+B(\xi B+\eta A)\right] \tag{1-4-43}$$

若 $\tan E>0$，则取 E 为 $0°\sim90°$（E 是东经）或 $180°\sim270°$（$360°-E$ 是西经）。

若 $\tan E<0$，则取 E 为 $90°\sim180°$（E 是东经）或 $270°\sim360°$（$360°-E$ 是西经）。

两个解中只有介于 P_1、P_2 点的经度之间的那个解是所需要的。

确定 E 之后，代入式（1-4-41）计算 $\tan N$。若 $\tan N>0$，N 为北纬；反之，$-N$ 为南纬。

对于航段 P_1P_2 沿经线的特殊情况，式（1-4-41）～式（1-4-43）不适用，求解如下：

此时

$$E_1=E_2，C=0，A=\sin E_1\sin(N_2-N_1)，B=\cos E_1\sin(N_2-N_1)$$

由式（1-4-38）得：

$$\tan E=A/B=\tan E_1，E=E_1$$

由式（1-4-40）计算 $\mathrm{d}y/\mathrm{d}N$，令其等于 0，得：

$$\sin N_0\cos N-\cos N_0\cos(E-E_0)\sin N=0$$

由此解出：

$$\tan N=\tan N_0/\cos(E-E_0) \tag{1-4-44}$$

若 $\tan N>0$，则 N 为北纬；反之，$-N$ 为南纬。

例 1-4-10：航段起点 P_1（涪陵）经纬度为 N29°42.2′，E107°22.6′，终点 P_2（老粮仓）经纬度为 N28°04.5′，E112°12.6′，航段外之点 P_0（花垣）经纬度为 N28°34.6′，E109°27.1′，计算得航段上切点花垣的经纬度为 N28°58.4′，E109°37.5′，航图上标注的是 N28°57.1′，E109°36.9′。

例 1-4-11：航段起点 P_1 经纬度为 N01°00.0′，E177°00.0′，终点 P_2 经纬度为 S01°0′，W178°00.0′，航段外之点 P_0 经纬度为 N01°00.0′，W179°0′，计算得航段上切 P_0 点的经纬度为 S00°22.8′，W179°33.1′。

例 1-4-12：航段起点 P_1 经纬度为 N01°00.0′，W002°00.0′，终点 P_2 经纬度为 S01°0′，E001°00.0′，航段外之点 P_0 经纬度为 N01°00.0′，E001°0′，计算得航段上切 P_0 点的经纬度为 S00°23.1′，E000°04.6′。

1.4.6　计算无风时等时点的经纬度

在制作 ETOPS 飞行计划时需要确定到相邻的两个航路备降机场的等时点 ETP_1、ETP_2 等的经纬度,无风时等时点就是到相邻的两个航路备降机场大圆距离相等的点 CP_1、CP_2 等,见图 1-4-14,下面推导计算这种点的经纬度的公式。设航段两端点的经纬度分别是 $P_1(N_1,E_1)$ 和 $P_2(N_2,E_2)$,机场 G 和机场 H 的经纬度分别为 (N_G,E_G),(N_H,E_H),设到机场 G 和 H 的等时点为 P,P 点的经纬度是 (N,E)。

把各点的经纬度按下述公式换算为球坐标 θ_1、φ_1、θ_2、φ_2、θ、φ、θ_G、φ_G、θ_H、φ_H:

$\theta=90°-N$(南纬代入负值)

$\varphi=E$(东经,E 取正值)

$\varphi=360°+E$(西经,E 取负值)

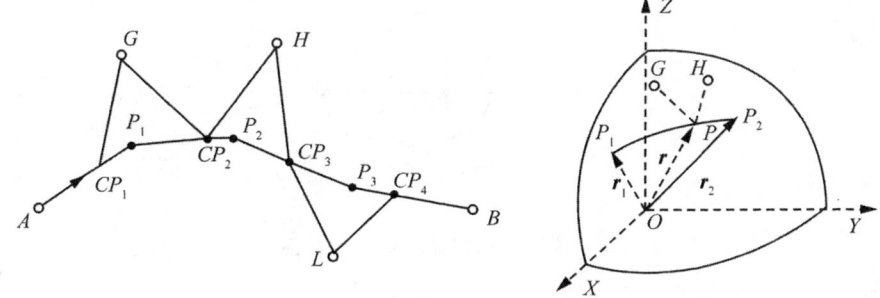

图 1-4-14　临界点(等时点)示意图

由于 P 到 G 和 H 的大圆距离相等,所以由式(1-4-1)得:

$$\sin\theta\sin\theta_G\cos(\varphi-\varphi_G)+\cos\theta\cos\theta_G=\sin\theta\sin\theta_H\cos(\varphi-\varphi_H)+\cos\theta\cos\theta_H \tag{1-4-45}$$

即

$$\cos N\cos N_G\cos(E-E_G)+\sin N\sin N_G=\cos N\cos N_H\cos(E-E_H)+\sin N\sin N_H$$

由三矢量共面条件:$\boldsymbol{r}\cdot\boldsymbol{r}_1\times\boldsymbol{r}_2=0$,得式(1-4-38)与式(1-4-39):

由式(1-4-38)得:

$$\sin\theta(A\cos\varphi-B\sin\varphi)+C\cos\theta=0$$

当航段 P_1P_2 不在赤道上时,$\cos\theta\neq0$,于是有:

$$\tan\theta=C/(B\sin\varphi-A\cos\varphi)$$

或

$$\tan N=(B\sin\varphi-A\cos\varphi)/C \tag{1-4-46}$$

由式(1-4-45)得:

$$\sin\theta[\sin\theta_G\cos(\varphi-\varphi_G)-\sin\theta_H\cos(\varphi-\varphi_H)]=\cos\theta(\cos\theta_H-\cos\theta_G)$$

$$\tan\theta[\sin\theta_G\cos(\varphi-\varphi_G)-\sin\theta_H\cos(\varphi-\varphi_H)]=\cos\theta_H-\cos\theta_G$$

将式(1-4-46)代入上式,得:

$$C[\sin\theta_G\cos(\varphi-\varphi_G)-\sin\theta_H\cos(\varphi-\varphi_H)]=(B\sin\varphi-A\cos\varphi)(\cos\theta_H-\cos\theta_G)$$

化简得:

$$\xi\sin\varphi=\eta\cos\varphi \tag{1-4-47}$$

式中:

$$\xi = B(\cos\theta_H - \cos\theta_G) - C(\sin\theta_G\sin\varphi_G - \sin\theta_H\sin\varphi_H)$$

$$\eta = A(\cos\theta_H - \cos\theta_G) + C(\sin\theta_G\cos\varphi_G - \sin\theta_H\cos\varphi_H)$$

①若 $\xi = 0$，$\eta \neq 0$，则 $\varphi = 90°$ 或 $270°$。

②若 $\xi = 0$，$\eta = 0$，则 φ 可取航段上的任一经度值，即航段上任一点到两个机场的大圆距离都相等。

③若 $\xi \neq 0$，则 $\tan\varphi = \eta/\xi$，如 $\tan\varphi > 0$，φ 取 $0° \sim 90°$、$180° \sim 270°$；如 $\tan\varphi < 0$，φ 取 $90° \sim 180°$、$270° \sim 360°$。

得出的两个角度应取介于航段两端点经度的那个，若两个角度都不在航段两端点经度之间，则表示到两个机场大圆距离相等的点不在该航段上（在其延长线上）。

设两端点经度 φ_1、φ_2（$0° \sim 360°$）中 $\varphi_1 < \varphi_2$，当 $\varphi_2 - \varphi_1 < 180°$ 时，取满足 $\varphi_1 \leq \varphi \leq \varphi_2$ 的一个解，当 $\varphi_2 - \varphi_1 > 180°$ 时，取满足 $0° \leq \varphi \leq \varphi_1$ 或 $\varphi_2 \leq \varphi \leq 360°$ 的一个解，见图 1-4-15。确定 φ 之后，由式（1-4-46）计算 $\tan N$。如 $\tan N > 0$，N 为北纬；如 $\tan N < 0$，N 为南纬。

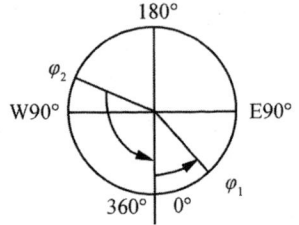

图 1-4-15　确定经度 φ 的示意图

特殊情况：

（1）航段在赤道上，此时 $A = 0$，$B = 0$，$C \neq 0$，由式（1-4-38）可得出 $N = 0$，若两个机场关于赤道对称，则 $\xi = 0$，$\eta = 0$，即上面的情况②，φ 可取航段上的任一经度值，即航段上任一点到两个机场的大圆距离都相等。若两个机场关于赤道不对称，但经度相同，如果经度 $= 0°$ 或 $180°$，则 $\xi = 0$，$\eta \neq 0$，即上面的情况①，$\varphi = 90°$ 或 $270°$，一般来说，这两个经度都不在航段中间，即航段上没有一点到两个机场的大圆距离相等。对于两个机场的其他位置一般属于上面的情况③。

（2）航段沿经线，$\varphi_1 = \varphi_2$，即 $E_1 = E_2$，此时 $C = 0$，由式（1-4-38）得 $\varphi = \varphi_1$，即 $E = E_1$。此时式（1-4-46）和式（1-4-47）不适用，可由式（1-4-45）求解 θ 或 N：

$$\tan\theta = (\cos\theta_H - \cos\theta_G)/[\sin\theta_G\cos(\varphi - \varphi_G) - \sin\theta_H\cos(\varphi - \varphi_H)] \qquad (1\text{-}4\text{-}48)$$

或由式（1-4-48）求解：

$$\tan N = [\sin\theta_G\cos(\varphi - \varphi_G) - \sin\theta_H\cos(\varphi - \varphi_H)]/(\cos\theta_H - \cos\theta_G)$$

若 $\tan N > 0$，N 为北纬；若 $\tan N < 0$，N 为南纬。

例 1-4-13：航段端点 P_1 经纬度为 N15°0′，E100°0′，P_2 经纬度为 N30°0′，E100°0′，机场 1 的经纬度为 N20°0′，E110°0′，机场 2 的经纬度为 N25°0′，E110°0′，航段上到两个机场大圆距离相等的点（ETP）的经纬度是 N22°11.5′，E110°0′。用计算出的 ETP 点的经纬度反向验算得：ETP 点到机场 1 的大圆距离为 1064.9 km，大圆航向为 101.4°，ETP 点到机场 2 的大圆距离为 1064.9 km，大圆航向为 71.0°。

例 1-4-14：航段端点 P_1 的经纬度为 S10°0′，W010°0′，P_2 的经纬度为 N05°0′，E010°0′，机场 1 的经纬度为 S15°0′，W005°0′，机场 2 的经纬度为 S10°0′，E008°0′，航段上到两个机场大圆距离相等的点（ETP）的经纬度是 S03°58.0′，W1°49.8′，用计算出的 ETP 点的经纬度反向验算得：ETP 点到机场 1 的大圆距离为 1274.4 km，大圆航向为 195.8°；ETP 点到机场 2 的大圆距离为

1274.5 km, 大圆航向为 135.6°。

1.4.7　计算航段终点经纬度

由航段起点经纬度、大圆航向及航段距离可计算航段终点经纬度。设起点的纬度为 N_1，经度为 E_1，大圆航向为 β，航段距离为 D，终点的纬度为 N，经度为 E，则由式（1-4-1）和式（1-4-3）可得以下方程组：

$$\sin N_1 \sin N + \cos N_1 \cos N \cos(E_1 - E) = \cos\omega \qquad (1\text{-}4\text{-}49)$$

$$\cos N_1 \sin N - \sin N_1 \cos N \cos(E_1 - E) = \cos\beta \sin\omega \qquad (1\text{-}4\text{-}50)$$

式中：$\omega = D/R$，R 为地球半径。不难由此方程组解出：

$$\sin N = \sin N_1 \cos\omega + \cos\beta \sin\omega \cos N_1 \qquad (1\text{-}4\text{-}51)$$

$$\cos(E_1 - E) = (\cos N_1 \cos\omega - \cos\beta \sin\omega \sin N_1)/\cos N \qquad (1\text{-}4\text{-}52)$$

由式（1-4-51）解出 N 之后，代入式（1-4-52）即可确定 E，N 为负值则表示是南纬。

如 $\beta = 0°$ 或 $180°$，则 $E = E_1$。

由于航段距离是大圆距离，所以 $|E_1 - E| \leqslant 180°$，故 $|E_1 - E|$ 应取反余弦主值区间的角，即：

$$|E_1 - E| = \theta = \arccos[(\cos N_1 \cos\omega - \cos\beta \sin\omega \sin N_1)/\cos N]$$

当 $\beta < 180°$ 时，$E_1 - E < 0°$，$E = E_1 + \theta$，反之 $E = E_1 - \theta$。

如 $E > 180°$，则取 $360° - E$ 为西经值，如 $-180° < E < 0°$，则取 $|E|$ 为西经值。

如 $E \leqslant -180°$，则取 $360° + E$ 为东经值。

例 1-4-15：航段起点经纬度为 N30°0′，E090°0′，大圆航向为 60°，航段距离为 500 km，试确定终点经纬度。

$\omega = D/R = 500/6368 = 4.4987°$

$\sin N = 0.5324237$，$N = 32.1694° = 32°10.16′$（北纬）

$\cos(E_1 - E) = 0.9967753$，$E_1 - E = 4.6025°$，$E = 94.6025° = 94°36.16′$

例 1-4-16：航段起点经纬度为 N04°0′，W003°0′，大圆航向为 150°，航段距离为 800 km，计算得终点经纬度为 S02°14.24′，E000°35.68′。

例 1-4-17：航段起点经纬度为 N02°0′，E176°0′，大圆航向为 120°，航段距离为 800 km，计算得终点经纬度为 S01°36.22′，W177°46.08′。

例 1-4-18：航段起点经纬度为 S02°0′，E177°0′，大圆航向为 50°，航段距离为 800 km，计算得终点经纬度为 N02°37.73′，W177°29.17′。

1.5　航路当量风和当量气温计算

1.5.1　当量风速计算

在使用简化图表制作飞行计划或把巡航段作为一段用平均燃油流量计算巡航油量时，对整个航路上的风和温度只能分别用一个风分量和温度来表示。如给出的是各航段上的风分量和温度,这时可使用下述的当量风和当量气温作为整条航线上的风分量和温度。使用当量风确定的空中距离与分段确定的空中距离之和相差很小,算出的油量之差也就很小,所以可用当量风作为整条航线上的风分量。对当量气温亦然。如图 1-5-1 所示,设已知一条航线上各航段的风矢量(如知道的是各航段端点的风矢量,可把航段两端点的风矢量的矢量平均作为这个航段上的风矢量)$V_{\mathrm{W}i}$,则可由下式计算出与这条航线的地面距离(即航程)D 对应的空中距离 D_{A}:

$$D_{\mathrm{A}} = \sum_i \frac{D_i V_i}{\sqrt{V_i^2 - (V_{\mathrm{W}i}\sin\theta_i)^2} + V_{\mathrm{W}i}\cos\theta_i} \tag{1-5-1}$$

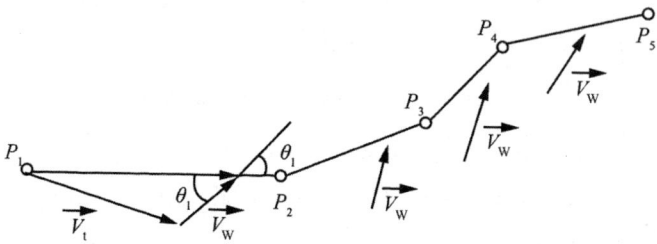

图 1-5-1　航路风示意图

式中:D_i——第 i 个航段的地面距离,$D = \sum_i D_i$,$\sum_i$ 表示对所有航段求和(下同);

　　　$V_{\mathrm{W}i}$——第 i 个航段的风速;

　　　θ_i——第 i 个航段的风矢量与该段航向的夹角;

　　　V_{t}——飞机真空速,设在整条航线上 V_{t} 为常数。

设对此航线存在一个"当量风速"W_{E},使得 $D_{\mathrm{A}} = D \times V_{\mathrm{t}} / (V_{\mathrm{t}} + W_{\mathrm{E}})$,即

$$\frac{DV_i}{V_i + W_{\mathrm{E}}} = \sum_i \frac{D_i V_i}{\sqrt{V_i^2 - (V_{\mathrm{W}i}\sin\theta_i)^2} + V_{\mathrm{W}i}\cos\theta_i} \tag{1-5-2}$$

由式(1-5-2)解出:

$$W_{\mathrm{E}} = \frac{D}{\sum_i \left[D_i / \left(\sqrt{V_i^2 - (V_{\mathrm{W}i}\sin\theta_i)^2} + V_{\mathrm{W}i}\cos\theta_i \right) \right]} - V_i \tag{1-5-3}$$

令 $V_{\mathrm{W}i} = V_{\mathrm{W}i}/V_{\mathrm{t}}$,$A$ 代表上式中的分母,则:

$$A = \sum_i \frac{D_i \left(\sqrt{V_{\mathrm{t}}^2 - (V_{\mathrm{W}i}\sin\theta_i)^2} - V_{\mathrm{W}i}\cos\theta_i \right)}{V_{\mathrm{t}}^2 - V_{\mathrm{W}i}^2}$$

$$= \frac{1}{V_{\mathrm{t}}} \sum_i D_i (1 + V_{\mathrm{W}i}^2) \frac{\sqrt{1 - (V_{\mathrm{W}i}\sin\theta_i)^2} - V_{\mathrm{W}i}\cos\theta_i}{1 - V_{\mathrm{W}i}^4}$$

$$A = \frac{1}{V_{\mathrm{t}}} \sum_i \left\{ D_i (1 + V_{\mathrm{W}i}^2) \times \left[1 - (V_{\mathrm{W}i}\sin\theta_i)^2/2 - V_{\mathrm{W}i}\cos\theta_i \right] \right\}$$

$$= \frac{1}{V_{\mathrm{t}}} \sum_i \left[D_i (1 - V_{\mathrm{W}i}\cos\theta_i) \right]$$

令 $W_i = V_{\mathrm{W}i}\cos\theta_i$,$\overline{W}_i = W_i/V_{\mathrm{t}}$,则:

$$A = \frac{1}{V_{\mathrm{t}}} \sum_i \left[D_i (1 - \overline{W}_i) \right] = \frac{1}{V_{\mathrm{t}}} \left[D - \sum_i (D_i \overline{W}_i) \right]$$

代入式(1-5-3),得:

$$W_E = \left[\frac{D}{D - \sum_i (D_i W_i)} - 1 \right] V_t = \frac{V_t \sum_i (D_i \overline{W}_i)}{D \left[1 - \sum_i (D_i W_i) / D \right]} = \frac{\sum_i (D_i W_i)}{D} \qquad (1\text{-}5\text{-}4)$$

定义当量风速:$W_E = \sum_i (D_i W_i)/D$,式中,W_i为各航段上的风在航线方向上的风分量,顶风为负。由式(1-5-4)可以看出,当量风速实为各航段风分量的加权平均。确定了一条航线的当量风速之后即可由下式将航线的地面距离 D 换算为近似的空中距离:

$$D_A = D \times V_t / (V_t + W_E) \qquad (1\text{-}5\text{-}5)$$

由式(1-5-5)定义的当量风与真空速无关,这对于制作飞行计划无疑是很方便的,以不同的速度巡航或不同的机型巡航速度不同时,不必重算当量风速。如式(1-5-1)中 $\theta_i = 0°$,则式(1-5-1)可简化为:

$$D_A = \sum_i \frac{D_i V_t}{V_t + V_{Wi}} = V_t \sum_i \frac{D_i}{V_t + V_{Wi}} \qquad (1\text{-}5\text{-}6)$$

这也是只考虑沿航路风分量计算空中距离的公式(忽略侧风的影响),使用手册中给出的地面距离与空中距离的换算图表都是只考虑沿航路风分量。对多条航线计算结果表明:使用式(1-5-4)、式(1-5-5)计算的空中距离与按式(1-5-6)计算的空中距离相差很小,而且与按式(1-5-1)计算的准确空中距离相比,相对误差一般也在 1% 以下,个别情况下为 2%~2.4%。

1.5.2　当量气温计算

对于航线上的气温,亦是用"当量气温(当量温差)"T_E来表示:

$$T_E = \sum_i (D_i \times T_i) / \sum_i D_i \qquad (1\text{-}5\text{-}7)$$

式中:T_i为第 i 个航段上的气温,是该航段两个端点的气温的算术均值,对式(1-5-7),T_i也可以是第 i 个航段上的气温与标准大气温度之差,T_E也就是与标准大气温度之差。

式(1-5-7)说明:某个航段的距离越长,该段的温度影响越大,在 T_E 中占的比重越大,因此以 D_i 加权平均计算整条航线上的当量气温。式(1-5-7)可按以下方法导出:假设耗油与距离成正比,同时也与温度和 ISA 之偏差成正比,用 K_d 表示单位距离耗油量,K_t 表示温度比 ISA 每高 1 ℃耗油量增加的百分比,T_i、T_E 都是与 ISA 之偏差。在非 ISA 时,耗油与 ISA 时耗油之差 $\Delta F = \sum_i (K_d \times D_i \times T_i \times K_t) = K_d \times K_t \times \sum_i (D_i \times T_i)$,如存在一个等效的航路温差 T_x,使按整条航线算的 ΔF 与上式相等,即 $\Delta F = (K_d \times \sum_i D_i) \times (K_t \times T_x)$ 与上式相等,显然:$T_x = \sum_i (D_i \times T_i) / \sum_i D_i$,这就是说,这个等效温差即当量温差(气温)。

第 2 章

常规飞行计划制作

执行航班任务前需要根据具体的气象条件、机场和飞机状况,按照民航有关的限制和规定,计算确定可携带的业载以及完成本次航班任务所需的飞行时间和燃油量,在保证飞行安全的同时提高营运经济性。

2.1 制作飞行计划使用的图表

飞行计划有详细、简化之分,制作简化飞行计划只需计算出总油量、轮挡油量、备份油量、航程油量、允许业载、轮挡时间、航程时间等主要参数。制作详细飞行计划还要给出到达各航路点的时间、消耗油量、各点的速度、航向等参数。制作详细飞行计划需要利用爬升、巡航、下降、等待、高度能力、机动能力、最佳高度等数值表。这种详细计划的计算工作量很大,可以编制程序计算。

2.1.1 飞行计划图表介绍

为了制作飞行计划,在飞机使用手册中都给出了有关的曲线图、数值表。不同飞机公司的手册给出的这些曲线图、数值表的形式是不同的,但只要掌握了制作飞行计划的方法,会使用一个公司的飞机使用手册,其他飞机公司的手册也容易看懂并学会使用。

附录 1 提供了波音公司 B737-800 飞机《飞行计划与性能手册》(FPPM)中与制作飞行计划有关的部分曲线图、数值表,同类图表只列出 1~2 张,具体包括:

附图 1-1　制作飞行计划的相关图表概述

附图 1-2　最佳高度图表

附图 1-3　LRC 巡航模式下航程油量和航程时间图表

附图 1-4　$M0.79$ 巡航模式下的航程油量和航程时间图表

无论是制作简化飞行计划还是详细飞行计划,都要用到这些图表。此外,为了制作一份完整的飞行计划,还需要知道地面滑行、防冰、APU、进近等耗油情况,有关数据可以在使用手册的"FLIGHT PLANNING ALLOWANCES"一节中找到。

以 B737-800(CFM56-7B26)为例,计算油量时所需的相关油量的数据包括:

(1)地面滑行每分钟耗油 25 lb(11.34 kg);

(2)APU 在地面正常工作时每小时耗油 225 lb(102 kg),在空中正常使用高度上每小时耗油查附录说明内容中"APU Operation During Flight"部分的图表;

(3)爬升油量中没考虑 10000 ft 以下表速不得大于 250 kt 的限制,如受到这个限制,爬升油量应增加 50 lb(22.5 kg);

(4)巡航中使用发动机防冰每小时耗油 100 lb(45 kg),使用发动机和机翼防冰每小时耗油 300 lb(140 kg);

(5)进近、着陆时襟翼放下状态每分钟耗油 130 lb(58.97 kg)。

有些机场缺少地面电源车、气源车,飞机在地面上只能使用自己的 APU 发电、供气,供飞行员起飞前检查设备及为空调包供气。在这种情况下,计算油量时应加上 APU 地面运行耗油量。

对于计算地面滑行耗油、APU 地面运行耗油、进近耗油,各公司可能会有自己的规定,例如,某航空公司规定按 APU 地面运行 60 min、起飞前地面滑行 20 min、进近 10 min 计算相应的油量。

制作简化飞行计划用的图表上标注有计算、绘制这些图表的条件,如爬升速度、下降速度、巡航速度等。在附图 1-6 LRC 阶梯巡航下的航程油量和航程时间图表上没注明巡航高度,巡航高度是按保持在最佳高度的±2000 ft 巡航来确定的,如图 2-1-1 所示。附图 1-7 LRC 改航油量和航程时间图表上的巡航高度是按最佳高度(航程长于 250 NGM)或按短距巡航高度图(航程短于 250 NGM)来选择的。每次阶梯爬升 2000 ft 可以使巡航高度更接近最佳高度,可以更省油。目前,在 FPPM 中给出的阶梯巡航下的航程油量和航程时间图表仍是按每次爬升 4000 ft 制作的。

制作简化飞行计划用的图表给出的油量和时间分别是航程油量(Trip Fuel)和航程时间(Trip Time),指从松刹车起飞直到在目标(或备降)机场接地所用的油量和时间。有些图表更适用于已知着陆重量求航程油量的情况,有些图表则更适用于已知起飞重量求航程油量的情况,读图的方法简单明了。

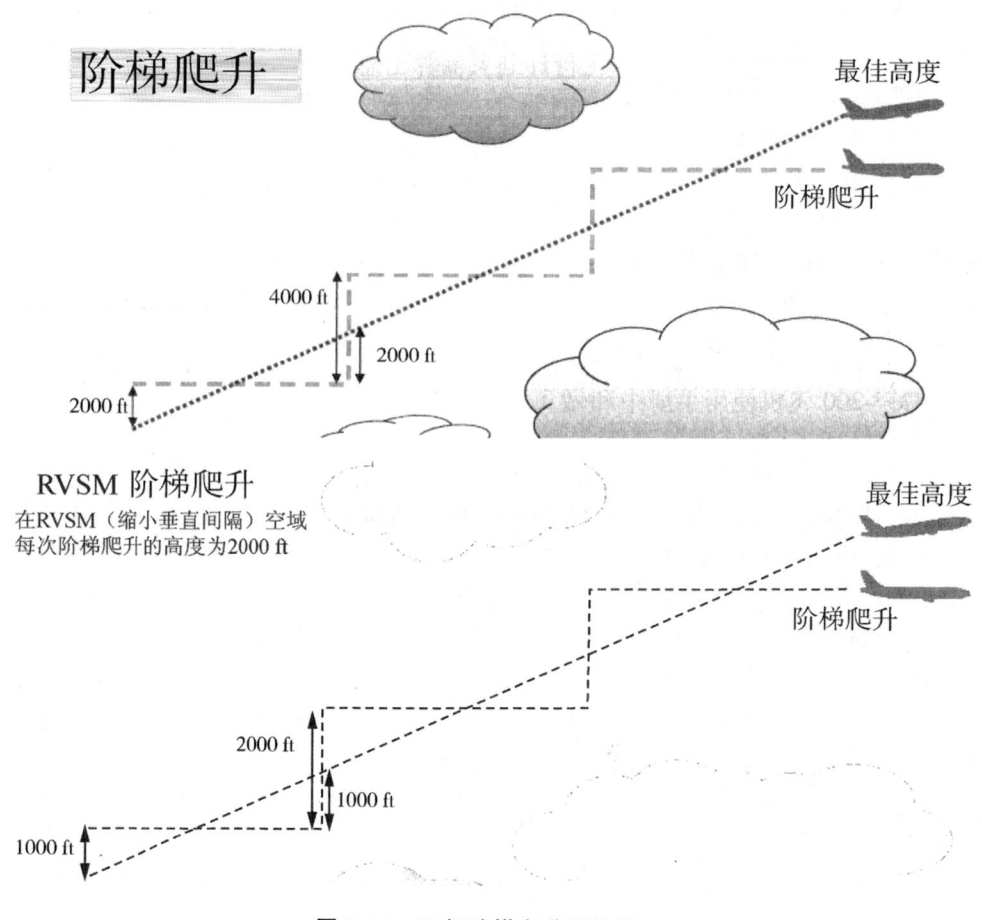

图 2-1-1　飞机阶梯爬升巡航图

在制作飞行计划时,比较方便的是从备降机场停机坪往起飞方向逐步计算,每算出一段油量后都要把它加到飞机重量上,然后向前计算。这样可以把后面一段的油量对前面一段耗油的影响考虑进去。

不同飞机公司给出的这些制作飞行计划用的图、表、格式、内容是不同的,主要区别如下:

(1)有的飞机公司给出的地面滑行每分钟耗油量、进近(襟翼放下)每分钟耗油量与飞机重量有关。

(2)大部分飞机公司的手册上没有短距巡航选择高度用的曲线。

(3)大部分飞机公司的手册上没有载运燃油分析曲线("TANKER ANALYSIS")。

(4)波音公司手册中的爬升数值表上有机场标高修正,爬升时间、油量分别包括起飞阶段的时间、油量,有的飞机公司手册中的爬升数值表中给出的爬升时间、油量不含起飞阶段用的时间、油量,使用这种数值表时要注意分别加上起飞用的时间、油量(手册中单独给出起飞用的时间、油量、距离),大部分飞机公司手册中的爬升数值表没有给出机场标高对爬升时间、油量的修正值。

波音公司手册中的爬升数值表没考虑 10000 ft 以下表速不得大于 250 kt 的限制(如 B737-800 的爬升规律是 280 kt/M0.78),有些公司手册中的爬升数值表考虑了这个限制(例如按照速度 250 kt/280 kt/M0.78 爬升),波音公司在手册上给出了受这种限制爬升时油量的增量。

(5)波音公司手册中的下降数值表中给出的下降用的时间、油量是从巡航高度下降、直接进近直到接地的时间、油量,而其他公司给的下降时间、油量是从巡航高度下降到 1500 ft 的时间及油量,不包含进近、着陆部分。

波音公司手册上的直接进近、着陆是指:从指定高度下降,在离接地点 15 NM 处放进近襟翼,过远台时放起落架和着陆襟翼,直接对着跑道方向过远台、近台、接地。

(6)巡航燃油流量与温度有关,波音公司手册上是按总温与标准大气时的总温的偏差来修正的,有的公司是按静温与 ISA 温度的偏差来修正的,还有的公司直接给出 ISA、ISA±5 ℃、ISA±10 ℃ 等情况下的燃油流量。

关于机场标高对爬升、下降的油量、时间、距离的一般修正方法,将在后面讨论详细飞行计划时介绍。

2.1.2　图表使用中的相关问题计算

2.1.2.1　爬升和下降性能中的机场标高修正

飞机使用手册中的爬升、下降性能的数值表都是针对机场气压高度为零的情况给出的,即给出的是由海平面机场起飞爬升到某一高度层所需要的油量、时间及飞过的水平距离和由某一高度层下降到气压高度为 1500 ft 时所需的油量、时间及飞过的水平距离(波音飞机的下降数据是直到接地的)。

当机场气压高度不为零时,应进行修正。一般来说,机场的标高并非其气压高度,两者之间的差别可达几百英尺,这种差别对机场分析(即确定最大起飞重量)的影响较大(尤其是对大型飞机),对确定爬升、下降的油量、时间、距离等影响不大,可以把标高作为气压高度。

设机场标高为 ELE,飞行高度为 FL,由图 2-1-2 不难看出,从标高为 ELE 的机场起飞爬升到巡航高度 FL 所需的油量 $F_{\text{ELE}\to\text{FL}}$、时间 $T_{\text{ELE}\to\text{FL}}$ 及飞过的水平距离 $D_{\text{ELE}\to\text{FL}}$ 计算如下:

$$F_{\text{ELE}\to\text{FL}} = F_{0\to\text{FL}} - F_{0\to\text{ELE}+1500\,\text{ft}} + F_{0\to1500\,\text{ft}}$$

$$t_{\text{ELE}\to\text{FL}} = t_{0\to\text{FL}} - t_{0\to\text{ELE}+1500\,\text{ft}} + t_{0\to1500\,\text{ft}}$$

$$D_{\text{ELE}\to\text{FL}} = D_{0\to\text{FL}} - D_{0\to\text{ELE}+1500\,\text{ft}} + D_{0\to1500\,\text{ft}}$$

这里近似认为从海平面机场起飞到离地 1500 ft 的油量、时间及距离分别等于在标高为 ELE 的机场起飞到离地 1500 ft 的油量、时间及距离。

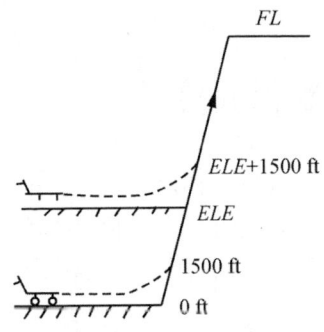

图 2-1-2　爬升过程中的标高修正示意图

对于下降而言(参见附图 1-13 中的下降性能图表),由巡航高度 FL 下降直到在标高为 ELE 的机场接地时的油量、时间及飞过的水平距离分别为(见图 2-1-3):

$$F_{\text{FL}\to\text{ELE}} = F_{\text{FL}\to0} - F_{\text{ELE}+1500\,\text{ft}\to0} + F_{1500\,\text{ft}\to0}$$

$$t_{FL \to ELE} = t_{FL \to 0} - t_{ELE+1500 \text{ ft} \to 0} + t_{1500 \text{ ft} \to 0}$$

$$D_{FL \to ELE} = D_{FL \to 0} - D_{ELE+1500 \text{ ft} \to 0} + D_{1500 \text{ ft} \to 0}$$

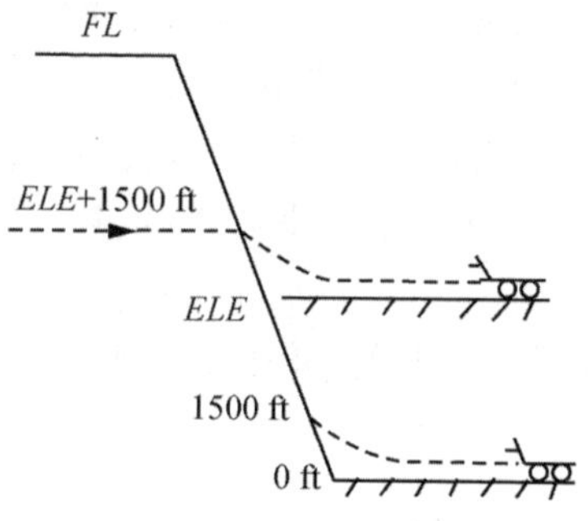

图 2-1-3　下降过程中的标高修正示意图

如手册中下降表中最后一行的油量、时间、距离是零,则表中每一行是由给定高度 FL 下降到离地 1500 ft 高的油量、时间、距离。由巡航高度 FL 下降到距标高为 ELE 的机场 1500 ft 高时的油量、时间及飞过的水平距离分别为:

$$F_{FL \to ELE} = F_{FL \to 1500 \text{ ft}} - F_{ELE+1500 \text{ ft} \to 1500 \text{ ft}}$$

$$t_{FL \to ELE} = t_{FL \to 1500 \text{ ft}} - t_{ELE+1500 \text{ ft} \to 1500 \text{ ft}}$$

$$D_{FL \to ELE} = D_{FL \to 1500 \text{ ft}} - D_{ELE+1500 \text{ ft} \to 1500 \text{ ft}}$$

对进近的油量、时间、距离应分别另外加上。对于使用波音手册下降数值表的情况,如不是直接进近、接地,在制作飞行计划时也应酌情适当增加进近时间。

2.1.2.2　风对爬升、下降过程中飞过的水平距离的影响

爬升、下降数值表中给出的水平距离均指无风时的距离(或者说是空中距离),一般爬升、下降中都是有风的,风速从地面到巡航高度是逐渐增大的。为求出爬升、下降过程中飞过的地面距离,一般近似认为爬升、下降过程中的风速为巡航高度上风速的 2/3,设巡航高度上的风速为 V_w,爬升、下降过程中飞过的空中距离为 DA(即上面计算出的距离),飞行时间为 t,则飞机爬升、下降中水平方向的平均空速为 DA/t,地速 $= DA/t \pm V_w \times 2/3$(顶风取负号),则飞过的地面距离 $D = (DA/t \pm V_w \times 2/3) \times t = DA \pm V_w \times t \times 2/3$,式中 t 以小时为单位。由于风是水平方向的,所以风不改变飞机的爬升、下降速率,不影响爬升、下降的油量及时间。

2.2　简化飞行计划的制作方法

常见的飞行计划有两种制作方法:第一种由备降机场停机坪开始向起飞机场方向往回推算;第二种由起飞机场按照最大允许起飞重量向备降机场停机坪推算。两种方法都需要结合飞行剖面进行计算。下面以简化飞行计划为例,介绍这两种方法的使用。

2.2.1　由备降机场停机坪开始往回推算

2.2.1.1　基本思路

已知飞机的使用空机重量(OEW),如果知道包括旅客、货物、行李以及邮件的实际业载重量(PL),则在备降机场停机坪时的无油重量 $ZFW = OEW + PL$。考虑航空公司酌情携带的燃油

（COF）一般不会被消耗掉，飞机在备降机场停机坪时的飞机重量 $W=ZFW+COF$。

如果想计算最大允许业载（MPL），则可以按以下方式处理：

假设在备降机场停机坪时 $ZFW=MZFW$，此时最大业载 $MPL=MZFW-OEW$。于是备降机场停机坪的飞机重量 $W=MZFW+COF$，然后由此重量开始往回推算，加上各阶段消耗油量，一直计算到起飞机场停机坪，在计算中应保证：

（1）无油重量≤最大无油重量（$ZFW \leqslant MZFW$）。

（2）备降机场着陆重量≤备降机场最大允许着陆重量（$LWA \leqslant MLWA$）。

（3）目的地机场着陆重量≤目的地机场最大允许着陆重量（$LWD \leqslant MLWD$）。

（4）起飞机场起飞重量≤起飞机场最大允许起飞重量（$TOW \leqslant MTOW$）。

（5）总油量≤油箱容量（$RPF \leqslant FTC$）。

在推算过程中，如有任何一个条件不满足，则应减少业载（PL）重新计算直到满足条件为止，最终得到所允许的业载（实际业载有可能减载了）及总油量等数据。

2.2.1.2 计算过程

如图 2-2-1 所示，以最典型的常规航线的飞行剖面（涡轮动力飞机）为例，从后往前制作飞行计划的详细步骤如下：

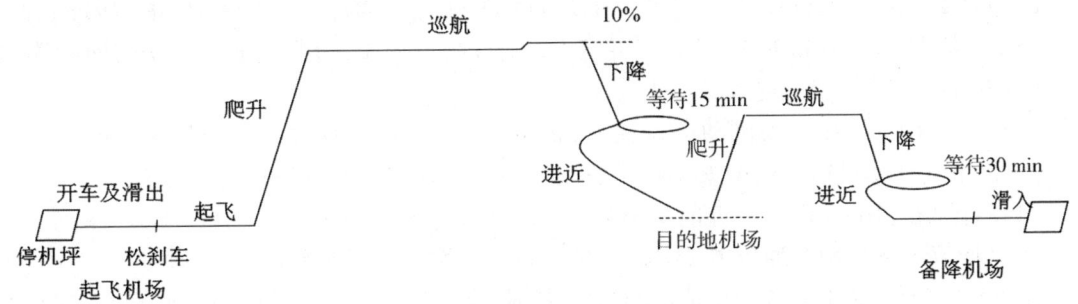

图 2-2-1 飞行剖面（从后往前推算）

（1）若主航段（起飞机场到目的地机场）或备降航段（目的地机场到备降机场）分段，则首先算出它们的当量风（W_E）和当量气温（T_E），用于后续计算。

（2）计算实际业载（PL）：$PL=$ 旅客重量＋货物重量＋托运行李重量＋邮件。

（3）计算无油重量（ZFW）：$ZFW=OEW+PL$。此时要检查是否满足 $ZFW \leqslant MZFW$。如果满足条件，则继续下一步；否则需要减少 PL，回到第（2）步，直至 $ZFW=MZFW$，然后继续下一步。

（4）计算在备降机场时的停机坪重量（$W_{停}$）：$W_{停}=ZFW+COF$。

（5）计算在备降机场滑入停机位的耗油（TIF）：$TIF=$ 滑行油耗率×滑入时间（TIT），滑行耗油率可通过机型 FPPM 手册获知。

（6）计算在备降机场的着陆重量（LWA）：$LWA=W_{停}+TIF$，此时要检查是否满足 $LWA \leqslant MLWA$。如果满足条件，则继续下一步；否则需要减少 PL，回到第（2）步，直至 $LWA=MLWA$，然后继续下一步。

（7）计算在备降机场机动进近耗油（$F_{进近备}$）：$F_{进近备}=$ 进近耗油率×备降机场进近时间（$T_{进近备}$），进近耗油率可通过机型 FPPM 手册获知。

（8）计算备降机场等待结束时的重量（$W_{结束}$），即在备降机场进近前的重量：$W_{结束}=LWA+F_{进近备}$。

（9）计算在备降机场等待的油量（HODF），即规章规定的最后储备油量。按照规章规定的最后储备油量的计算要求,$HODF$ 为等待 30 min 的油量。因为在等待过程中燃油流量(FF)是变量,因此要计算准确的等待油量可以采用迭代方法,为简便起见,采用平均燃油流量($FF_{平均}$)进行计算。

先根据 $W_{结束}$ 查等待油量表,得到全发燃油流量 FF_1,这是一个粗略的平均燃油流量。然后根据 FF_1 计算等待油量 $HODF_1 = FF_1 \times 30/60$,这也是一个粗略值。接着计算等待中的平均重量 $W_{平均} = (W_{结束} + W_{开始})/2 = W_{结束} + \frac{1}{2}HODF_1$。再根据 $W_{平均}$ 重新查等待油量表,得到全发燃油流量 FF,这就是等待中的平均燃油流量($FF_{平均}$)。最后计算等待油量 $HODF = FF_{平均} \times 30/60$。

(10)计算等待开始重量($W_{开始}$):$W_{开始} = W_{结束} + HODF$,此时要检查是否满足 $W_{开始} \leqslant MLWA$,如果满足条件,则继续下一步;否则需要减少 PL,回到第(2)步,直至 $W_{开始} = MLWA$,然后继续下一步。

(11)计算备降油量($DIVF$)和备降时间($DIVT$)。根据 $W_{开始}$ 查改航备降图表,可获得不含机动进近的备降油量($DIVF'$)和不含机动进近的备降时间($DIVT'$)。注意,$DIVF'$ 不含机动进近油量,$DIVT'$ 也不含机动进近时间。真实的备降油量和备降时间为:

$$DIVF = DIVF' + F_{进近备}, \quad DIVT = DIVT' + T_{进近备}$$

(12)计算在目的地机场着陆重量(LWD):$LWD = W_{开始} + DIVF' = LWA + HODF + DIVF$。此时要检查是否满足 $LWD \leqslant MLWD$。如果满足条件,则继续下一步;否则需要减少 PL,回到第(2)步,直至 $LWD = MLWD$,然后继续下一步。

(13)计算在目的地机场机动进近耗油($F_{进近}$):$F_{进近} =$ 进近耗油率 × 目的地机场进近时间($T_{进近}$),进近耗油率可通过机型 FPPM 手册获知,一般与在备降机场时一样。

(14)计算在目的地机场进近前重量($W_{进近前}$):$W_{进近前} = LWD + F_{进近}$。

(15)计算不可预期燃油(CF,Contingency Fuel)。按照以下过程计算 CF:

①首先计算目的地机场 15 min 的等待耗油($F_{15\text{ min}等待}$):根据目的地机场进近前重量 $W_{进近前}$ 查等待油量表,参考第(9)步迭代计算等待耗油 $F_{15\text{ min}等待}$,再计算目的地机场等待开始重量 $W_{目标开始} = W_{进近前} + F_{15\text{ min}等待}$。

②计算 10% 的航程油量($F_{10\%航程}$):根据目的地机场等待开始重量 $W_{目标开始}$ 查航程油量和航程时间图表,得到不含机动进近的航程油量 TRF' 和不含机动进近的航程时间 TRT';计算:$TRF = TRF' + F_{进近}$;$TRT = TRT' + T_{进近}$;$F_{10\%航程} = 10\% \times TRF$。

③计算不可预期燃油(CF),$CF = \max\{F_{15\text{ min}等待}, F_{10\%航程}\}$。

若 $F_{10\%航程} > F_{15\text{ min}等待}$,应返回第(14)步,以 $W_{进近前} + F_{10\%航程}$ 为基础,重新查航程油量和航程时间图表,重新计算不含机动进近的航程油量 TRF' 和不含机动进近的航程时间 TRT',并进行迭代计算,直至两次计算的 $F_{10\%航程}$ 结果的差值在一定的容差范围内即停止,最后的结果即为 $F_{10\%航程}$,$CF = F_{10\%航程}$,继续第(16)步;

若 $F_{10\%航程} < F_{15\text{ min}等待}$,$CF = F_{15\text{ min}}$,继续第(16)步。

注意:TRF' 不含机动进近油量,TRT' 也不含机动进近时间。

真实的航程油量和航程时间分别为:

$$TRF = TRF' + F_{进近}, \quad TRT = TRT' + T_{进近}$$

(16)计算起飞机场的起飞重量(TOW):$TOW = W_{进近前} + TRF' + CF$。此时要检查是否满足 $TOW \leqslant MTOW$。如果满足条件,则继续下一步;否则需要减少 PL,回到第(2)步,直至 $TOW = MTOW$,然后继续下一步。

（17）计算在起飞机场滑出耗油（TOF）：TOF ＝ 滑行油耗率×滑入时间（TOT），可通过查机型 FPPM 手册获知备降机场停机坪的滑行油耗率。

（18）计算在起飞机场的停机坪重量（RPW）：RPW ＝ TOW ＋ TOF。

（19）油量和时间汇总：

备降油量和备降时间：$DIVF$ ＝ $DIVF'$ ＋ $F_{进近备}$，$DIVT$ ＝ $DIVT'$ ＋ $T_{进近备}$。

航程油量和航程时间：TRF ＝ TRF' ＋ $F_{进近}$，TRT ＝ TRT' ＋ $T_{进近}$。

轮挡油量 ＝ 航程油量＋滑出耗油＋滑入耗油，即 BKF ＝ TRF ＋ TOF ＋ TIF。

轮挡时间 ＝ 航程时间＋滑出时间＋滑入时间，即 BKT ＝ TRT ＋ TOT ＋ TIT。

备份油量 ＝ 备降油量＋等待油量＋不可预期燃油＋酌情携带的燃油，即 $RESF$ ＝ $DIVF$ ＋ $HODF$ ＋ CF ＋ COF。

总油量 ＝ 轮挡油量＋备份油量，即 RPF ＝ BLF ＋ $RESF$。此时需要检查总油量是否超过油箱容量，即是否满足 $RPF \leqslant FTC$。如果满足条件，则继续下一步；否则需要减少 PL，回到步骤（2），直至 RPF ＝ FTC，然后继续下一步。

（20）检查：

①起飞重量－航程油量≤目的地机场最大允许着陆重量，即 TOW － TRF ≤ $MLWD$。

②检查备降机场等待开始重量＋不可预期燃油≤备降机场允许最大着陆重量，即 $W_{开始}$ ＋ CF ≤ $MLWA$。

两个条件中有任何一个不满足，则需要减载，回到步骤（2）重新计算；如果条件都满足，则继续下一步。

（21）验算：是否满足 RPF ＝ RPW － ZFW，即飞机总重减去飞机无油重量是否等于总油量。如果满足条件，则表明计算过程中求和无误；如果不满足条件，则表示在某一步骤的运算过程中出错，从步骤（1）开始检查，找出错误，直至满足条件。

（22）计算下降顶点的重量（W_{TOD}）：W_{TOD} ＝ $W_{进近前}$ ＋ $F_{下降}$。下降段耗油（$F_{下降}$）可以根据飞行高度层查下降性能数值表获得。

（23）计算下降顶点最佳高度（$HOPT_{下降}$）：根据 W_{TOD} 查飞机高度能力表和机动能力表，或者查询最佳高度图表，可以获得 TOD 点 $HOPT_{下降}$、H_{MCR}、$H_{1.3G}$。

（24）计算爬升顶点的重量（W_{TOC}）：W_{TOC} ＝ TOW － $F_{爬升}$。爬升段耗油（$F_{爬升}$）可以根据 TOW 查飞机爬升性能数值表获得。

（25）计算爬升顶点最佳高度（$HOPT_{爬升}$）：根据 W_{TOC} 查飞机高度能力表和机动能力表，或者查询最佳高度图表，可以获得 TOC 点 $HOPT_{爬升}$、H_{MCR}、$H_{1.3G}$。

（26）确定最佳飞行高度层。根据 $HOPT_{爬升}$ 和 $HOPT_{下降}$，以及航路走向和飞行高度层的分配原则，确定最佳飞行高度层。

2.2.2　由最大允许起飞重量往后算

2.2.2.1　基本思路

在计算之前应加业载未知，所以实际无油重量是未知的，因此只能由最大允许起飞重量（$MTOW$）开始计算，向后逐步推算出各阶段的耗油量及到达目的地机场及备降机场的重量，若出现 LWD ＞ $MLWD$ 或者 LWA ＞ $MLWA$，则减小起飞重量（TOW ＝ $MTOW$ － ΔW）重新计算。当全部计算得到的总油量（RPF）超过了油箱容量时，则应减小起飞重量并重新计算，直到总油量刚好等于油箱容量。

由起飞重量减总油量得无油重量（ZFW），应该满足 $ZFW \leq MZFW$，或者 $ZFW \leq OEW+PL$（如果给定了实际业载 PL），否则减少 TOW 重新计算。可以从 TOW 中减去（$ZFW-MZFW$）或 [$ZFW-(OEW+PL)$]，再次计算，直到计算出的 ZFW 近似等于 $MZFW$ 或（$OEW+PL$）为止。最后得到的 ZFW 减去 OEW 即能携带的业载（PL）。

2.2.2.2　计算过程

如图 2-2-2 所示，仍然以最典型的常规航线的飞行剖面（以涡轮动力飞机）为例，从前往后制作飞行计划详细步骤如下：

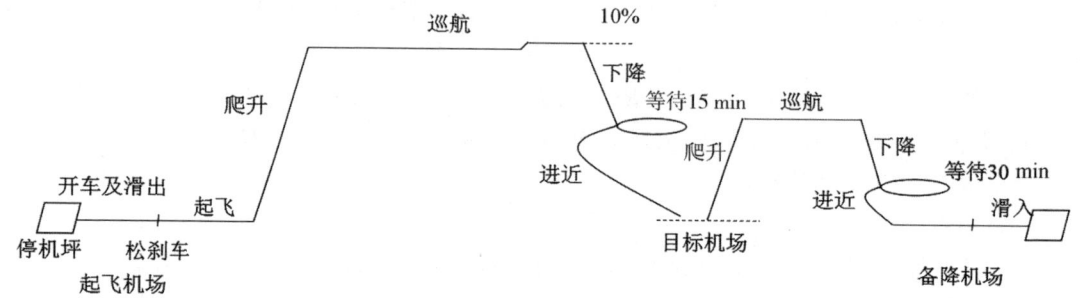

图 2-2-2　飞行剖面（从前往后推算）

（1）若主航段或备降航段分段，则先算出它们的当量风（W_E）和当量气温（T_E）。

（2）假设起飞重量为最大允许起飞重量，令 $TOW=MTOW$。

（3）计算不含机动进近的航程油量（TRF'）和不含机动进近的航程时间（TRT'）。

计算 TRF' 和 TRT' 的方法有两种，一种是辅助线法，另一种是迭代法。

辅助线法：按照巡航方式（LRC 或等马赫数），在航程油量表上作一条 $TOW=MTOW$ 的辅助线，然后根据 $MTOW$ 查得不含机动进近的航程油量 TRF'、不含机动进近的航程时间 TRT'；若巡航方式为阶梯爬升巡航，则直接根据 $MTOW$ 查得不含机动进近的航程油量 TRF'、不含机动进近的航程时间 TRT'。

迭代法：假设目的地机场进近前重量 $W_{进近前}=TOW=MTOW$，查航程图得 TRF'_1；

重新计算 $W_{进近前}=MTOW-TRF'_1$，再根据 $W_{进近前}$ 查航程图得 TRF'_2；

重新计算 $W_{进近前}=MTOW-TRF'_2$，再根据 $W_{进近前}$ 查航程图得 TRF'_3；

以此类推，不断迭代；

直至 $TRF'_n \approx TRF'_{n-1}$，迭代收敛，$TRF'=TRF'_n$。

（4）计算目的地机场进近前重量（$W_{进近前}$）：$W_{进近前}=MTOW-TRF'$。

（5）计算在目的地机场进近耗油（$F_{进近}$）：$F_{进近}=$ 进近油耗率×目的地机场进近时间，进近耗油率可通过查机型 FPPM 手册获知。

（6）计算航程油量（TRF）和航程时间（TRT）：$TRF=TRF'+F_{进近}$，$TRT=TRT'+T_{进近}$。

（7）计算不可预期燃油（CF）：

$F_{10\%航程}=10\% \times TRF$；

$F_{15\,min等待}$ 根据 $W_{进近前}$ 参考从备降机场停机坪往回推算的方法的步骤（9）迭代计算。

$CF=\max\{F_{15\,min等待}, F_{10\%航程}\}$。

（8）计算在目的地机场着陆重量（LWD）：$LWD=W_{进近前}-F_{进近}-CF$。检查是否满足 $LWD \leq MLWD$，如果满足条件，则继续下一步计算；否则回到步骤（2），需要减少 TOW，直至 $LWD=$

$MLWD$,然后继续下一步。

（9）计算不含机动进近的备降油量（$DIVF'$）和不含机动进近的备降时间（$DIVT'$）。

与步骤（3）计算方法类似，计算 $DIVF'$ 和 $DIVT'$ 有两种方法，一种是辅助线法，另一种是迭代法。假设飞机在备降机场等待开始的重量为 $W_{开始}$，以 $W_{开始}$ 作为备降航程油量和航程时间图表上的着陆重量。

①辅助线法：在备降航程油量和航程时间图表上作一条 $TOW_{备降}=LWD$ 的辅助线，然后根据 $TOW_{备降}$ 查得不含机动进近的备降油量（TRF'）和备降时间（TRT'）。

②迭代法：假设备降机场等待开始的重量 $W_{开始}=LWD$，查航程图得 $DIVF'_1$；

重新计算 $W_{进近前}=LWD-DIVF'_1$，再根据 $W_{进近前}$ 查航程图得 $DIVF'_2$；

重新计算 $W_{进近前}=LWD-DIVF'_2$，再根据 $W_{进近前}$ 查航程图得 $DIVF'_3$；

以此类推，不断迭代；

直至 $DIVF'_n \approx DIVF'_{n-1}$，迭代收敛，$DIVF'=DIVF'_n$。

（10）计算等待开始重量（$W_{开始}$）：$W_{开始}=LWD-DIVF'$。此时要检查是否满足 $W_{开始}+CF \leqslant MLWA$（考虑不可预期燃油 CF 没有消耗的情况）。如果满足条件，则继续下一步计算；否则回到步骤（2），需要减少 TOW，直至 $W_{开始}+CF=MLWA$，然后继续下一步。

（11）计算最后储备油量或者称为等待油量（$HODF$）：计算方法与从备降机场停机坪往回推算的方法的步骤（9）方法相同。因为在等待过程中燃油流量（FF）是变量，因此要计算准确的等待油量可以采用迭代法进行计算，为简便起见，采用平均燃油流量（$FF_{平均}$）进行计算。

先根据 $W_{开始}$ 查等待油量表，得到全发燃油流量 FF_1，这是一个粗略的平均燃油流量。然后根据 FF_1 计算等待油量 $HODF_1=FF_1 \times 30/60$，这也是一个粗略值。接着计算等待中的平均重量 $W_{平均}=(W_{开始}+W_{结束})/2=W_{开始}-\dfrac{1}{2}HODF_1$。再根据 $W_{平均}$ 重新查等待油量表，得到全发燃油流量 FF，这就是等待中的平均燃油流量（$FF_{平均}$）。最后计算等待油量 $HODF=FF_{平均} \times 30/60$。

（12）计算等待结束重量 $W_{结束}=W_{开始}-HODF$。

（13）计算在备降机场的进近耗油（$F_{进近备}$）：$F_{进近备}=$ 进近油耗率 × 备降机场进近时间，进近耗油率可通过查机型 FPPM 手册获知。

（14）计算备降油量和备降时间：

备降油量（$DIVF$）：$DIVF=DIVF'+F_{进近备}$；

备降时间（$DIVT$）：$DIVT=DIVT'+T_{进近备}$。

（15）计算在备降机场的着陆重量（LWA）：$LWA=W_{结束}-F_{进近备}$，检查是否满足 $LWA+HODF+CF \leqslant MLWA$。如果满足条件，则继续下一步计算；否则回到步骤（2），需要减少 TOW，直至 $W_{开始}+CF=MLWA$，然后继续下一步。

（16）计算在备降机场停机坪的重量（$W_{停}$）：$W_{停}=LWA-TIF$。计算备降机场滑入耗油（TIF）：$TIF=$ 滑行油耗率 × 滑入时间（TIT），滑行油耗率可通过查机型 FPPM 手册获知。

（17）计算起飞机场停机坪重量（RPW）：$RPW=TOW+TOF$。起飞机场滑出耗油 = 滑行油耗率 × 滑入时间，滑行油耗率可通过查机型 FPPM 手册获知。

（18）计算无油重量（ZFW）和最大业载（MPL）：无油重量为备降机场停机坪的重量减去酌情携带的燃油的重量，即 $ZFW=W_{停}-COF$。此时要检查是否满足 $ZFW \leqslant MZFW$。如果满足条件，则继续下一步；否则回到步骤（2），需要减少 TOW，直至 $W_{开始}+CF=MLWA$。计算最大业载（MPL）：$MPL=ZFW-OEW$。

（19）油量和时间汇总：

备降油量和备降时间：$DIVF = DIVF' + F_{进近备}$，$DIVT = DIVT' + T_{进近备}$。

航程油量和航程时间：$TRF = TRF' + F_{进近}$，$TRT = TRT' + T_{进近}$。

轮挡油量=航程油量+滑出耗油+滑入耗油，即 $BKF = TRF + TOF + TIF$。

轮挡时间=航程时间+滑出时间+滑入时间，即 $BKT = TRT + TOT + TIT$。

备份油量=备降油量+等待油量+不可预期燃油+酌情携带的燃油，即 $RESF = DIVF + HODF + CF + COF$。

总油量=轮挡油量+备份油量，即 $RPF = BLF + RESF$。此时需要检查总油量是否超过油箱容量，即是否满足 $RPF \leqslant FTC$。如果满足条件，则继续下一步；否则需要减少 TOW，回到步骤（2），直至 $RPF = FTC$，然后继续下一步。

（20）检查：

①起飞重量−航程油量≤目的地机场最大允许着陆重量，即 $TOW - TRF \leqslant MLWD$。

②检查备降机场等待开始重量+不可预期燃油≤备降机场允许最大着陆重量，即 $W_{开始} + CF \leqslant MLWA$。

两个条件中有任何一个不满足，则需要减载，回到步骤（2）重新计算；如果条件都满足，则继续下一步。

（21）验算：是否满足 $RPF = RPW - ZFW$，即飞机总重减去飞机无油重量是否等于总油量。如果满足条件，则表明计算过程中求和无误；如果不满足条件，则表示在某一步骤的运算过程中出错，从步骤（1）开始检查，找出错误，直至满足条件。

（22）计算下降顶点的重量（W_{TOD}）：$W_{TOD} = W_{进近前} + F_{下降}$。下降段耗油（$F_{下降}$）可以根据飞行高度层查下降性能数值表获得。

（23）计算下降顶点最佳高度（$HOPT_{下降}$）：根据 W_{TOD} 查飞机高度能力表和机动能力表，或者查询最佳高度图表，可以获得 TOD 点 $HOPT_{下降}$、H_{MCR}、$H_{1.3G}$。

（24）计算爬升顶点的重量（W_{TOC}）：$W_{TOC} = TOW - F_{爬升}$。爬升段耗油（$F_{爬升}$）可以根据 TOW 查飞机爬升性能数值表获得。

（25）计算爬升顶点最佳高度（$HOPT_{爬升}$）：根据 W_{TOC} 查飞机高度能力表和机动能力表，或者查询最佳高度图表，可以获得 TOC 点 $HOPT_{爬升}$、H_{MCR}、$H_{1.3G}$。

（26）确定最佳飞行高度层。根据 $HOPT_{爬升}$ 和 $HOPT_{下降}$，以及航路走向和飞行高度层的分配原则，确定最佳飞行高度层。

2.2.2.3　两种方法的取舍

对于较短的航线采用第一种方法由备降机场停机坪开始向起飞机场方向往回推算较好，一般计算出来的起飞重量不会超过最大允许起飞重量，可以避免迭代计算，适用于使用手册上所给出的制作简化飞行计划的图表。对于长航线则采用第二种方法由起飞机场按照最大允许起飞重量向备降机场停机坪推算较好，计算出来的着陆重量一般不会超过最大允许着陆重量，可以避免迭代计算，适用于手册上所给出的阶梯巡航的简化飞行计划图表。如果编制计算机飞行计划程序，可以只按一种方法来编制程序，一般来说第一种方法较为方便。

对于人工手算制作飞行计划，如由后向前推算得出的起飞重量超过最大允许起飞重量，可改为由前向后计算；反之，如由前向后推算得出的着陆重量超过最大允许着陆重量，可改为由后向前计算，以避免迭代计算。

2.3 详细飞行计划的制作方法

2.3.1 详细飞行计划的符号说明

详细飞行计划的制作步骤也是编制计算机软件采用的计算步骤。首先对制作飞行计划时所用的符号(如图 2-3-1 所示)说明如下：

APFD——目的地机场的进近油量

APFA——备降机场的进近油量

APUF——APU 耗油量

BKF——轮挡油量

BKT——轮挡时间

CFPM——主航段应急油量占航程油量的百分比(一般可取为 0)

CFPA——备降航段的应急油量占备降油量的百分比(一般可取为 0)

COF——酌情携带的燃油(公司备份油)

DCA——由目的地机场爬升到备降高度时飞过的地面距离

DCM——由起飞机场爬升到巡航高度时飞过的地面距离

DDA——由备降高度下降到备降机场上空 1500 ft 时飞过的地面距离

DDM——由巡航高度下降到目的地机场上空 1500 ft 时飞过的地面距离

DIVF——备降油量

DIVT——备降时间

EWM——主航段风分量(顶风为负)

EWA——备降航段风分量(顶风为负)

FAI——防冰用油

FTC——油箱最大油量

FCLA——备降爬升用油

FCLM——主航段爬升用油

FCRA——备降航段巡航油量

FCRM——主航段巡航油量

FDEA——备降航段下降油量

FDEM——主航段下降油量

HODF——等待油量

HODT——等待时间

MLWA——备降机场最大允许着陆重量

MLWD——目的地机场最大允许着陆重量

MTOW——目的地机场最大允许起飞重量

MZFW——飞机的最大无燃油重量

OEW——使用空机重

PL——业载

RA——由目的地机场到备降机场的航程

RM——由起飞机场到目的地机场的航程

RPF——停机坪油量(即起飞总油量)

RPW——停机坪重量(即开始滑出重量)

RFO——不可预期燃油

RESF——总备份油量

TAPA——备降机场上的进近时间

TAPD——目的地机场上的进近时间

TCLA——备降时的爬升时间

TCLM——主航段爬升到巡航高度的时间

TCRA——备降航段巡航时间

TCRM——主航段巡航时间

TDEA——备降航段下降时间

TDEM——主航段下降时间

TIF——滑入油量

TIT——滑入时间

TOF——滑出油量

TOT——滑出时间

TRF——航程油量

TRT——航程时间

WCA——备降段爬升顶点的重量

WCLA——在目的地机场复飞时的重量,即着陆重量

WDA——备降段开始下降时的重重

WH1——等待结束时的重量

WH2——等待开始时的重量

WTOC——主航段爬升顶点时的重量

WTOD——主航段开始下降点的重量

ZFW——无燃油重量

F154——对于 TY-154 之类飞机,要求总备份油量不少于某一最小值 RF_{min},如迭代开始算出的总备份油量 $RESF_0 < RF_{min}$,则令 $F154 = RF_{min} - RESF_0$,迭代,最后使 $RESF = RF_{min}$。

注：CFPM、CFPA 分别是考虑绕飞雷雨、飞机老化等因素在主航段和备降航所多加的油量占航程油量和备降油量的百分比。

2.3.2 制作详细飞行计划的计算步骤

下面以最典型的常规航线的飞行剖面(以涡轮动力飞机为例),介绍详细飞行计划的制作过程,重点介绍从备降机场停机坪往回推算至起飞机场的方法。为简便计算,把整个航路看成一段,如图2-3-1所示。

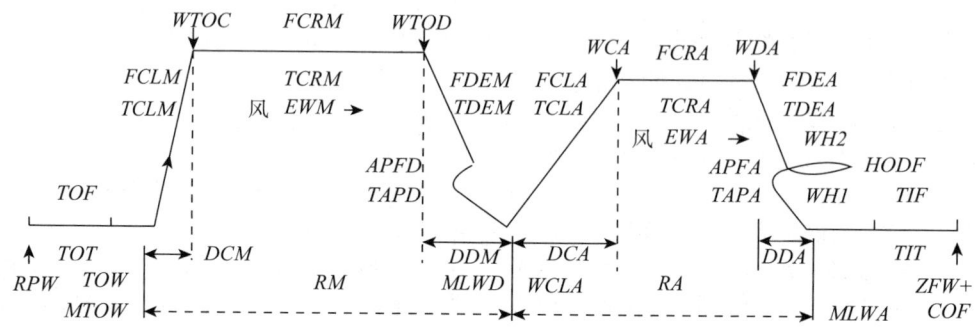

图2-3-1 飞行剖面上各段的油量与时间及飞机重量

(1)计算无燃油重量(ZFW):若给定了业载(PL),则$ZFW = OEW + PL$,ZFW应不大于$MZFW$,否则由$ZFW = MZFW$开始计算。

(2)计算目的地机场的进近油量($APFD$)、滑入油量(TIF)、防冰用油(FAI)、备降机场的进近油量($APFA$):

$$APFD = TAPD \times AFF$$
$$TIF = TIT \times TFF$$
$$FAI = FEI \times TIE + FEW \times TEW$$
$$APFA = TAPA \times AFF$$

式中:TFF——滑行每分钟耗油量;AFF——进近时每分钟耗油;FEI、FEW——使发动机防冰或同时使用发动机及机翼防冰的小时耗油量;TIE、TEW——使用发动机防冰与同时使用发动机及机翼防冰的时间。

(3)计算不可预期燃油(RFO):令$F154 = 0$,先设在主航段TOD点的飞机重量$WTOD = ZFW + COF + TIF + APFA + FAI$,根据此重量、巡航高度、巡航速度,初步确定$TOD$点的燃油流量$FFD$,同时确定$TOD$点之真空速$VT$,$TRTO = RM/(VT + EWM)$,不可预期燃油$RFO = \max\{10\%$ $(FFD \times TRTO)$,$F_{15\text{ min等待}}\}$。其中$F_{15\text{ min等待}}$根据WTOD计算。

(4)计算备降航段等待结束时的重量($WH1$)和等待开始时的重量($WH2$):等待结束时的重量为$WH1 = ZFW + COF + TIF + APFA + FAI + F154 + RFO$,先令$WH1$为等待的平均重量,确定等待中的平均燃油流量$FFH^{(1)}$(如需要,做温度修正),则$HODF^{(1)} = FFH^{(1)} \times HODT$,$WH2 = WH1 + HODF^{(1)}$,再以$(WH1 + WH2)/2$为等待中的平均重量确定燃油流量$FFH^{(2)}$……迭代计算出$HODF$,$WH2 = WH1 + HODF$。由于等待油量及防冰用油和$RFO$可能并未消耗,所以应使$WH2 \leq MLWA$,否则应减少$ZFW$,重新计算直到$|WH2 - MLWA| \leq \varepsilon$为止,$\varepsilon$是所选定的小正量,例如可取为5~10 lb,用于控制迭代精度。手算时只重复算一次即可。

(5)计算备降段下降顶点的重量(WDA):根据$WH2$及飞机使用手册中的下降数值表确定备降航段下降油量($FDEA$)、下降时间($TDEA$)和下降距离(DDA),$WDA = WH2 + FDEA$。

(6)迭代计算在目的地机场的着陆重量($WCLA$):利用使用手册中LRC巡航的燃油流量

及真空速 TAS 的数值表,利用数值积分确定备降航段巡航油量($FCRA$)、巡航时间($TCRA$)、巡航距离($DCRA$),并使用爬升数值表确定备降航段的爬升油量($FCLA$)、爬升时间($TCLA$)及爬升距离(DCA)。这是一个迭代过程:

①先假设巡航距离 $DCRA^{(1)} = RA - DDA$,计算 $FCRA^{(1)}$、$TCRA^{(1)}$。

②$WCLA^{(1)} = WDA + FCRA^{(1)}$,利用爬升数值表确定 $FCLA^{(1)}$、$TCLA^{(1)}$ 及 $DCLA^{(1)}$,则 $DCRA^{(2)} = RA - DDA - DCLA^{(1)}$。

③再计算 $FCRA^{(2)}$、$TCRA^{(2)}$。

④$WCLA^{(2)} = FCRA^{(2)} + WDA + FCLA^{(1)}$,再确定 $FCLA^{(2)}$、$TCLA^{(2)}$、$DCLA^{(2)}$。

……

迭代计算,直到 $|WCLA^{(K+1)} - WCLA^{(K)}| \leqslant \varepsilon$ 为止,取 $WCLA = WCLA^{(K+1)}$,K 为迭代次数。

若 APU 在工作并要求加应急油(即 $CFPA \neq 0$),还应在 $WCLA$ 加上这部分油。

手算时,巡航段可用平均燃油流量计算,也可用积分航程表计算,如有的话。

(7)检查着陆重量是否超限:若目的地机场着陆重量 $WCLA > MLWD$,则应减少业载 PL,回到步骤(1)重新计算 ZFW,直至 $|WCLA - MLWD| \leqslant \varepsilon$。

(8)检查备份油量是否满足 $WCLA - ZFW - TIF = RESF$,对于 TY-154M 之类有最小备份油量要求的飞机,若 $RESF < RESF_{min}$,可令 $F154 = F154 + 0.94 \times (RESF_{min} - RESF)$ 回到步骤(4)重新计算;若现在已受到 $MLWD$ 限制,则应相应减少 ZFW 再返回步骤(4)计算。迭代计算直到 $RESF$ 大于等于 $RESF_{min}$。

(9)计算主航段下降顶点的重量($WTOD$):利用飞机使用手册上的下降数值表及在目的地机场的着陆重量($WCLA$)计算主航段下降油量($FDEM$)、主航段下降时间($TDEM$)及下降时飞过的水平距离(DDM),$WTOD = WCLA + (1 + CFPM) \times FDEM + APFD + APUFDM$,其中 $APFD$ 为目的地机场机动油量,$APUFDM$ 为在目的地机场下降过程中的 APU 耗油。

(10)计算主航段航程油量(TRF)和航程时间(TRT):采用步骤(6)中的方法计算主航段巡航油量($FCRM$)、巡航时间($TCRM$)、爬升顶点重量($WTOC$)、爬升油量($FCLM$)、爬升时间($TCLM$)、爬升距离(DCM)、起飞重量(TOW)。注意,巡航方式应使用与指定的巡航方式相应的燃油流量数值表,在步骤(6)及步骤(10)中应注意温度的影响,温度不仅影响燃油流量,也影响真空速。还应注意加上 $APUF$ 及 $CFPM$ 的油量。此时 $TRT = TCRM + TCLM + TDEM + TAPD$,其中 $TAPD$ 为目的地机场进近时间。

(11)根据 TRT 及在 TOD 点的燃油流量,求出 RFO 返回步骤(4)重新计算。

(12)检查起飞重量是否超限:若步骤(10)计算的 $TOW > MTOW$,则应减少业载 PL,重新计算 ZFW,可以令 $ZFW = ZFW - 0.9(TOW - MTOW)$,返回步骤(4)重新计算,直到 $|TOW - MTOW| \leqslant \varepsilon$。

(13)计算起飞机场停机坪重量(RPW)和滑出油量(TOF):$RPW = TOW + TOF$,$TOF = TFF \times TOT$。

(14)计算油量:

①备降油量($DIVF$):$DIVF = APFA + (FDEA + FCRA + FCLA) \times (1 + CFPA)$,若 APU 工作,应加上 $APUF$,$APUF = APU$ 小时耗油量 $\times DIVT$。

②总备份油量($RESF$):$RESF = DIVF + HODF + COF + RFO + F154$(最后一项是针对 TY-154 类特殊机型)。

③航程油量:$TRF = (FCLM + FCRM + FDEM) \times (1 + CFPM) + APFD$,如 APU 工作应加上 $APUF$,$APUF = APU$ 小时耗油量 $\times TRT$。

④轮挡油量：$BKF = TRF + TOF + TIF$。

⑤总油量：$RPF = BKF + RESF$。

检查：若 RPF 大于油箱最大容量（FTC），则应减少业载（PL），重新计算 ZFW，可以令 $ZFW = ZFW - 5 \times (RPF - FTC)$。返回步骤（4）重新计算，直到 $|RPF - FTC| \leqslant \varepsilon$ 为止。

（15）计算业载（PL）：$PL = ZFW - OEW$，可以看出最后得出的允许业载可能小于最初给定的业载。根据计算过程可判定业载（PL）受何种限制。

（16）计算时间：

①备降时间（$DIVT$）：$DIVT = TCLA + TCRA + TDEA + TAPA$；

②航程时间（TRT）：$TRT = TCLM + TCRM + TDEM + TAPD$；

③轮挡时间（BKT）：$BKT = TRT + TIT + TOT$。

同时，也可得到主航段爬升到 TOC 时的时间、消耗的油量及飞过的距离，到下降点 TOD 的时间、剩余油量及飞过的距离等。

2.4 航线飞行计划制作实例

本节通过具体算例介绍人工制作飞行计划的方法，算例中所涉及的符号的含义如下：

A/I——防冰

APU——辅助动力装置

COF——酌情携带的燃油

ELE——机场标高

HOPT——最佳高度

LWA——在备降机场的着陆重量

MLWA——备降机场的最大允许着陆重量

LWD——在目的地机场的着陆重量

MLWD——目的地机场最大允许着陆重量

TOW——起飞重量

MTOW——最大允许起飞重量

TAXW——滑行重量，即停机坪重量（RPW）

KNOT——节（kt），海里/小时

FF——燃油流量，lb/h

FFI——ISA 时的燃油流量，lb/h

ISA——国际标准大气

HR——小时

ZFW——无油重量

MZFW——最大无油重量

NAM——空中距离，海里

NGM——地面距离，海里

OEW——使用空机重

PL——业载

MPL——最大允许业载，$MPL = MZFW - OEW$

TOC——爬升顶点

WTOC——飞机在爬升顶点重量

TOD——下降顶点，即开始下降点

WTOD——飞机在下降顶点重量

km/h——千米/小时

H:MM——时间，H 小时 MM 分

TAS——真空速

TAT——总温

ΔTt——实际总温与 ISA 时的总温之差

GS——地速

案例：

北京大兴（PKX）→ 成都天府（TFU），备降机场为昆明长水（KMG），已知条件如下：

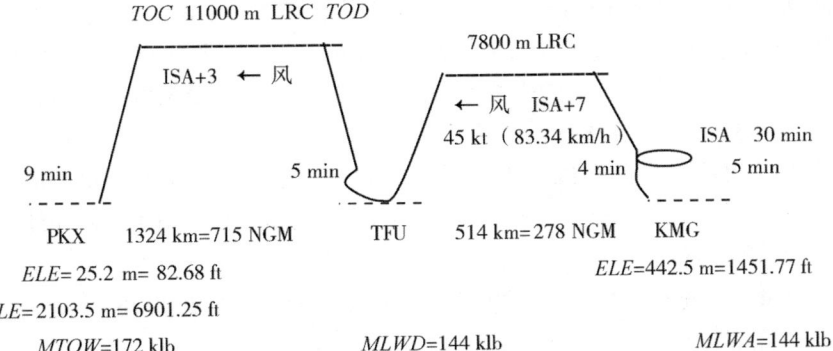

假设 PKX→TFU 分四个航段,各段距离分别是:125 km、351 km、434 km、414 km,各段风分量分别是:−106 km/h、−89 km/h、−86 km/h、−101 km/h,OEW = 42638 kg(94000 lb),$MZFW$ = 61688 kg(136000 lb),COF = 0,最大油箱容量为 45611 lb,旅客 130 人,旅客标准质量每人按 165 lb(75 kg)计,行李 3500 lb,货物 10000 lb,不使用 APU 和防冰,用简化图表制作飞行计划。[单位:m(米),km(千米),NGM(海里,地面距离),ft(英尺),lb(磅),klb(千磅),kg(千克),km/h(千米/小时),h(小时),min(分钟)]。由于航程较短,采用等高度巡航。

利用上述条件制作简化飞行计划和详细飞行计划。

2.4.1　航线简化飞行计划

简化飞行计划制作步骤如下:

首先计算北京至成都航路上的当量风,再开始制作飞行计划。

当量风 EW = [−106×125−89×351−86×434−101×414]/1324 ≈ −93.4 km/h ≈ −50.4 kt。

PL = 130×165+3500+10000 = 34950 lb = 15853 kg。

ZFW = 94000+34950 = 128950 lb < $MZFW$ = 136000 lb。

滑入:时间 = 5 min,耗油量 = 5×25 = 125 lb。

在备降机场进近:4 min,耗油量 = 4×130 = 520 lb。

等待结束重量 = 128950+125+520 = 129595 lb。

按 129595 lb 查得全发燃油流量为 4733 lb/h,等待平均重量为 129595+4733×0.5/2 ≈ 130778 lb。

按平均重量 130778 lb 查得全发燃油流量为 4771 lb/h,此即等待中的平均燃油流量,等待油量 = 4771×0.5 ≈ 2386 lb;等待开始重量 ≈ 129595+2386 = 131981 lb < $MLWA$ = 144000 lb,以等待开始重量作为 LWA 由附图 1-7 查得:

改航备降机场的油量 = 5000 lb,时间 = 0.87 h = 52.2 min。

在目的地机场着陆重量 LWD = 131981+5000 = 136981 lb < $MLWD$ = 144000 lb。

在目的地机场进近:5 min,耗油量 = 5×130 = 650 lb。

进近前重量 = 136981+650 = 137631 lb,由此可估算出 $WTOD$ ≈ 138000 lb,按 $WTOD$ 查得:

在 TOD 点 $HOPT$ = 368000 ft,最大巡航推力限制高度为 36800 ft,1.3G 过载限制高度为 39 480 ft。

巡航高度可选 11000 m(36089 ft)。

计算不可预期燃油:

（1）天府机场 15 min 等待耗油量 $F_{15\,min等待}$：按进近前重量 137631 lb 查得全发燃油流量为 5112 lb/h，等待平均重量 = 137631+5112×0.25/2 = 138270 lb；按平均重量 138270 lb 查得全发燃油流量为 5132 lb/h，此即等待中的平均燃油流量，$F_{15\,min等待}$ = 1283 lb；天府机场等待开始重量 $W_{天府机场开始}$ = 137631+1283 = 138914 lb。

（2）10%的航程油量：按天府机场等待开始重量 $W_{天府机场开始}$ 查航程油量图表，查得图表航程油量为 11600 lb，图表航程时间为 2.05 h = 123 min；$F_{10\%航程}$ = 10%×（11600+650）= 1225 lb。

（3）不可预期燃油 = $\max\{F_{15min等待}, F_{10\%航程}\}$ = 1283 lb。

起飞重量 TOW = 137631+1283+11600 = 150514 lb<$MTOW$ = 172000 lb。

滑出：9 min，耗油量 = 9×25 = 225 lb。

停机坪重量（开始滑行重量）$TAXW$ = 150514+225 = 150739 lb。

备降油量 = 520+5000 = 5520 lb，备降时间 = 0:52+0:04 = 56 min（加上进近油量和时间）。

备份油量 = 5520+2386+1283 = 9189 lb。

航程油量 = 11600+650 = 12250 lb，航程时间 = 2:03+0:05 = 2:08（加上进近油量和时间）。

轮挡油量 = 12250+125+225 = 12600 lb，轮挡时间 = 2:08+0:09+0:05 = 2:22。

起飞总油量 = 轮挡油量+备份油量 = 12600+9189 = 21789 lb<最大油箱容量 = 45611.4 lb

验算：$TAXW-ZFW$ = 150739-128950 = 21789 lb（表明求和无误）。

检查：

大兴机场起飞重量 TOW-航程油量 = 150514-12250 = 138264 lb<$MLWD$ = 144 klb。

昆明机场等待开始重量+不可预期燃油 = 131981+1283 = 133264 lb<$MLWA$ = 144 klb。

另外，可检查一下初始巡航高度是否满足要求：由起飞重量 150514 lb 查得：

爬升：时间 20 min，油量 = 3875 lb，飞过的距离 = 119.83 NAM = 108.42 NGM。

$WTOC$ = 150514-3875 = 146639 lb，在 TOC 点：$HOPT$ = 35400 ft = 10789 m。

最大巡航推力限制高度 = 35571 ft，1.3G 过载限制高度 = 38337 ft。

可见，选择 11000 m 巡航合适。

2.4.2　航线详细飞行计划

下面针对上例的条件制作详细飞行计划，这里所谓的详细是针对手算而言。后面给出有用计算机制作的详细飞行计划。

对于手算来说，如果逐个航段计算到各航路点的时间、油量等数据，工作量太大，因此，仍把北京至成都看为一段，用当量风来计算。北京至成都航路的当量风：

EW = [-106×125-89×351-86×434-101×414]/1324 ≈ -93.4 km/h ≈ -50.4 kt。

PL = 130×165+3500+10000 = 34950 lb = 15853 kg。

ZFW = 94000+34950 = 128950 lb<$MZFW$ = 136000 lb。

滑入：时间 = 5 min，耗油 = 5×25 = 125 lb。

LWA = 125+128950 = 129075 lb<$MLWA$ = 144000 lb。

在备降机场进近：4 min，耗油 = 4×130 = 520 lb。

等待油量的计算：等待结束重量 = 128950+125+520 = 129595 lb。

按 129595 lb 查得全发燃油流量为 4733 lb/h，等待平均重量 = 129595+4733×0.5/2 ≈ 130778 lb。

按平均重量 130778 lb 查得全发燃油流量为 4770 lb/h,此即等待中的平均燃油流量,等待油量 = 4770×0.5 = 2385 lb。

等待开始重量 = 129595+2385 = 131980 lb<$MLWA$ = 144000 lb。

下降数据的计算:

以 131980 lb 作为备降机场的着陆重量查下降数值表,得:

下降时间 = 21.26-9.04+4 = 16.22 min,下降油量 = 665-398+250 = 517 lb。

下降时飞过的空中距离 = 98.2-31.05+9 = 76.15 NM,飞过的地面距离 = 76.15-2/3×45×16.22/60 = 68.04 NM。

在备降段开始下降点重量 W = 131980+517 = 132497 lb。

备降航段巡航部分用平均真空速 TAS 及平均燃油流量 FF 计算,因为爬升段距离未知,所以要采用迭代法计算爬升和巡航两部分:

先按 140000 lb 和 9000 m(29 528 ft)由 LRC 巡航数值表查出:

ISA 时:TAS = 426 kt,M = 0.721,一发 FF = 2 678 lb/h。

ISA+7 时:ΔTt = 7×(1+0.2×0.721²) ≈ 7.73 ℃,TAS = 434 kt,GS = 479 kt,一发 FF ≈ 2678×(1+0.773×3%) ≈ 2740 lb/h。

先设巡航距离 = 278-68.04 = 209.96 NM,则飞行时间 = 209.96/479 ≈ 0.438 h,油量 ≈ 2740×2×0.438 ≈ 2400 lb。

先以 132497+2400 = 134897 lb 作为起飞(即复飞)重量查 ISA+10 的爬升数值表,得:

爬升时间 = 12-2+2 = 12 min,油量 = 2702-594+500 = 2608 lb。

爬升距离 = 66-2 = 64 NM,NGM = 64-45×2/3×12/60 = 58 NM。

巡航距离 = 278-68.04-58 = 151.96 NM,用上面的地速和 FF 粗算得:

巡航时间 = 0.317 h,油量 = 0.317×2740×2 ≈ 1737 lb。

巡航平均重量 = 132497+1737/2 ≈ 133366 lb,按 133366 lb 由 LRC 巡航数值表查出:

ISA 时:TAS = 414.8 kt,M = 0.702,一发 FF = 2538 lb/h。

ISA+7 时:ΔTt = 7×(1+0.2×0.702²) ≈ 7.69 ℃,TAS = 423 kt,GS = 378 kt,一发 FF ≈ 2538×(1+0.769×3%) ≈ 2597 lb/h。

由此可更准确地算出:

巡航时间 = 151.8/378 ≈ 0.402 h,油量 = 0.402×2597×2 ≈ 2088 lb。

在备降段的 TOC 点重量 $WTOC$ ≈ 132497+2088 = 134585 lb。

在目的地机场的着陆重量(即复飞重量)第二次近似值 = 134585+2608 = 137193 lb。

按 197785 lb 查出:

ISA+10 时:爬升时间 = 12 min,油量 = 2671 lb,距离 = 66 NAM = 59.6 NGM。

ISA 时:爬升时间 = 12 min,油量 = 2571 lb,距离 = 63 NAM = 56.7 NGM。

ISA+7 时:爬升时间 = 12 min,油量 = 2641 lb,距离 = 65 NAM = 58.7 NGM。

在目的地机场的着陆重量(即复飞重量)第三次近似值 = 134585+2641 = 137226 lb,这次得出的爬升数据和着陆重量与前一步的很接近,不再迭代。

LWD = 137226 lb<$MLWD$。

在目的地机场进近:5 min,耗油 = 5×130 = 650 lb,进近前重量 = 650+137226 = 137876 lb。

根据进近前重量 137876 lb 计算天府机场上空 1500 ft 的 15 min 等待耗油 $F_{15\,min等待}$,查得全发燃油流量为 5122 lb/h,等待平均重量 = 137876+5122×0.25/2 ≈ 138516 lb。

按平均重量 138516 lb 查得全发燃油流量为 5143 lb/h,此即等待中的平均燃油流量,$F_{15\,min等待}=1286$ lb;天府机场等待开始重量 $W_{天府机场开始}=137876+1286=139162$ lb。

下降数据的计算:

以 139162 lb 作为目的地机场着陆重量查下降数值表,得:

下降时间 $=23.6-5.8+4=21.8$ min,下降油量 $=706-289+250=667$ lb,飞过的空中距离 $=114$ NM,飞过的地面距离 $=114-2/3\times50.4\times21.8/60\approx101.8$ NM,在主航段开始下降点重量 $WTOD=139162+667=139829$ lb。

巡航和爬升段的迭代计算:

先按 11000 m $=36089$ ft ≈36000 ft 由 LRC 巡航数值表查出 140 klb:

ISA 时:$TAS=451$ kt,$M=0.786$,一发 $FF=2631$ lb/h。

ISA+3 时:$\Delta Tt=3\times(1+0.2\times0.786^2)\approx3.37$ ℃,$TAS=454.4$ kt,$GS=404$ kt,一发 $FF=2631\times(1+0.337\times3\%)\approx2658$ lb/h。

先设巡航距离 $\approx715-101.8=613.2$ NM,则飞行时间 $=613.2/404\approx1.52$ h,油量 $\approx2658\times2\times1.52\approx8080$ lb。

先以 $139829+8080=147909$ lb 作为起飞(即复飞)重量查 ISA 的爬升数值表,得:

爬升时间 $=20-2+2=20$ min,油量 $=3716-585+579=3710$ lb,爬升距离 $=114-0.1=113.9$ NM,$NGM=113.9-50.4\times2/3\times20/60=102.7$ NM。

巡航距离 $=715-101.8-102.7=510.5$ NM,用上面的地速和 FF 粗算得:

巡航时间 $=510.5/404\approx1.26$ h,油量 $\approx1.26\times2658\times2\approx6698$ lb,巡航平均重量 $\approx139829+6698/2=143178$ lb,按此重量由 LRC 巡航数值表查出:

ISA 时:$TAS=452$ kt,$M=0.788$,一发 $FF=2692$ lb/h。

ISA+3 时:$\Delta Tt=3\times(1+0.2\times0.788^2)\approx3.37$ ℃,$TAS=455.6$ kt,$GS=405$ kt,一发 $FF\approx2692\times(1+0.337\times3\%)\approx2719$ lb/h。

巡航时间 $=510.5/405\approx1.26$ h,巡航油量 $\approx1.26\times2719\times2\approx6852$ lb。

在主航段的 TOC 点重量 $WTOC\approx139829+6852=146681$ lb。

起飞重量第二次近似值 $=146681+3710=150391$ lb。

按 150391 lb 查出:

ISA+10 时:爬升时间 $=21$ min,油量 $=4010$ lb,距离 $=124.57$ NAM $=112.78$ NGM。

ISA 时:爬升时间 $=20$ min,油量 $=3810$ lb,距离 $=117.49$ NAM $=106.3$ NGM。

ISA+3 时:爬升时间 $=20$ min,油量 $=3870$ lb,距离 $=119.62$ NAM $=108.22$ NGM。

起飞重量第三次近似值 $=146681+3870=150551$ lb,按 150551 lb 查出:

ISA+10 时:爬升时间 $=21$ min,油量 $=4016$ lb,距离 $=124.86$ NAM $=113.05$ NGM。

ISA 时:爬升时间 $=20$ min,油量 $=3816$ lb,距离 $=117.75$ NAM $=106.5$ NGM。

ISA+3 时:爬升时间 $=20$ min,油量 $=3876$ lb,距离 $=119.88$ NAM $=108.47$ NGM。

起飞重量第三次近似值 $=146681+3876=150557$ lb $<MTOW$。

根据第三次的迭代的爬升油量,计算航程油量 $=3876+6852+667+650=12045$ lb。

10%的航程油量 $=12045\times10\%=1204.5$ lb,则计算不可预期燃油,不可预期燃油 $=\max\{F_{15\,min等待},F_{10\%航程}\}=1286$ lb。

滑出:时间 $=9$ min,耗油 $=9\times25=225$ lb。

滑行重量 $TAXW=150557+225=150782$ lb。

累加分段计算的结果：

备降油量 = 520+517+2085+2641 = 5763 lb，备降时间 = 4+16.22+0.317×60+12 = 51.24 min。

备份油量 = 5763+2385+1286 = 9434 lb = 4279 kg。

航程油量 = 3876+6852+667+650 = 12045 lb，航程时间 = 20+1.26×60+21.8+5 = 122.4 min。

轮挡油量 = 12045+125+225 = 12395 lb，轮挡时间 = 48.06+5+9 = 62.06 min。

起飞总油量 = 9434+12395 = 21829 lb ≈ 9901 kg < 最大油箱容量 = 20689 kg。

验算：$TAXW-ZFW$ = 150782−128950 = 21832 lb ≈ 9903 kg（表明求和无误）。

检查：

大兴机场起飞重量 TOW−航程油量 = 146681−12042 = 134639 lb < $MLWD$ = 144 klb。

昆明机场等待开始重量 + 不可预期燃油 = 131980+1286 = 133266 lb < $MLWA$ = 144 klb。

另外，同上例一样可根据计算中得到的 $WTOC$ 和 $WTOD$ 检查一下巡航高度是否合适，不再重复。

下面给出按本例的条件用计算机制作的两份飞行计划：第一份是用航路当量风把整条航线作为一段计算的；第二份是用各段的风分段计算的。这两份飞行计划的结果很接近，也和手动计算结果差不多。

```
    -= * == * ==B757-200(RB211-535E4)    飞行计划== * == * ==--
  航班：      机号：B××××    机长：    11-19-95    03:41:35
  PPP      (35.m)——CCC    (500.m)  备降：KKK      (1901.m)

      * * 主航段：290/0.78M  ——LRC    ——0.78M/290/250 KIAS * *
      高度=11000.m  温度=-54.  (℃)  ISA+3.  紧急油量/航程油量=.0%

    航段      地面距离          航路风分量(顶风为负)
    1      104.km(56.NM)        -106.km/h(-57.kt)
    2      347.km(187.NM)        -111.km/h(-60.kt)
    3      529.km(286.NM)        -116.km/h(-63.kt)
    4      687.km(371.NM)        -120.km/h(-65.kt)
    * * 备降航段：290/0.78M  ——LRC    ——0.78M/290/250 KIAS * *
      高度=9000.m  温度=-37.  (℃)  ISA+7.  紧急油量/备降油量=.0%

    航段 1                地面距离          航路风分量(顶风为负)
                      682.km(368.NM)      -76.km/h(-41.knot)
```

最大允许起飞重量：108862.kg　　飞行距离：1667.km　900.NM　　滑出时间：9.min

最大着陆重量(目)：89811.　　巡航高度：11000.m　36090.ft　　离场时间：0.min

最大着陆重量(备)：89811.　　巡航速度：LRC　　　风：-63.kt

滑入时间：5.min　　最大无油重量：83461.　　备降距离：682.km　368.NM

进近时间(目)：5.min　　使用空机重量：58040.　　巡航高度：9000.m　29528.ft

进近时间(备)：4.min　　最大结构业载：25421.　　巡航速度：LRC

风：-41.kt　　防冰预计使用：.0 h　　最大油箱容量：33253.

等待高度：458.m　　　1503.ft　　等待高度气温：ISA+0.

(γ= .779 kg/L)　　(距地面)　　飞行中APU：断开

滑行全重：99077.kg　218427. lb　轮挡油量：9433.kg　20797.lb　轮挡时间：2:45

起飞全重：98918.　218076.　　航程油量：9186.　20251.　　航程时间：2:31

着陆重量：89732.　197824.　　备降油量：4292.　9463.　　备降时间：1:8

无油重量：82988.　182957.　　等待油量：2363.　5210.　　等待时间：0:45

允许业载：24948.　55001.　　备份油量：6655.　14673.　　等待表速：208.kt

起飞油量：16089.　35470.　　(扣除滑行油后：15841.　34924.)　滑行油量：247 kg

起飞前APU地面工作：0.min,　　耗油：0 kg.　　不计此油量及滑行油量，

　　　　　　　　　　　　　　　　　　　　　　　　　　则总油量= 15841.kg

考虑飞机老化等因素航程油量与备　　　　　　　　　　公司备份油：0.kg

降油量各多加了　　　　　　0.%和0.%.

　　到爬升顶点：时间　距离 km　剩余油量 kg　到下降顶点：时间　距离 km　剩余油量 kg
　　　　　　28.　180.　13574.　　　　　　2:15　1499.　7423.

对计算出的起飞重量(98918.kg)：最佳巡航高度=35312.ft(10763.m)，

1.3G 过载限制高度=38812.ft(11830.m)，最大巡航推力限制高度=39541.ft(12052.m)

对巡航结束时的重量(90411.kg)：最佳巡航高度=36671.ft(11177.m)，

1.3G 过载限制高度=40275.ft(12275.m)，最大巡航推力限制高度=40868.ft(12456.m)

该飞行计划是把 PKX→TFU 看作一个航段,使用当量风计算的,下面的飞行计划是按题意把 PKX→TFU 分为四个航段(PKX→WP2、WP2→WP3、WP3→WP4、WP4→TFU,各段距离依次是 104 km、347 km、529 km、687 km,设真航迹为 0°,各段气象风角度也是 0°,风速依次是 106 km/h、111 km/h、116 km/h、120 km/h,使得各段顶风刚好是 106 km/h、111 km/h、116 km/h、120 km/h),使用各段的风分量计算的。

```
--= * = = * = = B757-200(RB211-535E4)   飞 行 计 划 = = * = = * =--
      航班：        机号:B××××    机长:11-19-95    03:49:45
```

PKX(35.m) -- TFU	(500.m) 备降：KMG	(1901.m)
最大允许起飞重量：108862.kg	飞行距离：1667.km 900.NM	滑出时间：9.min
最大着陆重量（目）：89811.	巡航高度：11000.m 36090.ft	离场时间：0.min
最大着陆重量（备）：89811.	巡航速度：LRC	滑入时间：5.min
最大无油重量：83461.	备降距离：682.km 368.NM	进近时间（目）：5.min
使用空机重量：58040.	巡航高度：9000.m 29528.ft	进近时间（备）：4.min
最大结构业载：25421.	巡航速度：LRC	防冰预计使用：.0 h
最大油箱容量：33253.	等待高度：458.m 1503.ft	等待高度气温：ISA+0.
（γ=.779 kg/L）	（距地面）	飞行中APU：断开

滑行全重：99078.kg 218430.lb	轮挡油量：9435.kg 20801.lb	轮挡时间：2:45
起飞全重：98919. 218079.	航程油量：9187. 20255.	航程时间：2:31
着陆重量：89732. 197825.	备降油量：4292. 9463.	备降时间：1:8
无油重量：82988. 182957.	等待油量：2363. 5210.	等待时间：0:45
允许业载：24948. 55001.	备份油量：6655. 14673.	等待表速：208.kt
起飞油量：16091. 35474.	（扣除滑行油后：15843. 34928.）	滑行油量：247 kg
起飞前APU地面工作：0.min,	耗油：0 kg.	不计此油量及滑行油量,
		则总油量= 15843.kg

考虑飞机老化等因素航程油量与备

降油量各多加了　　　　　　　　　0.%和0.%　　　　　　　公司备份油：0.kg

航路点 NDB VOR	经纬度	真空速 地速 （knot）	磁向 磁向 （度）	巡航高度 最佳高度 （m）	气象风 温度 （°,km/h）	累计距离 航段距离 （km）	时间 时间	剩余油量 油量 （kg）
PKX		352.	360.	35.	0./106.	0.	0:0	16091.
		314.	0.	10763.	ISA+3.	104.	0:11	1348.
滑行							0:9	159.
离场							0:0	0.
（入航）								
WP2		352.	360.	爬升	0./111.	104.	0:20	14583.
		313.	0.	10763.	ISA+3.	77.	0:8	1008.
爬升顶点		462.	360.	11000.	0./111.	181.	0:28	13575.
		402.	0.	10763.	ISA+3.	270.	0:22	1279.
WP3		462.	360.	11000.	0./116.	451.	0:49	12297.
		399.	0.	10849.	ISA+3.	529.	0:43	2479.
WP4		462.	360.	11000.	0./120.	980.	1:32	9817.
		397.	0.	11015.	ISA+3.	520.	0:42	2394.
下降始点		461.	360.	11000.	0./120.	1500.	2:15	7423.
		396.	0.	11177.	ISA+3.	167.	0:21	328.
进近							0:5	352.
TFU				500.		1667.	2:40	6744.
						（滑行：	0:5	88）

备降机场:							
TFU	323.	360.	500.	0./ 76.	0.	0:0	6744.
	296.	0.	11342.	ISA+7.	109.	0:12	1662.
爬升顶点	440.	360.	9000.		109.	0:12	5082.
	399.	0.	11342.		456.	0:37	2082.
下降始点	437.	360.	9000.		564.	0:49	3000.
	396.	0.	11490.		118.	0:15	267.
进近						0:4	281.
KMG			1901.		682.	1:8	2452.
					滑行:	0:5	88.

这两份飞行计划的结果非常接近,这表明制作简化飞行计划时使用当量风是可行的。这个例子中各航段的风相差不太大,即使相差很大,用航路当量风、把整条航线作为一段计算和用各段的风分量计算的结果也很接近。例如,把这个例子中四段的风速改为:-100 km/h、50 km/h、-120 km/h、-60 km/h,分段计算的结果:起飞油量=15512 kg,轮挡油量=8896 kg,轮挡时间=2:37;按当量风=-58.64 km/h 计算的结果为:起飞油量=15509 kg,轮挡油量=8854 kg,轮挡时间=2:36。在有侧风且侧风比较大时,用航路当量风把整条航线作为一段计算和用各段的风分量准确计算的结果会有些差别,侧风较小时,两者仍是很接近的。

第3章
特殊飞行计划制作

除了常规飞行计划外,实际运行时还会遇到各种其他的特殊飞行计划,例如规定起飞油量的飞行计划、无备降机场的燃油规定与飞行计划、目的地机场不能加油和只能部分加油的飞行计划、涡轮动力飞机的最小成本飞行计划、利用燃油差价的飞行计划、途中改航(全发改航)的飞行计划、公务机飞行计划、二次放行飞行计划等。这些特殊的飞行计划的计算方法或者使用的图表会有所不同。

3.1 规定起飞油量的飞行计划

在某些情况下,可能希望起飞油量不要少于某一数值。对于有这种要求的情况,首先按正常情况根据有关的加油规定计算起飞时应加的油量,如该油量多于希望的(规定的)最小起飞油量,则以计算出的应加油量为准,否则,应增加油量,重新计算。其具体算法是:在第一步最后算出的备降机场停机坪重量(即由备降机场停机坪开始往回推算时的重量)上增加一定的额外油量,例如增加规定的起飞油量和第一步算出的应加油量的差值的 0.9~0.95 倍(航程远取小值),然后往回推算到起飞机场,如这次算出的起飞油量仍小于规定的起飞油量,则再增加一些油量重新计算,如这次算出的起飞油量大于规定的起飞油量,则把额外油量减少一些重新计算,反复迭代,直到算出的起飞油量近似等于规定的起飞油量为止。在这种计算中如受到最大允许起飞重量或最大着陆重量或油箱容量限制,则应减少业载来满足各种限制。比较第一次和这次的飞行计划中的允许业载、轮挡油量、航程油量等即可得知为达到规定的起飞油量而造成的业载损失(业载也可能不减少)和(或)多耗的燃油。

例 3-1-1: 不规定起飞油量和规定起飞油量的飞行计划及其计算结果对比。

（1）不规定起飞油量的飞行计划（由最大结构业载开始计算）：

北京　　　　（25.2.m）——广州　　　（15.2.m）　备降:桂林　　　（173.6.m）		
航路爬升速度:280/M0.78	航路下降速度:M0.78/280/250 KIAS	
最大允许起飞重量:78244.kg	飞行距离:1662.km　897.NM	滑出时间:10.min
最大着陆重量(目):65317.	巡航高度:9800.m 32100.ft	离场时间:3.min
最大着陆重量(备):65317.	巡航速度:M0.79	滑入时间:5.min
最大无油重量:61688.	备降距离:272.km　147.NM	进近时间(目):5.min
使用空机重量:42638.	巡航高度:8400.m 27600.ft	进近时间(备):5.min
最大结构业载:19050.	巡航速度:LRC	防冰预计使用:.0 h
最大油箱容量:20689.	等待高度气温:ISA+0.	
(γ=.779 kg/L)	（距地面）	飞行中 APU:断开
滑行全重:70893.kg 156292.lb	轮挡油量:6073.kg　13389.lb	轮挡时间:2;23
起飞全重:70780.　156042.	航程油量:5903.　13014.	航程时间:2;08
着陆重量:64877.　143028.	备降油量:1425.　3142.	备降时间:0;29
无油重量:61688.　135997.	等待油量:1174.　2588.	等待时间:0;30
允许业载:19050.　41998.	备份油量:3189.　7030.	等待表速:221.kt
起飞油量:9093.　20046.	（扣除滑行油后:8923.　19672.）	滑行油量:170 kg
起飞前 APU 地面工作:0.min,耗油: 0 kg. 不计此油量及滑行油量,则总油量=8923 kg		
考虑飞机老化等因素航程油量与备降油量各多加了 0.%和 0.%. 公司备份油: 0.kg		
*业载受目的地机场最大着陆重量限制。		

（2）规定起飞油量为 11000 kg 的飞行计划（摘要）：

滑行全重:70893.kg 156292.lb	轮挡油量:6073.kg　13389.lb	轮挡时间:2;23
起飞全重:70780.　156042.	航程油量:5903.　13014.	航程时间:2;08
着陆重量:64877.　143028.	备降油量:1425.　3142.	备降时间:0;29
无油重量:61688.　135997.	等待油量:1174.　2588.	等待时间:0;30
允许业载:17143.　37793.	备份油量:5096.　11235.	等待表速:221.kt
起飞油量:11000.　24251.	（扣除滑行油后:10830.　23876.）	滑行油量:170 kg
起飞前 APU 地面工作:0.min,耗油: 0 kg. 不计此油量及滑行油量,则总油量=10830.kg		
考虑飞机老化等因素航程油量与备降油量各多加了 0.%和 0.%. 公司备份油: 0.kg		
为达到规定起飞油量 11000 kg,额外加油 1907 kg,多耗油 0 kg,业载减少 1907.kg		
*业载受目的地机场最大着陆重量限制。		

（3）如不规定起飞油量,计划业载为 18000 kg,计算结果为：

起飞油量=9003 kg,航程油量=5852 kg,允许业载=18000 kg(业载不受限制)。

如规定起飞油量=11000 kg,计划业载 18000 kg,计算结果同(2),此时为达到规定起飞油量 11000 kg,额外加油 1997 kg,多耗油 0 kg,业载减少 857 kg

*业载受目的地机场最大着陆重量限制。

3.2 无备降机场的燃油规定与飞行计划

本节所讲的无备降机场的飞行计划主要指的是无目的地备降机场的飞行计划。

3.2.1 无目的地备降机场的相关规定

一般情况下,按照仪表飞行规则签派飞机飞行前,都应当至少为每个目的地机场列出一个备降机场。但有时候当天气满足一定的要求时,也可以不选择目的地备降机场。无备降机场的情况有两种:一种是无须选择备降机场;另一种是孤立机场。这两种情况在 CCAR-121 中都有明确的规定。

3.2.1.1 无须选择备降机场的情况

对于国内和国际定期载客运行,当天气满足第 121.639、121.641 以及 121.642 条的规定时,可以不选择目的地备降机场。

第 121.639 条 仪表飞行规则国内定期载客运行的目的地备降机场

(a)按照仪表飞行规则签派飞机飞行前,应当在签派单上至少为每个目的地机场列出一个备降机场。当目的地机场和第一备降机场的天气条件预报都处于边缘状态时,应当再指定至少一个备降机场且在签派单上列出。

(b)如果目的地机场符合下列条件并且合格证持有人在飞机与运行控制中心之间建立了独立可靠的通信系统进行全程监控,则可以不选择目的地备降机场:

(1)如果天气实况报告、预报或者两者的组合表明,在飞机预计到达目的地机场时刻前后至少 1 小时的时间段内:

(i)机场云底高度至少在公布的最低的仪表进近最低标准之上 450 米(1500 英尺),或者在机场标高之上 600 米(2000 英尺),取其中较高值;

(ii)机场能见度至少为 5000 米。

(2)有独立可用的多条跑道,且其中一条跑道的仪表进近程序处在可用状态。

(c)按照本条规定选择的目的地备降机场的天气条件应当满足本规则第 121.643 条的要求。

第 121.641 条 国际定期载客运行的目的地备降机场

(a)按照仪表飞行规则签派飞机飞行前,应当在签派单上为每个目的地机场至少列出一个备降机场。但在下列情形下,如果在每架飞机与运行控制中心之间建立了独立可靠的通信系统进行全程监控,则可以不选择目的地备降机场:

(1)当预定的飞行不超过 6 小时,且相应的天气实况报告、预报或者两者的组合表明,在预计到达目的地机场时刻前后至少 1 小时的时间内,目的地机场的天气条件符合下列规定:

(i)机场云底高度符合下列两者之一:

(A)如果该机场需要并准许盘旋进近,至少在最低的盘旋进近最低下降高度(MDA)之上 450 米(1500 英尺);

(B)至少在公布的最低的仪表进近最低标准之上 450 米(1500 英尺),或者机场标高之上 600 米(2000 英尺),取其中较高者。

(ii)机场能见度(VIS)至少为 5000 米(或者 3 英里),或者高于计划使用的目的地机场仪表进近程序能见度最低标准 3200 米(或者 2 英里)以上,取其中较大者。

(2)有独立可用的多条跑道,且其中一条跑道的仪表进近程序处在可用状态。

(3)该次飞行是在前往无可用备降机场的特定目的地机场的航路上进行的,而且飞机有足够的燃油来满足本规则第 121.659 条的要求。

(b)按照本条规定选择的目的地备降机场的天气条件应当满足本规则第 121.643 条的要求。

3.2.1.2　孤立机场的情况

当从事国际定期载客运行时,如果目的地机场是孤立机场,按照第 121.642 条的规定也可以不用指定备降机场。

第 121.642 条　仪表飞行规则补充运行的目的地备降机场

(a)除本条(b)款规定外,当放行飞机按照仪表飞行规则进行补充运行时,应当在飞行放行单中至少为每个目的地机场列出一个备降机场。

(b)对于在国外飞行的航路上,当特定目的地机场无可用备降机场时,如果飞机装载了本规则第 121.659 条规定的燃油,在仪表飞行规则下可以不指定备降机场。

在 CCAR-121 的附录中对航线临界点和无可用备降机场的特定目的地机场(孤立机场)都有定义:

航线临界点(不可返回点):飞机能够从该点飞行到目的地机场以及特定飞行的可用航路备降机场的最后可能位置(地)点。

无可用备降机场的特定目的地机场(孤立机场):是指对于某一机型没有合适目的地备降机场的目的地机场。当从目的地机场决断高度/高或复飞点复飞备降至最近合适备降机场的所需燃油超过下列数值时,合格证持有人应当将该目的地机场视为无可用备降机场的特定目的地机场:

(a)对于涡轮发动机飞机,以等待速度在机场上空 450 米(1500 英尺)高度上在标准条件下飞行 90 分钟所需的油量;

(b)对于活塞发动机飞机,以等待速度在机场上空 450 米(1500 英尺)高度上在标准条件下飞行 75 分钟所需的油量。

3.2.2　无备降机场的加油规定

第 121.659 条　特定情况燃油要求

(a)特定情况下目的地备降机场燃油的计算:

(1)当不需要有目的地备降机场时,所需油量能够使飞机在目的地机场上空 450 米(1500 英尺)高度上在标准条件下飞行 15 分钟;

(2)预定着陆机场是一个孤立机场(无可用备降机场的特定目的地机场):

(i)能够以正常燃油消耗率在目的地机场上空飞行 2 小时的所需油量,包括最后储备燃油。

(ii)当按照本规则第 641 条(a)款第(2)项或第 642 条(b)款放行飞机前往孤立机场(无可用备降机场的特定目的地机场)时,需满足以下条件:

(1)在飞机与签派室之间建立了独立可靠的语音通信系统进行全程监控;

（2）必须为每次飞行至少确定一个航路备降机场和与之对应的航线临界点；

（3）除非气象条件、交通和其他运行条件表明在预计使用时间内可以安全着陆，否则飞往无可用备降机场的特定目的地机场的飞行不得继续飞过航线临界点。

（b）对于涡轮螺旋桨发动机飞机的国际定期载客运行或者包括有至少一个国外机场的补充运行，不可预期燃油不得低于以正常巡航消耗率飞往本规则第657条（b）款第（2）、（4）项规定的机场所需总时间的15%所需的油量，或者以正常巡航消耗率飞行60分钟油量，两者当中取其中较短的飞行时间。

（c）如果根据本规则121.657条计算的最低燃油不足以完成下列飞行，则应当要求额外燃油：

（1）假定在航路最困难临界点发动机发生失效或者丧失增压需要更多燃油的情况下，允许飞机在必要时下降并飞行到某一备降机场：

（i）以等待速度在该机场上空450米（1500英尺）高度上在标准条件下飞行15分钟；

（ii）在该机场进近并着陆。

（2）延程运行（EDTO）的飞机应当遵守经批准的延程运行临界燃油方案。

（3）满足上述未包含的其他规定。

对于未指定目的地备降机场时，是指按照到达目的地机场的预计着陆重量计算备降机场燃油量。

3.2.3　无备降机场的飞行计划

3.2.3.1　无须指定备降机场的情况

按照第121.659条（a）款（1）项的规定，当不需要有目的地备降机场时，备降燃油计算方法发生改变，即能够使飞机在目的地机场上空450米（1500英尺）高度上在标准条件下飞行15分钟，其余油量如滑行油量、航程油量、不可预期燃油、最后储备燃油、酌情携带的燃油等的算法不变。对于涡轮发动机飞机，在飞行计划剖面图中取消备降航段的计算。对于涡轮发动机飞机，最后储备燃油是以等待速度在机场上空450米（1500英尺）高度上在标准条件下飞行30分钟所需的油量。而对于活塞式发动机飞机，是以等待速度在机场上空450米（1500英尺）高度上在标准条件下飞行45分钟所需的油量。此时飞行计划剖面可以分别简化，如图3-2-1和图3-2-2所示。其飞行计划的计算方法与常规飞行计划方法相似，只不过缺少了备降油量的计算，计算过程在此不再重复。

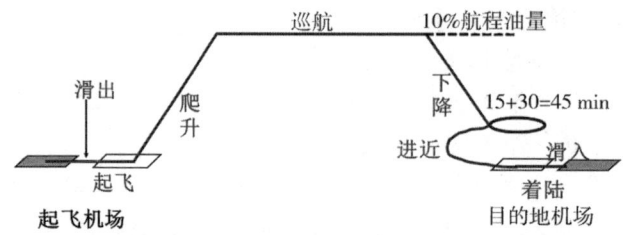

图3-2-1　无须指定备降机场的情况（涡轮发动机）

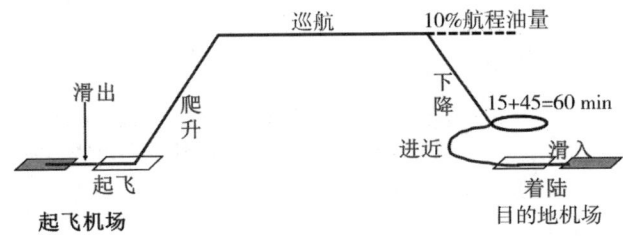

图 3-2-2　无须指定备降机场的情况(活塞式发动机)

3.2.3.2　孤立机场的情况

按照第 121.659 条(a)款(2)项的规定,在符合孤立机场的条件时,孤立机场不需要备降机场,此时按照上述条款计算备降油量,以正常燃油消耗率在目的地机场上空飞行 2 h 的所需油量,包括最后储备燃油。因此,除了备降燃油和最后储备燃油外,其余油量如滑行油量、航程油量、不可预期燃油、酌情携带的燃油等的算法不变。在飞行计划剖面图中,取消备降航段的计算,其飞行计划剖面可以简化,如图 3-2-3 所示。其飞行计划的计算方法与常规飞行计划方法相似,只不过缺少了备降油量和最后储备燃油的计算,计算过程在此不再重复。

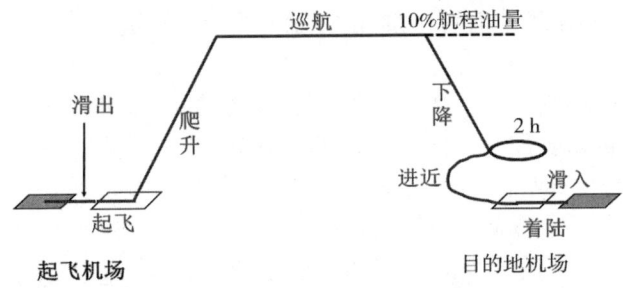

图 3-2-3　孤立机场的情况

3.3　目的地机场不能加油和只能部分加油的飞行计划

3.3.1　目的地机场不能加油的飞行计划

有时候会遇到有个别机场不能为飞机加油的情况,飞往这样的机场时飞机必须带上回程所需的油量。

如图 3-3-1 所示,假设飞机从机场 DEPART 出发飞到不能加油的机场 DEST1,其备降机场为 ALT1,正常情况下飞机在 DEST1 着陆并滑入停机坪上下旅客及装卸货物,然后滑到起飞线并起飞飞往目的地机场 DEST2(可以是 DEPART),其备降机场为 ALT2。在 DEST1 停机坪上剩余的油量应满足由 DEST1 起飞到 DEST2 备降 ALT2 的油量需求。注意,在 DEPART 起飞时的业载与在 DEST1 起飞时的业载可以是不同的。

这种情况下,其飞行计划的计算步骤如下:

(1)首先制作第二个航班的常规飞行计划。即从 ALT2 开始往回计算直到 DEST1 的停机坪为止,得到了从 DEST1 起飞到 DEST2 备降 ALT2 的飞行计划。

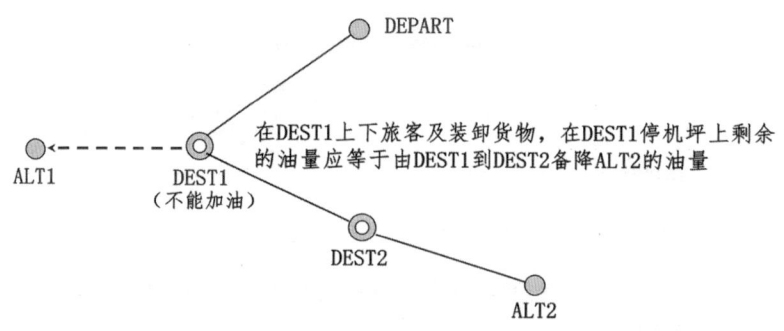

图 3-3-1　目的地机场不能加油

（2）然后制作第一个航班的常规飞行计划。方法是把步骤（1）计算出的总油量作为由 DEPART 到 DEST1 的备份油量,再由 DEST1 的停机坪往回算到 DEPART 的停机坪,这样就得到了由 DEPART 起飞到 DEST1 的飞行计划。

（3）一般情况下,步骤（2）得出的总油量大于由 DEPART 到 DEST1 再备降 ALT1 所需总油量（ DEST1 到 ALT1 比 DEST1 到 ALT2 距离近）;如果不能判断,则应再制作 DEPART→DEST1→ALT1 的飞行计划,再与步骤（2）的飞行计划比较,将总油量大的作为起飞应加的油量。当 DEST1 不繁忙,不用等待时可不加在 DEST1 的等待油量。

例如:

北京→拉萨,备降机场成都,然后拉萨→广州,备降机场桂林,即属于上面讨论的一般情况。

北京→拉萨,备降机场成都,然后拉萨→北京,备降机场太原,则是上面一般情况中的特例。

成都→拉萨,备降机场成都,然后拉萨→成都,备降机场重庆,亦是一种特例。

对上述案例的这些情况都应先制作回程（由拉萨起飞）的飞行计划,再制作飞往拉萨的飞行计划,如果回程长,需要带的回程油多,会由于拉萨机场的着陆重量限制（$LW = OEW + PL + $回程油 $\leq MLW$）,使由北京飞往拉萨时的业载过小,经济效益差,甚至亏本。因此,可以考虑由拉萨起飞时经停成都,这样由拉萨起飞时油量按拉萨至成都备降重庆（或昆明）机场计算,可使回程油减少,增大北京飞拉萨时的业载。

3.3.2　目的地机场只能部分加油的飞行计划

目前有些机场对在该机场着陆的飞机只能补充少量燃油,不能提供所需的全部燃油,对这种目的地机场只能部分加油的情况,可按下述步骤制作飞行计划。

考虑最一般的情况,如图 3-3-2 所示,假设从机场 DEPART 起飞飞往只能部分加油的机场 DEST1,其备降机场为 ALT1,由 DEST1 起飞执行飞往 DEST2 的航班,备降机场 ALT2。设在 DEST1 允许的加油量为 DF。

（1）先制作由 DEST1 飞往 DEST2,备降机场 ALT2 的飞行计划

计算出第二个航班的轮挡油量（$BKF2$）、备份油量（$RESF2$）、所需的起飞总油量（$RPF2$）及业载（$PL2$）等,这个飞行计划也是执行由 DEST1 飞往 DEST2 的航班所必需的。

对这种只能部分加油的情况,不能简单地套用目的地机场不能加油的做法（以 $RPF2-DF$ 作为在 DEST1 停机坪剩余油量往回推算）,因为由 DEPART 飞往 DEST1,备降机场 ALT1 的备份油量（$RESF1$）可能多于 $RESF$（$RESF = RPF2 - DF$）,如果简单地按 $RESF$ 计算有可能会导致飞机往 ALT1 时油量不足。因此,只有制作 DEPART 飞往 DEST1 备降机场 ALT1 的飞行计划才能知道 $RESF1$ 是否多于 $RESF$。

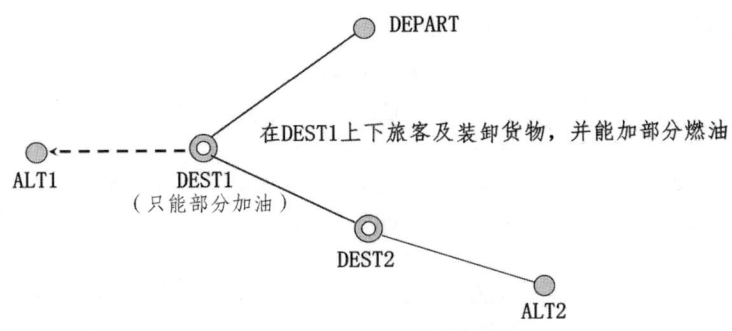

图 3-3-2　目的地机场只能部分加油

（2）制作由 DEPART 飞往 DEST1，备降机场 ALT1 的飞行计划

这个飞行计划按常规飞行计划方法制作，所使用的飞行剖面就是前面所用的制作正常飞行计划时用的飞行剖面。制作这个飞行计划时要保证飞机在 DEST1 时剩余油量不少于 $RESF$（$RESF=RPF2-DF$），以使在 DEST1 着陆、补充油量 DF 后总油量满足执行 DEST1 飞往 DEST2 备降机场 ALT2 所需的油量 $RPF2$，即 $RESF1+DF \geqslant RPF2$。步骤如下：

①先令在 ALT1 停机坪的重量 $W=OEW+PL+COF+F154+RFO$，由 ALT1 停机坪开始往回计算。式中：OEW 为使用空机重量；PL 为业载；COF 为酌情携带的燃油（公司备份油）；$F154$ 为使备份油等于所要求的数量需要补加的油量（$RESF1+DF=RPF2$），开始先令 $F154=0$；RFO 为不可预期燃油，最初开始计算时可取 $RFO=0$ 或先估算一个初值，在计算中通过迭代确定 RFO 及 $RESF1$ 和起飞油量等各项数据。

②当计算到 DEST1 时，如果着陆重量超过该机场的最大允许着陆重量（$LWD>MLWD$），则应减少业载由 ALT1 停机坪重新开始往回计算，直到在 DEST1 的着陆重量满足该机场的最大允许着陆重量限制（$LWD=MLWD$）为止。这时也就算出了备份油量 $RESF1$，$RESF1=RFO+COF+F154+DIVF+HLDF$。式中：$DIVF$ 为备降油量（包含在备降机场的进近油量）；$HLDF$ 为等待油量。如果 $RESF1 \geqslant RESF$，则由 DEST1 继续往回（DEPART）计算，完成这个飞行计划；否则，取 $F154=0.94 \times (RF-RF1)$，再返回 ALT1 停机坪重新开始计算，直到 $RESF1=RESF$ 为止，然后继续由 DEST1 往回（DEPART）计算，完成这个飞行计划。

③如果计算得到的 DEPART 的起飞重量超过该机场的最大允许起飞重量（$TOW>MTOW$），则应减少业载由 ALT1 停机坪重新开始往回计算，直到在 DEPART 的起飞重量满足该机场的最大允许起飞重量（$TOW=MTOW$）为止。这样就制作出了目的地机场只能部分加油的飞行计划。

从上述步骤可知，由 DEPART 飞到 DEST1 后的剩余油量 $RESF1$ 有可能多于（$RPF2-DF$），特别是在 DEST1 的备降机场比较远或在 DEST1 允许的加油量 DF 比较大时。当在 DEST1 允许的加油量 DF 很大时，（$TF2-DF$）很小甚至为负，这时制作的飞行计划和目的地机场能正常加油的飞行计划完全一样。对于目的地机场不能加油的情况，如果按目的地机场只能部分加油但允许加油量为 0 来制作飞行计划，则制作的飞行计划和按目的地机场不能加油制作的飞行计划一般相同。但在由 DEPART 飞往 DEST1 的备降距离远到使得备份油量 $RESF1$ 多于（$RPF2-0$）时，两个飞行计划不同，按目的地机场只能部分加油的方法制作飞行计划算出的起飞油量比按不能加油的方法算出的起飞油量多（后者是由 $RPF2$ 作为在 DEST1 的剩余油量往回推算的），这个飞行计划才是正确的。此外，这个计划也能给出备降航段的飞行数据。

下面结合案例进行说明。假设飞机由广州飞往拉萨，备降机场成都，再由拉萨飞往成都，备降机场重庆。

例 3-3-1: 由拉萨飞往成都,备降机场重庆的飞行计划。

```
－－＝＊＝＝＊＝＝ B737-700(CFM56-7B24A)　飞 行 计 划 ＝＝＊＝＝＊＝＝－－
        拉萨(3569.6.m) －－ 成都　　(512.4.m)　　　备降:重庆　(415.6.m)
最大允许起飞重量:69853.kg        飞行距离:1196.km　646.NM        滑出时间:9.min
最大着陆重量(目):58604.         巡航高度:10100.m 33100.ft      离场时间:0.min
最大着陆重量(备):58604.         巡航速度:LRC                   滑入时间:5.min
最大无油重量:55202.             备降距离:171.km　92.NM         进近时间(目):5.min
使用空机重量:39648.             巡航高度:5100.m　16700.ft      进近时间(备):4.min
最大结构业载:15554.             巡航速度:LRC                   防冰预计使用:0.h
最大油箱容量:20429.             等待高度气温:ISA+0.
(γ=.779 kg/L)                   飞行中 APU:断开
```

```
滑行全重:61484.kg 135548.lb    轮挡油量:3872.kg　8536.lb      轮挡时间:1:48
起飞全重:61382.　135323.       航程油量:3713.　8186.          航程时间:1:34
着陆重量:57528.　126826.       备降油量:935.　2061.           备降时间:0:21
无油重量:55202.　121698.       等待油量:1020.　2249.          等待时间:0:30
允许业载:15554.　34290.        备份油量:2326.　5128.          等待表速:208.kt
起飞油量:6180.　13624.         (扣除滑行油后:6021.　13274.)    滑行油量:159 kg
起飞前 APU 地面工作:60.min,耗油:107 kg.不计此油量及滑行油量,则总油量=5914.kg
考虑飞机老化等因素航程油量与备降油量各多加了 0.%和 0.%.公司备份油:0.kg
＊业载受最大着陆重量限制。
```

例 3-3-2: 由广州飞往拉萨,备降机场成都(拉萨不能加油)的飞行计划。

```
－－＝＊＝＝＊＝＝ B737-700(CFM56-7B24A)　飞 行 计 划 ＝＝＊＝＝＊＝＝－－
    航班:        机号:B××××    机长:        12-14-93    13:04:00
        广州(15.2.m) －－ 拉萨    (3569.6.m)  备降:成都      (512.4.m)
最大允许起飞重量:69853.kg        飞行距离:3003.km　1622.NM      滑出时间:9.min
最大着陆重量(目):58604.         巡航高度:11000.m 36100.ft      离场时间:0.min
最大着陆重量(备):58604.         巡航速度:LRC                   滑入时间:4.min
最大无油重量:55202.             备降距离:1196.km　646.NM       进近时间(目):5.min
使用空机重量:39648.             巡航高度:10100.m　33100.ft     进近时间(备):4.min
最大结构业载:15554.             巡航速度:LRC                   防冰预计使用:0.h
最大油箱容量:20429.             等待高度气温:ISA+0.
(γ=.779 kg/L)                   飞行中 APU:断开
```

```
滑行全重:67718.kg 149291.lb    轮挡油量:9177.kg　20232.lb     轮挡时间:3:57
起飞全重:67616.　149066.       航程油量:9018.　19881.         航程时间:3:43
着陆重量:58599.　129187.       备降油量:0.　0.                备降时间:0:0
无油重量:53416.　117761.       等待油量:0.　0.                等待时间:0:0
允许业载:12771.　28155.        备份油量:6180.　13624.         等待表速:206.kt
起飞油量:15198.　33506.        (扣除滑行油后:15039.　33155.)   滑行油量:159 kg
起飞前 APU 地面工作:0.min,耗油:0 kg. 不计此油量及滑行油量,则总油量=15039.kg
考虑飞机老化等因素航程油量与备降油量各多加了 0.%和 0.%. 公司备份油:0.kg
因在拉萨不能加油,起飞总油量按下列飞行所需计算:
广州→ 拉萨　 再起飞到 CTU　 备降 CKG
＊业载受目的地机场最大着陆重量限制。
```

例 3-3-3：由拉萨飞往广州，备降机场桂林(拉萨只能部分加油 7000 kg)的飞行计划。

```
－－＝＊＝＝＊＝＝ B737-700(CFM56-7B24A)　飞 行 计 划 ＝＝＊＝＝＊＝－－
　航班：　　　机号：B××××　机长：　　　　　　12-14-93　　13:10:49
　拉萨(3569.6.m) －－ 广州　　　(15.2.m)　备降：桂林　　　　(173.6.m)
```

最大允许起飞重量:69853.kg	飞行距离:3003.km　1622.NM	滑出时间:9.min
最大着陆重量(目):58604.	巡航高度:10700.m　35100.ft	离场时间:0.min
最大着陆重量(备):58604.	巡航速度:LRC	滑入时间:4.min
最大无油重量:55202.	备降距离:272.km　147.NM	进近时间(目):5.min
使用空机重量:39648.	巡航高度:8400.m　27600.ft	进近时间(备):4.min
最大结构业载:15554.	巡航速度:LRC	防冰预计使用:0.h
最大油箱容量:20429.	等待高度气温:ISA+0.	
(γ=.779 kg/L)	飞行中 APU:断开	

滑行全重:67217.kg 148187.lb	轮挡油量:8874.kg　19564.lb	轮挡时间:3:58
起飞全重:67115.　147962.	航程油量:8715.　19213.	航程时间:3:44
着陆重量:58400.　128749.	备降油量:1293.　2851.	备降时间:0:29
无油重量:55202.　121698.	等待油量:1034.　2279.	等待时间:0:30
允许业载:15554.　34290.	备份油量:3198.　7050.	等待表速:206.kt
起飞油量:11913.　26263.	(扣除滑行油后:11754.　25913.)	滑行油量:159 kg

起飞前 APU 地面工作:0.min，耗油: 0 kg. 不计此油量及滑行油量，则总油量 = 11754.kg
考虑飞机老化等因素航程油量与备降油量各多加了 0.% 和 0.%. 公司备份油:0.kg
＊业载受目的地机场最大着陆重量限制。

例 3-3-4：由广州飞往拉萨，备降机场成都(拉萨只能部分加油)的飞行计划

```
－－＝＊＝＝＊＝＝ B737-700(CFM56-7B24A)　飞 行 计 划 ＝＝＊＝＝＊＝－－
　航班：　　　机号:B××××　机长:　　　　　　12-14-93　　13:04:00
　广州(15.2.m) －－ 拉萨　　　(3569.6.m)　备降:成都　　　(512.4.m)
```

最大允许起飞重量:69853.kg	飞行距离:3003.km　1622.NM	滑出时间:9.min
最大着陆重量(目):58604.	巡航高度:11000.m　36100.ft	离场时间:0.min
最大着陆重量(备):58604.	巡航速度:LRC	滑入时间:4.min
最大无油重量:55202.	备降距离:1196.km　646.NM	进近时间(目):5.min
使用空机重量:39648.	巡航高度:10100.m　33100.ft	进近时间(备):4.min
最大结构业载:15554.	巡航速度:LRC	防冰预计使用:0.h
最大油箱容量:20429.	等待高度气温:ISA+0.	
(γ=.779 kg/L)	飞行中 APU:断开	

滑行全重:67718.kg 149291.lb	轮挡油量:9177.kg　20232.lb	轮挡时间:3:57
起飞全重:67616.　149066.	航程油量:9018.　19881.	航程时间:3:43
着陆重量:58599.　129187.	备降油量:3283.　7238.	备降时间:1:32
无油重量:53416.　117761.	等待油量:998.　2200.	等待时间:0:30
允许业载:13768.　30353.	备份油量:5183.　11426.	等待表速:206.kt
起飞油量:14200.　31305.	(扣除滑行油后:14041.　30955.)	滑行油量:159 kg

起飞前 APU 地面工作:0.min，耗油: 0 kg. 不计此油量及滑行油量，则总油量 = 15039.kg
考虑飞机老化等因素航程油量与备降油量各多加了 0.% 和 0.%. 公司备份油:0.kg
因在拉萨只能加油 7000.kg，起飞总油量按下列飞行所需计算:
在拉萨着陆后剩油 = 拉萨→CAN　　备降 KWL　　所需油量减可加油量计算
＊业载受目的地机场最大着陆重量限制。

从案例计算得知：

（1）如果按照目的地机场不能加油的方法，例 3-3-1 中由拉萨飞往成都，备降机场重庆的所需总油量为 6180 kg，在例 3-3-2 中由广州飞往拉萨在拉萨的剩余油量（备份油量）为 6180 kg，正好满足飞行要求。

（2）如果按照目的地机场拉萨能加部分燃油的方法，例 3-3-3 中拉萨飞广州，备降机场桂林所需的总油量为 11913 kg，减去拉萨机场可以加的 7000 kg 燃油，只需在拉萨剩油 4913 kg（11913−7000）即可；但是，在例 3-3-4 中由广州飞往拉萨，备降机场成都在拉萨的实际剩余油量（备份油量）是 5183 kg，即由广州飞到拉萨后应剩油 5183 kg，而不是所需的 4913 kg，满足拉萨飞广州，备降机场桂林的要求。

如果按照目的地机场不能加油的方法，例 3-3-3 中以 4913 kg 油量由拉萨飞往广州开始计算，起飞油量将偏少，若有特殊情况可能导致广州飞拉萨航段中飞机无法在拉萨降落，此时飞机在拉萨机场上空所剩的油量将无法满足飞机从拉萨备降成都。因此，不能将拉萨飞广州，备降机场桂林所需的油量减去拉萨机场可以加的油量而得到的剩余燃油直接往回推算拉萨飞广州的所需油量，在计算过程中仍需检查广州飞拉萨，备降机场成都的备份油量，以保证安全。

下面的例子用本节的方法研究了在第一目的地机场或者说是经停站 DEST1 不能加油或只能部分加油时，DEST1 的业载和起飞机场的业载相互影响的问题，这和经营策略有关。

例 3-3-5：业载分析

某 B737-700（CFM56-7B24A）型飞机由机场 A 起飞，经停机场 B，再飞往机场 C，备降机场 D，在机场 B 可以上下客和装卸货物，但不能加油，业载分析飞行剖面如图 3-3-3 所示。设在机场 A 的最大允许起飞重量为 105 t，在机场 B 的最大允许起飞重量为 95 t，在机场 B、C、D 的最大允许着陆重量是结构限制重量 89811 kg，最大无油重量为 83461 kg，使用空机重为 59360 kg，最大油箱容量为 33253 kg，机场 A、B、C、D 的标高分别为 500 m、3500 m、0 m、0 m。

A→B 的距离 = 1000 NGM = 1852 km，巡航高度 35000 ft，ISA+20，顶风 50 kt，以 LRC 巡航；

B→C 的距离 = 750 NGM = 1389 km，巡航高度 37000 ft，ISA+10，顶风 75 kt，以 LRC 巡航；

C→D 的距离 = 300 NGM = 556 km，巡航高度 35000 ft，ISA，无风，以 LRC 巡航。

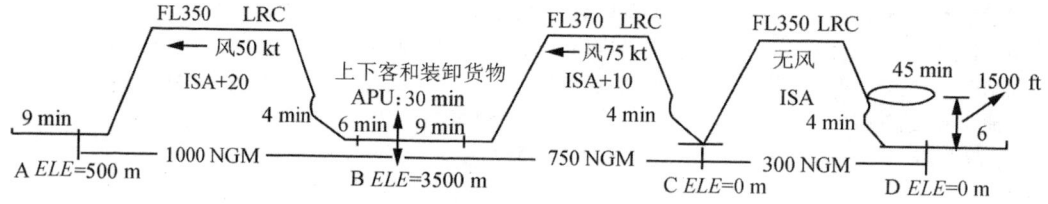

图 3-3-3　业载分析飞行剖面

在机场 A 的滑出时间为 9 min，离场时间为 0 min，在机场 B 的滑入时间为 6 min，滑出时间为 9 min，在停场期间用 APU 供电供气 30 min，在机场 D 的滑入时间为 6 min，进近时间一律为 4 min。

如由机场 A 起飞时携带的业载多，则受最大允许着陆重量或最大允许起飞重量限制，能带的油量就少，这将使从 B 起飞时能带的业载减少。如果机场 B 的客货是需要优先抢运的，这就会失去客源、货源，造成经济损失。反之，如果由机场 A 起飞时携带的业载少、带的油多，而 B 客货不足，这也会浪费运力。所以需要综合考虑 A、B 的货源，合理安排携带的业载和燃油。为此需要知道 B 处的业载和从 B 处起飞的油量的关系，以及 B 处的业载对从 A 处起飞携带的业载的影响。

解决方法是:假定一系列从 B 起飞时携带的业载 PL1、PL2……分别算出所需的起飞油量 BF1、BF2……,这就是从 A 起飞在 B 着陆时应剩余的油量。着陆重量最多只能等于最大允许着陆重量,由此重量开始向 A 计算即可得出起飞油量和从 A 起飞携带的业载,把计算结果列成表格,就可从表中根据计划在 B 处装的业载来确定从 A 起飞时应带的业载和油量。表 3-3-1 只给出计算的结果,详细过程略,数据是用计算机计算的,和人工查简化飞行计划图表计算的结果可能有些差别。如果打算从机场 B 运走的业载是 PL,利用右表很容易就可确定在机场 A 起飞时应加的油量和能装的业载。

表 3-3-1　业载与油量表　　　　　　　　　单位:kg

B→C→D						A→B					
PL	TOF	TOW	LW	BLKF	HOLDF	PL	TOF	TOW	LW	BLKF	RESF
6000	7642	53291	48873	4405	906	11306	14137	65091	58596	6495	7642
8000	7813	55461	50955	4486	937	11134	14308	65090	58596	6495	7813
10000	7990	57638	53038	4572	968	10962	14485	65095	58600	6495	7990
12000	8172	59820	55123	4662	1000	10780	14667	65094	58600	6495	8172
14000	8359	62007	57208	4757	1032	10593	14854	65094	58600	6495	8359
15293	8481	63422	58601	4821	1053	10470	14976	65094	58599	6495	8481

对于在机场 B 只能部分加油的情况,做法类似。例如,设在机场 B 能补充 2 t 燃油,计算时把在机场 B 着陆时应剩余油量减少 2 t 即可,其他算法不变。

3.4　涡桨动力飞机的飞行计划

对于涡桨动力飞机,关于其国内运行加油量的规定与涡轮动力飞机的一样,因此完全可以按上述方法制作涡桨动力飞机的飞行计划。而对于国际定期载客运行或者包括有至少一个国外机场的补充运行时,涡桨动力飞机的不可预期燃油计算方法有所不同,在第 121.659 条(c)款中有规定:不可预期燃油不得低于以正常巡航消耗率飞往本规则第 657 条(b)款第(2)、(4)项规定的机场所需总时间的 15%所需的油量,或者以正常巡航消耗率飞行 60 分钟的油量,两者当中取较短的飞行时间。其飞行计划剖面如图 3-4-1 所示,具体计算过程与常规飞行计划类似,不再重复。

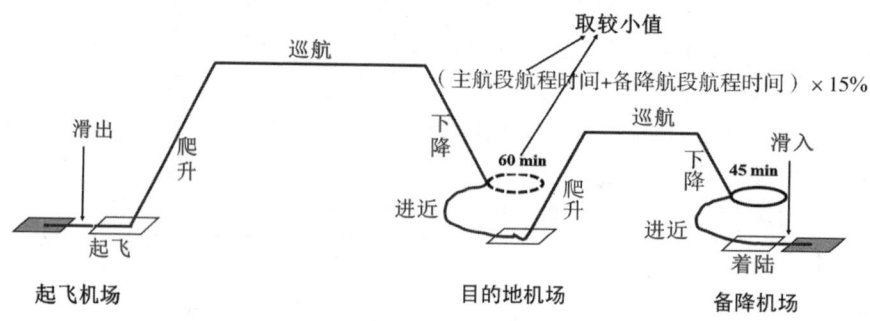

图 3-4-1　国际定期、补充运行飞行计划剖面(活塞式发动机飞机+涡桨动力飞机)

3.5　最小成本飞行计划

3.5.1　成本与成本指数

作为企业,航空公司应在完成客货运输任务的同时,力争创造最大的经济效益。为此,应尽量降低每一个航班的成本,包括直接使用成本和间接使用成本,其中应尽量降低飞机直接使用成本。

3.5.1.1　直接使用成本与燃油成本

飞机的直接使用成本(DOC,Direct Operation Cost)一般包括以下费用:

(1)燃油费用;

(2)机组费用:固定工资、飞行小时费等;

(3)乘务组费用:固定工资、飞行小时费等;

(4)维修费:飞机维修材料费和工时费、发动机维修材料费和工时费、杂项费用;

(5)机身保险费;

(6)飞机折旧费;

(7)起降费、停场费、通信导航费;

(8)旅客餐食费、住宿费、保险费;

(9)行李、货物的搬运保管费;

(10)机票、货物的代办费;

(11)公司的经营管理费和广告费;

(12)地面设备维修费、折旧费;

(13)贷款利息、租机费。

上述费用中除燃油费用以外只有(2)、(3)、(4)项与飞行时间(飞行速度)有关,且这三项中只有一部分和飞行时间有关。燃油费用仅是直接使用成本中的一部分,采用不同的爬升速度、巡航速度、下降速度飞行,则消耗的燃油和所用的飞行时间不同,航班成本也就不同。假设燃油成本(即燃油价格)为 C_f、与飞行时间有关的小时成本为 C_t、航班的轮挡油量为 F、轮挡时间为 T,则这个航班与飞行时间有关的成本 C(不是总成本)为:

$$C = C_f \times F + C_t \times T$$

　　燃油的消耗和飞行时间均与飞行速度有关,只要能选择合适的飞行速度,使这部分成本最小,就可以使航班总成本最小。

3.5.1.2　成本指数

　　所谓成本指数(Cost Index),是指小时成本与燃油成本(价格)的比值。由成本指数的概念可知成本指数为:

$$CI = \frac{\text{小时成本(美元/小时)}}{\text{燃油价格(美分/磅)}} = \frac{C_t}{C_f}, \text{单位}: \frac{\$/h}{\text{¢}/lb} = 100 \times \frac{lb}{h}$$

即 CI 的单位为(美元×磅)/(美分×小时)或 100 lb/h,为了不使成本指数的数值过大,燃油成本用美分/磅为单位。如果货币改用人民币,燃油成本以元/千克为单位,则:

$$CI = 0.0220462 \times \frac{\text{小时成本(元/小时)}}{\text{燃油价格(元/千克)}}$$

这样算出的 CI 的单位为(元×千克)/(元×小时)。

　　空客公司的小时成本以美元/分钟为单位,燃油价格以美元/千克为单位,则 CI 的单位为 kg/min,即

$$1 \text{ kg/min} = 1.3228(\text{美元×磅})/(\text{美分×小时})$$

这里的小时成本仅指那些与飞行时间长短有关的成本(不包括燃油费用),与财务部门使用的小时成本不同,财务部门的小时成本是把一年的总费用(前面介绍的 13 项费用)除以一年的总飞行小时数得出的,这样计算得到的小时成本中包括很多与飞行速度的快慢(飞行时间长短)无关的费用。

　　准确地确定与飞行时间长短有关的小时成本 C_t(燃油费用除外)是很难的,波音公司根据统计结果给出了与飞行时间相关的各类成本曲线,如图 3-5-1～图 3-5-3 所示,可以帮助用户确定与飞行时间长短有关的成本。由于各航空公司的具体营运状况具有差异性,小时成本只作为指导性材料,不能用于实际运行。

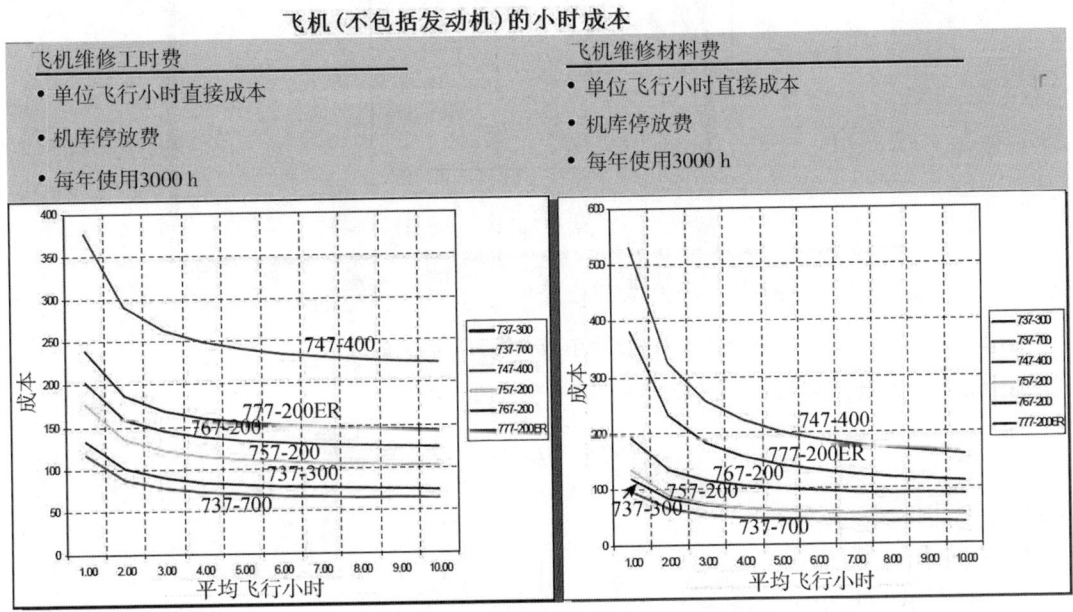

图 3-5-1　确定飞机维修费小时成本用的曲线

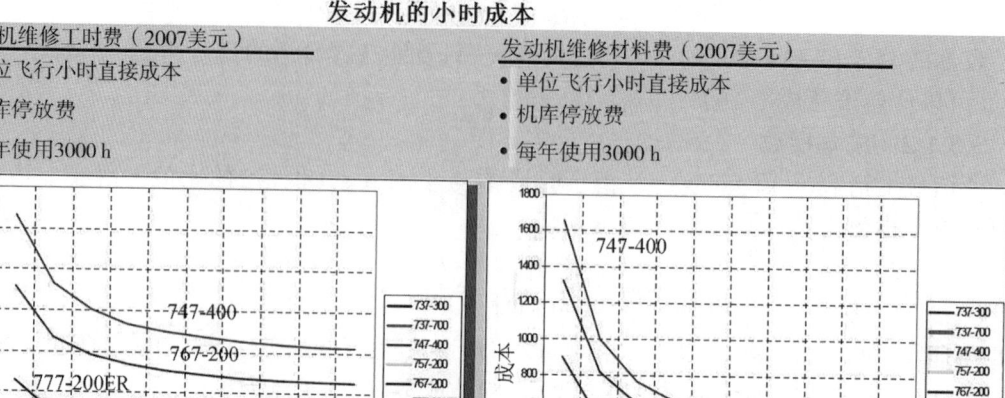

图 3-5-2　确定发动机维修费小时成本用的曲线

注意:这里未包括空勤组费用中与飞行时间有关的部分。

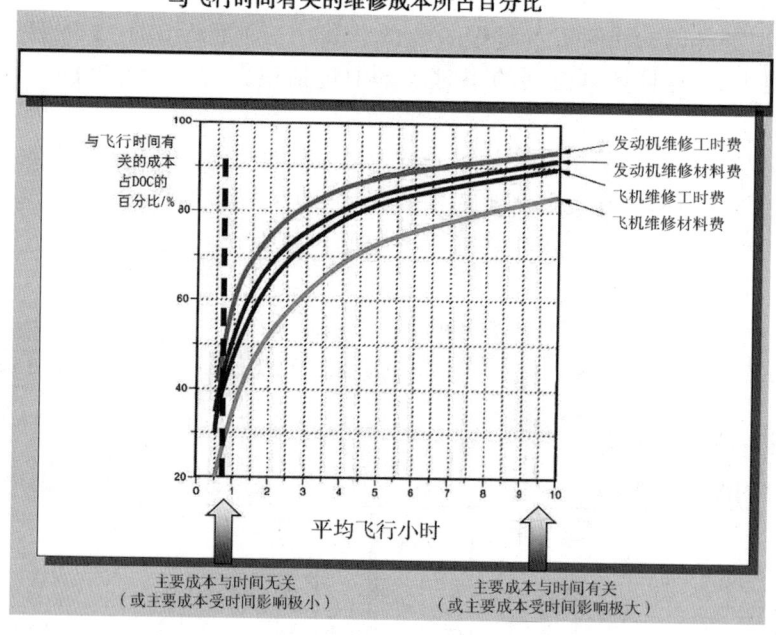

图 3-5-3　维修费中与飞行时间有关的成本

利用成本指数公式可以将成本 C 的计算公式改写为:

$$C = C_f \times \left(CI \times T + \frac{F}{100} \right)$$

式中:C_f——油价,美分/磅;

　　　F——轮挡油量,磅;

　　　T——轮挡时间,小时。

当 $CI = 0$,即不计时间成本时,最省油,则成本最小;当 CI 不为 0 时,应仔细选择各段飞行速度,使总成本最小。CI 越大,成本最小的飞行速度越快。

飞行员在飞行管理系统(FMS)中输入的 CI 越大,在选择经济飞行模式时,因受 FMS 控制,飞机的飞行速度越大。但是,不应该用改变 CI 的方法控制飞行速度,CI 应该是公司的财务等部门对飞机的各项成本经过仔细研究后确定的,飞行员在 FMS 中输入这个 CI,FMS 根据这个 CI 计算飞机的速度,并控制飞机以该速度飞行,使航班成本最小。FMS 不仅根据 CI 控制巡航速度,也控制爬升和下降速度,这些速度还与重量有关,尤其是在巡航中重量变化比较大,随重量的变化,使成本最小的巡航速度也应改变。此外,经济巡航速度还与航路风有关。如果不知道准确的 CI,在 FMS 中输入的 CI 过大或过小,在经济飞行模式下受 FMS 控制的飞行速度就会过大或过小,使航班成本提高,失去了使用 CI 的意义。

在 FMS 中输入的 CI 不必太精确,速度对 CI 不太敏感,即 CI 稍有变化,速度变化很小,航班成本变化不大。在不知道准确的 CI 时,宁可输入较小的值,也不要输入太大的值,在 CI 较大时,随 CI 的增加,航班成本增加得是比较快的。

3.5.1.3 经济巡航速度(马赫数)

如果飞机手册有在各高度、各重量下燃油里程(FM)或燃油流量(FF)随马赫数(M)变化的曲线或数值表分别如图 3-5-4 与图 3-5-5 所示,则可利用下述公式来计算在给定高度上使巡航成本最小的马赫数(即经济巡航马赫数),此马赫数与 CI、重量、风速、温度偏差有关,在 CI、风速、温度偏差一定时,此马赫数是重量的函数:$M=f(W)$。经济巡航马赫数可用以下方法计算:

$$\left(M+\frac{V_\mathrm{W}}{a}\right)\frac{\mathrm{d}FF}{\mathrm{d}M}-FF=100CI \quad 或 \quad \frac{V_\mathrm{W}}{FW}-\frac{M(Ma+V_\mathrm{W})}{FM^2}\frac{\mathrm{d}FM}{\mathrm{d}M}=100CI$$

式中:V_W——风速,顶风为负;a——巡航高度上的声速。

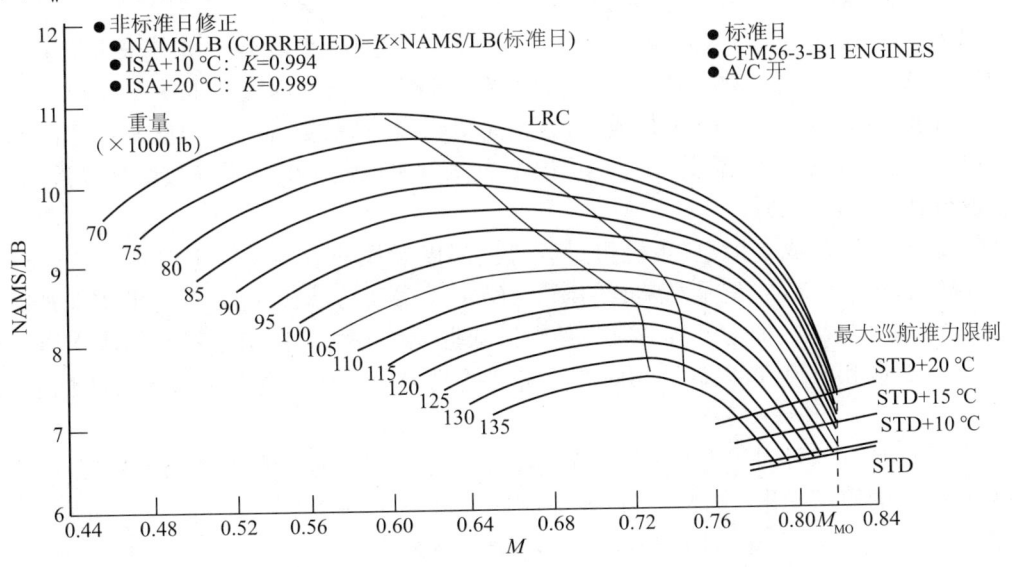

图 3-5-4 B737-300/CFM56-3-B1 31000 ft 的燃油里程曲线

利用上述公式,需要把燃油流量(FF)或燃油里程(FM)与飞行马赫数(M)的关系通过曲线拟合的方法用一个函数式(例如多项式)表示出来,这样才能计算导数 $\mathrm{d}FF/\mathrm{d}M$、$\mathrm{d}FM/\mathrm{d}M$,曲线拟合误差对导数计算的准确程度影响较大,经济巡航马赫数不易计算准确。更好的方法是利用经济巡航成本函数($ECCF$)计算经济巡航马赫数:

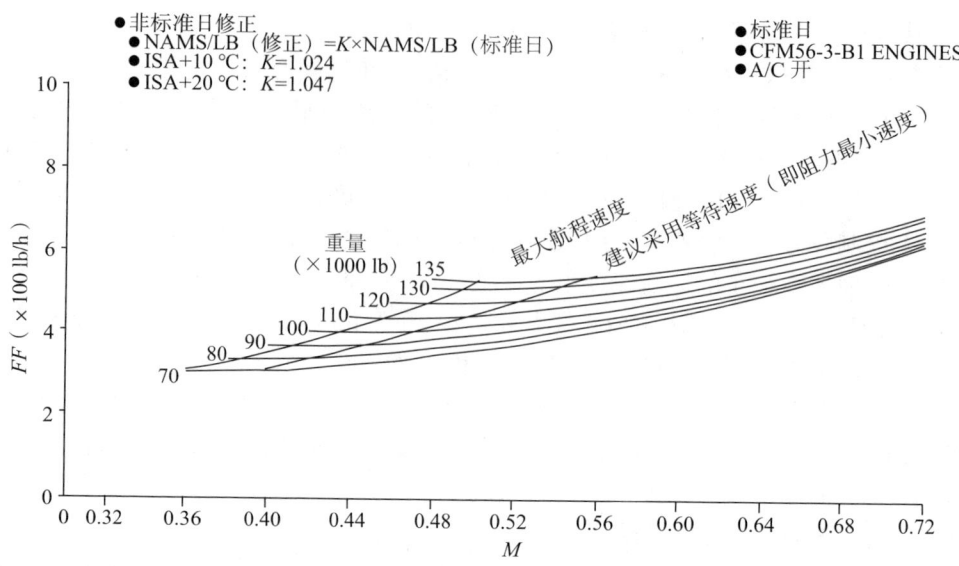

图 3-5-5 B737-300/CFM56-3-B1 20000 ft 的燃油流量曲线

$$ECCF = \frac{100CI + W_f}{Ma + V_W}$$

式中：$ECCF$（Economic Cruise Cost Function）——经济巡航成本函数（lb/NGM）；

$\quad\quad W_f$——全发（整架飞机的）燃油流量（lb/h）；

$\quad\quad M$——马赫数；

$\quad\quad a$——声速（kt）；

$\quad\quad V_W$——风速（kt），顶风为负，顺风为正。

$ECCF \times C_f$（单位：\$/NGM），即巡航 1 NGM 的总成本。使 $ECCF$ 最小，就是使成本最小。

$ECCF$ 与成本指数、马赫数、温度、风速和燃油流量有关，而燃油流量取决于飞机重量、马赫数、飞行高度与温度，因此，$ECCF$ 是成本指数、飞机重量、马赫数、高度、温度和风速的函数。在成本指数、飞机重量、高度、温度和风速给定时，使 $ECCF$ 取最小值的马赫数就是经济巡航马赫数，即最小成本巡航马赫数。

经济巡航马赫数与成本指数、飞机重量、高度、温度和风速有关。给定成本指数、飞机重量、高度、温度和风速，$ECCF$ 就是马赫数的函数，给定一系列马赫数，计算对应的 $ECCF$ 值，使 $ECCF$ 取最小值的马赫数，即为经济巡航马赫数。依据此种方法分别给定不同的成本指数、飞机重量、高度、温度和风速，可以计算相应的经济巡航马赫数。详细情况可参考飞机性能方面的教材。

通过上述计算可以得到和图 3-5-6 一样的结果。巡航时随重量的减少，FMS 控制马赫数按通过计算得到的规律变化，这样就可以使巡航成本最小。FMS 也控制飞机按经济爬升速度爬升（如图 3-5-7 所示）、按经济下降速度下降（如图 3-5-8 所示）。

波音机型的 FMC 给出的最大允许的经济爬升速度、经济巡航速度是 $V_{MO}-5$ kt 或 $M_{MO}-0.02$。$M_{MO}-0.02$ 是这样一个马赫数：在 V_{MO}/M_{MO} 的转换高度以上，按 M_{MO} 计算对应的修正空速 V_c，V_c-5 kt 对应的马赫数即最大允许经济巡航马赫数。对 B737-300/NG，FMC 限制的最大经济下降速度为 $V_{MO}-10$ kt $=340-10=330$ kt；对 B757-200/767/777/747，FMC 限制的最大经济下降速度为 $V_{MO}-16$ kt。空客机型 FMC 控制的最大经济速度为 $V_{MO}-10$ kt/$M_{MO}-0.02$。

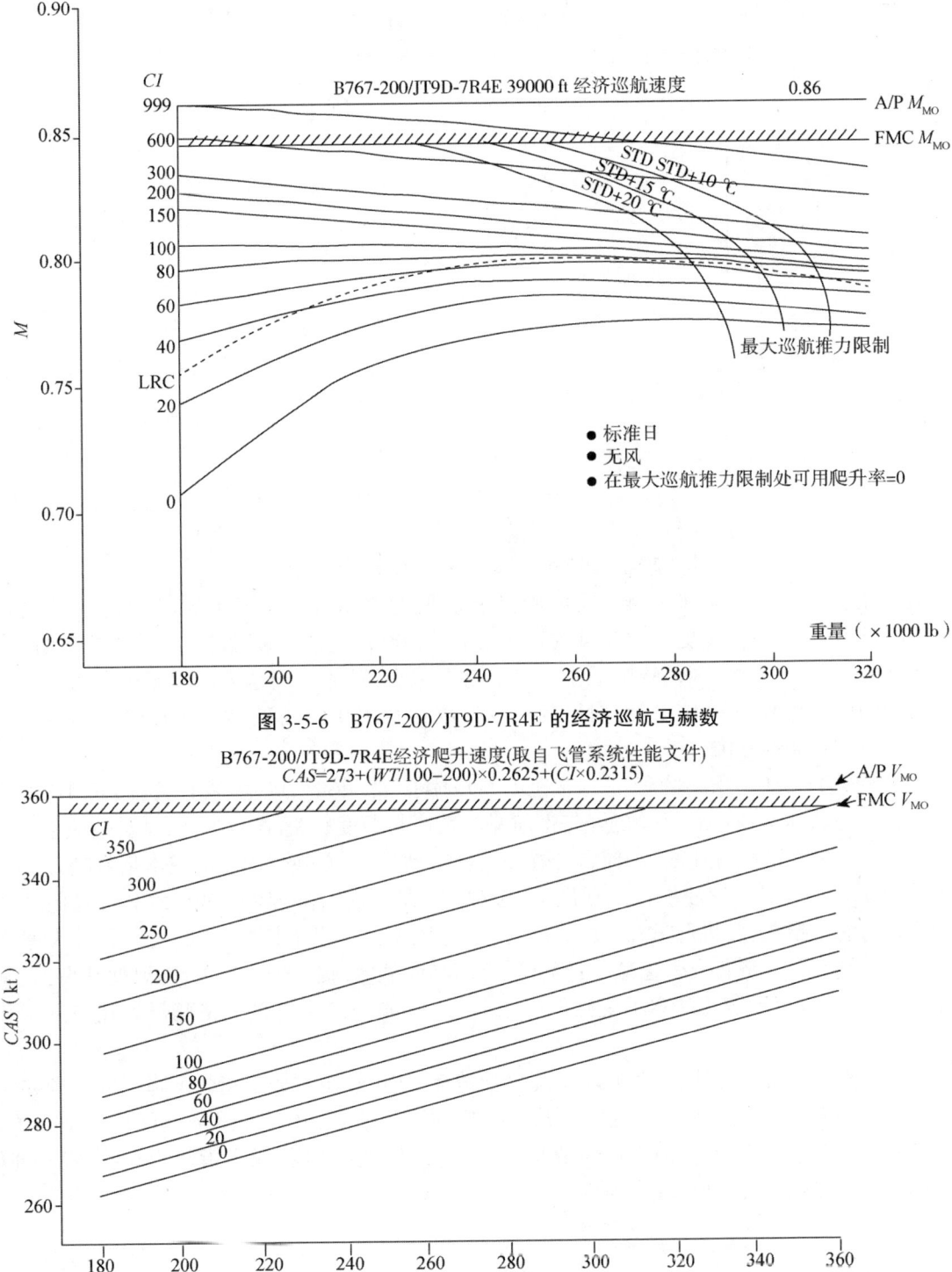

图 3-5-6　B767-200/JT9D-7R4E 的经济巡航马赫数

图 3-5-7　B767-200/JT9D-7R4E 的经济爬升速度

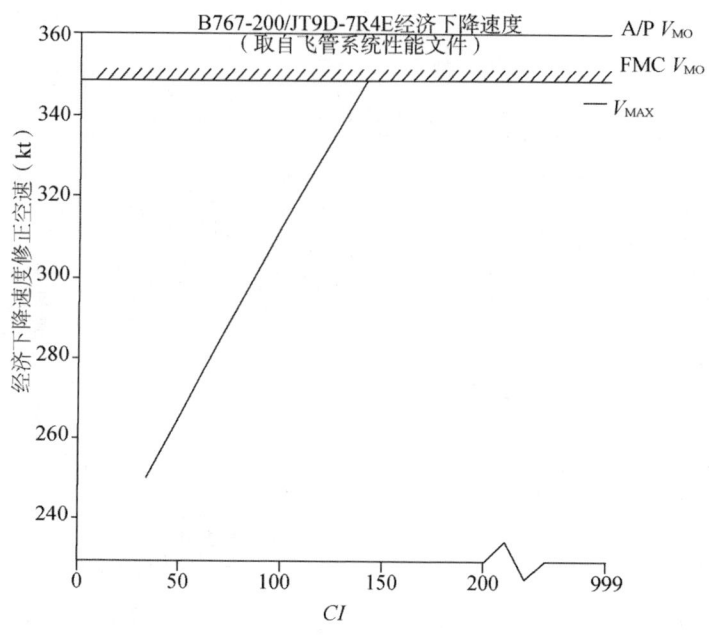

图 3-5-8　B767-200/JT9D-7R4E 的经济下降速度

　　风对经济巡航马赫数有影响,顶风使经济巡航马赫数增大(即顶风时为使巡航飞过1 NGM的成本最小,应该用更大的马赫数飞行),顺风使经济巡航马赫数减小。在低空,风的影响较大,随着高度的增加,风的影响减小。成本指数越小,风的影响越大。对于正常的巡航高度和常用的成本指数(和 LRC 速度对应的成本指数),每 100 kt 的顶(顺)风使经济巡航马赫数增大(减小)0.008~0.02。

　　空客飞机 FCOM 中给出的经济巡航速度与成本指数、重量、高度、风有关,如表 3-5-1 所示。空客性能软件 WINPEP 计算的经济巡航马赫数还与温度有关,温度升高,马赫数减小。

　　空客飞机的经济爬升速度:为使爬升燃油消耗减到最小,必须使用小的成本指数爬升。成本指数为 0 表示可以使燃油消耗减到最小,这对应着最大爬升率速度。WINPEP 计算的经济爬升表速与成本指数、飞机重量、高度、温度有关,与风无关。成本指数增加,经济爬升表速增加;飞机重量减小,经济爬升表速减小;爬升顶点的高度越低,爬升表速越大;随温度升高,表速增加(在平台温度以下表速基本不变),但变化不大。A340-313/642 的经济爬升表速与温度和风无关。

　　空客飞机的经济下降速度:为减小下降燃油消耗,使用较小的成本指数下降。成本指数为0 表示可以使燃油消耗减到最小,这将导致一个低的下降速度,此时的下降速度与机型有关,对 A320 系列机型是 250 kt。但在任何情况下,不得小于绿点速度,成本指数为 0 的经济下降速度如果小于绿点速度,则取绿点速度。

表 3-5-1　A340-300 经济巡航马赫数

COST INDEX = 0 kg/min								COST INDEX = 100 kg/min							
		FLIGHT LEVEL								FLIGHT LEVEL					
WEIGHT 1000 kg	WIND kt	310	330	350	370	390	410	WEIGHT 1000 kg	WIND kt	310	330	350	370	390	410
150	100.	.645	.680	.705	.725	.749	.777	150	100.	.783	.793	.803	.813	.821	.824
	50.	.667	.698	.720	.738	.761	.786		50.	.790	.799	.809	.818	.825	.826
	0.	.694	.719	.739	.756	.776	.794		0.	.798	.807	.816	.823	.827	.828
	−50.	.724	.743	.759	.774	.789	.802		−50.	.807	.814	.821	.827	.830	.829
	−100.	.753	.765	.777	.789	.801	.809		−100.	.815	.822	.828	.832	.832	.836
170	100.	.694	.712	.732	.757	.784	.804	170	100.	.792	.803	.813	.820	.821	.819
	50.	.710	.726	.745	.769	.791	.804		50.	.798	.808	.817	.823	.825	.821
	0.	.729	.744	.762	.782	.798	.804		0.	.805	.814	.822	.827	.826	.823
	−50.	.750	.764	.778	.794	.804	.807		−50.	.813	.820	.826	.829	.828	.825
	−100.	.769	.781	.793	.804	.810	.810		−100.	.820	.826	.830	.831	.829	.825
190	100.	.716	.736	.762	.787	.801		190	100.	.801	.811	.818	.820	.816	
	50.	.730	.749	.773	.793	.802			50.	.807	.816	.822	.822	.819	
	0.	.747	.765	.785	.800	.804			0.	.812	.820	.826	.825	.821	
	−50.	.766	.781	.796	.805	.806			−50.	.819	.825	.828	.826	.823	
	−100.	.783	.795	.805	.811	.809			−100.	.824	.829	.830	.828	.825	
210	100.	.738	.764	.788	.801			210	100.	.809	.816	.818	.815		
	50.	.751	.774	.794	.802				50.	.814	.820	.821	.817		
	0.	.767	.786	.800	.804				0.	.819	.824	.824	.820		
	−50.	.782	.797	.805	.806				−50.	.824	.827	.826	.822		
	−100.	.795	.806	.811	.809				−100.	.827	.829	.827	.824		
230	100.	.763	.788	.801				230	100.	.814	.817	.814			
	50.	.774	.794	.802					50.	.818	.819	.816			
	0.	.786	.800	.804					0.	.822	.823	.819			
	−50.	.796	.805	.806					−50.	.826	.825	.821			
	−100.	.805	.810	.809					−100.	.828	.827	.824			
250	100.	.785	.800					250	100.	.815	.813				
	50.	.792	.802						50.	.818	.816				
	0.	.799	.804						0.	.821	.819				
	−50.	.805	.807						−50.	.825	.821				
	−100.	.810	.809						−100.	.826	.823				
270	100.	.799						270	100.	.813					
	50.	.801							50.	.815					
	0.	.804							0.	.818					
	−50.	.807							−50.	.821					
	−100.	.810							−100.	.823					

	COST INDEX = 200 kg/min								COST INDEX = 300 kg/min					
		FLIGHT LEVEL								FLIGHT LEVEL				
WEIGHT	WIND	310	330	350	370	390	410	WEIGHT	WIND	310	330	350	370	390	410
1000 kg	kt							1000 kg	kt						
150	100.	.814	.821	.827	.832	.832	.830	150	100.	.831	.835	.839	.840	.837	.833
	50.	.820	.827	.831	.834	.834	.831		50.	.835	.839	.840	.840	.838	.834
	0.	.827	.831	.835	.837	.836	.832		0.	.839	.840	.840	.840	.839	.835
	−50.	.831	.836	.839	.840	.837	.833		−50.	.840	.840	.840	.840	.840	.836
	−100.	.837	.840	.840	.840	.838	.835		−100.	.840	.840	.840	.840	.840	.838
170	100.	.818	.825	.830	.831	.829	.825	170	100.	.833	.836	.837	.835	.831	.828
	50.	.823	.829	.832	.832	.829	.826		50.	.836	.839	.839	.836	.832	.829
	0.	.829	.833	.835	.834	.830	.827		0.	.840	.840	.840	.837	.833	.830
	−50.	.833	.836	.838	.835	.831	.828		−50.	.840	.840	.840	.838	.834	.831
	−100.	.839	.840	.839	.836	.833	.829		−100.	.840	.840	.840	.839	.836	.832
190	100.	.822	.828	.829	.828	.824		190	100.	.833	.835	.833	.830	.826	
	50.	.826	.830	.831	.828	.825			50.	.836	.837	.835	.831	.827	
	0.	.831	.833	.832	.829	.825			0.	.840	.839	.835	.832	.828	
	−50.	.834	.835	.834	.830	.826			−50.	.840	.840	.837	.833	.830	
	−100.	.838	.838	.835	.831	.828			−100.	.840	.840	.838	.834	.831	
210	100.	.826	.828	.827	.824			210	100.	.833	.832	.829	.825		
	50.	.829	.830	.828	.824				50.	.835	.833	.830	.826		
	0.	.832	.831	.828	.825				0.	.838	.834	.831	.827		
	−50.	.834	.833	.829	.826				−50.	.839	.836	.832	.829		
	−100.	.837	.834	.830	.827				−100.	.840	.837	.833	.830		
230	100.	.827	.826	.823				230	100.	.831	.829	.825			
	50.	.829	.827	.824					50.	.833	.829	.826			
	0.	.831	.828	.825					0.	.834	.830	.827			
	−50.	.832	.829	.825					−50.	.835	.831	.828			
	−100.	.833	.830	.826					−100.	.836	.833	.829			
250	100.	.825	.822					250	100.	.829	.825				
	50.	.827	.824						50.	.829	.826				
	0.	.828	.825						0.	.830	.826				
	−50.	.829	.825						−50.	.831	.828				
	−100.	.830	.826						−100.	.832	.829				
270	100.	.822						270	100.	.826					
	50.	.824							50.	.826					
	0.	.825							0.	.826					
	−50.	.826							−50.	.827					
	−100.	.826							−100.	.829					

3.5.2 最小成本飞行计划的计算方法

前文所述的经济巡航图表对于给定的成本指数和重量,能够提供最经济的爬升速度、巡航速度和下降速度,不能提供爬升、下降用的油量、时间、飞过的距离及巡航时的燃油流量,如果只有这种图表的话,是不能制作出最小成本飞行计划的。目前飞机使用手册上还没有提供与成本指数有关的制作飞行计划用的爬升、巡航、下降性能数值表,但可能会提供计算飞行性能用的软件,如波音公司给用户提供 INFLT 软件,用户可以用这种软件自己计算制作最小成本飞行计划所需的各种数据。

以 B757-200(RB211-535E4)为例,表 3-5-2、表 3-5-3、表 3-5-4 给出了成本指数为 100 的制作最小成本飞行计划所需的各种数值表,对经济爬升、巡航可以计算得到 ISA+0 ℃、ISA+5 ℃、ISA+10 ℃等各种温度偏差的数值表,这里仅给出了其中的一张。用这种软件可以计算得到给定的任何成本指数的数值表,其中的重量和高度的范围及间隔由用户自己决定,使用很方便。

表 3-5-2 为 B757-200 机型成本指数为 100 的经济爬升数据,经济爬升仍是一种等表速(修正空速)/等马赫数爬升,该马赫数即爬升顶点的重量对应的经济巡航马赫数。表中各松刹车重量(即每列)对应的爬升表速和马赫数都是不同的。

表 3-5-2　B757-200/RB211-535E4 成本指数为 100 的经济爬升数据

B757-200(RB211-535E4) ECONOMY CLIMB

COST INDEX = 100 $ · lb/¢ · HR　ISA+0 ℃　WIND 0 kt　TIME(min)/FUEL(lb)

DIST(NM)/ATAS(kt)

PRESS	BRAKE RELEASE WEIGHT(1000 lb)										
ALT(ft)	260	250	240	230	220	210	200	190	180	170	160
42000								23/5600	20/5000	18/4600	17/4200
								154/438	133/437	119/437	107/435
41000							24/5800	21/5200	19/4800	17/4400	16/4100
							156/437	135/436	121/435	110/435	100/434
40000						24/6000	21/5500	19/5000	17/4600	16/4300	15/3900
						156/436	137/434	123/433	111/433	103/433	94/432
39000				29/7300	24/6200	21/5700	19/5200	18/4800	17/4500	15/4100	14/3800
				196/438	156/435	138/433	124/432	113/431	104/431	96/431	88/430
38000			28/7200	24/6400	22/5900	20/5400	18/5000	17/4700	16/4300	15/4000	14/3700
			185/436	156/433	139/432	126/431	115/430	106/429	98/429	91/429	83/428
37000		28/7400	24/6600	22/6100	20/5600	19/5200	17/4800	16/4500	15/4200	14/3900	13/3600
		182/434	156/432	140/431	127/429	117/429	107/428	99/428	92/427	86/428	79/427
36000	27/7500	24/6800	22/6300	21/5800	19/5400	18/5000	16/4700	15/4400	14/4100	13/3800	12/3600
	178/433	156/431	141/429	129/428	118/427	109/427	101/426	94/426	87/426	82/426	75/425
35000	25/7000	23/6500	21/6000	19/5600	18/5200	17/4900	16/4600	15/4300	14/4000	13/3700	12/3500
	159/430	143/428	131/427	121/426	112/426	103/425	96/425	89/424	83/424	78/424	72/424
34000	23/6700	21/6200	20/5800	19/5400	17/5100	16/4800	15/4500	14/4200	13/3900	13/3700	12/3400
	146/427	133/426	123/425	114/424	106/424	98/423	92/423	85/422	79/422	74/423	69/422
33000	22/6400	20/6000	19/5600	18/5300	17/5000	16/4600	15/4400	14/4100	13/3800	12/3600	11/3300
	135/425	125/424	116/423	107/422	100/422	93/421	87/421	81/421	76/420	71/421	66/420
32000	21/6200	19/5800	18/5500	17/5100	16/4800	15/4500	14/4200	13/4000	12/3700	12/3500	11/3200
	127/422	117/421	109/421	102/420	95/420	89/419	83/419	78/419	72/418	68/419	63/418
⋮						⋮					
10000	6/2400	6/2200	6/2100	5/2000	5/1900	5/1800	5/1700	4/1600	4/1500	4/1400	4/1300
	19/347	18/347	17/347	16/347	15/347	14/347	14/346	13/346	12/346	11/346	10/315
5000	4/1700	4/1600	4/1500	4/1400	3/1400	3/1300	3/1200	3/1200	3/1100	3/1000	3/1000
	7/335	7/334	6/334	6/334	6/334	5/334	5/334	5/333	4/333	4/333	4/332
1500	3/1200	3/1100	3/1100	2/1000	2/1000	2/900	2/900	2/800	2/800	2/700	2/700

FUEL ADJUSTMENT FOR HIGH ELEVATION AIRPORTS	AIRPORT ELEVATION	2000	4000	6000	8000	10000	12000
EFFECT ON TIME AND DISTANCE IS NEGLIGIBLE	FUEL ADJUSTMENT	−200	−400	−600	−800	−1100	−1300

表 3-5-3 为 B757-200 机型成本指数为 100 的经济巡航数据。经济巡航马赫数与风有关，顺风使之减小，顶风使之增大，在最佳高度附近巡航时风对经济巡航马赫数的影响比较小，温度对经济巡航马赫数的影响更小。飞行管理计算机（FMC）计算、控制的经济巡航马赫数考虑了风和温度的影响。波音的 INFLT 软件在计算经济巡航马赫数时也可以考虑风和温度的影响。

表 3-5-3 B757-200/RB211-535E4 成本指数为 100 的经济巡航数据

B757-200 RB211-535E4 经济巡航

ECON($CI=100$)　　SPEED LIMIT = 0.847 MACH　　ISA+0 DEG C　　WIND 0 KTS

PRESS ALT 1000 ft (STD TAT)	WEIGHT 1000 lb												
	260	250	240	230	220	210	200	190	180	170	160	150	140
42 (−29)									1.70	1.66	1.62	1.59	1.56
									−11	−4	1		
									230	231	232	233	234
									.797	.800	.803	.806	.808
									3439	3272	3087	2942	2819
									457	459	460	462	464
⋮						⋮							
38 (−29)				1.74	1.70	1.66	1.63	1.60	1.58	1.55	1.53	1.52	1.50
				−17	−9	−3	2						
				251	252	253	254	254	255	256	256	257	257
				.793	.795	.798	.799	.801	.803	.805	.807	.808	.809
				4447	4117	3881	3682	3516	3375	3251	3153	3071	3051
				455	456	457	459	460	461	462	463	463	464
37 (−29)			1.73	1.69	1.66	1.63	1.60	1.58	1.55	1.53	1.52	1.50	1.49
			−15	−8	−2	3							
			257	258	259	259	260	261	261	262	262	263	263
			.793	.795	.797	.799	.801	.802	.804	.806	.807	.808	.809
			4591	4273	4040	3841	3674	3530	3402	3300	3215	3143	3082
			455	456	457	458	459	460	461	462	463	464	464
36 (−29)		1.72	1.69	1.66	1.63	1.60	1.58	1.55	1.53	1.52	1.51	1.49	1.48
		−13	−6	−1	4								
		263	264	265	265	266	266	267	267	268	268	269	268
		.793	.795	.797	.799	.800	.802	.803	.805	.806	.807	.807	.807
		4730	4427	4199	4001	3833	3688	3558	3451	3363	3288	3224	3157
		455	456	457	458	459	460	461	462	462	463	463	463
35 (−29)	1.72	1.68	1.65	1.62	1.60	1.57	1.55	1.53	1.52	1.51	1.49	1.48	1.47
	−11	−4	1	5									
	270	270	271	271	272	272	273	273	274	274	274	274	274
	.794	.795	.797	.798	.800	.801	.803	.804	.805	.806	.806	.806	.804
	4890	4603	4381	4183	4015	3868	3737	3626	3535	3457	3389	3324	3251
	457	458	459	460	461	462	463	463	464	464	465	465	464

CRUISE EPR LIMIT POWER SETTING = CRS

PRESS ALT 1000 ft	31	32	33	34	35	36	37	38	39	40	41	42
EPR LIMIT	1.69	1.71	1.72	1.73	1.74	1.75	1.75	1.74	1.74	1.74	1.74	1.74
TAT ℃ AND COLDER	−6	−8	−10	−12	−15	−17	−17	−17	−17	−17	−17	−17

REDUCE EPR BY 0.07 PER 10 DEGREES C TAT HOTTER THAN TABLE VALUE

表 3-5-4 为 B757-200 机型成本指数为 100 的经济下降数据,在成本指数比较大(例如 $CI \geq 100$)时,下降性能与成本指数的关系不大,基本上不随成本指数变化。

表 3-5-4　B757-200/RB211-535E4 成本指数为 100 的经济下降数据

B757-200　AIRPLANE WITH　RB211-535E4-"B"　经济下降　　　　TIME(min)/FUEL(lb)

DIST(NAM)/TAS(kt)

ECON/250	CI=100	WIND 0 kt	ISA+0	ECON/250	CI=300	WIND 0 kt	ISA+0
PRESS ALT	LANDING WEIGHT−lb			PRESS ALT	LANDING WEIGHT−lb		
ft	140000	170000	200000	ft	140000	170000	200000
42000	21.0/756	21.6/758	23.1/791	42000	21.0/810	21.4/753	23.0/789
	107/378	110/376	119/376		107/380	109/377	119/377
41000	20.2/687	21.2/751	22.7/784	41000	20.2/738	21.0/746	22.6/783
	101/375	107/374	117/374		102/377	106/375	116/375
40000	19.4/610	20.9/743	22.4/777	40000	19.4/653	20.7/739	22.3/776
	95/372	104/372	114/372		96/374	103/373	114/373
39000	18.6/521	20.5/736	22.0/770	39000	18.6/558	20.3/732	21.9/768
	89/369	102/370	111/371		90/370	101/371	111/371
37000	17.9/678	19.8/722	21.2/755	37000	17.7/674	19.7/719	21.2/754
	84/364	96/366	105/367		83/364	96/367	105/368
35000	17.3/667	19.2/709	20.6/741	35000	17.2/664	19.1/708	20.5/741
	80/359	91/361	100/363		79/360	91/362	100/364
29000	16.1/642	17.6/678	18.8/706	29000	16.1/643	17.6/679	18.8/707
	70/346	79/348	87/351		70/348	80/349	87/351
27000	15.7/635	17.2/670	18.3/696	27000	15.8/636	17.2/671	18.3/697
	67/342	76/344	82/346		68/343	76/345	83/346
25000	15.3/627	16.7/661	17.8/687	25000	15.3/627	16.7/661	17.8/687
	64/336	72/339	79/341		64/336	72/339	79/341
23000	14.8/617	16.1/650	17.1/674	23000	14.8/617	16.1/650	17.1/674
	60/330	67/332	73/334		60/330	67/332	73/334
21000	14.2/607	15.5/638	16.4/660	21000	14.2/607	15.5/638	16.4/660
	56/323	63/325	68/327		56/323	63/325	68/327
17000	13.2/585	14.2/612	15.0/631	17000	13.2/585	14.2/612	15.0/631
	48/308	54/309	58/311		48/308	54/309	58/311
15000	12.6/573	13.6/598	14.3/616	15000	12.6/573	13.6/598	14.3/616
	44/300	49/301	53/303		44/300	49/301	53/303
13000	12.1/561	12.9/584	13.5/599	13000	12.1/561	12.9/584	13.5/599
	41/292	45/293	48/294		41/292	45/293	48/294
11000	11.0/536	11.7/554	12.2/566	11000	11.0/536	11.7/554	12.2/566
	34/278	37/278	39/278		34/278	37/278	39/278
10000	10.0/511	10.6/526	10.9/535	10000	10.0/511	10.6/526	10.9/535
	29/271	31/271	33/271		29/271	31/271	33/271
9000	9.5/497	9.9/510	10.2/519	9000	9.5/497	9.9/510	10.2/519
	26/269	28/269	30/269		26/269	28/269	30/269
7000	8.3/467	8.7/477	8.9/484	7000	8.3/467	8.7/477	8.9/484
	21/265	22/265	23/265		21/265	22/265	23/265
5000	7.1/434	7.4/441	7.5/445	5000	7.1/434	7.4/441	7.5/445
	15/261	16/261	17/261		15/261	16/261	17/261
1500	5.0/370	5.0/370	5.0/370	1500	5.0/370	5.0/370	5.0/370
	6/72	6/72	6/72		6/72	6/72	6/72

制作最小成本飞行计划时要选择经济巡航高度,需要考虑三种高度:最佳高度、最大巡航推力限制高度和机动能力(过载)限制高度,如表3-5-5所示。这和成本指数有关,成本指数不同,则巡航速度不同,不能使用原来的 LRC 等巡航速度的数值表,必须用性能软件重新计算出给定成本指数下的这些数值表。在最大巡航推力限制高度上爬升率为0;最佳高度可以认为与温度偏差无关。有的机型按在最大巡航推力限制高度上剩余爬升率为 100 ft/min 计算推力限制高度,这取决于该机型的 FMC 计算规则。经济巡航的最佳高度本应是巡航成本最小的高度,但波音的 INFLT 软件是按以经济巡航速度巡航、燃油里程最大(最省油)确定的最佳高度。

所选择的巡航高度应在最佳高度附近、低于最大巡航推力限制高度,在巡航高度上应具有 $1.3g$ 的机动(过载)能力(如气流很平稳,可以放宽到 $1.2g$)。因此,所选择的巡航高度还应不超过 $1.3g$(气流平稳时 $1.2g$)过载限制高度。$1.2g$、$1.3g$ 对应的最大坡度(倾斜角)分别是 $33°$、$39°$。

表 3-5-5　B757-200(RB211-535E4) 成本指数为 100 的高度能力数据

重量	最大巡航推力限制高度(ft)				最佳高度(ft)				机动能力(过载)限制高度(ft)	
1000 lb	ISA+0	ISA+10	ISA+15	ISA+20	ISA+0	ISA+10	ISA+15	ISA+20	1.20g	1.30g
140	42000	42000	42000	42000	42000	42000	42000	42000	42000	42000
158	42000	42000	42000	42000	42000	42000	42000	42000	42000	42000
160	42000	42000	42000	42000	41911	41835	41832	41818	42000	42000
162	42000	42000	42000	42000	39327	39442	39499	39557	42000	42000
164	42000	42000	42000	42000	39709	39850	39920	39988	42000	42000
166	42000	42000	42000	42000	40124	40231	40284	40337	42000	42000
168	42000	42000	42000	42000	40497	40628	40694	40760	42000	42000
170	42000	42000	42000	42000	40526	40506	40507	40502	42000	42000
172	42000	42000	42000	42000	40265	40252	40246	40238	42000	42000
174	42000	42000	42000	42000	40005	39986	39988	39989	42000	42000
176	42000	42000	42000	41844	39749	39736	39728	39732	42000	42000
178	42000	42000	42000	41634	39496	39486	39483	39481	42000	42000
200	40871	40838	40235	39470	36915	36917	36915	36915	41796	40143
205	40385	40354	39764	39012	36371	36374	36373	36371	41282	39629
210	39913	39882	39303	38566	36089	36089	36089	36089	40781	39127
215	39452	39422	38855	38132	35681	35675	35665	35659	40291	38637
220	39003	38973	38418	37708	35165	35151	35146	35146	39812	38158
225	38565	38535	37991	37293	34634	34631	34623	34624	39344	37691
230	38137	38108	37574	36886	34132	34125	34120	34117	38886	37233
235	37718	37690	37165	36487	33646	33645	33645	33637	38439	36785
240	37308	37280	36764	36095	33187	33189	33182	33184	38000	36347
245	36906	36878	36370	35577	32699	32701	32699	32699	37571	35917
250	36512	36484	35952	35031	32243	32239	32240	32237	37150	35495

确定了成本指数后,制作最小成本飞行计划只要查相应的成本指数的各种数值表就可以了,和前面介绍的制作详细飞行计划的方法完全相同。由于计算量很大,应该利用计算机,为此应建立飞机性能数据库(或数据文件)。为建立数据库(数据文件),应使用 INFLT 软件提供的这些性能数据的磁带格式输出文件。表3-5-6、表3-5-7分别是 B757-200/RB211-535E4 经济爬升与经济巡航的磁带格式输出样本,这种格式输出的数据更加精确。巡航、下降等性能数据

都可以以这种磁带格式输出,通过这种格式的文本文件很容易建立飞机性能数据库。

表 3-5-6 B757-200(RB211-535E4)成本指数为 100 的经济爬升数据磁带格式

ISA	WEIGHT	ALT	TIME	FUEL	DIST	R/C	CARD
℃	lb	ft	min	lb	NM	ft/min	
0.0	140000.	1500.	1.681	579.856	0.000	5796.102	1
				⋮			
0.0	140000.	42000.	14.004	3548.109	89.460	1406.621	38
				⋮			
0.0	250000.	1500.	2.674	1125.474	0.000	3002.008	615
				⋮			
0.0	250000.	37000.	27.829	7376.369	182.135	184.964	645
				⋮			
15.0	140000.	1500.	1.681	579.856	0.000	5172.614	1
				⋮			
15.0	140000.	42000.	15.935	3953.206	106.865	1200.426	38
				⋮			
15.0	250000.	1500.	2.674	1125.474	0.000	2650.795	608
				⋮			
15.0	250000.	37000.	34.219	8713.371	236.692	99.255	638
				⋮			
20.0	140000.	1500.	1.681	579.856	0.000	4691.844	1
				⋮			
20.0	140000.	42000.	18.056	4275.765	124.270	1009.305	38
				⋮			
20.0	250000.	1500.	2.674	1125.474	0.000	2377.039	602
				⋮			
20.0	250000.	36000.	35.181	8891.806	245.177	233.904	631

表 3-5-7 B757-200（RB211-535E4）成本指数为 100 的经济巡航数据磁带格式

ALT	ISA	WEIGHT	MACH	IAS	EPRA/EPRI		N2	N1	FF/ENG	EGT	TAT	TAS	FM
ft	℃	lb		kt					lb/h	℃	℃	kt	NAM/1000 lb
33000	0.0	140000	0.797	283	1.45	1.72	0.0	0.0	3423.9	0	−22.1	463	67.660-50
33000	0.0	160000	0.803	286	1.47	1.72	0.0	0.0	3592.7	0	−21.6	467	65.013-50
33000	0.0	180000	0.805	286	1.49	1.72	0.0	0.0	3734.5	0	−21.5	468	62.650-50
33000	0.0	200000	0.803	286	1.52	1.72	0.0	0.0	3887.5	0	−21.6	467	60.072-50
33000	0.0	220000	0.801	285	1.55	1.72	0.0	0.0	4098.3	0	−21.8	466	56.857-50
33000	0.0	240000	0.799	284	1.59	1.72	0.0	0.0	4378.8	0	−21.9	465	53.069-50
33000	0.0	250000	0.798	284	1.61	1.72	0.0	0.0	4544.1	0	−22.0	464	51.063-50
35000	0.0	140000	0.804	274	1.47	1.74	0.0	0.0	3250.5	0	−26.0	464	71.323-50
35000	0.0	160000	0.806	274	1.49	1.74	0.0	0.0	3389.4	0	−25.9	465	68.568-50
35000	0.0	180000	0.805	274	1.52	1.74	0.0	0.0	3534.7	0	−26.0	464	65.618-50
35000	0.0	200000	0.803	273	1.55	1.74	0.0	0.0	3736.6	0	−26.2	463	61.899-50
35000	0.0	220000	0.800	272	1.60	1.74	0.0	0.0	4014.6	0	−26.3	461	57.418-50
35000	0.0	240000	0.797	271	1.65	1.74	0.0	0.0	4380.8	0	−26.6	459	52.425-50
35000	0.0	250000	0.795	270	1.68	1.74	0.0	0.0	4602.7	0	−26.7	458	49.801-50

利用上述给定成本指数的经济爬升、经济巡航、经济下降和经济巡航的高度能力数据建立机型性能数据库（可以包含多个成本指数的数据），就可以制作对给定成本指数的航班成本最小的飞行计划。

3.6　利用燃油差价的飞行计划

世界各地,包括国内各机场的燃油价格都不一样。当从油价低的机场飞往油价高的机场时,如果能够多带油使得在目的地机场不加油或少加油,则有可能节省燃油费用,当两地油价相差很大时,则会带来可观的经济效益。

例如,由机场 A 飞往机场 B,备降机场 C,然后由机场 B 飞往机场 D（或 A）,备降机场 E,机场 A 的油价 $P_T = 5000$ 元/吨,机场 B 的油价 $P_D = 6000$ 元/吨,假设不多带油时从 A 起飞的总油量为 14 t,消耗 9 t,到 B 时剩余 5 t（即备份油量）,从 B 起飞飞往 D（或 A）时所需的总油量为 15 t,则要在 B 买 10 t 油,需要 60000 元。假设在 A 多带 11 t 油（业载不变）,即起飞总油量为 25 t,路途消耗 10 t（多带的 11 t 油被消耗掉 1 t）,到达 B 时剩余 15 t 油,在 B 就不用再加油,从而节省了 10×6000−11×5000 = 5000 元燃油费用。

两地的油价相差越大,节省的燃油费用越多。若目的地机场油价比起飞机场油价高得不多,多带油可能反而不合算,尤其是航程比较长的情况。例如,上面的案例中如果 B 的油价为5100 元/吨,多带 11 t 油节省的燃油费用 = 10×5100−11×5000 = −4000 元,即多花了 4000 元。若 B 的油价为 5500 元/吨,多带 11 t 油节省的燃油费用 = 10×5500−11×5000 = 0 元。此时多带油不亏不盈,此时目的地机场油价称为保本油价（Break-even Fuel Price）,用 P_{DB} 表示。因此,只有目的地机场油价高于保本油价时多带油才能节省燃油费用。

为了利用燃油差价需要解决三个方面的问题：

（1）对一个具体航班来说能不能多带油？

（2）如果能,能多带多少油？

（3）多带油是否合算？ 能产生多大的经济效益？

此外,还可以进一步研究多带多少油使节省的燃油费用最多,为满足下一航班的需要应该多带多少油等问题。

3.6.1 利用燃油差价的经济效益计算

波音公司的飞机使用手册的"载运燃油分析（Tanker Analysis）"给出了几张如何利用燃油差价的图表,见附图 1-8 保本油价计算图表。"载运燃油分析"曲线的用处是:根据主航段的空中距离、巡航高度、不多带油时的着陆重量来确定多带的油在到达目的地机场时所消耗的百分比,根据多带的油在到达目的地机场时所消耗的百分比和起飞机场的油价来确定目的地机场的保本油价。保本油价图与机型无关,是按以下公式计算出来的:

$$P_{DB} = P_T / (1-X)$$

式中：P_T——起飞机场油价；

P_{DB}——目的地机场保本油价；

X——多带的燃油消耗的百分比。

利用燃油价差计算产生的经济效益的步骤如下:

3.6.1.1 先制作正常（不考虑多带燃油）的飞行计划

若：

①$TOW < MTOW$；且

②$LWD < MLWD$；且

③$LWA < MLWA$；且

④$RPF < FTC$。

这四个条件都满足,才能多带燃油；若任何一个条件不满足,则无法利用燃油差价,不能多带油。公式中：$MTOW$、$MLWD$、$MLWA$ 分别是由机场分析得出的在起飞机场的最大允许起飞重量、在目的地场及备降机场的最大允许着陆重量；FTC 是油箱可容纳的最大油量；TOW、LWD、LWA、RPF 分别是制作正常飞行计划时计算出的起飞重量、在目的地机场的着陆重量、备降机场的着陆重量（指开始等待时的重量）及起飞总油量（即停机坪油量）。

3.6.1.2 计算保本油价

在能多带油时,根据巡航高度、该高度上的风与温度、巡航速度、航程、飞机重量估算所飞的空中距离（NM）,然后确定多带的燃油消耗百分比 X,再确定目的地机场的保本油价 P_{DB}。当 $P_D > P_{DB}$ 时,多带燃油才能产生经济效益；否则,虽然可多带油,但由于路途消耗燃油过多,并不合算。

估算空中距离的方法有三种:

（1）利用简化飞行计划图表上的风修正部分；

（2）利用平均真空速 V_T 计算空中距离：$NAM = V_T \times NGM / (V_T + V_W)$；

（3）平均地速 $V_G = NGM / TRT$,平均真空速 $V_T = V_G - V_W$,$NAM = V_T \times TRT$,故

$$NAM = (V_G - V_W) \times TRT = NGM - V_W \times TRT$$

式中：V_W——风速，顺风为正；

TRT——航程时间（不计进近时间和离场时间）。

3.6.1.3 确定可多带的燃油量

计算：

① ΔF_T，$\Delta F_T = MTOW - TOW$；

② ΔF_D，$\Delta F_D = MLWD - LWD$；

③ ΔF_A，$\Delta F_A = MLWA - LWA$；

④ ΔF_R，$\Delta F_R = FTC - RPF$。

假设在起飞机场可以多带的油量为 $\Delta F'$，在目的地机场时所剩余的数量为 $\Delta F''$，$\Delta F''$ 必然小于 $\Delta F'$。

（1）当 ΔF_T 或 ΔF_R 最小时，令 $\Delta F' = \min(\Delta F_T, \Delta F_R)$，$\Delta F'$ 即起飞多带油量，则 $\Delta F'' = \Delta F' \times (1-X) < \Delta F_D$。

（2）当 ΔF_D 最小时，$\Delta F'' = \Delta F_D$，则 $\Delta F' = \Delta F_D / (1-X)$，此时：

① 若 $\Delta F'$ 不大于 ΔF_T 和 $\Delta F'$，ΔF_D 即起飞多带的油量在目的地机场剩余部分 $\Delta F''$，即 $\Delta F'' = \Delta F_D$；

② 若 $\Delta F'$ 大于 ΔF_T 或 ΔF_R，则 ΔF_T 或 ΔF_R 中较小者即起飞时所应多带的油量 $\Delta F'$，即 $\Delta F' = \min(\Delta F_T, \Delta F_R)$，此时应重新计算 $\Delta F''$，$\Delta F'' = \Delta F' \times (1-X)$。

（3）当 ΔF_A 最小时，一般情况下，备降机场的最大允许着陆重量 $MLWA$ 与目的地机场的最大允许重量 $MLWD$ 相等或稍小些，因而不起限制作用，即第③种情况很少发生。如果出现备降机场最大允许着陆重量起限制作用，ΔF_A 最小，可以利用载运燃油分析曲线确定在目的地机场飞往备降机场时多带的油的消耗百分比 Y。此时，先计算：$\Delta F''$，$\Delta F'' = \Delta F_A / (1-Y)$。

① 若 $\Delta F'' \leqslant \Delta F_D$，则计算 $\Delta F'$，$\Delta F' = \dfrac{\Delta F''}{1-X}$，此时：

如果 $\Delta F'$ 不大于 ΔF_T 及 ΔF_R，则前面计算的 $\Delta F'$、$\Delta F''$ 均成立。

如果 $\Delta F'$ 大于 ΔF_T 或 ΔF_R，则 ΔF_T 或 ΔF_R 中较小者即起飞时所应多带的油量 $\Delta F'$，即 $\Delta F' = \min(\Delta F_T, \Delta F_R)$，应重新计算 $\Delta F''$，$\Delta F'' = \Delta F' \times (1-X)$。

② 若 $\Delta F'' > \Delta F_D$，则 $\Delta F'' = \Delta F_D$，重新计算 $\Delta F'$，$\Delta F' = \Delta F_D / (1-X)$，此时：

如果 $\Delta F'$ 不大于 ΔF_T 及 ΔF_R，则前面计算的 $\Delta F'$、$\Delta F''$ 均成立。

如果 $\Delta F'$ 大于 ΔF_T 或 ΔF_R，则 ΔF_T 或 ΔF_R 中较小者即起飞时所应多带的油量 $\Delta F'$，即 $\Delta F' = \min(\Delta F_T, \Delta F_R)$，应重新计算 $\Delta F''$，$\Delta F'' = \Delta F' \times (1-X)$。

3.6.1.4 经济效益的计算：

经济效益可以利用以下公式进行计算：

$$¥ = \Delta F'' \times P_D - \Delta F' \times P_T$$

式中：$¥$——经济效益；

P_T——起飞机场油价；

P_D——目的地机场油价。

例 3-6-1：

北京大兴（PKX）→成都天府（TFU），备降机场昆明长水（KMG），已知条件如下：

$OEW = 42638$ kg $= 94000$ lb，$MZFW = 61688$ kg $= 136000$ lb，$PL = 15000$ kg $= 33069$ lb，$COF =$

0,最大油箱容量 $FTC = 45611$ lb,不使用 APU 和防冰,$P_T = 5000$ 元/吨,$P_D = 6000$ 元/吨。用简化图表确定利用燃油差价的效益。

北京大兴（PKX）→ 成都天府（TFU），备降机场 昆明长水（KMG），已知条件如下:

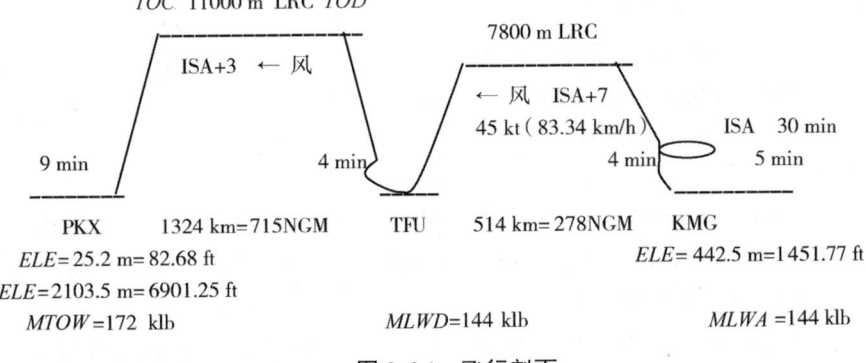

图 3-6-1　飞行剖面

解:

(1)制作不利用燃油差价的正常飞行计划:

在备降机场的 $ZFW = 94000 + 33069 = 127069$ lb

滑入:5 min,耗油量 $= 125$ lb,着陆重量 $= 127194$ lb

进近:4 min,耗油量 $= 520$ lb

等待结束重量 $= 127714$ lb,等待油量 $= 2355$ lb

等待开始重量 $LWA = 130069$ lb

由 DEST 飞到 ALT 的时间 $= 52$ min,油量 $= 4900$ lb

备降时间 $= 56$ min,备降油量 $= 5420$ lb

在 DEST 的着陆重量 $LWD = 134969$ lb,进近:4 min,耗油量 $= 520$ lb

进近前重量 $= 135489$ lb

计算不可预期燃油:

天府机场 15 min 等待耗油量 $F_{15\,min等待}$:按进近前重量 135489 lb 查得全发燃油流量为 5042 lb/h,等待平均重量 $= 135489 + 5042 \times 0.25/2 \approx 136119$ lb;按平均重量 136119 lb 查得全发燃油流量为 5063 lb/h,此即等待中的平均燃油流量,$F_{15\,min等待} = 1266$ lb;天府机场等待开始重量 $W_{天府机场开始} = 135489 + 1266 = 136755$ lb。

10%的航程油量:按天府机场等待开始重量 $W_{天府机场开始}$ 查航程油量图表,查得图表航程油量为 11400 lb,图表航程时间为 2.05 h $= 123$ min;$F_{10\%航程} = 10\% \times (11400 + 520) = 1192$ lb。

不可预期燃油 $= \max\{F_{15\,min等待}, F_{10\%航程}\} = 1266$ lb

由 DEPA 飞到 DEST 的时间 $= 2$:03,油量 $= 11400$ lb

航程时间 $= 2$:07,航程油量 $= 11920$ lb

起飞重量 $TOW = 148155$ lb,滑出:9 min,耗油量 $= 225$ lb

滑行重量 $= 148380$ lb,总油量 $= 21086$ lb

(2)判断能否多带燃油

检查:

$TOW = 148155$ lb $< MTOW = 172$ klb

$LWD = TOW - 航程燃油 = 148155 - 11920 = 136235$ lb $< MLWD = 144$ klb

$LWA = LWA +$ 不可预期燃油 $= 130069 + 1266 = 131335$ lb $< MLWA = 144$ klb

$RPF = 21086$ lb $< FTC = 45611$ lb

故可多带油,此时:

$\Delta F_\text{T} = MTOW - TOW = 23845$ lb

$\Delta F_\text{D} = MLWD - LWD = 144000 - 136235 = 7765$ lb

$\Delta F_\text{A} = MLWA - LWA = 144000 - 131335 = 12665$ lb

$\Delta F_\text{R} = FTC - RPF = 45611 - 21086 = 24525$ lb

（3）计算保本油价

由上面得到的 TOW 和 LWD 计算平均重量 $W \approx 142$ klb,由 LRC 巡航数值表查得平均真空速 $TAS = 455$ kt,则空中距离 $NAM = 715 \times 455/(455 - 50) \approx 803$ NM。［注:利用简化飞行计划图表上的风修正部分确定的空中距离 $NAM = 825$ NM。］

查附图 1-8 保本油价计算图表中的燃油价格率为 1.062,计算保本油价可得:$P_\text{DB} = P_\text{T} \times 1.062 = 5000 \times 1.062 = 5310$ 元/吨。

依据公式 $P_\text{DB} = P_\text{T}/(1 - X)$,可计算出多带的燃油消耗百分比 $X = 5.84\%$。

由于 $P_\text{D} = 6000 > P_\text{DB} = 5310$,因此多带燃油能产生经济效益。

（4）计算多带燃油产生的经济效益

由于 ΔF_D 最小,取 $\Delta F''' = \Delta F_\text{D}$,计算:$\Delta F' = 7765/(1 - 0.0584) \approx 8247$ lb,小于 ΔF_T 和 ΔF_R,即起飞时多带油量 $\Delta F' = 8247$ lb,着陆时剩余量 $\Delta F''' = 7765$ lb,可节省燃油费用:¥ $= (7765 \times 6000 - 8247 \times 5000)/2204.6 \approx 2429$ 元。

有些飞机制造商的手册上指出:载运燃油分析曲线对由于多带燃油而降低了巡航高度的情况是不适用的。另外,因为各航段的航向、风、温度都不同,再加之爬升、下降段的影响,所以对一个航班飞行估算的空中距离不会很准确,用这种方法计算出的结果是粗略的,而且只给出了能多带的油量和经济效益,没有提供多带油的飞行计划。此外,大部分飞机制造商提供的飞机使用手册没有这种载运燃油分析曲线图,因此有必要给出更一般的计算方法。

3.6.2 考虑燃油差价飞行计划的制作方法

3.6.2.1 制作不考虑多带燃油的正常飞行计划

对于航程较长的情况应使用阶梯巡航。若:$TOW < MTOW$、$LWD < MLWD$、$LWA < MLWA$、$RPF < FTC$ 同时成立,则可以多带燃油,否则不能多带燃油,燃油差价不能利用。$MTOW$、$MLWD$、$MLWA$ 分别是由机场分析得出的在起飞机场的最大允许起飞重量、在目的地机场及备降机场的最大允许着陆重量,FTC 为油箱容纳的最大油量。TOW、LWD、LWA、RPF 分别是制作正常飞行计划时计算出的起飞重量、在目的地机场的着陆重量、备降机场的着陆重量(指开始等待时的重量还应包括不可预期燃油)及总油量。

3.6.2.2 制作多带燃油的飞行计划

在可以多带燃油的情况下,设 $\Delta F = \min\{MTOW - TOW, MLWD - LWD, MLWA - LWA, FTC - RPF\}$,取 $\Delta F_\text{o} = \Delta F \times \xi$ 作为可以多带的燃油在备降机场所剩余部分。

式中:$\xi = \begin{cases} 1, & \text{当 } \Delta F = MLWA - LWA \text{ 时} \\ 0.93 \sim 0.95, & \text{当 } \Delta F = MLWD - LWD \text{ 时} \\ 0.90 \sim 0.93, & \text{当 } \Delta F = MTOW - TOW \text{ 或 } \Delta F = FTC - RPF \text{ 时} \end{cases}$

航程长,ξ 取小值。用计算机计算时可取 $\xi=1$,只不过多迭代几次而已。

以 $W = OEW + PL + CF$(正常飞行计划不可预期燃油)$+ COF + \Delta F$。从备降机场停机坪往前推算,制作多带燃油时的飞行计划,计算中仍应保证:

①$TOW < MTOW$,且

②$LWD < MLWD$,且

③$LWA < MLWA$,且

④$RPF < FTC$,

如任何一项不满足,则应减少 ΔF。再重新计算直到刚好满足上述限制为止。多带燃油的飞行计划中的业载应和不考虑多带燃油的正常飞行计划中的业载相同,不能靠减少业载来多带燃油,因为同样重量的业载比燃油能带来更多的经济效益。

因为多带了燃油,飞机有可能在较低的高度上巡航,这时应注意使用新高度上的风与温度来制作飞行计划。

3.6.2.3　计算可多带的燃油及产生的经济效益

设 $TOW1$ 是不考虑多带燃油的正常飞行计划中计算出的起飞重量,$TOW2$ 是多带燃油的飞行计划中计算出的起飞重量,则起飞时多带的燃油 $\Delta F' = TOW2 - TOW1$,设 $LWD1$、$LWD2$ 分别是正常飞行计划和多带燃油飞行计划算出的在目的地机场的着陆重量,则 $\Delta F'$ 在目的地机场剩余的燃油重量 $\Delta F'' = LWD2 - LWD1$,所产生的经济效益 \$ 为:

$$\$ = \Delta F'' \times P_D - \Delta F' \times P_T$$

若 \$ < 0,则不应多带燃油。航程较长且 P_D 比 P_T 高得不多时就会产生这种情况,即虽然可多带燃油,但由于路途消耗太多而不经济。

利用 $\Delta F'$ 和 $\Delta F''$ 可计算多带燃油消耗百分比,即

$$X = (\Delta F' - \Delta F'') / \Delta F' = 1 - \Delta F'' / \Delta F'$$

也可以计算出目的地机场的保本油价为:

$$P_{DB} = P_T \times \Delta F' / \Delta F'' = P_T / (1 - X)$$

经济效益计算中没有考虑多带燃油对航程时间的影响。若航程时间由于多带燃油而增加了 Δt(例如因巡航高度降低、风及温度的影响),设小时成本为 C_T,则计入时间成本后的经济效益:

$$\$ = \Delta F'' \times P_D - \Delta F' \times P_T - C_T \times \Delta t$$

这种算法可以利用简化飞行计划图表进行,也可以按详细飞行计划方法计算,后者工作量要大得多。为此应编制程序,由计算机计算。

下面给出一个利用简化飞行计划图表人工计算的例子并给出由计算机计算的结果,在程序中是按上面提到的公式计算经济效益的。

例 3-6-2:本例的条件即例 3-6-1 的条件。

解:

(1)首先制作不利用燃油差价的正常飞行计划:

在例 3-6-1 中已算出:

在 ALT 着陆(等待开始)重量 $LWA = 130069$ lb,在 DEST 的着陆重量 $LWD = 134969$ lb,起飞重量 $TOW = 148155$ lb,起飞总油量 $RPF = 21086$ lb。

因:$TOW < MTOW$,$LWD < MLWD$,$LWA < MLWA$,起飞总油量<油箱最大油量,故可多带油。

$\Delta F_T = MTOW - TOW = 23845$ lb

$$\Delta F_{\mathrm{D}} = MLWD - LWD = 144000 - 136235 = 7765 \text{ lb}$$

$$\Delta F_{\mathrm{A}} = MLWA - LWA = 144000 - 131335 = 12665 \text{ lb}$$

$$\Delta F_{\mathrm{R}} = FTC - RPF = 45611 - 21086 = 24525 \text{ lb}$$

ΔF_{D} 最小。

（2）制作燃油差价飞行计划

取 $\Delta F_{\circ} = \Delta F_{\mathrm{D}} \times 0.95 = 7377$ lb 作为初值，加到 ALT 的 ZFW 中，再次制作飞行计划：

在 ALT 停机坪飞机重量 $W = 127069 + 7377 = 134446$ lb

滑入：5 min，耗油量 = 125 lb，着陆重量 $LWD = 134571$ lb

备降进近：4 min，耗油量 = 520 lb

等待结束重量 = 135091 lb，等待油量 $HODF = 2477$ lb

等待开始重量 = 137568 lb

由 DEST 飞到 ALT 的时间 = 52 min，油量 = 5100 lb

备降时间 $DIVT = 56$ min，备降油量 $DIVF = 5620$ lb

在 DEST 的着陆重量 $LWD = 142668$ lb

目的地机场进近：4 min，耗油量 = 520 lb，进近前重量 = 143188 lb

计算不可预期燃油：

天府机场 15 min 等待耗油 $F_{15\,\mathrm{min}等待}$：按进近前重量 143188 lb 查得全发燃油流量为 5292 lb/h，等待平均重量 = $143188 + 5292 \times 0.25/2 \approx 143850$ lb；按平均重量 143850 lb 查得全发燃油流量为 5314 lb/h，此即等待中的平均燃油流量，$F_{15\,\mathrm{min}等待} = 1328$ lb；天府机场等待开始重量 $W_{天府机场开始} = 143188 + 1328 = 144516$ lb。

10% 的航程油量：按天府机场等待开始重量 $W_{天府机场开始}$ 查航程油量图表，查得图表航程油量为 12000 lb，图表航程时间为 2.05 h = 123 min；$F_{10\%航程} = 10\% \times (12000 + 520) = 1252$ lb。

不可预期燃油 = $\max\{F_{15\,\mathrm{min}等待}, F_{10\%航程}\} = 1328$ lb

由 DEPA 飞到 DEST 的轮挡时间 $BKT = 2:03$，轮挡油量 $BKF = 12000$ lb

航程时间 $TRT = 2:07$，航程油量 $TRF = 12520$ lb

起飞重量 $TOW = 156516$ lb，滑出：9 min，耗油量 = 225 lb

滑行重量 = 156741 lb，总油量 = $156741 - 127069 = 29672$ lb

以上计算中：$TOW < MTOW$，$LWD \approx MLWD$，$LWA < MLWA$，并且起飞总油量小于油箱最大油量，而且不难验证多带燃油后在 10200 m 高度巡航仍是没问题的。

对这次多带油的飞行计划：

起飞多带油 $\Delta F' = 156516 - 148155 = 8361$ lb

着陆时剩余 $\Delta F'' = 142668 - 134969 = 7699$ lb

多带的燃油消耗百分比 $X = 0.16\%$

$P_{\mathrm{DB}} = 5000/(1 - 0.0016) \approx 5008$ 元/吨

可节省燃油费用：¥ = $(7699 \times 6000 - 8361 \times 5000)/2204.6 \approx 1991$ 元

在燃油差价比较大，或者说燃油价格比（起飞机场油价 P_{T}/着陆机场油价 P_{D}）比较小时，一般是多带的油越多，能省的燃油费用越多，经济效益越好。这时只要按前述的方法计算即可。在燃油差价比较小，或者说燃油价格比 $P_{\mathrm{T}}/P_{\mathrm{D}}$ 接近 1 时，也有可能因多带的油太多使其消耗过大，使得能节约的燃油费用反而减少。

3.6.3　最佳多带油量和经济效益

对于多带油量的情况,并不是多带油量越多经济效益越好。应当确定使经济效益最好的最佳多带油量,多带的油超过此最佳值,不仅经济效益下降,而且飞机的高度能力、机动能力变差,也使起飞、着陆性能的安全余度减小(仍然是安全的)。

例 3-6-3:A340-300(CFM56-5-C4)在 9800 m 高度上以 LRC 速度巡航,航程为 5000 km,备降距离为 370 km,巡航高度为 7200 m。设是 ISA、无风,业载为 20000 kg,使用空机重为 135322 kg,最大允许起飞重量为 270 t,最大允许着陆重量为 190 t,P_T = 2500 元/吨,P_D = 2870 元/吨,滑出 9 min,滑入 5 min,目的地机场进近 5 min,备降机场进近 4 min,不用防冰和 APU,航程油量额外增加 2%,航线计算结果如表 3-6-1 所示。

表 3-6-1　起飞多带油量与经济效益(P_D = 2870 元/吨)

起飞重量(kg)	着陆重量(kg)	航程油量(kg)	起飞多带油量(kg)	消耗(%)	着陆剩余油量(kg)	经济效益(元)
200532	163020	37511	0	0	0	0
219035	179389	39646	18503	11.54	16369	720
225035	184651	40384	24503	11.72	21631	823
229327	188400	40927	28795	11.86	25380	851
231169	190000	41169	30637	11.94	26980	840

对本例,最佳多带油量为 28795 kg。当然,少带一些也可以,经济效益仅稍微减少而性能得以提高。

为了确定最佳多带油量,可按照前述方法进行迭代计算,从比较小的 ΔF_0(多带油量在备降机场剩余部分)开始计算,算出对应的效益,然后逐渐增加 ΔF_0 的值,直到经济效益达到最大为止。

(1)最佳多带油量与燃油差价的大小有关。燃油差价减小,则最佳多带油量减小。

例 3-6-3 中,若 P_D 减少为 2850 元/吨,则计算结果如表 3-6-2 所示,最佳多带油量为 21098 kg。

表 3-6-2　起飞多带油量与经济效益(P_D = 2850 元/吨)

起飞多带油量(kg)	消耗(%)	着陆剩余油量(kg)	经济效益(元)
18503	11.54	16369	393
19503	11.56	17248	398
21098	11.61	18648	401
24503	11.72	21631	390
30637	11.94	26980	301

(2)最佳多带油量与巡航高度有关。特别是对于因多带燃油而必须在较低的高度巡航的情况和不论是否多带油都在同一高度巡航的情况,差别更大,甚至结论相反。

例 3-6-3 的结论是:不论是否多带燃油都在 9800 m 高度巡航的情况,此时多带油有经济效益。如果不多带油巡航高度为 10400 m、多带油的巡航高度为 9800 m,其他条件同例 3-6-3,其计算结果如表 3-6-3 所示。

表 3-6-3　起飞多带油量与经济效益($P_D = 2870$ 元/吨)

起飞重量(kg)	着陆重量(kg)	航程油量(kg)	起飞多带油量(kg)	消耗(%)	着陆剩余油量(kg)	经济效益(元)
198671	163020	35650	0	0	0	0
231169	190000	41169	32498	16.98	26980	−3812

可见,此种情况下如多带油 32498 kg(受最大着陆重量限制)反而亏了,注意此时的多带油量的消耗百分比为 16.98%,比例 3-6-2 中的 11.94% 大得多。所以在波音飞机使用手册中给出的确定消耗百分比的图及空客给出的确定最佳起飞重量的图都不能用于因多带油而降低巡航高度的情况。

有时多带油量的消耗百分比可能随多带油量的增加而稍微减少(耗油仍是增加的),这是因为巡航高度比不多带油时的最佳高度低得多,随多带油量的增加,最佳高度和巡航高度越发接近。

在空客的飞机使用手册中给出了用于帮助用户确定多带油量的最佳数值的图,这种图与高度有关,如图 3-6-2 所示。仅有这种图是不够的,仍需制作飞行计划,具体步骤为:先制作不多带油的飞行计划,得出此时的起飞重量和起飞油量,根据起飞机场和目的地机场油价比值 P_T/P_D 和空中距离查出最佳起飞重量,此重量与不多带油的起飞重量之差即最佳多带油量。如果最佳起飞重量超过了最大允许起飞重量,则最佳多带油量=最大允许起飞重量−不多带油时的起飞重量;如果最佳多带油量大于油箱最大容量与不多带油时的起飞油量的差值,则油箱最大容量与不多带油时的起飞油量的差值为最佳多带油量,然后制作多带此油量时的飞行计划;如这次的着陆重量超过最大允许着陆重量,则应减少多带的油量,直到着陆重量等于最大允许着陆重量,此时的多带油量即最佳多带油量。

(3)最佳多带油量与下一航班所需油量有关。如确定的最佳多带油量使得着陆后剩余油量超过了下一航班所需总油量 FRD(先制作由着陆机场飞下一航班的飞行计划,算出 FRD),应减少多带的油量,使得着陆后剩余油量刚好等于 FRD,为此应把前面确定 ΔF 的公式改为:

$$\Delta F = \min \{ MTOW-TOW, MLWD-LWD, MLWA-LWA, FTC-RPF, FRD-RESF \}$$

式中的 RESF 是不多带油的飞行计划中的备份油量,在制作多带油的飞行计划时还应保证此时的"备份油量"(即在目的地机场着陆后的剩余油量)不大于 FRD。

为更准确地计算多带油产生的经济效益,还需要考虑:

(1)加油时要收取一定的手续费,加油越多,加油手续费越高(可能每单位油量的手续费会减少),这相当于机场油价提高了,可以把加油手续费加到机场的油价上。起飞机场和目的地机场的加油手续费可能是不同的。

(2)多带燃油会使维护成本提高,因为:

①多带燃油增加了起飞重量,减推力起飞时减推力的程度减小;

②由于多带油使巡航重量、着陆重量增大,所需的推力增大,发动机损耗增大,着陆后反推、刹车用得多,这些部件损耗增加,使得维修成本提高。

这种影响难以量化,但在计算多带油产生的经济效益时应该考虑。一般可在起飞机场燃油价格上增加一个增量或一个固定的百分比来考虑维修成本的增加。

(3)在国外机场加油时可能还有一个退税的问题,退税相当于起飞机场的油价降低了。

另外,在制作由着陆机场 B 执行下一航班的飞行计划时,如它的目的地机场 C 油价比起飞机场(即本次航班目的地机场 B)的油价高,可以考虑是否应由机场 A 多带油的问题,如图 3-6-3 所示。

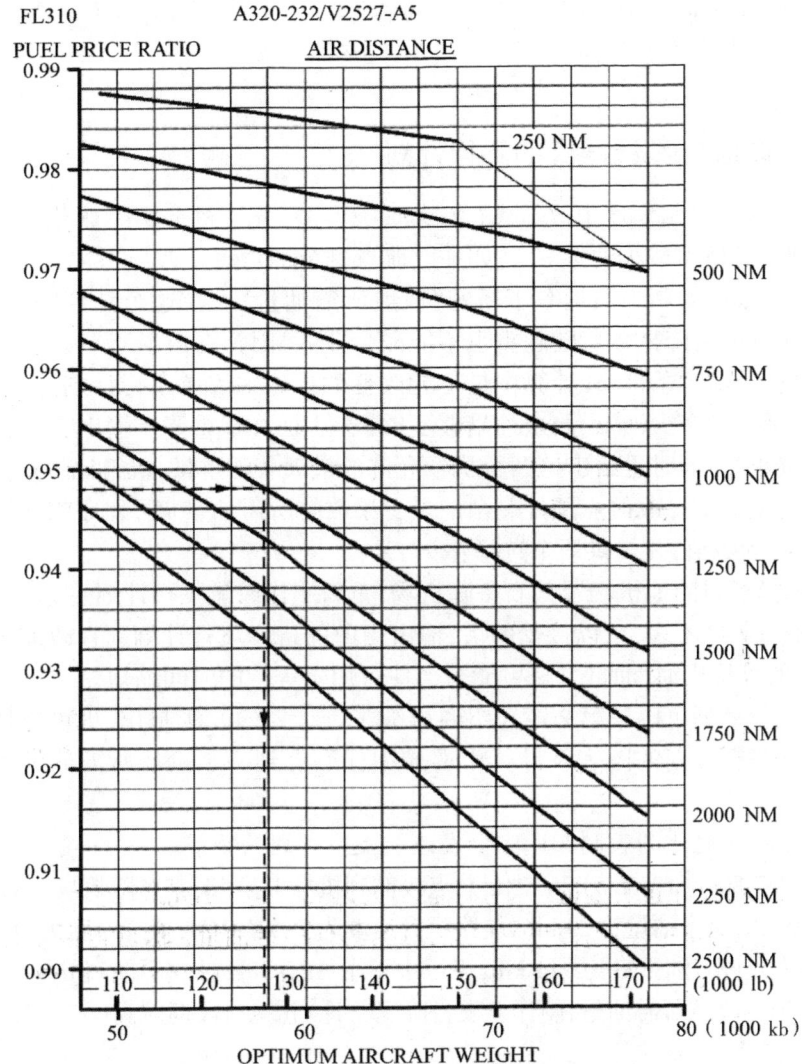

图 3-6-2　空客确定多带油量的最佳数值

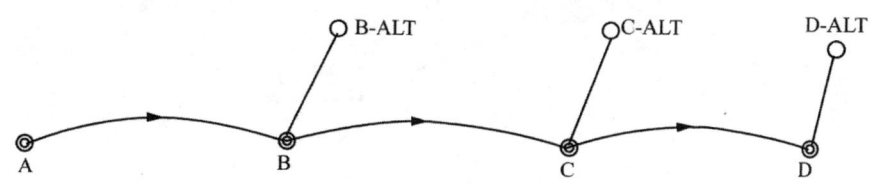

图 3-6-3　多机场利用燃油差价的选择

假设机场 A、B、C 的油价关系为 $P_A < P_B < P_C$，由 A 到 B 多带油消耗百分比为 a，由 B 到 C 多带油消耗百分比为 b。如果 $P_B \leqslant P_A/(1-a)$、$P_C > P_B/(1-b)$，则由 A 到 B 不用多带油，只考虑由 B 到 C 多带油；如果 $P_B > P_A/(1-a)$，$P_C < P_B/(1-b)$，由 A 到 B 可以多带油，则由 B 到 C 不用多带油，只考虑由 A 到 B 最佳多带油或满足由 B 到 C 所需油量要求。

例如：$P_A = 4000$ 元/吨，$P_B = 4800$ 元/吨，$a = 11\%$，$b = 14\%$。

$P_B > 4000/(1-11\%) \approx 4494$ 元/吨，如果 $P_C = 5200$ 元/吨 $< 4494/(1-14\%) \approx 5226$ 元/吨，只

考虑由 A 到 B 最佳多带油或满足由 B 到 C 所需油量要求;如果 $P_C > 5226$ 元/吨,则由 A 多带油时才考虑带上由 B 到 C 所需油量。

3.7 途中改航(全发改航)的飞行计划

如图 3-7-1 所示,在制作途中改航飞行计划时,程序首先按机上实际业载计算在改航点应剩余的油量(即改航所需最小油量 F_0),如在改航点剩余油量多于改航所需油量 F_0,程序进行迭代计算,直到在改航点剩余油量等于新的"改航所需油量"。在这个计算中保持业载不变,即使在目的地机场或备降机场着陆超重也不能减少业载,因为在改航点剩余油量和飞机上的业载是实际有的,是输入条件,在计算中是不能改变的。途中改航飞行计划计算的是:改航需要多少油量、实际消耗多少油量、剩余油量够不够、着陆是否超重等。输出有在目的地机场及备降机场的着陆重量,同时也给出这两个机场允许的最大着陆重量,机组比较这些重量就可以知道着陆是否超重。如在改航点预计剩余油量少于改航所需油量 F_0,程序将显示剩余油量及改航所需油量 F_0 的数值,并提供两个选择:

(1)重新选择新目的地机场及其新备降机场,重新制作改航飞行计划。

(2)如果油量欠缺不多,可以接受的话,输出当前的备降飞行计划。在改航飞行计划中也计算了 30 min 的等待用油,油量欠缺可能只会使可用于等待的时间缩短。

途中改航飞行计划可以按第 2 章介绍的详细飞行计划的方法制作,也可以使用 FPPM 手册中给出的简化飞行计划的图表及计算由巡航中任一点到一个机场所需油量和时间的图表制作。

下面是一个利用简化图表制作途中全发改航飞行计划的例子。

例 3-7-1:B737-800/CFM56-7B26 要由 P 点改航到机场 B 着陆,其备降机场为 C,预计在 P 点飞机重量为 120 klb,剩余油量为 23 klb,航路爬升速度为 280/M0.78,下降速度为 M0.78/280/250,在 B 及 C 的最大允许着陆重量为 144 klb,飞行中不用 APU 和防冰,公司备份油 $COF = 0$,其他条件见图 3-7-2。确定改航所需油量,制作全发备降飞行计划(主航段巡航高度为 37100 ft)。

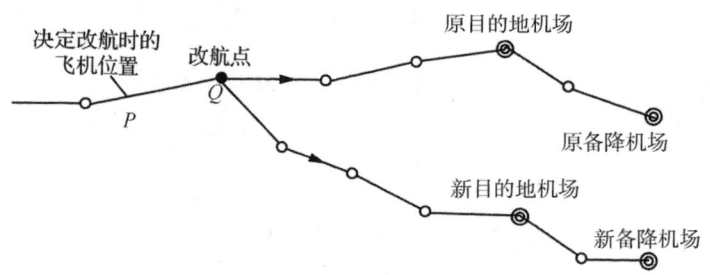

图 3-7-1 途中改航飞行计划

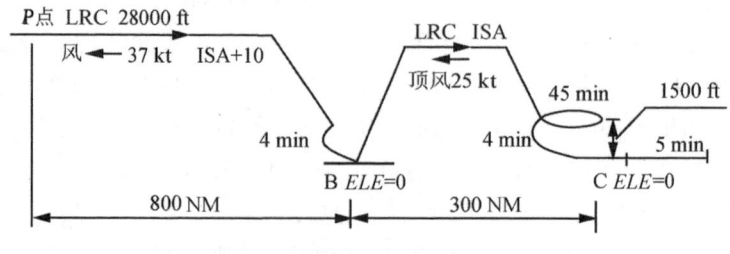

图 3-7-2 途中改航飞行计划

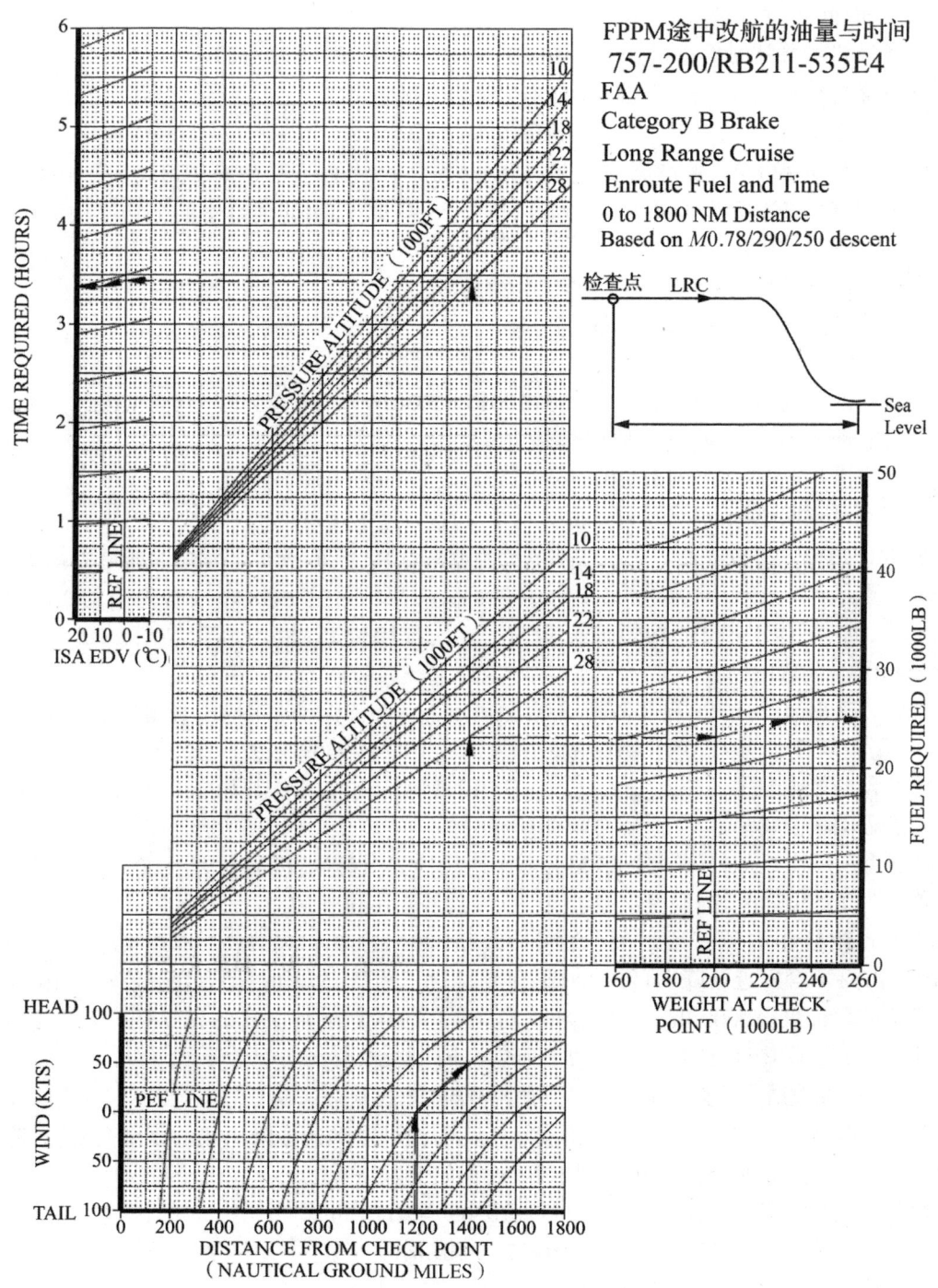

图 3-7-3　途中改航飞行计划图表

解：

（1）计算改航所需油量

$ZFW = 120000 - 23000 = 97000$ lb

先假设在备降机场停机坪剩余油量 = 0

在备降机场停机坪 $W = ZFW + 0 = 97000$ lb

滑入:时间 = 5 min,耗油 = 5×25 = 125 lb

在备降机场着陆时的重量 = 125+97000 = 97125 lb

在备降机场进近:4 min,耗油 = 4×130 = 520 lb

等待结束重量 = 520+97125 = 97645 lb

按 97645 lb 查得全发燃油流量为 3918 lb/h,等待平均重量 = 97645+3918×0.5/2 ≈ 98625 lb

按平均重量 98625 lb 查得全发燃油流量为 3940 lb/h,此即等待中的平均燃油流量,等待油量 = 3940×0.5 = 1970 lb

等待开始重量 = 97645+1970 = 99615 lb<$MLWA$

以等待开始重量作为 LWA 查备降图得:

备降机场的油量 = 4200 lb,时间 = 0.87 h = 52 min

在目的地机场着陆重量 LWD = 4200+99615 = 103815 lb<$MLWD$ = 144000 lb

在目的地机场进近:4 min,耗油 = 4×130 = 520 lb

进近前重量 = 520+103815 = 104335 lb

计算不可预期燃油:

机场 B 15 min 等待耗油 $F_{15\ min等待}$:按进近前重量 104335 lb 查得全发燃油流量为 4104 lb/h,等待平均重量 = 104335+4104×0.25/2 = 104848 lb;按平均重量 104848 lb 查得全发燃油流量为 4120 lb/h,此即等待中的平均燃油流量,$F_{15\ min等待}$ = 1030 lb;机场 B 等待开始重量 $W_{B开始}$ = 104335+1030 = 105365 lb。

10%的航程油量:按机场 B 等待开始重量 $W_{B开始}$ 查航程油量图表,查得图表航程油量为 9500 lb,图表航程时间 = 2.05 h = 123 min;$F_{10\%航程}$ = 10%×(9500+520) = 1002 lb。

不可预期燃油 = max$\{F_{15\ min等待}, F_{10\%航程}\}$ = 1030 lb。先按等待开始重量 $W_{B开始}$ 为 105365 lb 重新查航程图表查得航程油量为 9500 lb。

P 点重量一次近似值为 114865 klb,再次迭代计算,查得航程油量为 10000 lb,P 点重量二次近似值为 115365 lb。前后两次 P 点重量近似相同,不再迭代。

由 P 点改航 B、备降 C 所需最小油量 = 滑入油量(125 lb)+备降进近油量(520 lb)+等待油量(1970 lb)+不含进近备降油量(4200 lb)+目标进近油量(520 lb)+不可预期燃油(1030 lb)+不含进近航程油量(10000 lb) = 18365 lb = 8330 kg,即只要在 P 点剩余 18365 lb(8330 kg)油量就够改航了。实际剩油 23 klb,大于改航所需燃油 18365 lb,满足改航要求。

(2)计算在改航点实际消耗油量及到达备降机场的剩余油量

为计算改航重量为 120 klb 的飞机实际消耗的油量及到达备降机场的剩余油量,需要迭代计算。第一种方法是从备降机场往回推算,过程如下:

设备降机场的剩余油量 = 0.9×(23000−18365) = 4172 lb

则等待结束重量 = 4172+97645 = 101817 lb<$MLWA$

等待油量 = 4058×0.5 = 2029 lb

等待开始重量 = 101817+2029 = 103846 lb

改查备降图得:改航到备降机场的油量 = 4300 lb,时间 = 0.87 h ≈ 52 min

在目的地机场着陆重量 LWD = 4300+103846 = 108146 lb<$MLWD$

在目的地机场进近:4 min,耗油 = 4×130 = 520 lb

目的地机场进近前重量 = 520+108146 = 108666 lb

计算不可预期燃油:

（1）机场 B 15 min 等待耗油 $F_{15 \min 等待}$：按进近前重量 108666 lb 查得全发燃油流量为 4239 lb/h，等待平均重量 $=108666+4239\times0.25/2\approx109196$ lb；按平均重量 109196 lb 查得全发燃油流量为 4255 lb/h，此即等待中的平均燃油流量，$F_{15 \min 等待}=1064$ lb；B 机场等待开始重量 $W_{B开始}=108666+1064=109730$ lb。

（2）10%的航程油量：按机场 B 等待开始重量 $W_{B开始}$ 查航程油量图表，查得图表航程油量为 9700 lb，图表航程时间 $=2.05$ h $=123$ min；$F_{10\%航程}=10\%\times(9700+520)=1022$ lb。

（3）不可预期燃油 $=\max\{F_{15 \min 等待},F_{10\%航程}\}=1064$ lb

先按等待开始重量 $W_{B开始}=109730$ lb 查航程图表查得航程油量为 9700 lb

P 点重量一次近似值 $=119430$ lb，再次查得航程油量 $=10250$ lb

P 点重量二次近似值 $=119980$ lb，再次迭代得 P 点重量将几乎不变，故不再迭代。

现在得到在改航点 P 点重量近似等于飞机在 P 点的实际重量，所以也不用再返回备降机场重新计算了。

重量为 120 klb 的飞机由 P 点改航到在机场 B 着陆所需的油量 $=10250+520=10770$ lb，所需的时间 $=2.05$ h $+4$ min $=2:07$

加上滑入油量和时间后所需的油量 $=10895$ lb，所需时间 $=2:12$

由机场 B 备降机场 C 的备降油量 $=4300+520=4820$ lb，备降时间 $\approx0:52+0:04=0:56$，等待油量 $=2029$ lb

备份油量 $=$ 等待油量 $+$ 备降油量 $+$ 不可预期燃油 $+$ 在备降机场的剩余油量 $=2029+4820+1064+4172=12085$ lb

此时，总油量 $=12085+10895=22980$ lb，与飞机在 P 点的实际油量 23000 lb 相近。

也可按照以下的方法计算重量为 120 klb 的飞机实际消耗的油量及到达备降机场的剩余油量：从改航点往备降机场方向推算。

首先按实际重量 120 klb 查得由 P 点到机场 B 着陆所需的不含进近的航程油量为 9200 lb，航程时间 $=2.4$ h，进近前重量 $=120000-9200=110800$ lb。

计算不可预期燃油：

（1）机场 B 15 min 等待耗油 $F_{15 \min 等待}$：按进近前重量 110800 lb 查得全发燃油流量为 4306 lb/h，等待平均重量 $=110800-4306\times0.25/2\approx110262$ lb；按平均重量 110262 lb 查得全发燃油流量为 4288 lb/h，此即等待中的平均燃油流量，$F_{15 \min 等待}=1072$ lb。

（2）10%的航程油量：按实际重量 120 klb 查得由 P 点到机场 B 着陆所需油量 9200 lb，所需时间 $=2:24$，$F_{10\%航程}=10\%\times(9200+520)=972$ lb。

（3）不可预期燃油 $=\max\{F_{15 \min 等待},F_{10\%航程}\}=1072$ lb

进近 4 min，进近耗油 520 lb

由 P 点到机场 B 着陆所需航程油量 $=9200+520=9720$ lb，时间 $=2:24+0:04=2:28$

在机场 B 着陆重量 $=120000-9720-1072=109208$ lb$<MLWD$

由机场 B 到机场 C 的不含进近备降油量 $=4500$ lb，备降时间 $=0:52$

在机场 C 进近 4 min，进近耗油 520 lb

备降油量 $=4500+520=5020$ lb，备降时间 $=0:52+0:04=0:56$

等待开始时的重量 $=109208-4500=104708$ lb$<MLWA$

按等待开始时的重量 $=104708$ lb 计算的等待燃油流量 $=4116$ lb/h，平均等待重量 $=104708-4116\times0.5/2=103679$ lb，平均燃油流量 $=4084$ lb/h，等待油量 $=2042$ lb，等待结束时的

重量 = 104708－2042 = 102666 lb。

着陆时重量 = 102666－520 = 102146 lb

在机场 C 停机坪重量 = 102146－125 = 102021 lb

剩余油量 = 102021－97000 = 5021 lb

3.8 公务机飞行计划

3.8.1 公务飞行的申请

3.8.1.1 公务飞行的定义

《中华人民共和国民用航空法》第一百四十五条:"通用航空,是指使用民用航空器从事公共航空运输以外的民用航空活动,包括从事工业、农业、林业、渔业和建筑业的作业飞行以及医疗卫生、抢险救灾、气象探测、海洋监测、科学实验、教育训练、文化体育等方面的飞行活动。"通用航空企业的经营项目划分为甲、乙、丙三类,其中甲类包括陆上石油服务、海上石油服务、直升机机外载荷飞行、人工降水、医疗救护、航空探矿、空中游览、公务飞行、私用或商用飞行驾驶执照培训、直升机引航作业、航空器代管业务、出租飞行、通用航空包机飞行。

公务飞行的名称源于英文 Business Aviation,直译为商务航空,目前国际民用航空组织也无正式定义,但在对世界各国民用航空情况统计时,一直将其划为通用航空活动进行统计。公务飞行一部分是由用户租用公务机进行的,一部分是由用户自购公务机进行的,但共同点是单一用户、不定时、无固定目的地、无客票,其服务方式与要求、运营步骤与过程是一致的;为防止利用公务飞行开展加班包机,规定航空器的座位数是必要的;因此对公务飞行定义为,是通用航空活动的一种方式,系指使用民用航空器按单一用户(企业、事业单位、政府机构、社会团体或个人)确定时间和始发地、目的地,为其商业、事务、行政等活动提供的无客票飞行服务。通常使用 30 座(含)以下的民用航空器(轻型飞机、旋翼航空器、滑翔机除外)。大型跨国公司的业务扩展到全球很多地方,公司人员因公务需要租用公务机外出执行任务;或者大型公司(非航空运输企业)自己出钱购买飞机,成立机队用以运送内部职工或客户,运送急需的零部件等。这类间接为商务提供服务的航空活动被称为公务航空。

3.8.1.2 公务飞行计划的申请

国内涉及公务机运营的法规主要有《小型商业运输和空中游览运营人运行合格审定规则》(CCAR-135-R3)和中华人民共和国交通运输部令 2022 年第 3 号《一般运行和飞行规则》(CCAR-91-R4)两部法规。不论是包机、专机,还是加班飞行,一切飞行都应当预先申请并经过批准后方可执行,未经批准的飞行预报不得执行。《民用航空预先飞行计划管理办法》(CCAR-73)总则第四条规定:"航空营运人进行民用航空飞行活动,其预先飞行计划应当获得批准;未获得批准的,不得实施飞行。航空营运人应当按照批准的预先飞行计划实施飞行;取消获得批准的预先飞行计划,应当及时向预先飞行计划的批准部门备案。"

关于公务飞行的申请程序,CCAR-73 对加班和不定期飞行预先飞行计划申请、受理和批准有相关规定:

(1)申请。中国航空营运人在同一飞行管制区内飞行的加班和不定期飞行预先飞行计划,向该飞行管制区飞行管制部门所在地的民航地区空管局办事单位申请,在新疆管理局辖区内飞行的,向新疆空管局空管处申请;外国国家领导人或者重要客人乘坐的外国专机、包机预先飞行计划应当向国家外交主管部门提出申请。

(2)申请资料。一般情况下,预先飞行计划的申请资料应包括如下内容:

①航空器所有人、经营人及其联系方式;

②航空器型号、型别;

③经营人两字和三字代码、航班号、无线电通话和通信呼号;

④机载电子设备,是否装有机载防撞系统和航路、航线有特殊要求的机载电子设备;

⑤航空器的最大起飞重量和最大着陆重量;

⑥起降地点、起降时刻、班期、航线走向、飞行高度和进出中国飞行情报区的航路点代码及时刻;

⑦航班性质;

⑧计划起止日期;

⑨其他需要说明的事项。

(3)批准。中国航空营运人加班和不定期飞行预先飞行计划的申请由中国民航局空管局和民航地区空管局以 AFTN 电报、SITA 电报或者传真电报的形式批复。

(4)批准时限。一般情况下,受理单位不应迟于预先飞行计划开始执行之日前至少 2 个工作日做出决定并通知申请人。

关于公务飞行的补充说明:外国航空营运人和中国港澳台地区航空营运人在中国境内机场起飞或者降落的公务飞行,应当于飞行前至少 7 个工作日以 SITA 电报、航空固定业务电报或者中国民航局接受的其他方式提出预先飞行计划申请。申请应当包括以上第(2)条申请资料和下列内容:①机组成员和旅客名单、出生日期、护照号及国籍;②境内接待单位名称、地址及联系办法。

中国航空营运人进行加班和不定期飞行,使用的航空器型号和型别、飞行高度以及航线走向不超出现行规定范围的,应当于飞行前至少 3 个工作日以传真、SITA 电报、航空固定业务电报或者中国民航局接受的其他方式提出预先飞行计划申请;使用的航空器型号和型别、飞行高度以及航线走向超出现行规定范围的,应当于飞行前至少 5 个工作日以传真、SITA 电报、航空固定业务电报或者中国民航局接受的其他方式提出预先飞行计划申请。紧急情形下的加班和不定期飞行应当说明理由,预先飞行计划申请的时限不受限制。

3.8.2　公务飞行的油量规定和飞行剖面

3.8.2.1　公务飞行油量规定

《小型商业运输和空中游览运营人运行合格审定规则》(CCAR-135-R3)的第 135.351 条对公务飞行的油量相关规定如下。

第 135.351 条　燃油要求

(a)飞机应当携带可以安全完成计划飞行和改航所需的足够的可用燃油。

(b)携带的可用燃油量应当至少基于:

(1)下列数据:

(i)从燃油消耗监控系统中获得的特定飞机的当前数据(如可用)。

(ii)采用飞机制造商提供的数据(如果没有特定飞机的数据)。

（2）计划飞行的运行条件,包括：

（i）飞机的预计重量。

（ii）航行通告。

（iii）天气实况和预报。

（iv）空中交通服务程序、限制及预期的延误。

（v）维修保留项或者构型偏离的影响。

（c）飞行前对所需可用燃油的计算应当包含：

（1）滑行燃油：考虑到起飞机场的当地条件和辅助动力装置(APU)的燃油消耗,起飞前预计消耗的燃油量。

（2）航程燃油：考虑到本条(b)款(2)项的运行条件,允许飞机从起飞机场或者从重新放行点飞到目的地机场着陆所需的燃油量。

（3）不可预期燃油：为补偿不可预见因素所需的燃油量。根据航程燃油方案使用的燃油消耗率计算,占计划航程燃油或者飞行中重新放行点5%的所需燃油,但在任何情况下不得低于以等待速度在目的地机场上空450米(1500英尺)高度上在标准条件下飞行5分钟所需的燃油量。

（4）目的地备降机场燃油,包含：

（i）当需要有目的地备降机场时,飞机所需的燃油应当能够：

（A）在目的地机场复飞；

（B）爬升到预定的巡航高度；

（C）沿预定航路飞行；

（D）下降到起始进近定位点；

（E）在目的地备降机场进近并着陆。

（ii）当需要两个目的地备降机场时,根据本条(c)款(4)项(i)目计算的所需燃油量,能够使飞机飞行到需要更多备用燃油的目的地备降机场。

（iii）当不需要有目的地备降机场时,所需油量能够使飞机在目的地机场上空450米(1500英尺)高度上在标准大气条件下飞行15分钟。

（iv）当预定着陆机场是一个孤立机场时：

（A）对于活塞式发动机飞机,飞行45分钟的所需油量与巡航高度层上消耗的计划飞行时间的15%所需油量之和(包括最后储备燃油)或者飞行2小时的所需油量,取其中较小者。

（B）对于涡轮发动机飞机,能够以正常巡航燃油消耗在目的地机场上空飞行2小时的所需油量(包括最后储备燃油)。

（5）最后储备燃油：使用到达目的地备降机场,或者不需要目的地备降机场时,到达目的地机场的预计重量计算得出的燃油量：

（i）对于活塞式发动机飞机,按照局方规定的速度和高度条件飞行45分钟所需的油量。

（ii）对于涡轮发动机飞机,以等待速度在机场上空450米(1500英尺)高度上在标准大气条件下飞行30分钟所需的油量。

（6）额外燃油：所需燃油的补充,即如果根据本条(c)款(2)项至(5)项计算的最低燃油不足以：

（i）假定在航路最困难临界点发动机发生失效或者丧失增压需要更多燃油的情况下,允许飞机在必要时下降并飞行到某一备降机场。

（A）以等待速度在该机场上空 450 米(1500 英尺)高度上在标准条件下飞行 15 分钟。

（B）进近并着陆。

（ii）允许进行延长改航时间运行(EDTO)的飞机遵守局方批准的临界燃油方案。

（iii）满足上述未包含的其他规定。

（7）酌情燃油：机长自行决定携带的额外燃油。

（d）仅当机上可用的燃油按照要求符合本条(c)款(1)项至(6)项的要求，方可开始飞行(如适用)。仅当机上可使用的燃油按照要求符合本条(c)款(2)项至(6)项的要求(如适用)，方可从重新放行点继续飞行。

3.8.2.2　公务机飞行计划剖面

常规公务机的飞行计划剖面如图 3-8-1 所示。

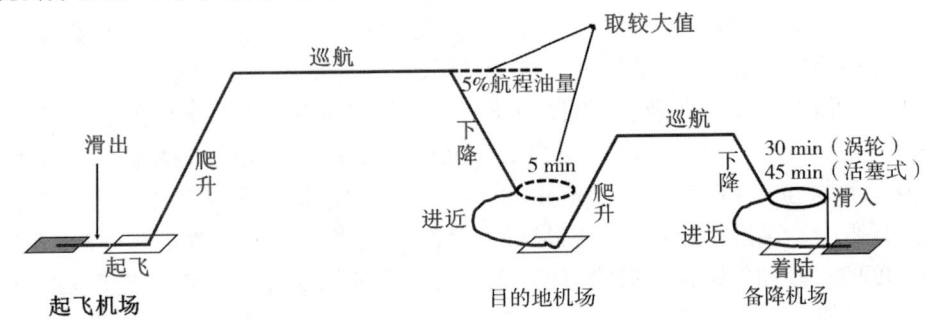

图 3-8-1　常规公务机的飞行计划剖面

图 3-8-2 是豪客 800XP 飞行手册提供的飞行计划剖面，并附有以此图为依据的简明飞行计划实例。剖面图中将等待位置放在了目的地机场进行计算(后文将根据飞行实例计算结果与将等待放在备降机场计算结果进行比较分析)，而且没有目的地机场的进近部分，而是将其视为附加备份油量中的二次进近油量，在目的地机场等待开始重量处，自主选择加或不加。

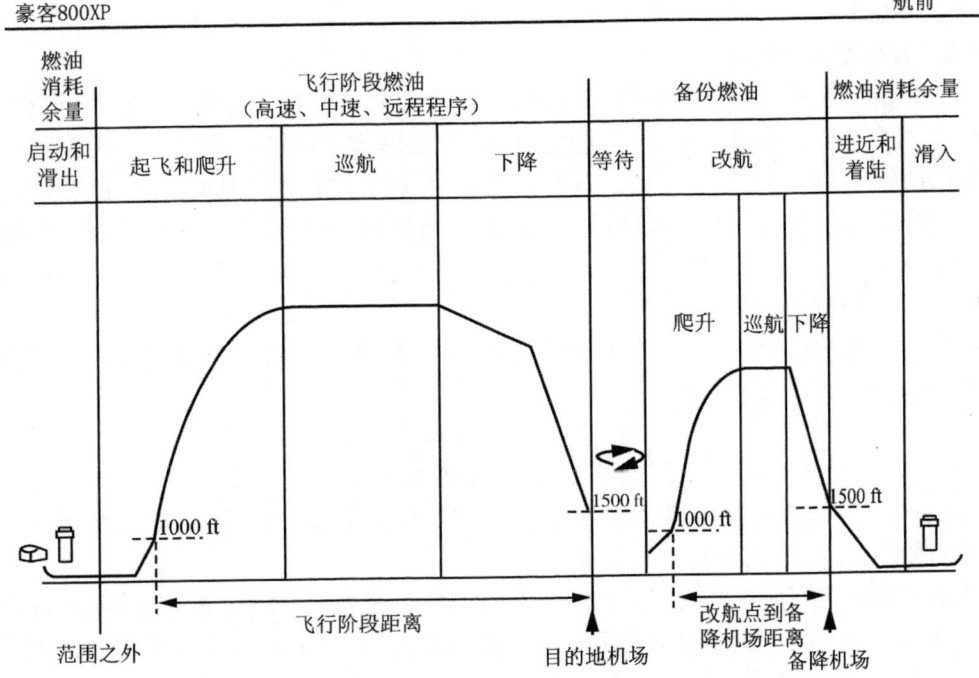

图 3-8-2　豪客 800XP 飞行计划剖面

3.8.3　公务飞行航线选择和相关性能计算

3.8.3.1　航线选择

在美国,通用航空发达,小型机场众多。公务航空在通往没有固定航班的中小城市时显得尤为方便快捷,这也正是大型公共航空运输所无法比拟和替代的。正因为公务飞行的不确定性,就会有飞往尚无已开通航线的中小型机场的业务需求,针对这种情况,在制作飞行计划时就要选择航路和进行可行性分析。在我国,各地区的经济发展程度不同,西北内陆地区相对落后,东南沿海地区经济相对发达,集中了国内绝大部分的公务航空需求。我国公务飞行所通往的地区城市大多已有成熟航线和大型机场,所需要考虑的问题相对较少,然而对于航路的选择同样也得考虑安全等相关可行性问题。同时也会存在飞往中小机场的业务需求,而且随着我国西部大开发战略的实施,西北地区的经济活动活跃,这一市场需求只会越来越大。

当前我国空域尚未完全开放,可供民用飞机使用的空域仅为有限的航路,所以航线选择即为航路的选择,选择的空间较小,需要做的工作也相对较少。航线的选择要求距离最短,考虑空中交通流量问题,避开拥挤、繁忙的航路,同时要求开始巡航重量条件下一发停车的改平高度高于航路最低安全高度。当出现航路最低安全高度高于一发停车改平高度时,找出航路上最低安全高度的航段和关键性障碍物,计算飞机飞行到该位置时重量和该重量下的改平高度,如果计算出的改平高度高于障碍物高度,则航路安全,可以飞行通过,否则要选择该航段的替代航路进行绕飞。由于篇幅有限,这里就不做详细叙述。

CCAR 第 121.637 条起飞备降机场(a)款第(1)项规定,对于双发动机飞机,备降机场与起飞机场的距离不大于飞机使用一发失效的巡航速度在静风条件下飞行 1 h 的距离。而在CCAR-135 中,均未发现与之类似的规定条款。当前涉及公务机运行的民航法规中,尚无关于双发喷气式公务机航路一发失效时备降机场和双发延程飞行备降机场的相关规定,所以当前在喷气式公务机制作飞行计划时不予考虑。

3.8.3.2　性能限重计算

公务飞行为不定期飞行,没有固定的航线和起降机场,在每次飞行前都要计算起降机场分析数据。最大允许起降重量是机场分析得出的结果,是考虑了机场、气象条件、飞机性能、结构限制、跑道道面限制后得出的最小值。这些数据是在飞行实施之前根据起降机场数据(机场标高、跑道长度、跑道坡度、温度、风速风向和障碍物情况)从飞行手册上查图表或编制软件计算得出的。

3.8.3.3　等效静风距离的计算

在巡航阶段或改航阶段,除了零风速情况外,必须估算用于确定所需燃油的等效静风距离。等效静风距离可以通过下式计算得出:

$$等效静风距离 = \frac{实际距离 \times 真空速}{真空速 - 风速}$$

式中,距离单位为 NM;风速单位为 kt,顶风为正值,顺风为负值;真空速是指巡航真空速或平均静风飞行阶段速度,单位为 kt。

首先根据实际飞行阶段距离,通过查表,在对应的程序表(高速、中速或远程)中得到近似静风速度飞行阶段时间。用飞机无油重量加备份油量,得到一个近似着陆重量。查实际大气条件(温度、ISA 或±10 ℃等)和巡航高度(8000~43000 ft)所对应的表,得到静风飞行阶段所用时间。

然后计算巡航真空速,或近似静风速度(平均静风飞行阶段速度)。巡航真空速通过下式计算:

$$巡航真空速 = \frac{飞行阶段实际距离}{静风飞行阶段时间}$$

最后根据等效静风距离公式,计算等效静风距离。

3.8.3.4　巡航高度的选定和最佳高度的确定

民用飞机巡航高度的选定,在遵守空中交通管制规定的前提下,主要考虑节省燃油、缩短飞行时间、降低成本,以取得最大的经济效益。公务航空以快捷、高效、灵活和舒适的特点区别于大型公共运输航空,但其经济性就难以兼顾了,也没有得到足够的重视。然而公务飞行在时间和效益之间亦可寻求一种平衡。

大型公共运输飞行因其固定的航班时刻、航线和追求最大经济效益的特性,在飞行高度和速度等参数的选择上,在安全的前提下以降低成本为首要目标。其参数选定的方法已被广泛研究并形成了一套成熟的体系和可靠的现行方法,而且大型飞机制造厂商提供的飞行手册都给出了最佳高度和速度选择的数据和方法。公务飞行因自身的特殊性,其经济性没有得到足够的重视,也少有这方面的相关研究。以豪客 800XP 型公务机为例,其飞行手册所提供的数据资料并没有类似大型飞机的最佳高度选择的图表数据,而巡航也只有高速程序、中速程序和远程程序三种固定速度方式,但也没有各种选择参考的条件依据。

下面将从燃油里程入手,针对豪客 800XP 机型最佳巡航高度选择进行初步探讨分析。

设飞机飞行一小段距离 ΔR 要消耗燃油 ΔW,ΔR 为空中飞行距离,则表示消耗单位燃油所飞过的距离即燃油里程,可用下式表示:

$$\frac{nam}{lb} = -\frac{\Delta R}{\Delta W}$$

式中:$\frac{nam}{lb}$ 为消耗每磅燃油所能飞行的海里数。

ΔR 以 NM (海里)为单位;

ΔW 以 lb (磅)为单位;

公式中的负号是因为 ΔW 为飞机重量减小值,即燃油消耗量用负值表示,而燃油里程必为正值,因而加负号。

燃油里程越大,说明消耗单位燃油所飞行的距离越远,或者说飞行相同距离所消耗的燃油越少。仅从燃油消耗方面考虑,飞机飞行时燃油里程越大,飞行过程越省油,经济效益越高。由此可见,最佳巡航高度即为燃油里程最大的巡航高度,巡航高度选择就是在空中交通管制许可的条件下尽量选择燃油里程最大的气压高度。

豪客 800XP 型公务机飞行手册中给出了标准大气条件下一定气压高度不同巡航重量的"巡航时每磅燃油英里数"曲线图(图中已换算为 NM)。现以 22000 lb 巡航重量查图,得出数据(气压高度单位:ft;燃油里程单位:NM/lb),见表 3-8-1 与图 3-8-3。

表 3-8-1　22000 lb 巡航不同气压高度燃油里程数据

气压高度(ft)	20000	25000	31000	33000	35000	37000	39000	41000	43000
远程(NM/lb)	0.230	0.260	0.294	0.306	0.311	0.324	0.330	0.330	0.326
高速(NM/lb)	0.214	0.237	0.259	0.218	0.230	0.246	0.269	0.295	0.319

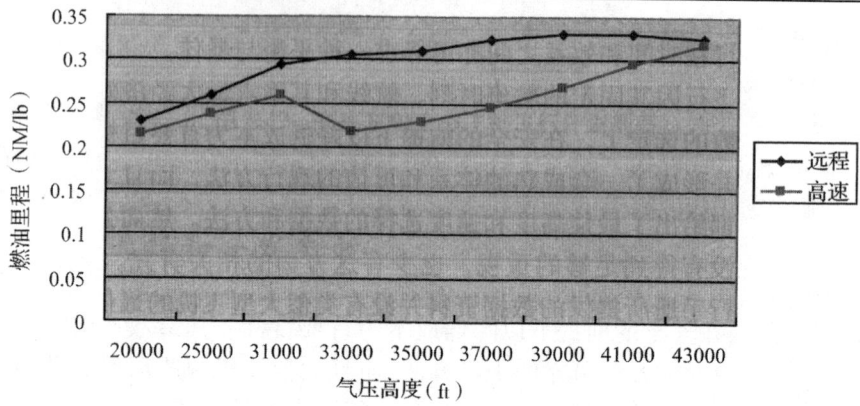

图 3-8-3　22000 lb 燃油里程–气压高度分析折线图

现从飞行手册中高速程序、中速程序和远程程序的性能数据中读出总燃油流量和真空速，由燃油里程与速度和燃油流量的关系可知：

$$\frac{nam}{lb} = V \cdot \frac{1}{W_F}$$

以豪客 800XP 飞行手册中性能数据中给出的总燃油流量和真空速计算燃油里程。选定 22000 lb 巡航重量、标准大气条件，计算得出燃油里程，见表 3-8-2 和图 3-8-4。

表 3-8-2　22000 lb 巡航不同气压高度燃油里程数据

气压高度(ft)	远程巡航(NM/lb)	中速巡航(NM/lb)	高速巡航(NM/lb)
1000	0.149	—	
3000	0.155	—	—
5000	0.163	—	0.154
7000	0.170	—	0.160
9000	0.178	0.168	0.151
11000	0.187	0.175	0.153
13000	0.195	0.183	0.163
15000	0.204	0.191	0.171
17000	0.214	0.200	0.180
19000	0.224	0.209	0.187
21000	0.235	0.218	0.195
23000	0.250	0.228	0.203
25000	0.259	0.237	0.208
29000	0.275	0.256	0.194

续表

气压高度(ft)	远程巡航(NM/lb)	中速巡航(NM/lb)	高速巡航(NM/lb)
31000	0.294	0.259	0.205
33000	0.305	0.268	0.217
35000	0.316	0.281	0.230
37000	0.324	0.295	0.243
39000	0.333	0.305	0.268

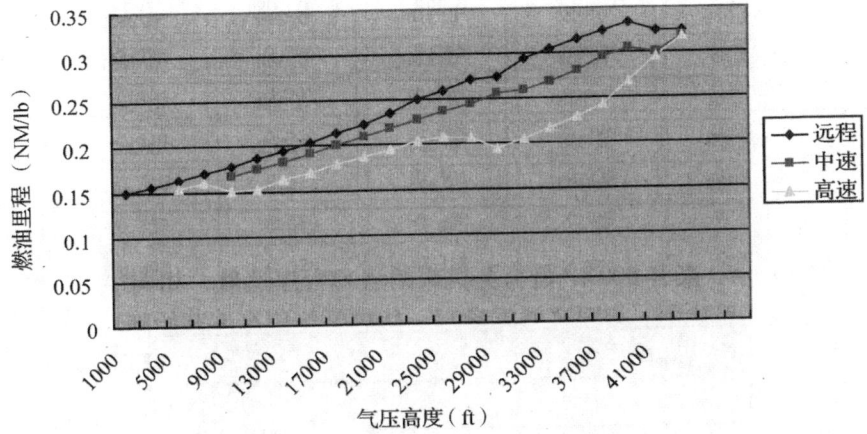

图 3-8-4　22000 lb 燃油里程–气压高度分析折线图

对比通过飞行手册读图所得数据和性能参数计算所得的燃油里程数据,考虑读图误差后,数据基本吻合,两种途径所得数据的折线图对应部分呈现相同的特征,如图 3-8-3 和图 3-8-4 所示。以 22000 lb 重量巡航时,在低高度随着气压高度的增加燃油里程总体上逐渐增加。在较高气压高度时,对于远程程序,燃油里程逐渐增加,在 39000 ft 高度达到最大值;对于中速程序,燃油里程先逐渐增加,在 39000 ft 高度达到一个峰值,在 41000 ft 高度略微下降,在 43000 ft 时又有所增加;对于高速程序,燃油里程从 9000 ft 开始增加,在 25000 ft 达到一个峰值后下降到 31000 ft 的更小值,随后持续增加。

以上对全气压高度 22000 lb 重量巡航时燃油里程数据的分析显示,在较低气压高度时燃油里程总体上随气压高度的增加而增加。下面将对不同重量以不同程序在较高气压高度巡航时,对性能数据计算得出的燃油里程进行分析并绘制对应的曲线图。

(1)远程巡航(标准大气温度,见表 3-8-3 和图 3-8-5)

表 3-8-3　远程巡航不同重量高度燃油里程数据　　　　　　　　　　　(单位:NM/lb)

	18000 lb	20000 lb	22000 lb	24000 lb	26000 lb	27000 lb
17000 ft	0.224	0.219	0.214	0.209	0.202	0.199
19000 ft	0.235	0.229	0.224	0.218	0.212	0.209
21000 ft	0.247	0.241	0.235	0.228	0.222	0.219
23000 ft	0.263	0.256	0.25	0.243	0.235	0.232
25000 ft	0.272	0.266	0.259	0.259	0.252	0.244

续表

	18000 lb	20000 lb	22000 lb	24000 lb	26000 lb	27000 lb
27000 ft	0.284	0.277	0.271	0.262	0.254	0.251
29000 ft	0.289	0.282	0.275	0.267	0.259	0.255
31000 ft	0.309	0.302	0.294	0.285	0.277	0.273
33000 ft	0.323	0.314	0.305	0.296	0.283	0.276
35000 ft	0.332	0.322	0.316	0.299	0.285	0.278
37000 ft	0.350	0.338	0.324	0.308	0.289	0.278
39000 ft	0.361	0.348	0.333	0.308	0.283	0.271
41000 ft	0.378	0.359	0.324	0.302	0.282	—
43000 ft	0.388	0.357	0.324	—	—	—

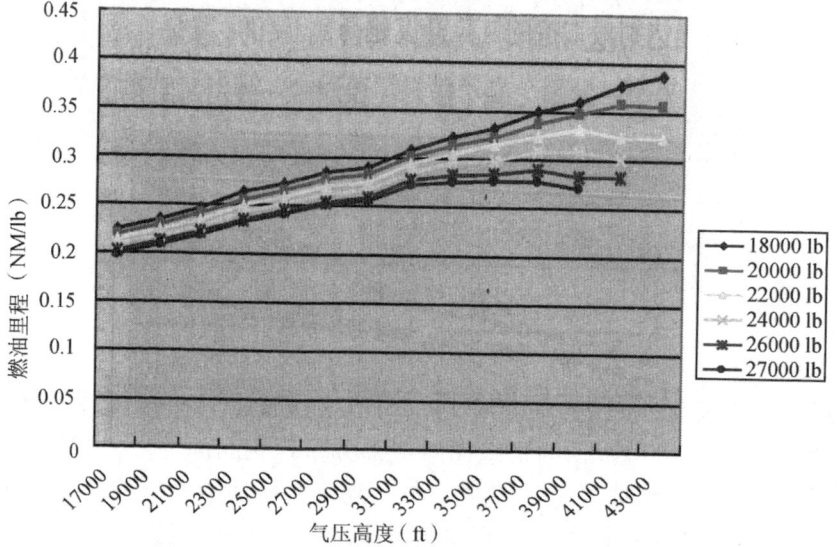

图 3-8-5　远程巡航燃油里程–气压高度分析曲线图

如图 3-8-5 所示,豪客 800XP 型公务机采用远程程序巡航:相同气压高度下,巡航重量越小,燃油里程越大。相同巡航重量时,从较低气压高度开始随着高度的增加燃油里程逐渐增加,其中在 31000 ft 时增速相对较小;在达到较高气压高度时燃油里程达到最大,随后减少。此外,巡航重量越大,燃油里程达到最大值的气压高度越高(其中 18000 lb 巡航重量在极限气压高度 43000 ft 时仍未出现最大值)。

(2)中速巡航(标准大气温度,见表 3-8-4 和图 3-8-6)

表 3-8-4　中速巡航不同重量高度燃油里程数据　　　　　　　　　(单位:NM/lb)

	18000 lb	20000 lb	22000 lb	24000 lb	26000 lb	27000 lb
17000 ft	0.206	0.203	0.200	0.196	0.192	0.191
19000 ft	0.219	0.213	0.209	0.205	0.201	0.199
21000 ft	0.225	0.222	0.218	0.214	0.210	0.208
23000 ft	0.235	0.232	0.228	0.214	0.220	0.218
25000 ft	0.244	0.241	0.237	0.233	0.228	0.226
27000 ft	0.253	0.249	0.245	0.240	0.235	0.235

续表

	18000 lb	20000 lb	22000 lb	24000 lb	26000 lb	27000 lb
29000 ft	0.259	0.256	0.256	0.246	0.240	0.238
31000 ft	0.263	0.259	0.259	0.246	0.239	0.235
33000 ft	0.282	0.276	0.268	0.259	0.250	0.245
35000 ft	0.300	0.291	0.281	0.271	0.259	0.252
37000 ft	0.319	0.308	0.295	0.280	0.261	0.252
39000 ft	0.346	0.323	0.305	0.278	0.262	0.259
41000 ft	0.350	0.329	0.300	0.289	0.281	—
43000 ft	0.357	0.326	0.318	—	—	—

如图 3-8-5 所示,豪客 800XP 型公务机以中速程序巡航时,随气压高度增加燃油里程总体上逐渐增加,当达到 31000 ft 气压高度时燃油里程减小(18000 lb 重量时增速减缓);随之燃油里程继续增加,到一定值时开始减少,随后又开始增加。不难发现,巡航重量越大,燃油里程达到最大值的气压高度越高。

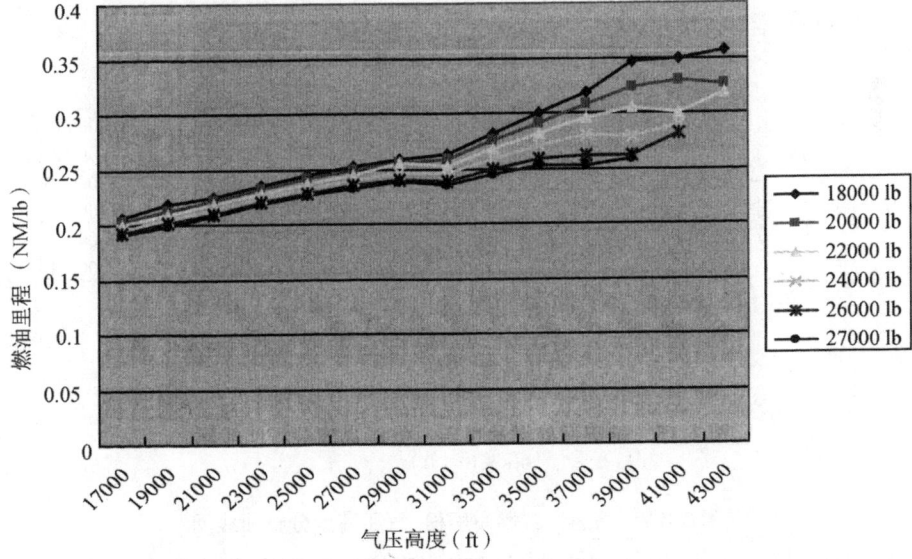

图 3-8-6　中速巡航燃油里程−气压高度分析曲线图

(3)高速巡航(标准大气温度,见表 3-8-5 和图 3-8-7)

表 3-8-5　高速巡航不同重量高度燃油里程数据　　　　　　　　　　　　　　　　　　(单位:NM/lb)

	18000 lb	20000 lb	22000 lb	24000 lb	26000 lb	27000 lb
17000 ft	0.181	0.180	0.180	0.178	0.192	0.191
19000 ft	0.190	0.189	0.187	0.186	0.201	0.199
21000 ft	0.198	0.197	0.195	0.194	0.210	0.208
23000 ft	0.206	0.204	0.203	0.201	0.220	0.218
25000 ft	0.212	0.210	0.208	0.206	0.228	0.226
27000 ft	0.211	0.209	0.207	0.204	0.235	0.233
29000 ft	0.200	0.197	0.194	0.194	0.236	0.238

续表

	18000 lb	20000 lb	22000 lb	24000 lb	26000 lb	27000 lb
31000 ft	0.194	0.207	0.205	0.205	0.237	0.233
33000 ft	0.225	0.218	0.217	0.216	0.215	0.215
35000 ft	0.239	0.230	0.230	0.228	0.228	0.227
37000 ft	0.249	0.246	0.243	0.243	0.242	0.241
39000 ft	0.272	0.270	0.268	0.266	0.262	0.259
41000 ft	0.300	0.298	0.295	0.289	—	—
43000 ft	0.311	0.326	0.319	—	—	—

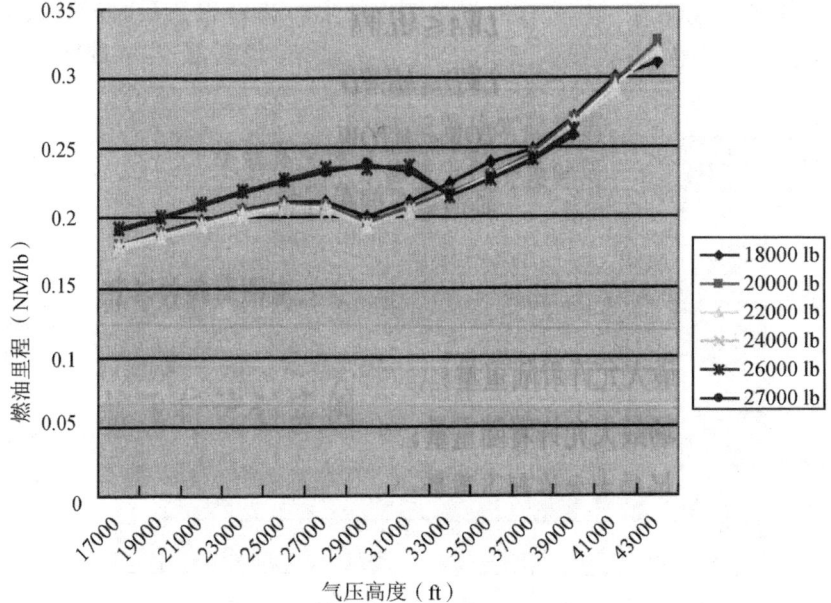

图 3-8-7 高速巡航燃油里程–气压高度分析曲线图

如图 3-8-7 所示,豪客 800XP 型公务机以高速程序巡航时随着气压高度的增加,燃油里程开始随之增加,达到一个峰值后开始下降,降到一个极小值后又转而增加。以 24000 lb 及更小重量巡航时在 25000 ft 气压高度时燃油里程达到峰值,在 29000 ft 气压高度时达到极小值;而以 26000 lb 及更大重量巡航时在 29000 ft 气压高度时燃油里程达到峰值,在 31000 ft 气压高度时达到极小值。

高速巡航时,在空中交通管制允许的前提下,应合理选择燃油里程较大的气压高度,并注意避开燃油里程达到极小值的气压高度;在燃油里程达到极小值气压高度附近,燃油里程相同或相当时,为减少爬升耗油,应选择较小的高度巡航。

3.8.4 公务机飞行计划的计算方法

普通的飞行计划制作方法有两种:(1)由备降机场停机坪开始往回推算,适用于已知实际业载和较短的航线飞行情况;(2)由最大允许起飞重量往后算,适用于长航线飞行和由最大允许起飞重量计算最大业载的情况。

公务飞行在我国被划归为加班、不定期飞行,每次都要在飞行前 3 个工作日向所在地的空管部门申请,通过审批以后才能执行,所以每次飞行都要提前申请计划。公务飞行以飞行小时数计费,飞行的客货重量基本可以确定不存在更大业载能带来更大经济效益的情况。公务机的最大业载小,而且基本确定,飞行前都限定了最大旅客数和允许携带行李重量,如果计算业载和燃油后的起降重量超过最大允许重量,则不宜通过减少旅客和行李的途径解决。

一些公务机(如豪客 800XP 型)制造厂商的飞行手册给出的燃油装载计划实例也只有从备降机场停机坪往回推算一种方法。综合分析可知,公务飞行计划适用于由备降机场停机坪开始往回推算的方法。

如果知道实际业载重量 PL(旅客和货物重),则在备降机场停机坪无油重量(ZFW)= 营运空重(OEW)+业载重量(PL),在备降机场停机坪飞机重量 $W=ZFW+$公司备份油量,然后由此重量开始往回推算,加上各阶段消耗油量,一直算到起飞机场停机坪。在计算中还应保证:

$$ZFW \leqslant MZFW$$
$$LWA \leqslant MLWA$$
$$LWD \leqslant MLWD$$
$$TOW \leqslant MTOW$$
$$RPF \leqslant FTC$$

式中:$MLWA$——备降机场最大允许着陆重量;

$\quad MLWD$——目的地机场最大允许着陆重量;

$\quad MTOW$——起飞机场最大允许起飞重量;

$\quad LWA$、LWD、TOW——制作飞行计划中计算出的在备降机场、目的地机场的着陆重量和起飞机场的起飞重量。

如有一个条件不满足,应减少业载重新计算,直到满足条件为止,通过计算结果就得到了所允许的业载和起飞总油量等数据。

3.9 利用二次放行的飞行计划

本节只考虑涡轮动力发动机的飞机,按照 CCAR-121 的规定,航班的起飞总油量由以下六部分组成:

(1)滑行油量。

(2)航程油量:从起飞机场到目的地机场的油量(包括进近、着陆用油)。

(3)不可预期燃油:为补偿不可预见因素,按计划航程燃油的 10% 计算的所需燃油,但在任何情况下不得低于以等待速度在目的地机场上空 450 m(1500 ft)高度在标准大气条件下飞行15 min 所需的油量。

(4)备降油量:从目的地机场到备降机场所需的油量。

(5)最后储备燃油:有目的地备降机场时按预计到达备降机场重量,不需要目的地备降机场时按到达目的地机场的预计着陆重量,以等待速度在相应机场上空 450 m(1500 ft)高度在标准大气条件下飞行 30 min 所需的油量。

(6)酌情携带的燃油:合格证持有人决定携带的附加燃油。

上述第（3）项不可预期燃油主要是为应对领航误差、航路天气预报误差和空中交通延误等情况而加的，而随着领航技术、天气预报技术、空中管制设备及技术的改进，这一部分燃油被消耗的可能性大大减少。因此，远程飞行的飞机在到达目的地机场后会剩下大量燃油。

按航程油量10%计算的不可预期燃油（或叫作航线应急油）的数量不小。例如，B747-400由洛杉矶飞往北京，航程时间约12 h，航程油量大约120 t，不可预期燃油约12 t；B787-8由北京飞往洛杉矶，航程时间约11 h，航程油量大约50 t，不可预期燃油约5 t。这部分燃油如能尽量少加，则可大大增加业载，每吨业载相当于10名旅客，其收益是相当可观的；如果没有旅客或货物可增加，那么飞机重量可减少，巡航所耗油量可减少，同样也能提高经济效益。

那么能不能减少不可预期燃油（航线应急油），而又不违反CCAR-121的规定确保飞行安全呢？答案是肯定的，利用二次放行（Redispatch，简称二放）就能做到这一点。航行中的放行（即二次放行）一方面要遵守有关飞行放行的规定，以确保飞行安全；另一方面又能减少到达目的地机场时过多的剩余燃油，从而提高经济性。飞机越大，航线越长，不可预期燃油越多，利用二次放行所能获得的效益就越大。因此，国际航线普遍采用二次放行。

注：按照《航空承运人不可预期燃油政策优化与实施指南》，经局方批准，航空公司可选择按小于10%的系数（即按航程油量×X，X<10%）计算不可预期燃油，但不得小于3%，按4%、3%计算不可预期燃油时，不能利用二次放行。

3.9.1　二次放行的依据和基本思想

3.9.1.1　二次放行的法规依据
第121.651条　初始签派或者放行、重新或者更改签派或者放行

（a）合格证持有人可以指定任一经批准用于该型飞机的正常使用机场、临时使用机场或者加油机场，作为初始签派或者放行的目的地机场。

（c）飞机在航路上飞行时，任何人不得擅自更改初始签派或者放行单上指定的初始目的地机场或者备降机场。如果确有必要改变为另外的机场时，则该机场应当是经批准用于该型飞机的，并且在重新签派或者更改签派或者放行单时，应当符合本规则第121.173条和第121.621条至第121.675条的相应要求。

第121.657条　燃油量要求

（d）除非机上可使用的燃油按照要求符合本条（b）款除滑行燃油以外的要求，否则不得从飞行中重新签派点继续飞往目的地机场。

第121.651条（a）项表明在初始签派放行中指定的机场可以不是最终的目的地机场，而可以是运行规范中获批的正常、临时使用或者加油机场。这是我们在二次放行中选择初始目的地机场的依据所在。

第121.651条（c）项明确了签派放行不一定在地面进行，空中同样也可以进行更改初始目的地或备降机场的第二次签派放行。

第121.651条（a）、（c）这两个条款为正常情况下二次放行的全过程提供了法规依据。

第121.657条对飞行中重新签派放行的油量要求做了相关规定，对二次放行的加油量提供了法律依据。

3.9.1.2　二次放行的基本思想

设起飞机场为 A,目的地机场为 B,备降机场为 E,在目的地机场 B 之前选择一个机场 C(该机型能起降的机场)作为由 A 起飞的最初目的地机场,即初次放行目的地机场,对于机场 C 选择一个备降机场 D,见图 3-9-1。

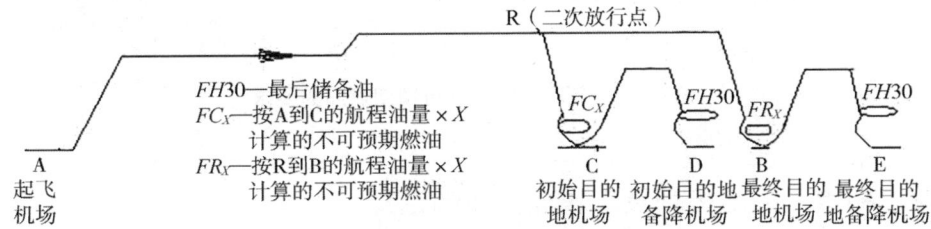

图 3-9-1　二次放行飞行计划的飞行剖面

在起飞机场 A 起飞时,起飞总油量按从 A 飞往 C 备降 D 计算,此油量应符合 CCAR-121 的油量规定,其中包括了由 A 到 C 的航程油量$\times X$($5\% \leqslant X \leqslant 10\%$)的不可预期燃油 FC_X。如果在 A 到 C 的航路中 FC_X 没有被消耗,在去往 C 的下降点或稍前的一点 R 检查机载剩余油量,并计算由 R 飞到 B 的备降机场 E 的所需总油量 F_{RE}:

$$F_{RE} = F_{RB} + X \times F_{RB} + F_{BE} + HODF + COF$$

式中:F_{RB}——R 到 B 的航程油量;

$\quad X \times F_{RB}$——不可预期燃油,不小于在 B 上空 1500 ft 标准大气条件下等待 15 min 的油量;

$\quad F_{BE}$——B 到 E 的备降油量;

$\quad HODF$——E 机场最后储备油量(等待燃油量);

$\quad COF$——酌情携带的燃油。

如果机载所剩油量大于 F_{RE},符合 CCAR-121 的规定,飞机可以不在 C 着陆,而是在 R 点再次签派放行到最终目的地机场 B。

如果机载所剩油量小于 F_{RE},则飞机必须在 C 着陆,补充燃油后再从 C 飞到最终目的地机场 B。R 点称为二次放行点。在二次放行点飞行员要与机场 C 进行地面通信联系,通知地面你的决定。由于 A 到 C 的距离小于 A 到 B 的距离(设 B 和 C 到各自备降机场距离相同),A 到 C 的航程油量和相应的不可预期燃油要少,所以采用二次放行的方法起飞油量可以减少,这就可以增加业载或减少起飞重量,至于能增加多少业载和经济效益则取决于 C 及 R 的位置,也与距离两个备降机场的远近等有关,这将在后面讨论。另外,机场 C 与 D 都必须是该机型能起降的机场,因为有可能会降落在这两个机场。

由上所述,二次放行的中心思想是利用一般不会被消耗的不可预期燃油作为由二次放行点到目的地机场的所需油量。按照 CCAR-121 的燃油政策,二次放行的方法既适用于国际航线,也适用于国内航线。

如图 3-9-2 所示,若航程比较短或者确定不可预期燃油(航程油量$\times X$)的系数 X 比较小,以致航程油量$\times X \leqslant F15$,其中 $F15$ 为以等待速度在目的地机场上空 1500 ft 高度在 ISA 下飞行 15 min 的燃油量,这时不可能利用二次放行。因为即使不可预期燃油完全未被消耗,在 R 点剩余油量也是不够的,如要加额外油量使得在 R 点能二次放行,所需增加的油量至少是 R 到 P 的油量(假设 B、C 到各自备降机场的距离相同),而这样得到的总油量就是不用二次放行 A 到 B,备降机场 E 的油量了。故利用二次放行的前提是基于不可预期燃油为 X 的航程燃油($4\% < X \leqslant 10\%$)。

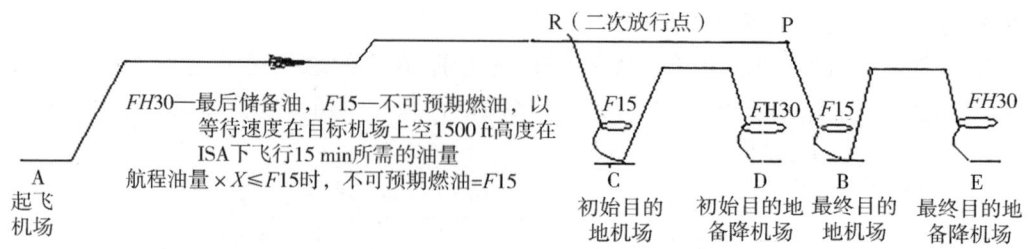

图 3-9-2　二次放行对不可预期燃油的要求

3.9.2　初始目的地机场及二次放行点的选择

不同的初始目的地机场和二次放行点的选择,其最终的效果是不同的。下面分几种情况讨论,不考虑滑行油量和酌情携带的燃油(它们对所讨论的问题没有影响)。

3.9.2.1　初始目的地机场在到最终目的地机场的航线上且它们到备降机场的距离相同

(1)初始目的地机场和二次放行点的选择

如图 3-9-3 所示,C 机场位于 A 机场和 B 机场之间,C 机场到其备降机场 D 与 B 机场到其备降机场 E 的距离相同。

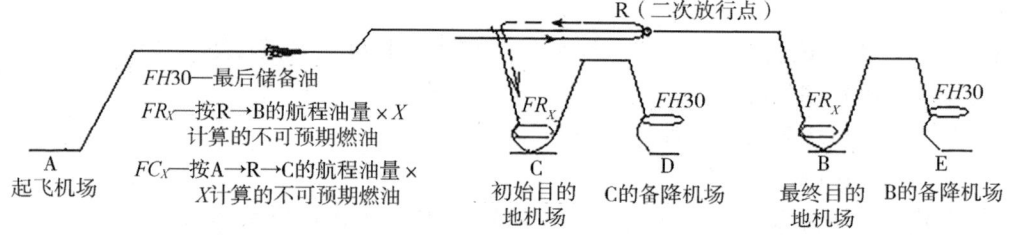

图 3-9-3　初始目的地机场在到最终目的地机场的航线上

用 $TOF1$ 表示由 A 机场起飞到 C 机场备降 D 机场的总油量,即初次放行应加的油量:

$$TOF1 = \begin{cases} \text{A 至 C 的航程油量 } F_{AC} \\ \text{A 至 C 的航程油量} \times X \text{ 的油量 } XF_{AC}(\text{即 A 至 C 的不可预期燃油}) \\ \text{C 至 D 的改航油量(备降油量)} F_{CD} \\ \text{在 D 等待 30 min 的油量(最后储备油量)} HODF_{CD} \end{cases}$$

用 $TOF2$ 表示采用二次放行方法由 A 机场起飞到最终目的地机场 B 备降 E 机场实际消耗的总油量:

$$TOF2 = \begin{cases} \text{A 至 B 的航程油量 } F_{AB} \\ \text{R 至 B 的航程油量} \times X \text{ 的油量 } XF_{RB}(\text{即 R 至 B 的不可预期燃油}) \\ \text{B 至 E 的改航油量(备降油量)} F_{BE} \\ \text{在 E 等待 30 min 的油量(最后储备油量)} HODF_{BE} \end{cases}$$

当初始目的地 C 和最终目的地 B 到它们各自的备降机场 D、E 的距离相同时,近似假设它们的备降油量和最后储备油量相同,即 $TOF1$ 和 $TOF2$ 中第 3、4 项分别相等。

如果以到初始目的地机场 C 的下降点作为二次放行点 R,则 C 一旦确定,R 也被确定。若 C 机场选择得适当,按照 $TOF1$ 加油从 A 机场起飞后,假设 A 至 R 的油量的 X 这部分应急油量未被消耗,则在 R 点剩油 $F1$ 为:

$$F1 = \begin{cases} R \text{ 至 C 的航程油量 } F_{RC} \\ A \text{ 至 C 的航程油量} \times X \text{ 的油量 } XF_{AC}(\text{即 A 至 C 的不可预期燃油}) \\ C \text{ 至 D 的改航油量(备降油量)} F_{CD} \\ \text{在 D 等待 30 min 的油量(最后储备油量)} HODF_{CD} \end{cases}$$

若从二次放行点 R 重新签派放行到 B 机场备降 E 机场,按照 CCAR-121 的规章要求油量为 F2:

$$F2 = \begin{cases} R \text{ 至 B 的航程油量 } F_{RB} \\ R \text{ 至 B 的航程油量} \times X \text{ 的油量 } XF_{RB}(\text{即 A 至 C 的不可预期燃油}) \\ B \text{ 至 E 的改航油量(备降油量)} F_{BE} \\ \text{在 E 等待 30 min 的油量(最后储备油量)} HODF_{BE} \end{cases}$$

若 $F1 \geq F2$,则可以在 R 点重新签派放行到 B 机场备降 E 机场。考虑 $F1 = F2$ 的情况,因为 $F_{CD} = F_{BE}$, $HODF_{CD} = HODF_{BE}$,则:

$$F_{RC} + XF_{AC} + F_{CD} + HODF_{CD} = F_{RB} + XF_{RB} + F_{BE} + HODF_{BE}$$

也可以转换成:

$$F_{AR} + F_{RC} + XF_{AC} + F_{CD} + HODF_{CD} = F_{AR} + F_{RB} + XF_{RB} + F_{BE} + HODF_{BE}$$

也就是:

$$F_{AC} + XF_{AC} + F_{CD} + HODF_{CD} = F_{AB} + XF_{RB} + F_{BE} + HODF_{BE}$$

即

$$TOF1 = TOF2$$

因此,二次放行时,C 机场的位置满足 $TOF1 = TOF2$,即 C 机场应该按 $TOF1 = TOF2$ 来选择。

(2)最佳目的地机场和最佳二次放行点

对于给定的 A、B、E 三个机场:

①$TOF2$ 仅取决于二次放行点 R 的位置;

②当二次放行点 R 在 A 与 C 之间某点时,$TOF1$ 仅取决于 C 的位置;

③若 R 在飞过 C 之后的 CB 段之间(见图 3-9-3),则 $TOF1$ 取决于 C 和 R 的位置。

现在首先考虑 R 在 A 与 C 之间且取为到 C 的下降点的情况,如图 3-9-4 所示。

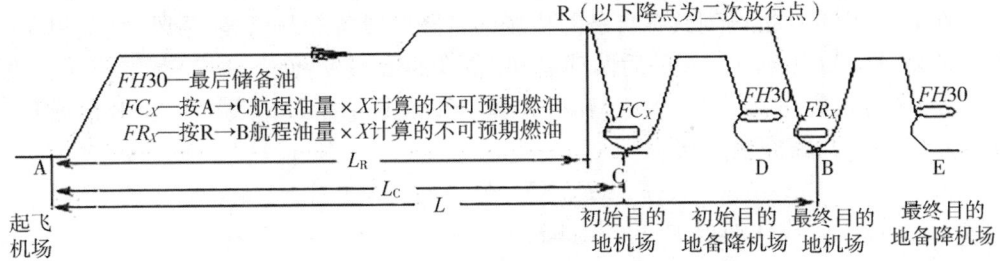

图 3-9-4　二次放行点位于 C 机场下降点的情况

设 L 为 A 至 B 的航程,L_C 为 A 至 C 的航程,L_R 为 A 至 R 的航程。一般下降段为 $(2\% \sim 2.5\%)L$。设不使用二次放行时由 A 至 B,备降机场 E 的起飞总油量为 TOF,其中不可预期燃油(即 A 至 B 的航程油量 $\times X$ 的油量)为 F_X。

$$TOF = \begin{cases} \text{A 至 B 的航程油量 } F_{AB} \\ \text{A 至 B 的航程油量} \times X \text{ 的油量 } XF_{AB}(\text{即 A 至 C 的不可预期燃油}) \\ \text{B 至 E 的改航油量(备降油量)} F_{BE} \\ \text{在 E 等待 30 min 的油量(最后储备油量)} HODF_{BE} \end{cases}$$

下面分析当二次放行点 R 在不同位置时,初始放行油量($TOF1$)与二次放行($TOF2$)的变化情况:

R 点对 $TOF2$ 的影响:

①$L_R = 0$,即不使用二次放行时,此时 $TOF2 = TOF$。

②随 L_R 的增加,XF_{RB} 逐渐减小,其余油量不变,$TOF2$ 逐渐减小。

③当 $L_R \approx L$,即完全不加不可预期燃油(即航线应急油)时,$TOF2 = TOF - F_X$。

R 点对 $TOF1$ 的影响:

①$L_R = 0$,$L_C \approx 0$,此时 $F_{AC} \approx 0$,$XF_{AC} \approx 0$。$TOF1$ 只剩下 $HODF_{CD}$ 和 F_{CD}。

②随 L_R 增加,L_C 也增加(R 是到 C 的下降点),使 $TOF1$ 增加。

③当 $L_C = L$ 时,$TOF1 = TOF$。

$TOF1$ 和 $TOF2$ 随 R 点(C 机场)位置的变化趋势如图 3-9-5 所示。

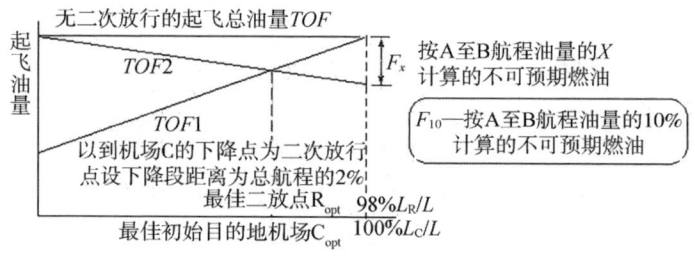

图 3-9-5 $TOF1$ 和 $TOF2$ 随 C 机场(R 点)变化的情况

如图 3-9-5 所示,当初始目的地机场 C 选得靠近 A 时,$TOF1 < TOF2$,为使在二次放行点 R 能放行到最终目的地机场 B,需要按 $TOF2$ 加油;当初始目的地机场 C 选得靠近 B 时,$TOF1 > TOF2$,要按 TOF1 加油。可见,C 选得靠近 A 或靠近 B 都不好。当初始目的地机场 C 的位置选在使 $TOF1 = TOF2$ 的位置时,所需的起飞总油量最小,能增加的业载 ΔPL 最多。

初始目的地机场 C 的最佳位置可按以下方法近似确定:

假设在 C 机场和 D 机场到其各自备降机场的备降航段备降距离、航路风、气温相同[这将使其备份油量,即 $TOF1$ 和 $TOF2$ 的后两项之和(备降油量+最后储备油量)基本相同],飞行时间、耗油量与飞行距离均成正比,令 F_{AC} 表示 A 至 C 的航程油量,F_{CB} 表示 C 至 B 的航程油量。为使 $TOF1 = TOF2$,只需满足:

$$F_{AC} + F_{AC} \times X = F_{AB} + F_{RB} \times X$$

改用距离表示,因下降段耗油很少,可以忽略,上式可以写成:

$$L_{AC} \times (1 + X) = L + L_{CB} \times X$$

经过整理得:

$$L_{AC} \times (1 + X) = L + (L - L_{AC}) \times X$$
$$L_{AC} \times (1 + 2 \times X) = L \times (1 + X)$$

则 C 机场的最佳位置:

$$L_{AC} = L \times (1 + X) / (1 + 2 \times X)$$

一般下降点距机场距离一般在 120 NM 左右,对于远程航线(如 L_{AC} 在 5000 NM 左右),下降点距机场约为 $2.5\%L_{AC}$,对于较短的航线(例如 2000 NM 左右的国内航线),下降点距机场约为 $5\%L_{AC}$。按照上面得到的 C 机场最佳位置公式进行计算,初始目的地机场 C 和二次放行点(下降点)R 的近似最佳位置见表 3-9-1。

表 3-9-1 初始目的地机场和二次放行点的近似最佳位置

X	$\xi = L_{AC}/L$	L_R/L(下降段 = $2.5\%L_{AC}$)	L_R/L(下降段 = $5\%L_{AC}$)
10%	91.6%	89.1%	86.6%
9%	92.4%	89.9%	87.4%
8%	93.1%	90.6%	88.1%
7%	93.9%	91.4%	88.9%
6%	94.6%	92.1%	89.6%
5%	95.5%	93.0%	90.5%
4%	96.3%	93.8%	91.3%
3%	97.2%	94.7%	92.2%

按照中国民用航空局飞行标准司的《航空承运人不可预期燃油政策优化与实施指南》(AC-121-FS-136)的要求,按照 4%、3% 计算不可预期燃油时,不能利用二次放行。从表 3-9-1 可以看出,按照 4%、3% 计算不可预期燃油(此油量本就不多)时,初始目的地机场很接近最终目的地机场,采用二次放行获益不大。

根据 CCAR-121 的规定,如果按 10% 航程油量计算不可预期燃油,初始目的地机场 C 在距起飞机场 A 的 91.6% 的总航程处,或者说二次放行点 R(下降点)在距起飞机场 A 约 89% 的总航程处,使 $TOF1 = TOF2$,此时起飞总油量最少,业载增加得最多。此时能增加的业载大约是不使用二次放行时 10% 航程油量(不可预期燃油)的 86%。这一点可以解释如下:从上面的分析及图 3-9-4 看,采用二次放行可以少加 91% 的不可预期燃油。但如果把少加的 91% 的不可预期燃油全部改为业载,则二次放行时在主航段及备降航段都要多耗油,多耗的油量等于业载增量 $\Delta PL \times 5\%$,或更多(这取决于备降距离的远近,备降距离越远,多耗的燃油越多)。因此,可多加的业载 ΔPL 应使 $\Delta PL \times 105\% = 91\% \times F_{10}$,即最多可增加业载:

$$\Delta PL = 86.6\% \times F_{10} = 8.66\% \times TRF_{AC}$$

式中:TRF_{AC} 为 A 至 C 的航程油量;

F_{10} 为 A 至 C 的不可预期燃油,$F_{10} = 10\%TRF$。

当二次放行点 R 距离大于或小于 $89\%L$ 时,ΔPL 都要减小,其变化如图 3-9-6 所示。

如果按 5% 航程油量计算不可预期燃油,初始目的地机场 C 在距起飞机场 A 为 95.5% 总航程处,或者说二次放行点 R(下降点)在距起飞机场 A 约 93% 总航程处,使 $TOF1 = TOF2$,起飞油量最少,业载增加得最多。此时采用二次放行可少加约 95% 的不可预期燃油 F_5,则可以增加业载,增加业载会导致多耗油。多耗的油量 = 业载增量 $\Delta PL \times 5\%$,由 $\Delta PL \times 105\% = 95\% \times F_5$,可知最多可增加业载:

$$\Delta PL \approx 90\% \times F_5 = 4.5\%TRF$$

当二次放行点 R 距离大于或小于 $93\%L$ 时,ΔPL 都要减小。

上述结论是由 CCAR-121 的油量规定决定的,与机型无关。但它只对业载受最大允许起

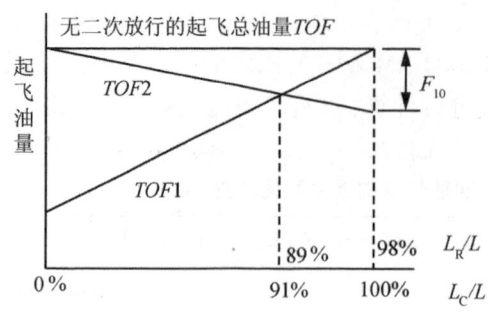 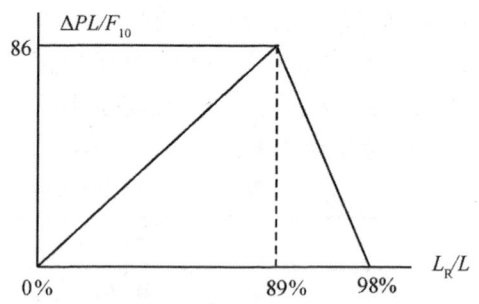

图 3-9-6　二次放行点 (C 机场位置) 与业载增量之间的关系

飞重量限制、下降段为 2.5% 总航程、多耗的油量 = 业载增量 ×5% 的情况适用,对受最大允许着陆重量限制以及不用二次放行时业载受油箱容量限制的情况是不适用的。

　　如果航线并非一条直线,而是如图 3-9-7 所示的折线,当 C 在 A 到 B 的航路上且 C 和 B 的备降距离相同时,上述结论同样成立。实际情况可能没有一个初始目的地机场恰好在上述最佳位置,往往有几个初始目的地机场可供选择,这就产生两个问题:一是对每个机场来说二次放行点应选在何处,二是哪个机场最好 (即能使增加的业载最多)。

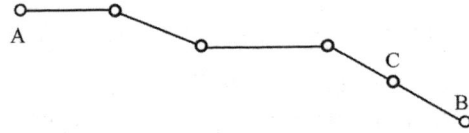

图 3-9-7　二次放行航线

3.9.2.2　初始目的地机场 C 在 A 至 B 的航路上,但距机场 A 太近 (假设备降距离相同)

　　如图 3-9-8 所示,机场 C 位于机场 A 和机场 B 之间,但机场 C 比较靠近起飞机场 A。机场 C 到其备降机场 D 与机场 B 到其备降机场 E 的距离相同。

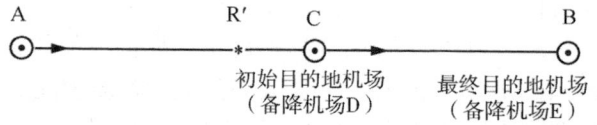

图 3-9-8　初始目的地机场在主航路上但距起飞机场太近

　　在机场 C 距机场 A 太近时,如果仍用机场 C 的下降点 R' 作为二次放行点,则 $TOF1 < TOF2$,如图 3-9-9 所示,为满足二次放行要求,初始放行就要按 $TOF2$ 来加油。$TOF2 = A$ 至 B 的油量 + R' 至 B 的油量的 X (不可预期燃油) + B 至 E 的改航油量 + 在 E 的等待油量。$TOF2$ 比 $TOF1$ 大得多,这样一来在机场 C 的着陆重量可能超过最大允许着陆重量,同时能增加的业载也少得多,如图 3-9-9 所示。

　　对于这种情况,就不能再采用机场 C 的下降点作为二次放行点了,应该在初始目的地机场 C 之后的某一个位置选择二次放行点 R,如图 3-9-10 所示,即应按 A→C→R→C 来制作初次放行计划,算出起飞油量 $TOF1$。二次放行点 R 的位置应选在:

$$L_{AR} + L_{RC} = \xi \times L$$

即

$$L_{AC} + L_{CR} + L_{RC} = \xi \times L$$

式中:ξ 为表 3-9-1 的 L_{AC}/L。

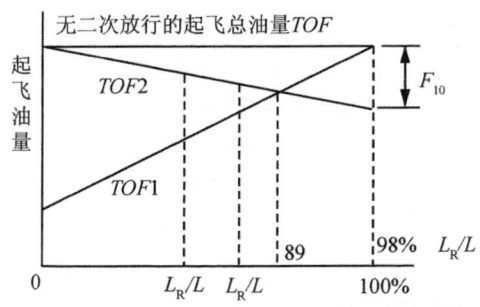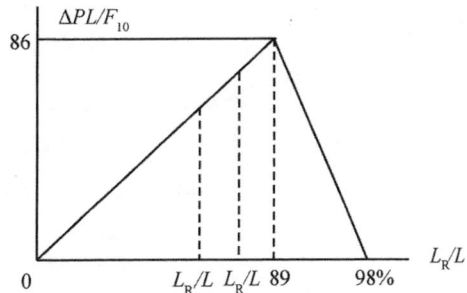

图 3-9-9　初始目标基础机场靠近起飞机场

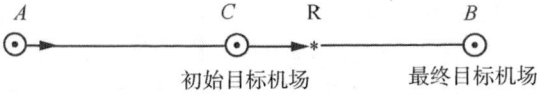

图 3-9-10　二次放行点选择在初始目的地机场之后

这样算出的油量与按 A→C→R→C′ 算出的油量是一样的,如图 3-9-11 所示。这时由于 L_R/L 小于表 3-9-1 中的值,故 $TOF2>TOF1$,$TOF2=$ A 至 B 的航程油量+R 到 B 的航程油量的 X+B 到备降机场 E 的备降油量+备降机场 E 的等待油量(最后储备油量)。为满足二次放行要求,初始放行要按 $TOF2$ 加油,这时能增加的业载按 L_R/L,参考图 3-9-9 确定。

```
A              C     R    C′    B
⊙━━━▶       ⊙  ━▶ *  ◯    ⊙
```

图 3-9-11　二次放行点选择在初始目的地机场之后的最佳位置

3.9.2.3　最初目的地机场 C 不在 A 至 B 的航路上(假设备降距离相同)

如图 3-9-12 所示,机场 C 位于机场 A 和机场 B 之间但偏出了航路,机场 C 到其备降机场 D 与机场 B 到其备降机场 E 的距离相同。

```
                        ⊙C
                          ╲
A                          R          B
⊙━━━━━━━━━━━━━━━★━━━━━━⊙
```

图 3-9-12　二次放行点不在 A 至 B 的航路上

同第二种情况类似,按 A→R→C 制作初次放行计划,计算 $TOF1$,R 应选择在:

$$L_{AR}+L_{RC}=\xi\times L$$

式中:ξ 与计算不可预期燃油的 X 有关,即表 3-9-1 第 2 列的 L_{AC}/L。

按 L_{AR}/L 查图 3-9-6 确定可增加的业载。

为确定第一、第二和第三种情况的二次放行点位置,参照图 3-9-13,可用下面的公式计算二次放行点到起飞机场的距离:

$$\sqrt{(D-B)^2+C^2}+D=\xi\times A$$

则二次放行点到起飞机场距离公式:

$$D=\frac{(\xi A)^2-B^2-C^2}{2\times(\xi A-B)}$$

按 $X=10\%$ 计算不可预期燃油,$\xi=91.6\%$,按 5% 计算不可预期燃油,$\xi=95.45\%\approx95.5\%$。当 $A=10000$ km,$B=7000$ km 和 8000 km,$C=500$ km 时,计算结果如表 3-9-2 所示。

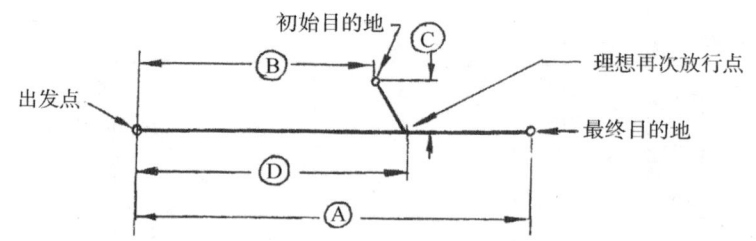

图 3-9-13　二次放行点位置计算

表 3-9-2　二次放行点的计算结果

B	A = 10000 km　　C = 500 km			
	X = 10%		X = 5%	
	ξ = 91.6%	ξ = 91%	ξ = 95.5%	ξ = 95%
7000 km	D = 8022 km	D = 7990 km	D = 8226 km	D = 8200 km
8000 km	D = 8472 km	D = 8436 km	D = 8694 km	D = 8667 km

实际航线不是直线,由 C 向 AB 作垂线的办法也许行不通(见图 3-9-14)。可以估计一个初始目的地机场偏离主航路的距离 C 和初始目的地机场沿主航路到起飞机场的距离 B,然后按二次放行点的距离公式计算二次放行点 R 沿航路到起飞机场的距离 $L_{AR} = D$,按此距离可确定二次放行点 R 在主航路的位置。由 L_{AR}/L 查业载增量图可确定能增加的业载 ΔPL。由 R 至 C 的航线应是 ATC 允许的航路,未必是直线。

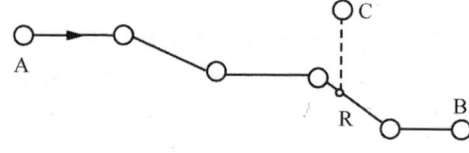

图 3-9-14　确定二次放行点的特殊情况

3.9.2.4　初始目的地机场太接近最终目的地机场(备降距离相同)

如图 3-9-15 所示,机场 C 位于机场 A 和机场 B 之间,但机场 C 比较靠近最终目的地机场 B。机场 C 到其备降机场 D 与机场 B 到其备降机场 E 的距离相同。

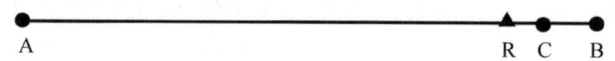

图 3-9-15　初始目的地机场太接近最终目的地机场

对于这种情况,可把到初始目的地机场的下降点作为二次放行点 R,按 L_R/L 查业载增量图确定业载增量 ΔPL。这时初次放行的起飞油量 $TOF1 > TOF2$,当二次放行点剩余油量大于二次放行所需时,把二次放行点 R 向 A 移动,虽可能使在 R 点剩油等于二次放行所需,但不能更多地增加业载,即使把二次放行点 R 取在最佳处也无用。

当航路上有若干个机场可作为初始目的地机场时,应对每个机场确定其二次放行点位置(见图 3-9-16),算出相应的业载增量 ΔPL,与最大 ΔPL 相应的那个机场即最好的初始目的地机场,对应的二次放行点即最佳二次放行点。

3.9.2.5　备降距离对二次放行点位置的影响

前面四种情况讨论的前提是初始目的地机场和最终目的地机场到其各自的备降距离相

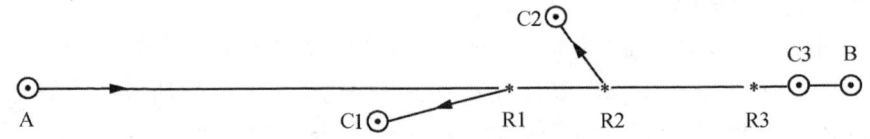

图 3-9-16　最佳初始目的地机场的确定

等、备份油量相同,但实际上往往备降距离不等。如图 3-9-17 所示,设初始目的地机场到其备降机场距离为 L_{CD},最终目的地机场到其备降机场距离为 L_{BE}。若 $L_{BE} > L_{CD}$,按常规方法确定的二次放行点 R 的余油就不够了,这时应把二次放行点 R 向最终目的地机场 B 移动大约 $\frac{1}{3}(L_{BE}-L_{CD})$,到 R′点;若 $L_{BE} < L_{CD}$,按常规方法确定的二次放行点 R 的余油过多,二次放行点 R 可以向 A 点方向移动约 $\frac{1}{3}(L_{CD}-L_{BE})$。因为通过这样移动二次放行点的位置,初次放行机场的航程大约变化 $\frac{2}{3}|L_{BE}-L_{CD}|$,再加上巡航 $\frac{2}{3}|L_{BE}-L_{CD}|$ 的油量的 X 油量,以及 TOF2 中由 R′到 B 的不可预期燃油的变化,差不多可以抵消备降距离不等的影响。

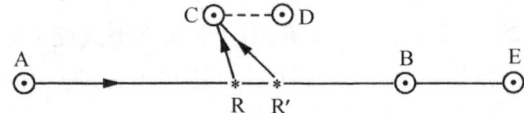

图 3-9-17　备降距离对二次放行点的影响

3.9.2.6　风对二次放行点位置的影响

在某些情况下,风对初始目的地机场机二次放行点的选择是有影响的。如果航路风在整个航程中大小不变,且二次放行点在初始目的地机场之前,则风对初始目的地机场及二次放行点的选择没什么影响。然而,实际上航路风沿航路是变化的,顶风相当于使航段变长,顺风相当于使航段变短,风会影响初始目的地机场和二次放行点的选择。例如,两个备降航段的距离相等,但风分量方向相反,假设最终目的地机场到其备降机场为顶风,那么初始目的地机场或二次放行点的位置应该从常规方法得出的理想位置再向最终目的地机场靠近一些,也可以估算出两个备降航段的空中距离,按照前文备降距离对二次放行点的影响的公式修正二次放行点的位置。

对于二次放行点在初始目的地机场之前、在二次放行点前后的航路风分量变化比较大的情况,在选择初始目的地机场和二次放行点时也应该考虑对风的影响做一些修正。例如,在接近最终目的地机场的航段上如果顶风风量变大,初始目的地机场和二次放行点应更靠近最终目的地机场一些,可以从制作出的飞行计划中给出的额外油量(EXTRA)来判断位置是否合适,或从二次放行点剩余的油量和二次放行所需油量的差值(未加额外油量之前)来判断。

3.9.3　二次放行飞行计划的制作方法

下面是编制计算机飞行计划软件中所用的方法,手算时可参照执行。假设初始目的地机场及二次放行点已选好:

（1）制作由起飞机场到最终目的地机场的飞行计划

以给定的业载（或最大业载）从最终目的地机场的备降机场往回计算（具体算法见第 2 章内容），这时不可预期燃油按由二次放行点到最终目的地机场的航程油量的 X 计算，同时要与在目的地机场上空 1500 ft 标准大气等待 15 min 的油量相比取大者。记下在二次放行点的剩余油量 RF，这是为二次放行所必需的油量。计算结束后得到的允许业载可能由于受最大允许起飞重量限制、最大允许着陆重量限制或最大油箱容量限制而减少。

（2）制作由起飞机场到初始目的地机场的飞行计划

以第（1）步得到的业载从初始目的地机场的备降机场开始往回计算，制作由起飞机场到初始目的地机场的飞行计划，这次的不可预期燃油按由起飞机场到初始目的地机场的航程油量的 X 计算（与在初始目的地机场上空 1500 ft 标准大气等待 15 min 的油量相比取大者）。这次计算要保证在二次放行点所剩油量 $RF^* \geq RF$，如这次计算得到的在二次放行点剩余油量 $RF^* < RF$，则把所差部分的 0.94 倍（$0.94 \times |RF - RF^*|$）加到 ZFW 上，从初始目的地机场的备降机场再次往回计算，再检查在二次放行点的剩余油量 RF^* 是否等于 RF，如仍然不等，再重复这个过程直到 $RF^* = RF$ 为止（实际计算按两者之差少于 10 lb 来控制迭代次数）。这个计算中业载也可能因受最大允许起飞重量限制、最大允许着陆重量限制或最大油箱容量限制而减少。

（3）检查第（1）步和第（2）步算出的业载是否相等

①如果第（2）步算出的业载和第（1）步算出的业载相等或相差不超过 10 lb，则结束计算，开始显示、输出飞行计划。输出的初次放行计划中的起飞总油量和允许业载就是这次航班应加的总油量和可携带的业载，飞行计划中的额外油量如大于 0，则表示初次放行所需油量少，在二次放行点剩余油量少于为二次放行所需油量，为满足二次放行需求则需要增加初次放行的油量，而额外油量的数值是为使在二次放行点剩油等于二次放行所需油量而多加的油量在到达初始目的地机场的备降机场后剩余部分。如额外油量为 0，那么在二次放行点剩余油量可能多于或刚好等于所需油量。在输出的初次放行计划中的二次放行点那一行可以查出剩余油量，在输出的二次放行计划中可查到二次放行所需油量，如在二次放行点剩余油量多于所需油量，则表明初始目的地离起飞机场太远，或其备降机场太远（把二次放行点向起飞机场移动并不能减少起飞油量，只能早些判断能否飞往最终目的地）。

②如果第（2）步算出的业载 PL_2 小于第（1）步算出的业载 PL_1：

若额外油量为 0，则以业载 PL_2 重复第（1）步的计算［可以以第（2）步的起飞油量为规定的起飞油量进行计算］。这次算出的业载与开始计算时的业载必定相同，而二次放行所需油量减少。对于以业载 PL_2 开始计算的情况，计算结束，输出飞行计划。

若额外油量大于 0，则以业载 $PL_4 = PL_1 - 0.85 \times (PL_1 - PL_2)$ 重复第（1）步的计算，再以业载 PL_4 重复第（2）步的计算。这次算出的业载 PL_5 有可能小于 PL_4，如果 $|PL_4 - PL_5| \leq 10$ lb，则停止计算，开始显示、输出飞行计划。否则，以业载 $PL_6 = PL_4 - 0.85 \times (PL_4 - PL_5)$ 分别重复第（1）步和第（2）步的计算，直到两次算出的业载之差小于 10 lb 为止，然后开始显示、输出飞行计划。一般要迭代 3~4 次。

第（3）步中已对输出的飞行计划做了部分说明，下面再补充一点。额外油量的大小反映了初始目的地机场及其备降机场和二次放行点选择的水平。额外油量越小越好（指在二次放行点剩余油量和二次放行所需油量相差越小越好），为做到这一点，关键是仔细选择初始目的地机场和备降机场及二次放行点。

如果初始目的地机场离起飞机场比较近，二次放行点应选在飞过了初始目的地之后的地方，但可能受航线结构限制,空管部门不允许由 R→C 或 R→P→C 这种飞行,如图 3-9-18 所示。这样二次放行点只有选在 P 点前,算出的额外油量就比较大,节省燃油、增加业载的效果就差些。

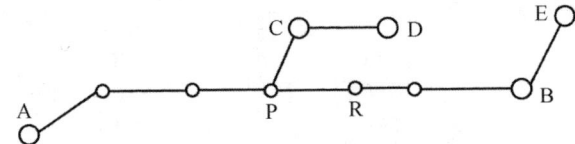

图 3-9-18　二次放行点位置受限制的情况

按上文计算出的额外油量是指为使在二次放行点剩余油量等于二次放行所需油量而多加的油量在到达初始目的地机场的备降机场后的剩余部分。而 SITA 报文中的飞行计划给出的额外油量(Extra Fuel)是指为使在二次放行点剩余油量等于二次放行所需油量而在起飞机场多加的部分。两者算出的总油量应该是相同的。

3.9.4　二次放行飞行计划的实例

前文介绍的确定选择初始目的地机场和最佳二次放行点的方法是近似的,再加上各航段的航向、风、温度都是不同的,对于实际航线也可能找不到合适的初始目的地机场,所选的二次放行点不一定是最佳的,所以一般情况下算出的到初始目的地机场的起飞油量 $TOF1$ 和到最终目的地的起飞油量 $TOF2$ 不相等,即在二次放行点剩油一般不等于所需油量。如 $TOF1>TOF2$,即剩油大于所需,则 $RPF=TOF1$;如 $TOF1<TOF2$,即在二次放行点剩油少于两次放行所需油量,这时要通过加额外油量的方法使起飞油量 $TOF1$ 增加到等于 $TOF2$,一般情况下额外油量都大于零。

例 3-9-1:三亚—哈尔滨(备降长春)二次放行飞行计划(见图 3-9-19)

B747-400/PW4056　B2427　O　PNLWT(使用空机重) 450000 lb　PAY(业载)= 120000 lb

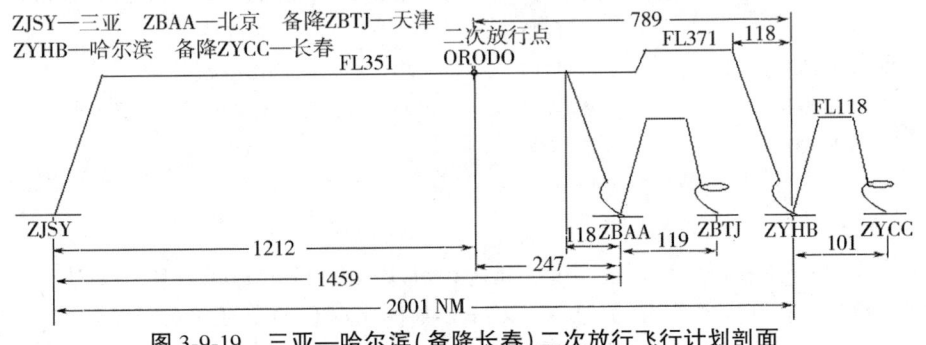

图 3-9-19　三亚—哈尔滨(备降长春)二次放行飞行计划剖面

PLAN 0039　　　　　　　　　ZJSY TO ZYHB 74FF　LRC/F IFR　06/07/20

NONSTOP COMPUTED 0347Z　FOR ETD 0200Z　PROGS　0518UK　B2427 LBS

ETD-预计离场时间

UKMO(United Kingdom Met Office)5 日 18 点的气象预报数据

SICHUAN AIRLINES　　　　　　　　　　　CALLSIGN:

I CERTIFY THAT THIS FLIGHT IS RELEASED/DISPATCHED IN ACCORDANCE

WITH ALL APPLICABLE ICAO REGULATIONS

DISPATCHER HE   SIGN. . . . .

PILOT IN COMMAND   SIGN. . . . .

BLOCK IN . . . .   ON . . .   T/O . . .   T/O FUEL . . .

BLOCK OUT . . .   OFF . . .   LDG . . .   LDG FUEL . . .

TOTAL . . . .   TOTAL . . .   TOTAL . . .   FUEL USED . . .

REASON FOR DELAY . . . .

COMPANY NOTES:

MTOW 870000   MLW 652000   MZFW 610000   OPCG

	PLAN	0039			0040			0041		
	PD/RCLR	ZJSY/ZYHB			ORODO	/ZYHB		ZJSY/ZBAA		
		FUEL	TIME	DIST	FUEL	TIME	DIST	FUEL	TIME	DIST
PA ZYHB		098342	0416	2001	030341	0143	0789 ZBAA	074759	0310	1459
AL ZYCC		008497	0022	0101	008436	0022	0101 ZBTJ	010294	0025	0119
HLD		009820	0030		009738	0030		010096	0030	−最后储备油
RES		009835	0030		004941	0015		007476	0022	−不可预期燃油
B/RC					067269	0233	−从起飞到二次放行点的油量和时间 航程油量×10%			
ETOP		000000	0000		000000	0000		000000	0000	
XTR		001000	0003		001000	0003		019100	0057	−酌情燃油/额外油量
TXI		000000			000000			000000		
TOT		127494	0541		121725	0526		121725	0523	

利用二次放行可少加油 = 127494−121725 = 5769 lb   *TOW* 减少 5769 lb（业载不变）

TOW		697494			691725			691725		
LDW		599152			594115			616966		
ZFW		570000			570000			570000		
PAY		120000			120000			120000		

注释说明 1：

RCLR—Reclear，二次放行；PA—Point of Arrival；B/RC—Burn-off to Reclear Point

PLAN 0039 是不用二次放行的 ZJSY—ZYHB 备降 ZYCC 的普通飞行计划

TOW = 697494 lb

LDW = 599152 lb = 697494−98342 = 570000+8497+9820+9835+1000

PLAN 0040 是以 ORODO 为二次放行点计算的 ZJSY—ZYHB 备降 ZYCC 的二次放行计划

TOW = 691725 lb

LDW = 594115 lb = 691725−67269−30341 = 570000+8436+9738+4941+1000

ZYHB 剩余油量 = 594115−570000 = 24115 lb

ZYCC 剩余油量 = 24115−8436 = 15679 = 9738+4941+1000

不可预期燃油、酌情燃油都是在备降机场的油量

PLAN 0041 是以 ZBAA 为初始目的地的 ZJSY—ZBAA 备降 ZBTJ 的初次放行计划

XTR = 19100 lb 是酌情燃油 = 1000 lb + 为二次放行所加的额外油量 18100 lb

TOW = 691725 lb = 570000 + 121725

LDW = 616966 lb = 691725 − 74759 = 570000 + 10294 + 10096 + 7476 + 19100

ZBAA 剩余油量 = 616966 − 570000 = 46966 lb

ZBTJ 剩余油量 = 46966 − 10294 = 36672 lb = 10096 + 7476 + 19100

不可预期燃油、酌情燃油、额外油量都是在备降机场的油量。

X=10%时起飞油量与机场C的关系

L_R/L = 1212/2001 × 100% ≈ 60.6% , $\Delta F/F_{10}$ = 61/98 × 100% ≈ 62%

本例计算结果:$\Delta F/F_{10}$ = 5769/9835 × 100% ≈ 58.7%

FUEL BURN ADJUSTMENT FOR 1000 LBS INCR/DECR IN TOW:0105 LBS/N/A　　　（N/A 不适用）

ROUTE AVG WIND M012　　MXSH 06/DOSKU　　　　　　（MXSH——最大风切变）

ROUTE AVG TEMP M42　　FLIGHT LEVEL　351/ZUH　　　341/POU　　　351/ZHO

　　　　　　　　　　　　　　　　　341/OBMEP　　351/BUMDU　　371

　　　　　　　　　　ACT. FLT LEVEL. . . . . . .

ALT.LEVEL	ETE	WIND	FUEL	（ETE——预计航程时间）
351	0416	M011	097577	
	HHMM			
331	0414	M008	097771	

REASONS FOR EXTRA FUEL . . . .

			MSA——最低安全高度　　TTK——真航迹　　W/C——风分量				
	MSA	TTK	DIST	FL	W/C	TIME	FUEL
ALTERNATE　−1　ZYCC	040	193	0101	118	P011	00.22	8436

ETOPS ENTRY　N2459.8E11326.3　　　　ETOPS EXIT　N4109.9E11743.6

---

　*　MOST CRITICAL MSA　09000 FEET AT BUMDU　*

DEPARTURE ATIS:　　　　　　　　　　　　　　　　（ATIS——自动天气通播）

. . . .

. . . .

ATC CLEARANCE:

. . . .

. . . .

ROUTE DESCRIPTION:COMPANY ROUTE ZJSY-ZYHB

ZJSY POR9YD PORAP W222 SAMAS G221 BIGRO R200 ZUH R473 POU W90 TEPID W22 YIN A461 ZHO B208 OBMEP W56 LEBUN W57 BEKDO W157 VYK G212 RUSBO W204 IGPUP IGP09A ZYHB

注释说明 2：

T——温度,℃（省略负号）；WIND——前 2 位为风向（10°），后 3 位为风速（节）；S 或 WS——风切变；GRS——地速；MSA——最低安全高度（100 ft）；MCS——磁航迹（°）；DST——航段距离（NM）；DSTR——剩余距离（NM）；ETE——预计航段时间（小时/分）；CUM——累计时间（小时/分）；FU——航段耗油（100 lb）；FR——剩余油量（100 lb）；FF/E——一发燃油流量（lb/h）。

ZJSY ELEV   00095FT

CPT	FLT	T	WIND	TAS	MSA	MCS	DST	DSTR	ETE	CUM	FU	FR	FF/E
FREQ	AWY	S	GRS										

DOSTA..   .. ..   082 031.4 018 1983 ./.. ./.. .. ..
      POR9YD..

PORAP..   .. ..   082 026.6 050 1933 ./.. ./.. .. ..
      POR9YD..

NUMKU..   .. ..   079 026.6 055 1878 ./.. ./.. .. ..
      W222..

TOC   351.. ....   ..   019 016.4 003 1875 0/20 0/20 168 1049 .. .
      W222..

ZGZU FIR   GUANGZHOU

SAMAS 351 40 07038 504 019 016.4 023 1852 0/02 0/22 011 1038 5860
      W222   03   484

GIVIV 351 40 07035 504 019 051.9 041 1811 0/06 0/28 020 1018 5828
      G221   03   472

VHHK FIR   HONG KONG

FIR   351 40 07032 504 054 052.1 032 1779 0/04 0/32 016 1002 5816
      G221   03   474

ZGZU FIR   GUANGZHOU

FIR   351 40 07032 504 054 052.1 018 1761 0/02 0/34 009 0993 5809
      G221   03   474

BIGRO 351 40 07032 504 054 052.1 007 1754 0/01 0/35 003 0990 5807
      G221   03   474

TOMUD 351 39 07030 505 054 069.5 044 1710 0/05 0/40 021 0968 5802
      R200   03   475

ADBIN 351 39 07029 505 047 069.6 017 1693 0/03 0/43 008 0960 5790

R200　　03　476

BOKAT 351 39 07028 505 071 070.0 011 1682 0/01 0/44 005 0955 5779
　　　　R200　　02　477

KIBAS 351 39 07028 505 071 069.0 015 1667 0/02 0/46 007 0947 5774
　　　　R200　　02　477

ZUH　351 39 07027 505 051 070.8 014 1653 0/02 0/48 007 0941 5768
116.7　R200　　02　478

VIBOS 341 36 06023 507 051 344.8 025 1628 0/02 0/50 010 0930 5267
　　　　R473　　02　503

POU　341 36 07022 507 070 344.6 025 1603 0/03 0/53 012 0919 5792
114.1　R473　　02　504

TEPID 351 39 07023 505 070 021.9 033 1570 0/04 0/57 017 0902 6363
　　　　W90　　02　490

YIN　351 39 07021 504 076 005.0 039 1531 0/05 1/02 018 0884 5726
113.5　W22　　02　495

BUBDA 351 38 06015 505 082 004.3 100 1431 0/12 1/14 046 0837 5714
　　　　A461　　02　496

LIG　351 38 02010 505 087 004.8 107 1324 0/13 1/27 049 0788 5676
112.4　A461　　03　495

PAVTU 351 38 35017 505 073 007.9 027 1297 0/04 1/31 013 0776 5653
　　　　A461　　03　489

LUMKO 351 38 34020 505 073 007.5 023 1274 0/02 1/33 011 0765 5643
　　　　A461　　03　487

AKUBA 351 38 33023 505 073 007.5 016 1258 0/02 1/35 007 0758 5634
　　　　A461　　02　486

ZHWH FIR　　　　　　　　　　　　　　　　　　　WUHAN
DAPRO 351 38 32023 505 073 008.2 030 1228 0/04 1/39 014 0744 5623
　　　　A461　　02　488

LKO　351 37 31023 506 070 008.1 039 1189 0/05 1/44 018 0726 5629
115.8　A461　　02　491

HOK   351 37 29030 506 070 028.2 093 1096 0/11 1/55 041 0685 5605
116.0   A461   01   506

ESDOS 351 37 28042 505 039 009.2 024 1072 0/03 1/58 011 0674 5582
     A461   01   500

OBLIK 351 38 29046 504 039 009.5 037 1035 0/04 2/02 017 0657 5555
     A461   02   490

ZHO   351 39 29048 503 022 009.7 080 0955 0/10 2/12 036 0621 5520
115.5 A461   05   488

OBMEP 341 37 29047 504 014 326.5 012 0943 0/02 2/14 004 0617 4520
     B208   05   464

OKTOX 351 40 29053 502 018 009.9 038 0905 0/04 2/18 019 0598 5979
     W56   03   481

DOSKU 351 41 30063 501 053 010.2 098 0807 0/13 2/31 045 0553 5437
     W56   06   474

VADMO 351 41 30069 501 053 010.7 012 0795 0/01 2/32 006 0547 5419
     W56   04   469

ORODO 351 42 30069 500 053 010.0 006 0789 0/01 2/33 003 0545 5400   二次放行点
     W56   04   466
ZBPE FIR   BEIJING
FIR   351 43 30070 499 053 031.5 003 0786 0/00 2/33 001 0543 5398
     W56   05   487

LEBUN 351 43 30070 499 053 031.5 091 0695 0/12 2/45 040 0503 5363
     W56   05   487

BEKDO 351 45 30060 496 038 055.1 042 0653 0/05 2/50 017 0486 5308
     W57   04   510
AVBOX 351 46 30056 495 021 022.0 024 0629 0/03 2/53 011 0475 5281
     W157   03   476

VYK   351 46 30052 495 053 021.9 034 0595 0/04 2/57 015 0460 5270
112.7 W157   03   476

SOTMU 351 47 30045 494 053 026.6 039 0556 0/05 3/02 017 0443 5245
     G212   03   483

BUMDU 351 49 30032 492 090 027.2 058 0498 0/07 3/09 025 0418 5196
　　　G212　　04　　486

MUDAM 371 53 30034 490 090 028.9 021 0477 0/03 3/12 012 0406 6603
　　　G212　　04　　481

UKDUM 371 54 31032 489 086 066.4 053 0424 0/06 3/18 022 0385 5107
　　　G212　　03　　500

ZYSH FIR　　　　　　　　　　　　　　　　　　SHENYANG
FIR 　371 55 32044 488 086 062.5 002 0422 0/00 3/18 001 0384 5106
　　　G212　　01　　489

TGO 　371 55 32044 488 086 062.5 205 0217 0/25 3/43 085 0299 5049
116.3 G212　　01　　489

DULEX 371 53 32035 489 023 064.5 039 0178 0/05 3/48 016 0283 5036
　　　G212　　02　　490

OTABO 371 52 32030 490 023 064.1 053 0125 0/07 3/55 022 0262 5034
　　　 G212　　02　　493

TOD 　371 52 31027 490 023 064.5 007 0118 0/00 3/55 003 0259 5029
　　　G212　　02　　497

RUSBO.. 　.. ....　.. 　023 064.5 004 0114 ./.. ./.. .. 　.. .. ..
　　　G212..

IGPUP.. 　.. ....　.. 　023 094.0 010 0104 ./.. ./.. .. 　.. .. ..
　　　W204..

LEGAG.. 　.. ....　.. 　023 064.9 030 0074 ./.. ./.. .. 　.. .. ..
　　　IGP09A..

DUBIK.. 　.. ....　.. 　028 065.7 017 0057 ./.. ./.. .. 　.. .. ..
　　　IGP09A..

HB510.. 　.. ....　.. 　021 065.9 012 0045 ./.. ./.. .. 　.. .. ..
　　　IGP09A..

HB504.. 　.. ....　.. 　021 073.6 017 0028 ./.. ./.. .. 　.. .. ..
　　　IGP09A..

HB503.. 　.. ....　.. 　021 050.4 007 0021 ./.. ./.. .. 　.. .. ..
　　　IGP09A..

```
LS..  .. ....   ..  023 048.6 007 0014 ./.. /.. ..  .. .. . .
445.0   IGP09A..
```

```
ZYHB..  .. ....   ..  023 049.7 014 0000 0/21 4/16 018 0241 .. .
ELEV   00457FT
```

ZJSY	N18181E109248	DOSTA	N18294E109368	PORAP	N19150E109588
NUMKU	N20050E110229	SAMAS	N20303E110297	GIVIV	N20572E111033
FIR	N21186E111300	FIR	N21298E111441	BIGRO	N21342E111496
TOMUD	N21515E112328	ADBIN	N21581E112493	BOKAT	N22023E113000
KIBAS	N22083E113145	ZUH	N22133E113280	VIBOS	N22375E113197
POU	N23013E113114	TEPID	N23321E113231	YIN	N24114E113249
BUBDA	N25521E113278	LIG	N27389E113310	PAVTU	N28063E113334
LUMKO	N28289E113349	AKUBA	N28453E113360	DAPRO	N29155E113384
LKO	N29544E113415	HOK	N31195E114258	ESDOS	N31433E114281
OBLIK	N32198E114315	ZHO	N33399E114393	OBMEP	N33493E114303
OKTOX	N34277E114339	DOSKU	N36051E114433	VADMO	N36168E114445
ORODO	N36225E114450	FIR	N36256E114468	LEBUN	N37479E115353
BEKDO	N38156E116147	AVBOX	N38389E116227	VYK	N39117E116343
SOTMU	N39480E116513	BUMDU	N40428E117169	MUDAM	N41020E117267
UKDUM	N41299E118267	FIR	N41313E118292	TGO	N43336E122118
DULEX	N43563E122557	OTABO	N44273E123549	RUSBO	N44336E124071
IGPUP	N44347E124211	LEGAG	N44523E124557	DUBIK	N45020E125153
HB510	N45087E125289	HB504	N45167E125509	HB503	N45219E125570
LS	N45270E126027	ZYHB	N45375E126151		

```
FIRS   ZGZU/0022   VHHK/0032   ZGZU/0034   ZHWH/0139   ZBPE/0233
FIRS   ZYSH/0318                              进入各情报区的时间(hhmm)
```

ENROUTE WIND AND TEMPERATURE SUMMARY

FL /	10000	29000	31000	33000	35000	37000
POS	Wind Tem WS	Wind Tem WS	Wind Tem WS	Wind Tem WS	Wind Tem WS	Wind Tem WS

风向风速 温度 风切变

DOSTA	20010P13-01	09028M26-02	08032M31-02	08035M36-02	08041M41-04	08050M46-04
PORAP	20008P13-01	09028M26-02	08031M31-02	08035M36-03	08041M41-04	08049M46-04
NUMKU	20012P13-01	09026M25-02	08030M30-03	08035M36-03	08040M40-03	08047M45-04
SAMAS	21014P13-02	09025M25-03	08028M30-02	08033M35-03	07038M40-03	07044M45-03
GIVIV	21015P12-01	09023M25-03	08026M30-02	08030M35-03	07035M40-03	07041M45-03
BIGRO	21016P12-01	09020M25-03	08024M30-02	08027M35-02	07032M40-03	07039M45-03
TOMUD	22016P12-01	09018M25-03	08022M30-03	07025M34-02	07030M39-03	07036M45-03
ADBIN	22016P12-01	09017M25-02	08021M30-02	07024M34-02	07029M39-03	07034M44-03
BOKAT	22016P12-01	09017M25-02	08020M30-02	07024M34-02	07028M39-02	07033M44-03
KIBAS	22016P12-01	09016M25-02	08020M30-02	07023M34-02	07028M39-02	07032M45-02

ZUH	22017P12-01	09016M25-02	08019M30-02	07023M34-02	07027M39-02	07032M45-02
VIBOS	22017P12-01	08015M25-03	07018M29-02	07022M34-02	07026M39-02	07030M44-02
POU	22018P13-01	08014M25-03	07017M29-02	07020M34-02	07024M39-02	07028M44-02
TEPID	22020P13-01	08012M24-02	07015M29-02	07020M34-02	07023M39-02	07026M44-02
YIN	22021P13-01	08011M24-03	06014M29-02	07018M34-02	07021M39-02	06024M44-02
BUBDA	23020P14-02	10008M24-03	07010M28-02	07012M33-01	06015M38-02	05017M44-02
LIG	25020P13-02	21004M23-01	25004M28-02	34006M33-03	02010M38-03	01016M44-03
PAVTU	24023P12-01	27009M23-01	29009M28-03	33012M33-04	35017M38-03	35023M43-03
LUMKO	24025P12-01	28011M23-01	30012M28-03	32016M33-03	34020M38-03	34025M43-03
AKUBA	24026P12-00	30014M23-01	31015M28-02	32019M33-03	33023M38-02	33027M43-02
DAPRO	24025P12-00	30015M23-02	31017M27-02	32020M32-02	32023M38-02	33027M43-02
LKO	24022P13-00	30016M23-01	30018M27-01	31020M32-02	31023M37-02	32025M43-02
HOK	17013P13-01	28024M22-02	29026M27-02	29029M32-01	29030M37-01	30030M42-03
ESDOS	09010P12-02	27033M23-03	28040M27-03	28043M32-01	28042M37-01	28041M43-01
OBLIK	08014P12-03	27036M24-03	28039M28-02	28042M33-03	29046M38-02	29045M43-01
ZHO	07014P11-03	28040M25-02	28043M29-01	28045M34-02	29048M39-05	30052M43-02
OBMEP	06009P11-02	28044M26-01	28046M30-02	29047M35-02	29050M40-03	30055M44-03
OKTOX	03008P11-02	28043M26-01	28046M31-03	29049M35-03	29053M40-03	30059M44-03
DOSKU	32008P11-02	29044M28-03	29050M33-03	29056M37-04	30063M41-06	31069M44-04
VADMO	30007P10-02	30053M30-03	30058M34-03	30063M38-03	30069M41-04	31076M44-04
ORODO	30007P10-02	30053M30-03	30058M34-03	30063M38-03	30069M42-04	31076M44-04
LEBUN	31010P09-03	30046M32-02	30052M36-04	30061M40-05	30070M43-05	31078M45-05
BEKDO	32012P08-02	30044M33-02	30048M37-02	30053M42-03	30060M45-04	30070M47-06
AVBOX	32012P08-01	30047M33-03	30051M38-01	30052M43-01	30056M46-03	30064M48-05
VYK	32011P07-01	30041M33-03	30046M38-02	30048M43-02	30052M46-03	30058M48-03
SOTMU	33011P07-01	29026M35-02	29031M39-03	30038M44-04	30045M47-03	30051M49-04
BUMDU	34013P07-01	29016M35-01	29019M40-02	30025M45-03	30032M49-04	30040M52-04
MUDAM	34019P06-01	29015M36-01	29017M41-01	30020M46-02	30026M50-04	30034M53-04
UKDUM	34022P06-01	30018M36-01	30020M41-01	30022M46-02	31026M51-03	31032M54-03
TGO	35025P07-01	34037M37-02	33040M42-02	33044M47-02	32045M51-02	32044M55-01
DULEX	35023P06-01	34047M38-01	34045M43-02	34041M47-03	33038M50-04	32035M53-02
OTABO	35021P04-02	34042M38-02	33040M42-02	33037M46-02	32033M49-02	32030M52-02
RUSBO	35021P04-01	33040M37-03	33036M42-02	32033M45-03	31030M49-02	31027M52-02
IGPUP	35021P04-01	33038M37-02	32035M41-02	32031M45-02	31028M49-03	30026M52-01
LEGAG	34021P04-01	33035M37-03	32031M41-03	31028M45-03	30025M48-02	29024M52-02
DUBIK	34021P04-02	32033M36-05	31028M40-03	30025M44-03	29023M48-02	28022M52-02
HB510	34020P04-01	30035M36-03	29030M40-03	28027M44-03	27024M48-02	27022M52-01
HB504	33019P04-01	27037M36-03	27032M40-02	26029M44-02	26026M48-02	26023M51-01
HB503	33018P03-01	24038M36-03	24034M40-02	24031M44-02	25028M48-02	25024M51-02
LS	33018P03-01	23039M36-02	23035M40-02	24032M44-02	24028M48-02	25025M51-03
ZYHB	33017P03-01	22040M36-02	22037M40-02	22033M44-02	23030M48-03	24025M51-03

RECLEAR FLIGHT PLAN　ORODO / ZBAA

CPT	LAT	LONG	MSA	TTK	DIST	CPT-CHECKPOINT
CG	N39048	E117225	079	149	0082	

ZBTJ   N39074   E117207   035   332   0037

ETOPS ENTRY N2459.8E11326.3   ETOPS EXIT   N3230.8E11432.6

----

    *   MOST CRITICAL MSA 08700 FEET AT LIG   *

DEPARTURE ATIS：

· ·

· ·

ATC CLEARANCE：

· ·

· ·

ROUTE DESCRIPTION：COMPANY ROUTE ZJSY-ZBAA

ZJSY POR9YD PORAP W222 SAMAS G221 BIGRO R200 ZUH R473 POU W90 TEPID
W22 YIN A461 ZHO B208 OBMEP W56 LEBUN W57 BEKDO W157 AVBOX AVB9ZA
ZBAA

ZJSY ELEV 00095FT
CPT   FLT T WIND TAS MSA MCS   DST DSTR ETE CUM FU   FR   FF/E
FREQ AWY   S   GRS
ORODO 351 42 30069 500 053 010.0 006 0247 0/01 2/33 003 0545 5400
      W56   04   466
ZBPE FIR   BEIJING
FIR   351 43 30070 499 053 031.5 003 0244 0/00 2/33 001 0543 5398
      W56   05   487

LEBUN 351 43 30070 499 053 031.5 091 0153 0/12 2/45 040 0503 5363
      W56   05   487

TOD   351 45 30060 496 038 055.1 035 0118 0/04 2/49 015 0489 5308
      W57   04   510

BEKDO.. .. .. 038 055.1 007 0111 ./.. ./.. .. ..
      W57..

AVBOX.. .. .. 021 022.0 024 0087 ./.. ./.. .. ..
      W157..

AA443.. .. .. 053 019.6 025 0062 ./.. ./.. .. ..
      AVB9ZA..

```
AA442..  .. ....  ..   053 000.3 027 0035 ./.. ./.. .. .. . ...
    AVB9ZA..


AA441..  .. ....  ..   053 000.1 006 0029 ./.. ./.. .. .. . ...
    AVB9ZA..


ZBAA..  .. ....  ..   079 023.8 029 0000 0/21 3/10 019 0470 .. .
ELEV   00115FT
```

ZJSY	N18181E109248	DOSTA	N18294E109368	PORAP	N19150E109588
NUMKU	N20050E110229	SAMAS	N20303E110297	GIVIV	N20572E111033
FIR	N21186E111300	FIR	N21298E111441	BIGRO	N21342E111496
TOMUD	N21515E112328	ADBIN	N21581E112493	BOKAT	N22023E113000
KIBAS	N22083E113145	ZUH	N22133E113280	VIBOS	N22375E113197
POU	N23013E113114	TEPID	N23321E113231	YIN	N24114E113249
BUBDA	N25521E113278	LIG	N27389E113310	PAVTU	N28063E113334
LUMKO	N28289E113349	AKUBA	N28453E113360	DAPRO	N29155E113384
LKO	N29544E113415	HOK	N31195E114258	ESDOS	N31433E114281
OBLIK	N32198E114315	ZHO	N33399E114393	OBMEP	N33493E114303
OKTOX	N34277E114339	DOSKU	N36051E114433	VADMO	N36168E114445
ORODO	N36225E114450	FIR	N36256E114468	LEBUN	N37479E115353
BEKDO	N38156E116147	AVBOX	N38389E116227	AA443	N39034E116297
AA442	N39300E116256	AA441	N39363E116246	ZBAA	N40044E116359

```
FIRS   ZGZU/0022   VHHK/0032   ZGZU/0034   ZHWH/0139   ZBPE/0233
```

进入各情报区的时间(hhmm)

ENROUTE WIND AND TEMPERATURE SUMMARY

FL /	10000	29000	31000	33000	35000	37000
POS	Wind Tem WS	Wind Tem WS	Wind Tem WS	Wind Tem WS	Wind Tem WS	Wind Tem WS

风向风速 温度 风切变

```
ORODO 30007P10-02 30053M30-03 30058M34-03 30063M38-03 30069M42-04 31076M44-04
LEBUN 31010P09-03 30046M32-02 30052M36-04 30061M40-05 30070M43-05 31078M45-05
BEKDO 32012P08-02 30044M33-02 30048M37-02 30053M42-03 30060M45-04 30070M47-06
AVBOX 32012P08-01 30047M33-03 30051M38-01 30052M43-01 30056M46-03 30064M48-05
AA443 32011P07-01 30044M33-03 30048M38-01 30049M43-01 30053M46-03 30059M48-03
AA442 32010P07-01 30033M34-03 30038M39-03 30043M43-03 30049M47-03 30055M49-04
AA441 33009P07-01 29026M34-03 30032M39-04 30039M44-04 30046M47-04 31052M49-04
ZBAA  33009P07-02 29019M35-02 29025M40-04 30035M44-05 30043M47-04 31049M49-04
```

DESTINATION ATIS：

· ·

· ·

ABNORMAL AFFAIR REPORT：

TYPE		OPERATION （ ）		SAFETY （ ）
TIME OF INCIDENT				
LOCATION OF INCIDENT				
PHASE	PARKED （ ） PUSHBACK （ ） TAXI OUT （ ） TAKE OFF （ ） CLIMB （ ） CRUISE （ ） DESCENT （ ） APPROACH （ ） LANDING （ ） TAXI IN （ ）			
CAPTAIN'S NARRATIVE				

（FPL-B2427-I

-B744/H-SJ1J2J3WXY/SU1

-ZJSY0200

-K0933S1070 POR9YD PORAP W222 SAMAS G221 BIGRO R200 ZUH/K0939S1040

R473 POU/K0935S1070 W90 TEPID W22 YIN A461 ZHO/K0933S1040 B208

OBMEP/K0930S1070 W56 LEBUN W57 BEKDO W157 VYK G212

BUMDU/K0907S1130 G212 RUSBO W204 IGPUP IGP09A

-ZYHB0416 ZYCC

-DOF/200707 REG/B2427

EET/ZGZU0022 VHHK0032 ZGZU0034 ZHWH0139 ZBPE0233 ZYSH0318

RIF/ORODO W56 LEBUN W57 BEKDO W157 AVBOX AVB9ZA ZBAA

RMK/AGCS EQUIPPED

-E/0526 P/TBN）

END OF JEPPESEN DATAPLAN

REQUEST NO. 42

例 3-9-2：浦东→洛杉矶（备降拉斯维加斯）的二次放行飞行计划（业载受 *MTOW* 限制的情况，见图 3-8-20）

B747-400/PW4056 B2427 *BOW* =450 klb *MTOW* =870 klb *MLW* =652 klb *MZFW* =610 klb

PLAN 0115　　　　　　　ZSPD TO KLAX 74FF　　LRC/F IFR　　28/08/20

NONSTOP COMPUTED 0759Z　FOR ETD 0200Z　PROGS　2800NWS　B2427 LBS

ETD——预计离场时间

NWS,美国 National Weather Service 中心 28 日 00 点的气象预报

OMNI AIR INTERNATIONAL

FLIGHT NUMBER R5XX　　　　　　　　　　　　（不可预期燃油按 10%航程油量计算）

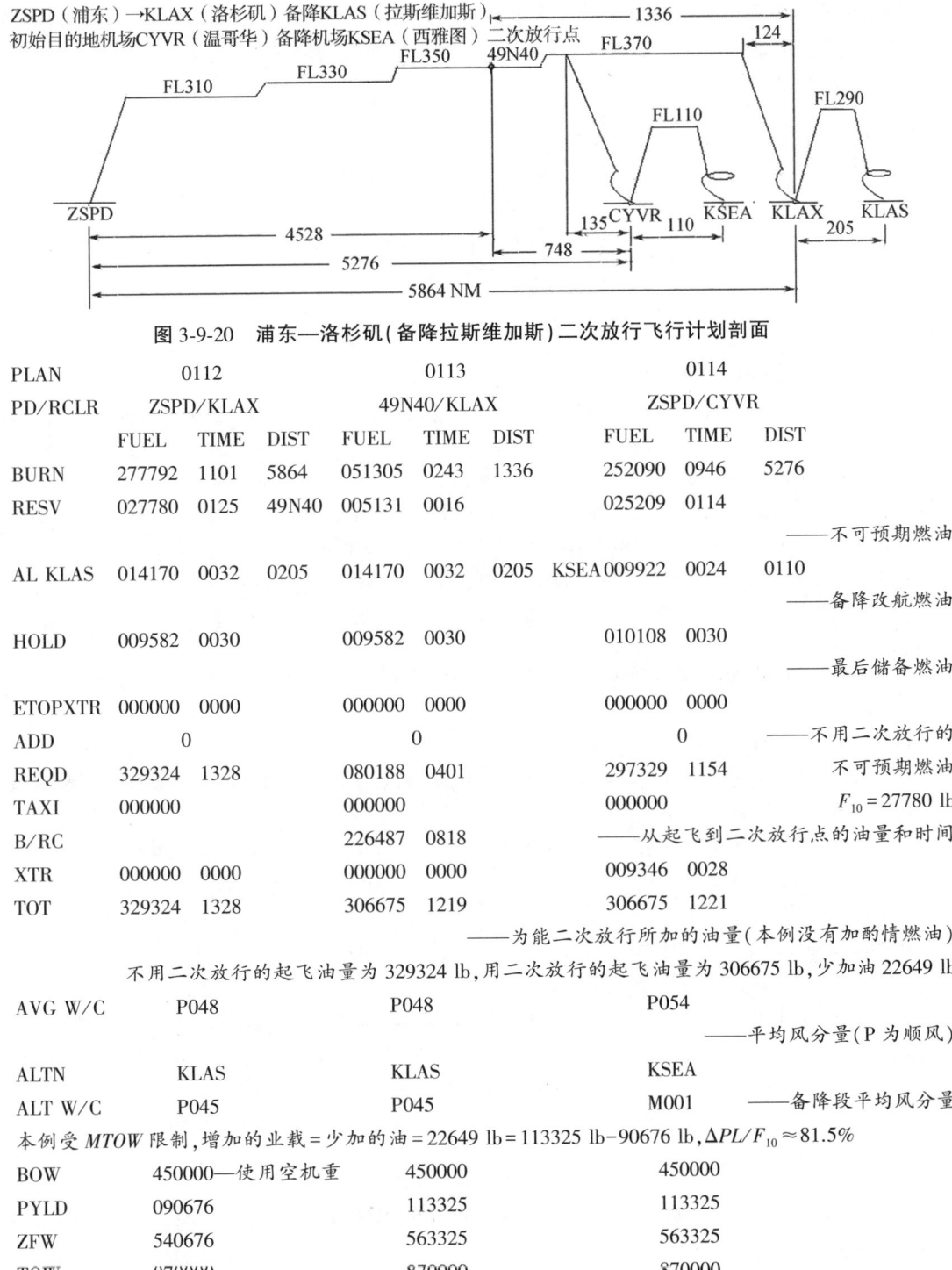

图 3-9-20　浦东—洛杉矶(备降拉斯维加斯)二次放行飞行计划剖面

PLAN	0112			0113			0114		
PD/RCLR	ZSPD/KLAX			49N40/KLAX			ZSPD/CYVR		
	FUEL	TIME	DIST	FUEL	TIME	DIST	FUEL	TIME	DIST
BURN	277792	1101	5864	051305	0243	1336	252090	0946	5276
RESV	027780	0125	49N40	005131	0016		025209	0114	
								——不可预期燃油	
AL KLAS	014170	0032	0205	014170	0032	0205	KSEA 009922	0024	0110
								——备降改航燃油	
HOLD	009582	0030		009582	0030		010108	0030	
								——最后储备燃油	
ETOPXTR	000000	0000		000000	0000		000000	0000	
ADD	0			0			0		——不用二次放行的
REQD	329324	1328		080188	0401		297329	1154	不可预期燃油
TAXI	000000			000000			000000		$F_{10} = 27780$ lb
B/RC				226487	0818				——从起飞到二次放行点的油量和时间
XTR	000000	0000		000000	0000		009346	0028	
TOT	329324	1328		306675	1219		306675	1221	

　　　　　　　　　　　　　　　　　　　　——为能二次放行所加的油量(本例没有加酌情燃油)

　　不用二次放行的起飞油量为 329324 lb,用二次放行的起飞油量为 306675 lb,少加油 22649 lb

AVG W/C	P048	P048	P054

　　　　　　　　　　　　　　　　　　　　　　——平均风分量(P 为顺风)

ALTN	KLAS	KLAS	KSEA
ALT W/C	P045	P045	M001

——备降段平均风分量

本例受 MTOW 限制,增加的业载 = 少加的油 = 22649 lb = 113325 lb−90676 lb, $\Delta PL/F_{10} \approx 81.5\%$

BOW	450000—使用空机重	450000	450000
PYLD	090676	113325	113325
ZFW	540676	563325	563325
TOW	870000	870000	870000
LDW	592208	592208	617910

注释说明:

　　PLAN 0112 在备降机场 KLAS 的剩余油量 = 592208−14170−540676 = 37362 lb = 9582+27780

　　PLAN 0113 在备降机场 KLAS 的剩余油量 = 592208−14170−563325 = 14713 lb = 9582+5131

PLAN 0114 在备降机场 KSEA 剩余油量＝617910－9922－563325＝44663 lb＝10108＋25209＋9346

不可预期燃油、为二次放行所加的额外油量（及酌情燃油）都是在备降机场停机坪的油量。

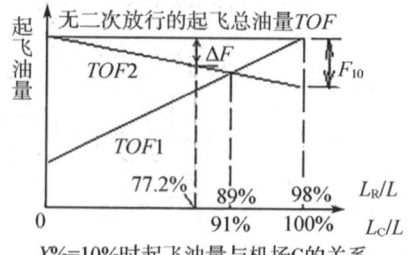

$X\%=10\%$ 时起飞油量与机场C的关系

$$L_R/L=4528/5864\times100\%\approx77.2\%$$

$$\Delta F/F_{10}=77.2/98\times100\%\approx78.8\%,本例计算结果\ \Delta F/F_{10}=22649/27780\times100\%\approx81.5\%$$

ZSPD SUR91D SURAK G455 LAMEN A593 ONIKU Y60 POTET A593 FUE Y60
ISAKY Y28 DGC V28 OLIVE Y28 SANDA V28 CUE Y28 BIWWA V28 KCC Y88
DAIGO Y886 ADNAP OTR7 EMRON..44E60..FIR..47E70..48N80..49N70..
51N60..51N50..49N40..FIR..ZANNG..MITOH..FIR..ENI..FIR..BURGL IRNMN2
KLAX

FL   310/GOMAR 330/DAIGO 350/49N40 370

LRC 290 11/05 2836
　　　　以 LRC 速度、FL290 等高度巡航,ZSPD—KLAX 的航程时间为 11 h 05 min,航程油量为 283600 lb
LRC 250 11/19 2964
　　　　以 LRC 速度、FL250 等高度巡航,ZSPD—KLAX 的航程时间为 11 h 19 min,航程油量为 296400 lb

FLT PLN PROOF
DEST＿＿＿＿＿＿＿＿　　　DIST＿＿＿＿＿＿＿＿＿＿　　　WIND＿＿＿＿＿＿

FLT TIME＿＿＿＿＿＿＿＿＿＿　　FUEL＿＿＿＿＿＿＿＿＿＿＿

ATIS＿＿＿＿＿＿＿＿＿＿＿＿＿＿＿＿＿＿＿＿　　　　　　　（自动天气通播）
ATC CLEARANCE ＿＿＿＿＿＿＿＿＿＿＿＿＿＿＿＿＿

每个航路点（CHKPNT,检查点）有三行数据,各字段对应的中文如下：
TROP:对流层顶高度,1000 ft;油量:100 lb;时间:小时.分;温度:℃;速度:kn;风向:10°;距离:NM;燃油流量:lb/h。

```
CHKPNT    LAT    LONG    TEMP    TROP    LEG    REMN
                                         DIST   DIST                    FUEL
-----------  ---------  ---------  ---------  ---------   ----------  ----------        USED    FLOW
                   MAG                                                            REMN    /ENGS
FREQ   WIND   CRS   HDG   AIRWAY   TAS   GS   TIME   TIME   REMN   /ENGS
----------  ----------  ---------  ---------  ----------  ----------  ---------  ---------  ----------        ----------
FL/SR   TTM   TC   TH                              L/O FUEL _____
```

注释：
航路点　纬度　经度　温度　TROP　　　　　　　　航段距离 剩余距离 航段耗油
频率或FIR　风向风速　磁航迹 磁航向 航路代号 真空速 地速 航段时间 剩余时间 剩余油量 一发燃油流量
高度层/风切变 累计时间 真航迹 真航向　　　　　　机组记录剩余油量_____

```
ZSPD      N31 08.7 E121 47.6                       5864
   .               .               .           11.01  3067


DE35R     N31 09.7 E121 47.2 P026   51    001   5863   004
          19007 347 346 SUR91 200 206 0.01 11.00  3063  ...
CLB/2     0001        341 340


PD501     N31 20.9 E121 42.6 P016   51    012   5851   062
          19014 347 347 SUR91 200 211 0.03 10.57  3001  ...
CLB/1     0004        341 341


HSH       N31 22.1 E121 50.8 P013   51    007   5844   014
114.4     19014 087 087 SUR91 270 274 0.01 10.56  2988  ...
CLB/0     0005        080 081


PD508     N31 27.0 E122 00.4 P010   51    010   5834   014
          20014 065 067 SUR91 364 374 0.02 10.54  2974  ...
CLB/1     0007        059 061


ALDAP     N31 37.5 E122 22.2 P004   51    021   5813   028
          20013 067 069 SUR91 386 396 0.03 10.51  2945  ...
CLB/1     0010        061 063


MATNU     N31 39.6 E122 38.0 M001   51    014   5799   020
          21012 088 090 SUR91 386 393 0.02 10.49  2925  ...
CLB/1     0012        081 083


EMSAN     N31 40.7 E122 46.5 M004   51    007   5792   008
          21011 088 090 SUR91 386 393 0.01 10.48  2918  ...
```

CLB/1   0013   081 083

SURAK   N31 46.4 E123 29.5 M011   52   037 5755   040
          21008 088 090 SUR91 386 391 0.06 10.42   2877   ...
CLB/1   0019   081 083

LAMEN   N31 36.6 E124 00.0 M019   52   028 5727   020
RKRR/FIR   24007 117 118 G455 386 387 0.03 10.39   2857   ...
CLB/2   0022   110 111

TOC   N31 45.6 E124 39.6 M030   53   035 5692   025
          21004 082 082 A593 516 519 0.05 10.34   2832   ...
310/1   0027   075 075

SADLI   N31 50.0 E124 59.9 M030   53   018 5674   010
          310 21004 082 082 A593 516 519 0.02 10.32   2822   07443
310/1   0029   075 075

PONIK   N32 00.4 E125 47.0 M030   53   041 5633   024
          310 13007 082 083 A593 516 512 0.04 10.28   2798   07456
310/0   0033   075 076

NIRAT   N32 03.9 E126 03.5 M030   53   014 5619   008
          310 14009 083 084 A593 516 512 0.02 10.26   2790   07438
310/1   0035   076 077

ONIKU   N32 11.7 E126 39.3 M030   54   031 5588   018
RJJJ/FIR 310 14011 083 083 A593 516 512 0.04 10.22   2772   07422
310/1   0039   075 076

POTET   N32 16.8 E127 02.4 M030   54   020 5568   012
          310 14012 082 083 Y60 516 510 0.02 10.20   2760   07410
310/2   0041   075 076

AZUKI   N32 24.1 E127 34.9 M030   54   028 5540   016
          310 13013 082 083 A593 516 509 0.03 10.17   2744   07395
310/1   0044   075 076

GOMAR   N32 27.7 E127 51.5 M030   54   015 5525   009
          310 13014 083 084 A593 516 508 0.02 10.15   2735   07386
310/1   0046   076 077

FUE   N32 40.1 E128 49.6 M035   54   051 5474   034
115.8   330 13015 083 084 A593 512 504 0.06 10.09   2701   08379

330/1　0052　　　076 077

ISAKY	N33 12.4 E129 39.2 M035	54		053	5421	031		
	330 14018 059 061 Y60	512 512	0.06	10.03		2670	07431	
330/0	0058		052 054					

DGC	N33 40.6 E130 23.4 M035	54		046	5375	027		
114.5	330 15019 060 063 Y28	512 512	0.06	09.57		2644	07397	
330/1	0104		053 055					

ONGHA	N33 47.3 E130 47.2 M035	54		021	5354	012		
	330 14017 079 081 V28	512 506	0.02	09.55		2631	07375	
330/1	0106		071 073					

SWE	N33 51.4 E131 01.8 M035	53		013	5341	008		
113.85	330 14017 079 081 V28	512 506	0.02	09.53		2624	07366	
330/0	0108		071 073					

UBE	N33 56.1 E131 17.0 M035	53		014	5327	008		
110.8	330 14016 077 080 V28	512 506	0.01	09.52		2616	07353	
330/0	0109		070 072					

KATTA	N34 00.2 E131 41.2 M035	53		021	5306	012		
	330 14016 085 088 V28	512 504	0.03	09.49		2604	07338	
330/0	0112		078 080					

MARCO	N34 04.8 E132 08.8 M036	53		023	5283	013		
	330 14016 086 088 V28	510 501	0.03	09.46		2590	07303	
330/1	0115		079 080					

MASKU	N34 08.1 E132 16.2 M036	53		007	5276	004		
	330 13016 070 072 V28	510 505	0.01	09.45		2586	07290	
330/1	0116		062 064					

HGE	N34 26.0 E132 55.4 M036	53		037	5239	021		
117.9	330 13016 069 071 V28	510 505	0.04	09.41		2565	07278	
330/0	0120		061 063					

WASYU	N34 28.3 E133 23.0 M036	53		023	5216	013		
	330 13015 091 093 V28	510 499	0.03	09.38		2551	07254	
330/1	0123		084 085					

OYE	N34 45.0 E133 50.1 M036	53		028	5188	016		
111.0	330 12015 061 063 V28	510 505	0.03	09.35		2535	07240	

330/1	0126		053 055							
OLIVE	N34 45.3 E134 27.0 M036	53		030	5158		017			
	330 12014 097 098 V28	510	498	0.04	09.31		2518		07216	
330/2	0130		089 090							
ASANO	N34 47.6 E134 38.8 M036	53		010	5148		006			
	330 11015 085 086 Y28	510	497	0.01	09.30		2512		07200	
330/1	0131		077 078							
HYOGO	N34 51.5 E134 59.7 M036	53		018	5130		010			
	330 11014 085 086 Y28	510	498	0.02	09.28		2502		07193	
330/2	0133		077 078							
SANDA	N34 55.8 E135 21.7 M036	53		019	5111		011			
	330 10014 085 086 Y28	510	497	0.02	09.26		2491		07180	
330/2	0135		077 078							
CUE	N35 01.0 E135 49.6 M036	53		023	5088		013			
117.1	330 10014 085 086 V28	510	497	0.03	09.23		2477		07167	
330/1	0138		077 078							
ADGUN	N35 03.7 E136 01.3 M036	53		010	5078		006			
	330 10013 081 083 Y28	510	498	0.01	09.22		2472		07150	
330/1	0139		074 075							
BIWWA	N35 07.0 E136 15.4 M036	53		012	5066		007			
	330 10013 082 083 Y28	510	498	0.02	09.20		2465		07148	
330/1	0141		074 075							
RITTO	N35 10.0 E136 29.0 M036	53		012	5054		007			
	330 10013 083 084 V28	510	498	0.01	09.19		2458		07137	
330/1	0142		075 076							
KCC	N35 15.9 E136 54.9 M036	52		022	5032		013			
114.2	330 10012 082 083 V28	510	499	0.03	09.16		2445		07126	
330/1	0145		074 075							
SWING	N35 23.4 E137 15.5 M036	52		018	5014		010			
	330 10011 074 075 Y88	510	501	0.02	09.14		2435		07116	
330/1	0147		066 067							
NAKTU	N35 28.0 E137 27.8 M036	52		011	5003		006			
	330 09010 073 074 Y88	510	501	0.01	09.13		2429		07108	

330/2　　0148　　　　065 066

SUGAL	N35 35.6 E137 47.7 M036	52	018	4985	010		
	330 09010 073 073 Y88	510 501	0.02	09.11	2419	07095	
330/2	0150　　　　065 065						

TENRU	N35 40.2 E138 00.4 M036	52	011	4974	006		
	330 09009 074 074 Y88	510 501	0.02	09.09	2412	07086	
330/2	0152　　　　066 066						

PIGAP	N35 47.4 E138 19.0 M036	52	017	4957	010		
	330 08009 072 073 Y88	510 501	0.02	09.07	2403	07074	
330/2	0154　　　　065 065						

OSPAN	N35 53.8 E138 35.8 M036	52	015	4942	008		
	330 07008 073 073 Y88	510 502	0.02	09.05	2394	07066	
330/2	0156　　　　065 065						

TEPEX	N36 00.2 E138 53.0 M036	52	015	4927	008		
	330 07008 073 073 Y88	510 502	0.01	09.04	2386	07050	
330/2	0157　　　　065 065						

GYODA	N36 09.8 E139 18.7 M036	52	023	4904	013		
	330 06007 073 073 Y88	510 503	0.03	09.01	2373	07041	
330/2	0200　　　　065 065						

AKAGI	N36 23.5 E139 41.9 M036	53	023	4881	013		
	330 05006 062 062 Y88	510 504	0.03	08.58	2360	07024	
330/2	0203　　　　054 054						

JD	N36 29.2 E139 51.8 M036	53	010	4871	006		
389.0	330 04007 062 062 Y88	510 503	0.01	08.57	2355	07012	
330/3	0204　　　　054 054						

DAIGO	N36 44.7 E140 21.0 M036	53	028	4843	016		
	330 01009 064 063 Y88	510 504	0.03	08.54	2339	07006	
330/1	0207　　　　056 055						

BEKPO	N36 54.6 E142 28.9 M041	53	103	4740	062		
	350 34015 092 090 Y886	505 508	0.13	08.41	2277	07613	
350/2	0220　　　　084 082						

PABBA	N37 00.2 E143 59.8 M041	53	073	4667	041		
	350 34018 093 091 Y886	505 510	0.08	08.33	2236	07119	

350/2   0228   085 083

DOVIX   N37 02.1 E144 15.5 M041   53   013   4654   007
           350 33020 089 086 Y886   505 511 0.02 08.31   2229 07082
350/2   0230   081 079

ADNAP   N37 11.8 E145 40.0 M042   53   068   4586   037
           350 33020 089 086 Y886   504 512 0.08 08.23   2192 07023
350/1   0238   081 079

EMRON   N38 06.4 E150 00.0 M042   53   214   4372   114
           350 29016 081 080 OTR7   504 517 0.25 07.58   2077 06911
350/1   0303   074 073

44E60   N44 00.0 E160 00.0 M040   51   575   3797   292
           350 28037 055 052..   506 527 1.05 06.53   1786 06681
350/0   0408   049 046

FIR   N45 08.6 E163 21.9 M042   45   159   3638   068
KZAK/FIR 350 28124 068 058..   503 596 0.16 06.37   1718 06572
350/2   0424   063 054

47E70   N47 00.0 E170 00.0 M042   45   299   3339   128
           350 28124 068 058..   503 596 0.30 06.07   1590 06368
350/2   0454   063 054

48N80   N48 00.0 W180 00.0 M044   40   411   2928   161
           350 28142 079 073..   501 629 0.39 05.28   1429 06148
350/8   0533   078 072

49N70   N49 00.0 W170 00.0 M045   33   403   2525   168
           350 26072 075 075..   499 571 0.43 04.45   1261 05963
350/0   0616   078 078

51N60   N51 00.0 W160 00.0 M046   40   405   2120   155
           350 23122 062 068..   498 607 0.40 04.05   1106 05806
350/2   0656   069 075
51N50   N51 00.0 W150 00.0 M048   43   379   1741   149
           350 26080 075 077..   496 575 0.39 03.26   0957 05640
350/5   0735   086 088

RECLEAR POINT REQD 080188   FOB_____ TIME_____ DISP_____
                              VIA_____ CAPT SIGN_____

DEST WX　–　　　　　　　　　　　　　　　（目的地机场天气报告）

ALTN WX　–　　　　　　　　　　　　　　　（备降机场天气报告）

49N40　　　N49 00.0 W140 00.0 M050　42　　405　1336　155　　　　　二次放行点
　　　　　　350 29080 089 087..　　493 572　0.43 02.43　0802　05484
350/1　　　0818　　　103 101

FIR　　　　N44 58.2 W131 31.5 M056　42　　423　0913　169
ADIZ/FIR　370 29049 105 107..　　487 535　0.47 01.56　0633　05372
370/3　　　0905　　　122 123

ZANNG　　N44 05.6 W129 58.7 M056　42　　085　0828　034
　　　　　　370 29049 105 107..　　487 535　0.10 01.46　0599　05350
370/3　　　0915　　　122 123

MITOH　　N41 47.4 W126 49.9 M054　40　　196　0632　080
KZSE/FIR　370 26028 118 122..　　489 506　0.23 01.23　0518　05188
370/2　　　0938　　　134 137

FIR　　　　N40 13.0 W124 44.0 M052　44　　134　0498　056
KZOA/FIR　370 24012 119 120..　　491 494　0.16 01.07　0462　05162
370/2　　　0954　　　134 135

ENI　　　　N39 03.2 W123 16.5 M052　44　　097　0401　040
112.3　　　370 24012 119 120..　　491 494　0.12 00.55　0422　05144
370/2　　　1006　　　134 135

FIR　　　　N35 44.1 W120 10.1 M047　50　　248　0153　102
KZLA/FIR　370 24026 126 131..　　496 498　0.30 00.25　0320　05134
370/3　　　1036　　　142 145

BURGL　　N35 40.0 W120 06.5 M047　50　　005　0148　002
　　　　　　370 24026 126 131..　　496 498　0.01 00.24　0318　05134
370/3　　　1037　　　142 145

TOD　　　　N35 19.8 W119 51.6 M045　50　　024　0124　010
　　　　　　370 24060 136 144 IRNMN　498 500 0.02 00.22　0308　05119
370/2　　　1039　　　148 156

DOUIT　　　N35 17.3 W119 49.5 M043　50　　003　0121　000
　　　　　　DSC 24057 136 144 IRNMN　489 491 0.01 00.21　0307　...
DSC/2　　　1040　　　148 156

CROWY   N34 54.9 W119 28.6 M024   50   028   0093   002
          DSC 23026 130 133 IRNMN   432 4290.03 00.18   0306   …
DSC/3   1043   142 145

MUPTT   N34 45.0 W119 19.7 M014   50   012   0081   001
          DSC 22019 131 133 IRNMN   450 4440.02 00.16   0305   …
DSC/1   1045   143 145

MDOTS   N34 39.1 W119 14.6 M008   50   007   0074   000
          DSC 21014 132 133 IRNMN   415 4080.01 00.15   0305   …
DSC/1   1046   144 145

GRIPR   N34 33.8 W119 10.0 M002   50   006   0068   000
          DSC 20013 132 133 IRNMN   433 4250.01 00.14   0304   …
DSC/1   1047   144 145

BIKNG   N34 26.6 W119 03.6 P004   50   009   0059   001
          DSC 19012 132 134 IRNMN   376 3640.01 00.13   0304   …
DSC/1   1048   144 146

RUNNN   N34 20.6 W118 58.4 P008   50   007   0052   001
          DSC 19014 132 134 IRNMN   336 3220.01 00.12   0303   …
DSC/1   1049   144 146

IRNMN   N34 15.3 W118 53.6 P010   50   007   0045   001
          DSC 19017 131 134 IRNMN   327 3120.01 00.11   0302   …
DSC/1   1050   143 146

SYMON   N34 09.9 W118 48.6 P014   49   007   0038   001
          DSC 19020 130 133 IRNMN   289 2770.02 00.09   0301   …
DSC/4   1052   142 145

BAYST   N34 01.8 W118 39.8 P021   49   011   0027   002
          DSC 20016 126 128 IRNMN   280 2770.02 00.07   0300   …
DSC/3   1054   138 140

JUUSE   N34 01.2 W118 33.2 P023   49   006   0021   001
          DSC 20013 084 085 IRNMN   273 2760.01 00.06   0299   …
DSC/1   1055   096 097

CLIFY   N34 00.6 W118 27.4 P023   49   005   0016   001
          DSC 21006 085 086 IRNMN   206 2100.01 00.05   0298   …
DSC/1   1056   097 098

DAHJR	N34 01.6 W118 21.5 P023	49	005	0011	002			
	DSC 22006 067 066 IRNMN	207	2130.02	00.03	0296	...		
DSC/2	1058	078 078						
GADDO	N34 02.0 W118 17.9 P024	49	003	0008	002			
	DSC 23006 071 070 IRNMN	185	1920.01	00.02	0294	...		
DSC/1	1059	082 082						
KLAX	N33 56.5 W118 24.5 P023	49	008	0000	005			
	DSC 26007 213 201 IRNMN	185	1790.02	00.00	0289	...		
DSC/2	1101	225 213						

ORIGIN	PVG/ZSPD	PUDONG	N3108.7 E12147.6	
DESTINATION	LAX/KLAX	LOS ANGELES INTL	N3356.5 W11824.5	
ALTERNATE	LAS/KLAS	MC CARRAN INTL	N3604.8 W11509.0	

```
( FPL-R5XX-I
-B744/H-SJ1J2J3WXY/SU1
-ZSPD0200
-N0516F310 SUR91D SURAK G455 LAMEN/N0516F310 A593 ONIKU Y60 POTET
 A593 GOMAR/N0512F330 A593 FUE Y60 ISAKY Y28 DGC V28 OLIVE Y28
 SANDA V28 CUE Y28 BIWWA V28 KCC Y88 DAIGO/N0505F350 Y886 ADNAP
 OTR7 EMRON/M085F350 DCT 44N160E 47N170E 48N180W 49N170W 51N160W
 51N150W 49N140W/N0487F370 DCT ZANNG DCT MITOH DCT ENI DCT BURGL
IRNMN2
-KLAX1101 KLAS
-DOF/200829 REG/B2427
 EET/RKRR0022 RJJJ0039 EMRON0303 160E0408 KZAK0424 170E0454
 180W0533 170W0616 160W0656 150W0735 140W0818 ADIZ0905 ZANNG0915
 KZSE0938 KZOA0954 KZLA1036
 RIF/49N40 VIDKU SHARI PEKAA SHARK1 CYVR RMK/AGCS EQUIPPED
-E/1219 P/TBN )
```

		MSA	TTK	DIST	FL	W/C	TIME	FUEL
ALTERNATE	- 1 KSEA	113	161	110	110	M001	00.24	9922

I CERTIFY THAT THIS FLIGHT IS RELEASED IN ACCORDANCE WITH ALL
APPLICABLE FAR 121 REGULATIONS INCLUDING CERTIFICATION

121.443 .445 .687 .689

DISPATCHER		BOW..450000	...MIN.REQ..297329
CAPT	F/O		F/E
CAPT SIGN			
OMNI FLT NBR	.. .. LEG	ZSPD TO CYVR..	

PER 49 CFR 1544 AND OMNI AIR INTERNATIONAL SECURITY MANUAL. I CERTI

COMPLIANCE WITH ALL SECURITY ISSUES/ CONCERNS FOR FLIGHT
HAVE BEEN RESOLVED AND COMMUNICATED TO THE IN-FLIGHT SECURITY
COORDINATOR(ISC).

DATE:......................................

OPERATIONS REPRESENTATIVE(NAME and SIGNATURE):....................
ISC: （ISC——飞行中的安全协调员）

STATION COPY   STATION COPY
OMNI AIR INTERNATIONAL

I CERTIFY THAT THIS FLIGHT IS RELEASED IN ACCORDANCE WITH ALL
APPLICABLE FAR 121 REGULATIONS INCLUDING CERTIFICATION
121.443 .445 .687 .689
FLIGHT NUMBER R5XX   747-400F/B2427   IFR
DISPATCHER   BOW..450000   ...   MIN.REQ..297329
CAPT   F/O   F/E
CAPT SIGN
OMNI FLT NBR   .. .. .. LEG   ZSPD TO CYVR.. ALTN KSEA..

PER REGULATORY REQUIREMENTS AND THE OMNI AIR INTERNATIONAL SECURITY
MANUAL.   I CERTIFY COMPLIANCE WITH ALL SECURITY ISSUES/ CONCERNS
FOR FLIGHT HAVE BEEN RESOLVED AND COMMUNICATED TO THE IN-FLIGHT
SECURITY COORDINATOR(ISC).

DATE:......................................

OPERATIONS REPRESENTATIVE(NAME and SIGNATURE):....................
ISC:

WINDS/TEMPERATURES ALOFT FORECAST

FD DATA BASED ON 2800NWS

	12000	18000	24000	30000	34000	39000	45000
						风向 风速 温度	
DE35R	2014P08	2111M02	2309M13	2509M27	2205M37	2107M51	1616M65
PD501	2013P08	2111M02	2309M13	2509M27	2206M37	2208M51	1206M65
HSH	2013P08	2111M02	2310M13	2609M27	2206M37	2208M51	1306M65
PD508	2014P08	2111M02	2309M13	2609M27	2206M37	2107M51	1108M65
ALDAP	2015P08	2111M02	2308M13	2708M27	2105M37	3206M51	0706M65
MATNU	2015P08	2112M03	2307M13	2906M27	3003M37	3105M51	0505M65
EMSAN	2015P08	2111M03	2307M13	2906M27	3003M37	3205M51	0505M65
SURAK	2015P08	2011M03	2108M13	2905M27	2903M37	3505M51	0505M65
LAMEN	2016P08	2011M03	1908M14	3004M27	0204M37	0304M51	0506M65
SADLI	1914P08	1811M03	1808M14	2305M27	1204M37	0805M51	1006M65
PONIK	1913P08	1711M03	1711M14	1307M27	1407M37	1206M51	1409M64
NIRAT	1912P08	1610M03	1715M14	1510M27	1308M38	1107M51	1411M64
ONIKU	1912P08	1610M03	1717M14	1511M28	1209M38	1211M51	1413M64
POTET	1911P08	1712M02	1717M14	1413M28	1210M38	1420M51	1414M64
AZUKI	2011P08	1714M02	1617M14	1414M28	1210M38	1208M51	1314M64
GOMAR	2010P08	1714M02	1617M14	1415M28	1212M38	1209M51	1314M64
FUE	2009P08	1712M02	1617M14	1417M28	1314M38	1313M51	1314M64
ISAKY	1908P08	1711M02	1616M14	1419M28	1418M38	1416M51	1315M64
DGC	1908P08	1811M01	1616M14	1519M28	1419M38	1416M51	1315M64
ONGHA	1906P08	1811M01	1715M14	1517M28	1417M38	1416M51	1313M64
SWE	2005P08	1810M01	1714M14	1517M28	1417M38	1416M51	1214M64
UBE	2005P08	1810M01	1714M14	1516M28	1417M38	1416M51	1214M64
KATTA	1905P09	1809M01	1613M14	1416M28	1416M38	1416M51	1214M64
MARCO	1806P09	1709M01	1612M14	1415M28	1317M38	1316M51	1214M64
MASKU	1806P09	1609M01	1612M14	1415M28	1317M38	1316M51	1114M64
HGE	1706P09	1609M01	1611M14	1414M28	1316M38	1316M51	1114M64
WASYU	1707P09	1608M01	1611M14	1313M28	1316M38	1317M51	1114M64
OYE	1707P09	1507M01	1610M14	1212M28	1316M38	1217M51	1113M64
OLIVE	1706P09	1407M01	1609M14	1111M28	1216M38	1217M51	1013M64
ASANO	1606P09	1307M01	1509M14	1011M28	1116M38	1117M51	1014M64
HYOGO	1505P09	1107M01	1509M14	1011M28	1116M38	1017M51	0914M64
SANDA	1505P09	1007M01	1408M14	0910M28	1116M38	1017M51	0914M64
CUE	1505P09	1007M01	1308M14	0910M28	1015M38	0917M51	0813M64
ADGUN	1505P09	0907M01	1308M14	1010M28	1014M38	0917M51	0813M64
BIWWA	1505P09	0907M01	1309M14	1010M28	1014M38	0917M51	0813M64
RITTO	1505P09	1007M01	1309M14	1110M28	1013M38	0917M51	0713M64
KCC	1405P09	0708M01	1309M14	1110M28	1013M38	0817M51	0713M64
SWING	1405P09	0707M01	1409M14	1209M28	0912M38	0816M51	0714M64
NAKTU	1404P09	0707M01	1409M14	1208M28	0811M38	0716M51	0714M64
SUGAL	1404P09	0707M01	1508M14	1308M28	0810M38	0715M51	0713M64

```
TENRU   1404P09 0706M01 1508M14 1307M28 0710M38 0615M51 0613M64
PIGAP   1404P09 0706M01 1607M14 1207M28 0709M38 0614M51 0612M64
OSPAN   1504P10 0606M01 1706M14 1206M28 0609M38 0514M51 0512M64
TEPEX   1603P10 0505M01 1705M14 1105M28 0508M38 0415M51 0513M64
GYODA   2703P10 0405M01 1804M14 1104M29 0408M38 0315M51 0411M64
AKAGI   2902P10 0305M01 2103M14 1003M29 0307M38 0216M51 0311M64
JD      2903P10 0206M01 2204M14 1005M29 0108M38 0117M51 0211M64
DAIGO   2903P10 0106M01 2506M14 0307M29 0009M38 0116M51 0112M64
BEKPO   2905P10 0006M01 2707M14 3410M29 3414M38 3521M51 0015M63
PABBA   3404P10 3504M01 2911M14 3214M29 3417M39 3523M51 3413M62
DOVIX   3304P10 3005M01 2813M14 3116M29 3319M39 3422M51 3211M62
ADNAP   3204P09 3007M01 2814M14 3018M29 3219M39 3320M51 3010M61
EMRON   2909P09 2814M02 2821M15 2820M30 2916M40 2914M50 2610M61
44E60   2943P08 2846M04 2848M15 2837M27 2837M38 2838M50 2840M63
47E70   2640M02 2772M08 2798M17 7712M29 7822M40 7928M52 7811M60
48N80   2846M05 2753M14 2886M24 7817M35 7832M42 7842M51 2988M54
49N70   2540M07 2547M15 2555M29 2663M43 2671M45 2666M45 2654M49
51N60   2349M03 2360M14 2396M25 7311M35 7320M44 7318M55 2469M55
51N50   2448P00 2450M10 2557M21 2574M35 2579M45 2683M58 2661M64
49N40   2838M02 2853M11 2966M23 2874M37 2979M47 3084M60 3058M62
VIDKU   2932M04 2853M11 2964M23 2966M37 2972M48 2977M61 2859M60
SHARI   2829P00 2745M11 2855M23 2856M38 2962M48 2863M60 2751M59
PEKAA   2726P01 2643M11 2752M24 2756M38 2859M49 2759M59 2654M58
FOCHE   2726P01 2547M11 2654M24 2561M39 2666M49 2667M59 2656M58
ANVAP   2625P00 2549M11 2555M25 2561M39 2468M49 2569M59 2656M57
CASDY   2625P00 2549M11 2456M24 2460M39 2465M49 2568M59 2656M57
URMIX   2625P01 2549M11 2456M24 2460M39 2463M49 2569M59 2655M57
ATUKI   2625P01 2548M11 2456M24 2458M39 2460M49 2566M60 2656M57
YVR     2525P01 2547M11 2455M24 2457M39 2458M49 2564M60 2556M58
PEMSU   2525P01 2547M10 2454M24 2455M39 2456M49 2562M60 2556M58
GOREG   2525P01 2547M10 2454M24 2455M39 2456M49 2562M60 2556M58
CYVR    2525P00 2547M11 2455M24 2457M39 2458M49 2563M60 2556M58
```

END OF JEPPESEN DATAPLAN
REQUEST NO.   115

第 4 章

延程运行

延程运行是指在飞机计划运行的航路上至少存在一点到任一延程运行可选备降机场的距离超过飞机在标准条件下静止大气中以经批准的一台发动机不工作时的巡航速度飞行 60 min 对应的飞行距离(以两台涡轮发动机为动力的飞机)或超过 180 min 对应的飞行距离(以多于两台涡轮发动机为动力的载客飞机)的运行。在中国民用航空局飞行标准司的咨询通告《延程运行和极地运行》(AC-121-FS-2019-009R2)中，延程运行(ETOPS, Extended Range Operations)和延伸航程运行(EDTO, Extended Diversion Time Operations)同义。对于 EASA，ETOPS 是 Extended Twins Operations 的缩写；对于 FAA，则是 Extended Operations 的缩写，它代表延展运行，不仅适用于双发飞机，还适用于两发以上的飞机(不包括两发以上的货机)。国际民用航空组织修订 ETOPS 规则后，将 ETOPS 更名为 EDTO，即延长改航时间的运行，适用两发及两发以上的飞机(包括两发以上的货机)。ETOPS 在确定航程时(即确定该飞行时间时)，假设飞机在标准条件下静止大气中以经批准的一台发动机不工作时的巡航速度飞行。EDTO 要求当双发飞机的一台发动机或主要系统发生故障时，飞机能在剩余一台发动机工作的情况下，以一发巡航速度(在标准条件下的静止大气中)，在规定时间内飞抵最近的备降机场(改航机场)。比如，获得"120 分钟 EDTO"就是指飞机在单发失效的情况下飞往备降机场所规定的时间不能超过 120 min。EDTO 主要应用于穿越沙漠、海洋还有极地的飞行，因为此时可供选择的备降机场较少。如果没有 EDTO 能力，意味着飞机需要选择尽量靠海岸线的航路飞行，保证可用备降机场在规定范围内，以确保安全。EDTO 的目的是在保证高水平安全性的前提下为航空承运人获取更多效益，使双发飞机不受先前条件限制，可与三发和四发飞机一样续航。通过 EDTO，航空公司可以根据自己的条件使用双发飞机开辟更多的直飞航线，获得更大的效益；航空旅客也可缩短旅行时间和获得更经济的票价，成为 EDTO 的直接受益者。因此，EDTO 已成为航空公司日常运营中的重要工作。

4.1 延程运行的历史发展和优势

4.1.1 延程运行的历史发展

早在 1936 年,FAA 就制定了类似于今天 FAR121 第 161 款的规定,当时的规定要求所有飞机(不论几发飞机)的飞行都要在距一个合适机场 100 mi(英里)以内的区域中,即所飞的航线上任意一点都必须在距某个合适机场 100 mi 以内的区域中,这 100 mi 是当时的许多飞机在一发不工作时在 60 min 内所能飞过的距离。1953 年,FAA 根据 20 世纪 50 年代初使用的活塞式发动机的可靠性制定了 FAR121 第 161 款的"60 分钟法则"——双发飞机一发故障后只允许飞行 60 min,也即一发故障后必须在 60 min 内飞到一个备降机场着陆。其目的是将飞机一发故障后另一发再发生故障的概率减小到可接受的水平(此概率与一发故障后的飞行时间有关,时间越短,概率越小),也就是把双发失效造成灾难性事故的可能性降低到可接受的水平。随着发动机技术的进步,活塞式发动机逐渐被可靠性和安全性更好的喷气式发动机代替,动力装置的可靠性有了很大提高,而且喷气式发动机的尺寸及推力的大小对故障率没有什么影响。1964 年,FAA 修改了相关条例,装配三发或四发发动机的飞机无须再遵循"60 分钟法则"的限制。航空管理机构和工业界认识到机体、航空电子、推进系统技术的发展提供了延长一发故障后飞行时间的可能性。

20 世纪 80 年代初,ICAO 组建了一个 ETOPS 研究组来考查双发喷气式飞机延程运行的可能性,该研究组确定了一些应当满足的标准来确保双发延程运行有很高的安全性,其结果是提出了对 ICAO 附件 6 的修正建议:如果能满足专门的 ETOPS 安全性标准,双发涡轮飞机在一发故障后的飞行时间允许超过 60 min,即可以实施延程运行,否则应以 60 min 为限。FAA 也在20 世纪 80 年代初对 ETOPS 进行了开创性的工作,于 1985 年 6 月 6 日发布了咨询通告AC120-42,确定了可以实施 120 min 延程运行的标准,即在满足该标准时,允许双发飞机在一发故障后可以飞行 120 min,如果再满足专门的附加标准,可以增加 15%,即一发故障后可以飞行 138 min。实施 120 min 延程运行时,北大西洋中还有一小块不能允许的飞行区域,如能实施 138 min 延程飞行则北大西洋刚好都是允许飞行区域,不必申请 180 min 延程运行,延程运行的时间越长,批准的条件越严格,这就是规定 138 min 标准的原因。

同一时期,其他一些民航管理机构也公布了自己的 ETOPS 标准,如英国民航局、法国民航总局、加拿大运输部、澳大利亚运输部等,其他许多国家则遵照 ICAO 附件 6 中的 ETOPS 标准。1988 年 12 月 30 日,FAA 公布了咨询通告 AC120-42A,取代了原来的 AC120-42,在 AC120-42A中对 ETOPS 的有关概念、批准 ETOPS 应考虑的因素及批准实施 75 min、120 min、180 min 延程飞行的标准做了规定。AC120-42A 基本上成了各国民航局批准 ETOPS 的准则。1993 年,JAA(欧洲联合适航性管理机构)综合欧洲各国的规则及 FAA 标准的优点制定了自己的标准——咨询材料汇编 AMJ120-42。所有这些 ETOPS 标准基本上都是要确保双发飞机延程飞行至少要像目前的三发、四发飞机一样安全和可靠。

2000 年 3 月,FAA 以政策函件(Policy Letter)EPL20-1 的形式批准了 B777 飞机在北太平洋区域[北纬 40°以北的太平洋区域,包括北太平洋空中交通服务航路,以及公布的位于日本和北美之间的太平洋组织航迹系统(PACOTS)的航迹]的 207 min 延程飞行,即在满足一些额外要求时可以在 180 min 延程飞行的基础上再增加 15%。因为在北太平洋区域俄罗斯和阿留申群岛的一些机场可能在恶劣天气下不能用作 ETOPS 备降机场,180 min 的允许飞行区域不能覆盖北太平洋区域,207 min 的允许飞行区域才能覆盖整个北太平洋区域。但是否批准只能依据每次飞行的具体情况确定,即由于政治或军事原因、火山活动、机场暂时的条件、机场天气条件低于签派要求或出现其他天气事件,导致飞机无法在 180 min 的改航时间内抵达一个延程运行备降机场时才能实施 207 min 的延程运行。为此,应在签派或飞行放行单中规定以最多 207 min 改航时间能飞抵的最近的可用延程运行备降机场,应考虑空中交通服务部门的首选航迹,飞机时限性最严格的延程运行关键系统以及货舱和行李舱的火情抑制系统的批准时间应当至少为 222 min(207 min+15 min),并且飞机必须配备卫星通信设备。

2001 年 3 月,FAA 又制定了极地运行要求,无论是几发的客机或货机都必须满足相关的要求才能在北极、南极区域飞行。2003 年 11 月,JAA ETOPS 标准被 EASA 采用,并发布为 EASA AMC 20-6,在技术要求方面无更新。

2007 年 1 月 16 日,FAA 公布了新的 ETOPS 条例[Extended Operations (ETOPS) of Multi-Engine Airplanes],此条例于 2 月 15 日生效。2007 年前,ETOPS 代表 Extended-range Operations with Two-engine Airplanes(双发飞机延伸航程运行),现在按照最新的 ETOPS 条例,缩写词 ETOPS 代表 Extended Operations (延程运行),另一个可替代的也更精确地描述了 ETOPS 含义的缩写是 EDTO(Extended Diversion Time Operations,延长改航时间运行),这两个缩写的含义相同。新条例使关于延程飞行的咨询通告(1988 年 12 月 30 日的 AC120-42A)上升为法规。以前的咨询通告仅针对双发飞机,延程时间被限制到 180 min(除 B777 在北太平洋航线被允许增加到 207 min 外),新条例允许双发飞机延程超过 180 min,对三发、四发客机,改航时间超过 180 min 的也被视为延程运行,要满足适用的 ETOPS 条例。此外,对计算由 ETP(等时点)到备降机场的临界油量的规定等方面也做了修改。新条例中也包括了极地运行政策。2008 年 6 月 13 日,FAA 公布了 AC120-42B"ETOPS and Polar Operations",取代了原先的 AC120-42A。在 AC120-42B 中对延程运行和极地运行提供了详尽的指南。

澳大利亚民航安全局(CASA)于 2007 年 7 月颁布了新的延程运行规则。国际民用航空组织(ICAO)、欧洲联合航空管理局(JAA)/欧洲航空安全局(EASA)和其他各方也对延程运行条例进行了类似的更新。

欧洲航空安全局(EASA,European Aviation Safety Agency)于 2010 年 12 月 23 日发布了 Revision 2 to AMC 20-6(Extended Range Operation with Two-Engine Aeroplanes ETOPS Certification and Operation),完成了对 ETOPS 指南的更新,和 FAA 的规定进行了协调。EASA 只规定了双发飞机的延程运行,对三、四发飞机没有考虑延程运行问题。

2017 年,ICAO 对附件 6 第 I 部分进行了第 36 次修订,出版的《延长改航时间运行(EDTO)手册》中用新术语 EDTO(延长改航时间运行)替代之前所用术语 ETOPS(双发飞机延伸航程运行)。

中国民用航空局于 2010 年 1 月 4 日公布了 CCAR-121-R4,对其中和延程运行有关的条例进行了修改,使其与 FAR-121 和 AC120-42B 一致。中国民用航空局依据 CCAR-121 W 章下发了咨询通告《延程运行和极地运行》(AC-121-FS-2019-009R2),为合格证持有人实施延程运行和极地运行提供了指南、监督和指导。2021 年 3 月 15 日,中国民用航空局对公布的 CCAR-121-R7 中与延程运行有关的条例又进行了修改。2021 年 3 月 15 日,中国民用航空局对公布的 CCAR-121-R7 中对与延程运行有关的条例进行了修改;2024 年 4 月 13 日,又一次对延程运行有关条例进行了修改,并把名称统一为"EDTO"。

4.1.2 延程运行的优势

延程运行在保证安全性的前提下,可以使公司获得更大的经济效益,其优势显而易见:

(1)可以大大节省运营成本。相比四发飞机,双发延程运行喷气机飞行时每座每英里运营成本低 5% ~9%,使用燃油量少。双发飞机获得 ETOPS 资格后,可以代替三发、四发飞机承担客运量比较小的航线运输任务,提高上座率,从而提升经济效益。

(2)更多的直飞机会。自 1985 年 120 min ETOPS 条例出台后,航空公司可以有效地将 B757、B767、A300、A310 等双发喷气式飞机投入跨大西洋航线的飞行中。这些飞机载客量较少,航空公司可以通过增加航班为出行的人提供更多的选择,航空公司也可在欧洲和北美之间开通更多的点到点的新航线,这些都为航空公司节省了资金,降低了经济风险。

如果双发飞机没有被允许实施延程运行,则图 4-1-1 所示的圆内区域是允许飞行区域,图中 A—P_1—P_2—P_3—B 航路是允许的,这个航程距离即合法距离。而 A—B 的直飞航路(如大圆航线)是不允许的,因为有个别航段不满足条例要求。这一条例限制了双发飞机的飞行区域,可能使飞行距离不得不延长,使油耗增加。例如,从纽约到西班牙首都马德里(见图 4-1-2)的飞行即为这种情况,正是这种"60 min 法则"限制了双发飞机的越洋飞行。

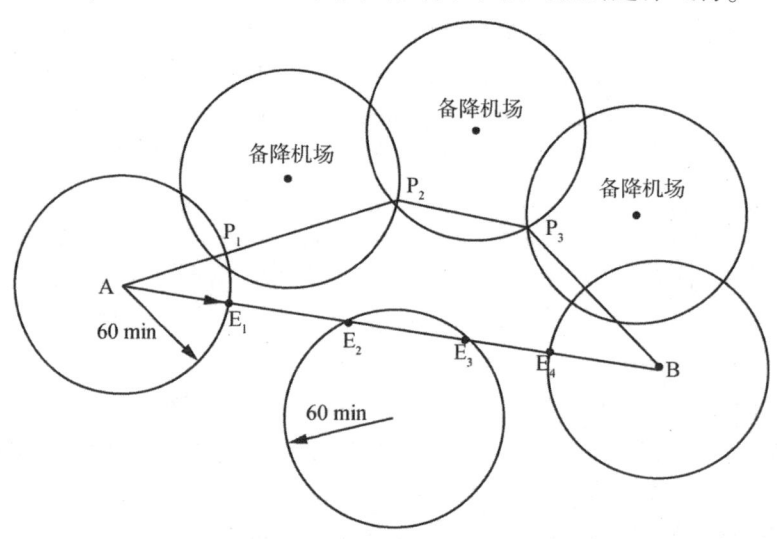

图 4-1-1　按"60 min 规则"确定的允许飞行区域——圆内区域

如果被批准实施 120 min 的延程运行,就可以从纽约沿大圆航线直飞马德里。从纽约至伦敦的航线亦是这种情况,见图 4-1-3,如能实施 120 min 的延程运行,就能沿大圆航线直飞伦敦,与 60 min 的非延程运行比较,营运者用 A310-300 可节省高达 2.4 t 的燃油或增加等值的业载,也减少了中途备降机场的数目,提高了效率。

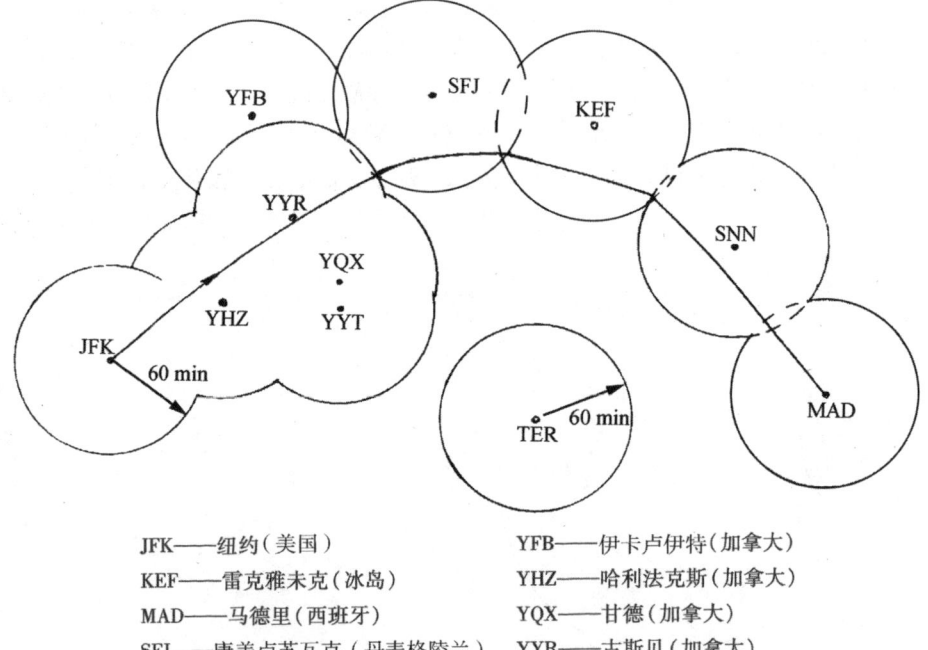

JFK——纽约（美国）　　　　　　YFB——伊卡卢伊特（加拿大）

KEF——雷克雅未克（冰岛）　　　 YHZ——哈利法克斯（加拿大）

MAD——马德里（西班牙）　　　　 YQX——甘德（加拿大）

SFJ——康盖卢苏瓦克（丹麦格陵兰）　YYR——古斯贝（加拿大）

SNN——香农（爱尔兰）　　　　　 YYT——圣约翰斯（加拿大）

TER——特塞拉（葡萄牙）

图 4-1-2　纽约—马德里航线示意图

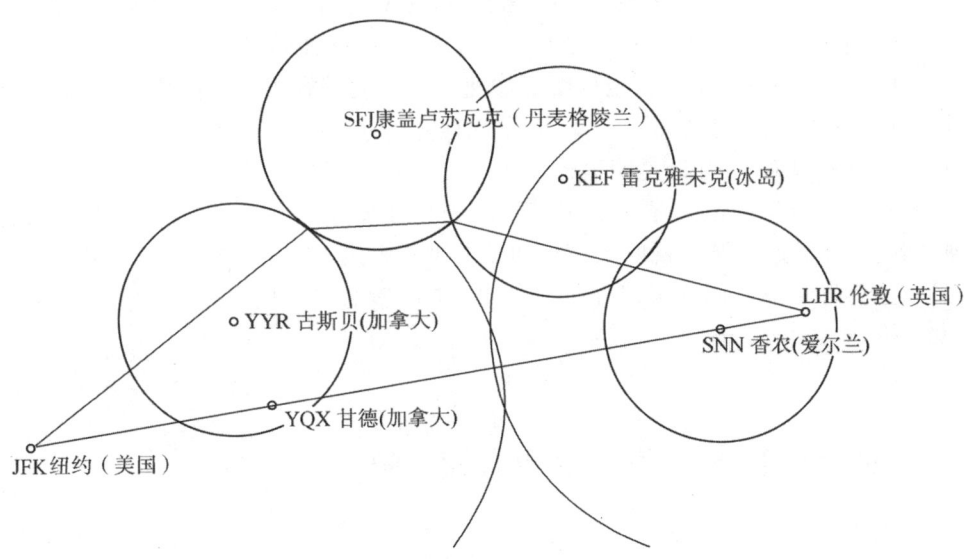

纽约至伦敦水平航迹：在60 min圆内（半径435 NM）

图 4-1-3　纽约—伦敦航线示意图

　　如果用户（航空公司）满足延程运行的要求，被批准实施 120 min 甚至 180 min 的延程运行，即一发故障后允许飞行 120 min 或 180 min，则允许飞行区域大为扩大，就可以飞一条较短的航线甚至直飞航线。

　　从内罗毕至新加坡的航线（见图 4-1-4）用 60 min 改航时间是不能实现的，这是因为没有足够的航路备降机场。然而，如将改航时间延长到 120 min，就可使营运者在这条航线上使用双发喷气机,不然就只好用较大的三发和四发飞机。如果这种航线客流量小,使用双发喷气机比用大的三发和四发飞机的经济效益要好。对于那些高密度航线,如北大西洋航线,使用双发喷气机可增加飞行频率(班次),对于相同的客流量,用大飞机飞,航班次数或者说航班密度就小,因为增加航班密度必然使每个航班的上座率或者说载运比减少,使经济效益下降,甚至亏本。用较小的双发飞机就可以增加航班密度,这既方便了旅客,也提高了航空公司经营的灵活性。

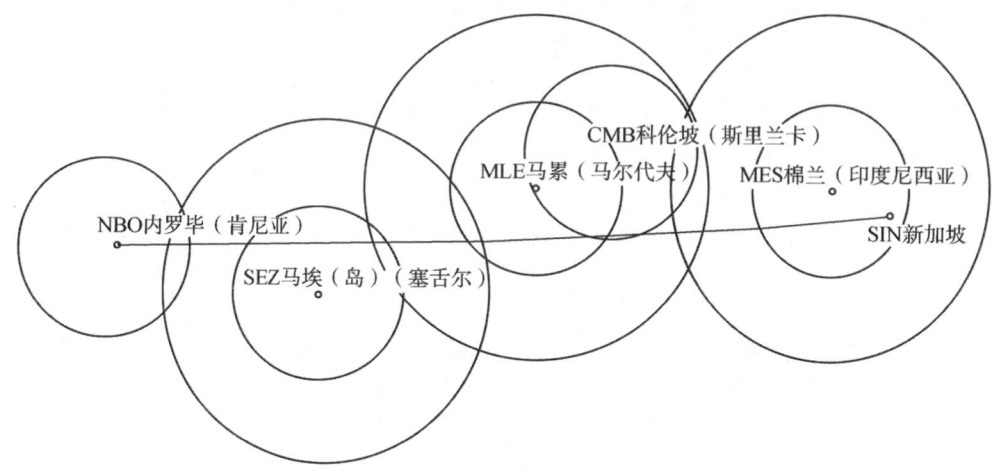

内罗毕—新加坡航线,只有用120 min转场时间才有可能实现

图 4-1-4　内罗毕—新加坡航线示意图

(3)提高了公司机队的灵活性。以往三发、四发飞机均为远程客机,因此这些飞机就只能用来执行远程航线,而不适用于国内的中程干线的飞行,这种局限性不利于提高公司机队的灵活性。ETOPS 的成功应用也在一定程度上加速了三发远程客机市场的萎缩,并减少了四发远程客机的市场份额,因为过去需要三发、四发远程客机才能执飞的航线现在可以由双发客机取代,这对于航空公司来说,好处是明显的。

(4)机队的机场适应性更强。某些国外机场等级较低,不能提供重型客机着陆的跑道宽度和强度等,要开辟这种航线就要用较小的双发飞机来执飞。

(5)安全性和可靠性同样可以保证。如 B777 机型的发动机安全性甚至已经超过某些三发、四发飞机。

4.2　延程运行的有关规定和名词术语

4.2.1　延程运行的有关规定

4.2.1.1　CCAR-121 规定

《大型飞机公共航空运输承运人运行合格审定规则》(CCAR-121-R8)中对延程运行的航路类型和飞机限制有明确的规定。

第 121.157 条　飞机的航路类型限制

(a)除经依据本规则 W 章的规定得到局方批准且满足合格证持有人运行规范许可的条件下,合格证持有人不得使用以涡轮发动机为动力的飞机实施以下运行:

(1)延程运行;

(2)在北极区域内的运行;

(3)在南极区域内的运行。

(b)合格证持有人用于延伸跨水运行的陆上飞机,应当是按照《运输类飞机适航标准》(CCAR-25)中的水上迫降规定审定合格或者被批准为适合于水上迫降的飞机。

CCAR-121 的 W 章延程运行与极地运行的第 121.711-720 条对延程运行做了规定。

第 121.711 条　总则

(a)合格证持有人除经局方批准外不得实施以下运行:

(1)在飞机计划运行的航路上至少存在一点到任一延程运行可选备降机场的距离超过飞机在国际标准大气和静止空气条件以经批准的一台发动机不工作时的巡航速度飞行 60 分钟对应的飞行距离(以两台涡轮发动机为动力的飞机)或超过 180 分钟对应的飞行距离(以多于两台涡轮发动机为动力的载客飞机)的运行;

(2)在北极区域内的运行;

(3)在南极区域内的运行。

(b)本章还规定了获得极地运行批准的各项要求。

(c)只有当机体发动机组合获得中国民航颁发的延程运行型号设计批准时,合格证持有人才有资格实施延程运行。延程运行型号设计批准通常在相应的机体发动机组合的《飞机飞行手册》(AFM)、型号合格证(TC)或补充型号合格证(STC)中给出。

(d)为纠正运行中出现可能危及所要求的可靠性水平时,局方可以在任何时候要求修改构型、维修与程序标准。局方将根据需要采取行动,要求对构型、维修与程序标准进行修订,以达到和保持所需的可靠性水平。修订之前的构型、维修与程序标准被认为不再适合于继续从事延程运行。

(e)延程运行的批准通过颁发或修改合格证持有人的运行规范的方法进行。

(f)对于每一次延程运行(EDTO),签派放行单中应当包含该次延程运行(EDTO)最大改航时间。

第 121.713 条　附加要求

(a)对于从航路上一个点飞往一个航路备降机场的运行超过 60 分钟的情况,合格证持有人必须确保:(新增条款)

(1)对于所有飞机:

(i)指定了航路备降机场;

(ii)向飞行机组提供指定航路备降机场的最新资料,包括工作状况与气象条件。

(2)对于双发涡轮驱动实施延程运行(EDTO)飞机,除非根据最新的资料评估,指定的航路备降机场气象条件在预计使用时间能够达到或高于合格证持有人运行规范批准的机场运行最低标准,否则不得超过本规则第 121.711 条(a)款第(1)项规定的时间飞行。如果经确定的任何条件不允许飞机在预计的使用时间在该机场安全进近并着陆,必须确定一个备用的运行方案。

(b)合格证持有人应当为所有已获得延程运行(EDTO)批准的飞机制定签派放行程序,以防止将飞机被签派到改航时间超过经局方批准的延程运行(EDTO)关键系统时间限制能力的航路上。(新增条款)

(c)对于超过 180 分钟的延程运行(EDTO)和极地运行,该次运行所指定的每个延程运行(EDTO)指定备降机场和极地运行改航机场,都应当具备足以保障乘客和机组生存需求的条件。

(d)合格证持有人应当在下列程序和大纲中增加本条(a)款至(c)款的内容,以保证整体安全水平:(新增条款)

(1)运行控制与飞行签派程序;

(2)运行程序;

(3)训练大纲。

第121.714条 延程运行(EDTO)的通信设备

如果实施超过180分钟的延程运行(EDTO),满足本规则第121.97条(c)款所规定的通信系统应当能提供基于卫星的语音通信,通话质量与固定电话相当。该系统应当能够在飞行机组和空管人员之间以及飞行机组和合格证持有人之间提供通信。合格证持有人在规划延程运行(EDTO)时,还应当考虑改航至延程运行(EDTO)备降机场所需的可能的航线和海拔高度,以确定基于卫星的语音通信设备是否可用。如果存在无法利用基于卫星的语音通信设备的情况,或者因其质量低劣而无法进行语音通信,则应当有另一种通信系统作为替代。

第121.715条 延程运行备降机场:救援与消防服务

(a)除本条(b)中的规定之外,在签派或飞行放行单上列明的每一个延程运行备降机场,均需能够提供下列救援和消防服务(RFFS):

(1)如果延程运行时间在180分钟内,每个指定的延程运行备降机场应当能够提供等效于或高于国际民航组织规定的第4类救援和消防服务(RFFS)的要求。

(2)如果延程运行时间超过180分钟,每个指定的延程运行备降机场应当能够提供等效于或高于国际民航组织规定的第4类救援和消防服务(RFFS)的要求。此外,飞机还应当保持在距能够提供等效于或高于国际民航组织规定的第7类救援和消防服务(RFFS)的合适机场的一定距离内,以便能在依据其延程运行批准的改航时间内飞抵该合适机场。

(b)如果无法直接在一个机场利用本条(a)要求的救援和消防设备与人员,只要该机场通过当地获得消防增援与消防服务能力后,能够达到本条(a)的要求,合格证持有人仍然可在签派或飞行放行单上将该机场列为备降机场。在改航的航路运行过程中如果当地资源可以及时被告知,30分钟的增援响应时间应当是充足的。在改航飞机飞抵备降机场时,增援设施与人员应当可用,另外,只要改航飞机需要救援和消防服务,这些增援设施与人员就应当始终处于随时可用状态。

第121.716条 手册内容

合格证持有人的手册中应当包含下列内容:

(a)对于所从事的延程运行,支持所有飞行阶段的性能数据。

(b)对于载客运行:

(1)如果实施超过180分钟的延程运行,应当制订每一相应延程运行备降机场具体的旅客航程恢复计划;

(2)对于在北极区域和南极区域的运行,应当制订每一相应延程运行改航机场具体的旅客航程恢复计划。

第121.718条 延程运行型号设计批准依据

除2015年2月17日之前制造的安装两台以上发动机的载客飞机和仅用于实施飞行时间少于或等于75分钟的延程运行的双发飞机之外,合格证持有人只能在下列条件下实施延程运行:飞机获得延程运行型号设计批准且每架用于延程运行的飞机均符合其构型、维修和程序文件的要求。延程运行型号设计批准通常在相应的机体发动机组合的《飞机飞行手册》(AFM)、型号合格证(TC)或补充型号合格证(STC)中给出。

第121.719条 延程运行持续适航性维修方案(CAMP)

为实施延程运行,合格证持有人应当制定并遵照由局方批准的运行规范中持续适航维修方案的要求对用于延程运行的机体发动机组合实施维修。合格证持有人可以通过对飞机制造厂推荐的维修方案,或对现在已获批准的持续适航维修方案增补维修要求来制定适用于延程运行的持续适航维修方案。

本书只摘录了第121.719条的部分内容,这个持续适航维修方案应当包括的内容详见规章CCAR-121.719条规定的内容。

第121.720条 机组成员与签派人员培训要求

合格证持有人应当对其机组成员和签派人员就其在合格证持有人的旅客航程恢复计划中的职责和作用进行培训。

4.2.1.2 咨询通告《延程运行和极地运行》

依据CCAR-121 W章制定了咨询通告《延程运行和极地运行》(AC-121-FS-2019-009R2),为合格证持有人实施延程和极地运行提供指南,也为局方对合格证持有人的延程运行或极地运行的批准和持续监督检查提供指导。该咨询通告从延程运行的规章要求、运行要求、申请、审查、验证、批准等方面做了详细的规定。

4.2.1.3 延程运行规定国内与国际上的差异

关于ETOPS的规范,不同的机构在某些要求上也存在差异。

(1)CMP方面要求的差异

FAA的构型、维修和程序(CMP,Configuration, Maintenance, and Procedure)文件的原始版本是固定的。可靠性检测过程所要求的任何后续修订都是通过适航指令来发布的。但EASA的CMP的原始版本不是固定的。如果出现安全问题,或由于飞机技术的改进、规则或参考文件的修改导致内容需要进行改进,可以随时启动CMP修订,意味着EASA的CMP总是最新的。但ETOPS批准是由适航当局对现运行飞机的可靠性的持续监测,不是永远授权的,随着可靠性监测的进行可能会导致机身或发动机的ETOPS标准发生变化。我国《延程运行和极地运行》的咨询通告CMP中所包含的要求通常是在机体/发动机组合的延程运行初始型号设计批准时由局方确定的。CMP文件通常由飞机制造厂公布并保持其有效性。飞机制造厂可能在初始制定的延程运行要求基础上持续发布CMP的修订版。

(2)救援和消防服务(RFFS)方面的差异

在救援和消防服务(RFFS)方面,CCAR的规章条例与FAA的规范相同,但和ICAO、EASA有些许差异。ICAO对于航路EDTO备降机场的救援和消防服务的要求是:若运营人可以提前30 min发出通知,则最大审定起飞质量超过27000 kg的飞机的机场救援和消防服务等级应达到4级,或者所有其他飞机的机场救援和消防服务等级应达到1级;若运营人不能提前30 min发出通知,则可接受的救援和消防服务保护可能比飞机救援和消防服务等级低两个等级。EASA发布的RFFS类型与ICAO的第四类等同,对于ETOPS的航路备降机场可提前30 min通知机场提供RFFS。

(3)通信方面的差异

在ETOPS的通信方面,CCAR-121.714、FAA-121.99条和ICAO 3.6.2.1的规章内容相差不大。其中在咨询通告《延程运行和极地运行》中也强调超过180 min的延程运行,要求两套语音通信系统中的一套必须是SATCOM VOICE。CCAR-121.714条的目的是确保机组具备随时可用的语音通信,其质量应相当于固定电话。如果飞机已经装有一套 SATCOM VOICE 系统,另一套语音通信系统不一定是 SATCOM VOICE。FAA 对于超过 180 min 的 ETOPS 强制要求卫星通信。EASA 要求双发飞机实施 ETOPS 时,运营人应确保飞机具有能够在正常和计划的应急高度可以与适当的地面站进行联系的通信设备。对于有语音通信设施的 ETOPS 航路,应提供语音通信。对于超过 180 min 的所有 ETOPS,应安装基于语音或数据链路的可靠通信技术设备。如果语音通信设施不可用,且无法进行语音通信或语音通信质量较差,则应确保使用替代系统进行通信。

总体而言,关于 ETOPS 的相关规章在 ICAO、FAA 和 EASA 三个机构中有些许不同,可从适应性、飞机认证、极地运行和旅客航程恢复计划四个方面总结,如表 4-2-1 所示。

表 4-2-1   延程运行规定差异性

差异项目	ICAO	FAA	EASA
适应性	ICAO EDTO 标准已经取代 ICAO 附件 6 中现有的 ETOPS 标准。EDTO 标准允许基于推进系统可靠性和当前在役双发飞机的整体运行安全性对双发飞机进行"无限制"改航时间操作。另外,ICAO 还为运行 EDTO 的三发和四发飞机引入了类似的措施,比如选择合适航路备降机场和监控航路时考虑的飞机限时系统等,且对于三发和四发飞机的 EDTO 操作没有维护要求。对于双发飞机,EDTO 阈值和最大改航距离以单发不工作的巡航速度计算,而对于两发以上的飞机必须使用全发工作的巡航速度计算阈值和改航距离。对于超过 180 min 的运行,具体可基于各个国家政策,再依据风力和温度预测调整改航速度	自 2007 年以来,FAA 引入了自己的 ETOPS 原则和标准,它们也适用于多发飞机的运行。对于双发飞机,在静止空气中标准条件(ISA)下,从合适机场以一发不工作(OEI)速度飞行 60 min。对于三发和四发飞机,在 ISA 条件下静止空气中从合适机场以 OEI 速度飞行 180 min。FAA 还引入了"非限制性" ETOPS 概念,即授权改航时间超过 180 min 的延程运行,其中最大改航时间和最大改航距离由飞机制造商根据最大限制系统的能力予以证明	2010 年前,EASA 的 ETOPS 法规仅适用于超过 60 min 阈值(以一发不工作速度计算的距离)的双发飞机。2010 年,EASA 修订了 ETOPS 的规定,允许最长改航时间超过 180 min 的飞机运营。但 EASA 还没有发布任何有关两发以上飞机的 ETOPS 运行的具体规则。为确定如何保证两发以上飞机的 ETOPS 运行内容,EASA 审查了其 ETOPS 的规定,审查后的 ETOPS 规定草案被命名为"远程操作"(LROPS),但此草案已被取消。目前 EASA 将继续审查 ETOPS 标准并相应地更新现行规则
飞机认证	在 EDTO 航线上运行的双发飞机需要获得 EDTO 的认证。与 FAA 不同的是,ICAO 的 EDTO 标准不要求两发以上的飞机获得任何 EDTO 认证。换言之,只要最大改航时间不超过最大限制系统的能力,任何获得型号设计批准的三发或四发飞机均可在 EDTO 航路上运行,但是超过阈值的三发或四发飞机的运行需要获得 EDTO 运行批准。在 ICAO 文件中,双发飞机的阈值参考值为60 min,两发以上飞机的阈值参考值为 180 min	在 ETOPS 航线上运行的双发飞机需要专门的 ETOPS 认证。在 2015 年 2 月之前(即自 FAA ETOPS 条例生效之日起八年内)生产的两发以上的飞机无须获得 ETOPS 认证即可在 ETOPS 航线上运行	在 ETOPS 航线上运行的双发飞机需要专门的 ETOPS 认证,两发以上的飞机不需要专门的认证。和 ICAO EDTO 一样,如果最大的改航时间不超过最大限制系统的能力,则两发以上的飞机也可以进行延程运行

续表

差异项目	ICAO	FAA	EASA
极地运行	在国际民用航空组织附件 6 的标准或相关指导材料中关于极地运行的要求并未具体述及;但一些国家已针对极地运行的授权确定了其他额外要求,在这些额外要求中也有可能适用于 EDTO 考虑的事项	FAA 将北纬 78°以北和南纬 60°以南所涵盖的区域定义为极地地区。FAA ETOPS 条例要求所有申请穿越南、北极地区的航线的运营商需遵守极地运营要求,例如制定避免燃油结冰策略和监测机组或太阳耀斑活动	EASA 在现有的运行规则中已经讨论了极地运行,除了额外补充的全货物作业外,在极地运行的每个合格证持有人都必须制订旅客航程恢复计划。该计划还应特别考虑极端寒冷天气的可能性、有限的旅客设施,以及立即启动旅客航程恢复计划的需要
旅客航程恢复计划	ICAO EDTO 标准中未提及旅客航程恢复计划	对于超过 180 min 的 ETOPS,运营人必须为每个 ETOPS 备降机场制订旅客航程恢复计划。该计划应验证机场基础设施的可接受性和服务能力,同时还应该考虑医疗、人身安全保障和通信等方面。此外,机场方面还应包含乘客和机组人员的疏散工作。48 h 内的旅客航程恢复计划是可以接受的,这样的计划应当能够满足所有关于为旅客和机组提供照顾和保证安全的要求	EASA 的规定中除极地运行外,不涉及专门的旅客航程恢复计划,因为它被认为不在其职责范围之内

4.2.2　延程运行的名词术语

关于延程运行的相关术语参考《延程运行和极地运行》(AC-121-FS-2019-009R2),包括:

(1)经批准的一台发动机不工作的巡航速度:是指合格证持有人选定且经局方批准的在飞机使用限制范围内的一个速度,用于:

①计算一台发动机不工作时所需燃油;

②确定在延程运行中飞机能否在批准的最大改航时间内飞抵延程运行指定备降机场。

(2)门限时间(Threshold Time):在标准条件下静止大气中以经批准的一台发动机不工作时的巡航速度飞行 60 min 对应的飞行航程(以时间表示)(以两台涡轮发动机为动力的飞机)或 180 min 对应的飞行航程(以时间表示)(以多于两台涡轮发动机为动力的载客飞机)。

(3)延程运行可选备降机场(Suitable ETOPS Alternate):对于特定延程运行航线,不考虑临时状况,列入合格证持有人运行规范的可选的航路备降机场。这些机场必须满足 CCAR-121.197 条规定的着陆限制要求。它可能是下列两种机场之一:

①经审定适合大型飞机公共航空运输承运人所用飞机运行的,或等效于其运行所需安全要求的机场,但不包括只能为飞机提供救援和消防服务的机场。

②对民用开放的可用的国内外军用机场。如果某军用机场满足合格证持有人安全运行的

基本要求,其军方主管部门以某种形式宣布向民用航空提供紧急情况下备降的服务支持,合格证持有人已经获得该机场运行的必要资料并且向局方证明可以在延程运行期间随时与该机场运营人之间建立可靠的通信联系,则可以将该军用机场列为延程运行可选备降机场。

(4)延程运行指定备降机场(Designated ETOPS Alternate):是指列入了合格证持有人的运行规范,并且考虑到当时的状况,在签派或飞行放行时预计可以供延程运行改航备降使用的,在签派或飞行放行中指定的航路备降机场。这一定义适用于飞行计划,并不限制机长在最终改航备降决策时根据实际情况选择其他的备降机场。

(5)延程运行区域(ETOPS Area):对于以两台或两台以上涡轮发动机为动力的飞机,延程运行区域是超过其门限时间才能抵达一个延程运行可选备降机场的区域。

(6)延程运行航线(ETOPS Route):是指计划航路上,包括灵活航路,至少有一点处在延程运行区域中的航线。在这样的航线上实施延程运行需要获得局方的批准,并在运行规范中列明。特定的延程运行航线是通过起飞机场和目的地机场以及两者之间的航路来确定的。

(7)延程运行航段(ETOPS Segment):是指计划航路上处在延程运行区域中的部分。一条延程运行航线上可能存在多段延程运行航段。每一段延程运行航段都是由前后两个延程运行指定备降机场来确定的。

(8)延程运行进入点(EEP,ETOPS Entry Point):是指延程运行航路上进入延程运行航段的进入点。

(9)延程运行等时点(ETP,ETOPS Equal-time Point):延程运行航路中的一点,考虑到预计飞行高度和预报风的影响,自该点以经批准的一台发动机不工作的巡航速度飞向相邻两个延程运行指定备降机场的计划飞行时间是相等的。

(10)延程运行退出点(EXP,ETOPS Exit Point):是指延程运行航路上退出延程运行航段的退出点。

(11)批准的最大改航时间(Maximum Authorized Diversion Time):为了延程运行航路计划之用,经局方批准在合格证持有人运行规范中列明的延程运行可使用的最大改航时间。在计算最大改航时间所对应的飞行距离时,假设飞机在标准条件下静止大气中以经批准的一台发动机不工作的巡航速度飞行。对于某特定机体/发动机组合,批准的最大改航时间对应的是经局方批准的最大改航距离。

(12)最早预计到达时刻(Earliest ETA):对于每一延程运行指定备降机场,假设飞机飞抵前一个相关等时点后以经批准的一台发动机不工作的巡航速度直线飞抵该机场的时刻。

(13)最晚预计到达时刻(Latest ETA):对于每一延程运行指定备降机场,假设飞机飞抵下一个相关等时点后以经批准的一台发动机不工作的巡航速度直线飞抵该机场的时刻。

(14)指定备降机场的改航备降关注时间段:是指从最早预计到达时刻开始,至最晚预计到达时刻之间的时间范围。

(15)燃油关键点(Fuel Critical Point):延程运行航线各等时点中,所需临界燃油量大于根据正常备份油量计算出的飞行计划中在该点的预计剩余燃油量且差值最大,或者,所需临界燃油量等于或小于根据正常备份油量计算出的飞行计划中在该点的预计剩余燃油量且差值最小。

(16)临界燃油量(Critical Fuel):假设飞机在燃油关键点一台发动机失效,按照合格证持有人延程运行临界燃油量的相关要求,飞抵延程运行指定备降机场所需的最少燃油量。

(17)旅客航程恢复计划(Passenger Recovery Plan):在航班改航备降之后,合格证持有人应当恰当地安置旅客和机组,保障旅客和机组的基本生存条件,满足旅客和机组的基本需要,并尽快安排旅客恢复其前往原定目的地的旅程。针对上述整个过程,合格证持有人所制定的方案就是旅客航程恢复计划。

(18)延程运行合格人员:是指圆满完成了合格证持有人的延程运行培训,为合格证持有人从事延程运行相关工作的人员。

(19)延程运行关键系统(ETOPS Significant System):是指包括发动机在内的飞机系统,其失效或发生故障时会危及延程运行安全,或危及飞机在延程运行改航备降时飞行和着陆的安全。延程运行关键系统被分为一类和二类延程运行关键系统。

①一类延程运行关键系统(ETOPS Group 1 Significant System)为:

(ⅰ)对飞机发动机数量所提供的安全冗余度产生直接影响的,具有"失效后安全"特征的系统;

(ⅱ)在发生故障或失效时可能导致空中停车、推力控制丧失或其他动力丧失的系统;

(ⅲ)能在一台发动机失效时提供额外的安全冗余度,进而显著提高延程运行改航备降过程中安全水平的系统;

(ⅳ)在一台飞机发动机不工作的飞行高度上保持长时间运行必不可少的系统。

②二类延程运行关键系统(ETOPS Group 2 Significant System)是除一类延程运行关键系统之外的延程运行关键系统。二类延程运行关键系统的失效不会导致航空器飞行性能的丧失或客舱环境问题,但可能导致航空器返航或改航备降。

(20)时限系统:是指那些预计在导致飞机改航备降的最临界的情况出现后,为了保障改航备降的安全运行必须持续工作,且具有最高连续工作时间限制的飞机系统,如货舱抑火系统。

对于延程运行的飞行时间有以下要求:

(a)对于不超过 180 min 的延程运行,在标准条件下静止大气以经批准的一台发动机失效巡航速度备降至计划的延程运行指定备降机场所需的时间,不能超过该飞机时限最严格的延程运行关键系统(包括货舱抑火装置)所规定的最长时限减去 15 min。

(b)对于双发或多于两台发动机的飞机超过 180 min 的延程运行,在正常的全发巡航高度,修正了风和温度的影响,以全发运行的巡航速度备降至计划的延程运行指定备降机场所需的时间,不能超过该飞机抑火系统的最大合格审定时限减去 15 min。

(c)对于双发飞机超过 180 min 的延程运行,在正常的一台发动机失效巡航高度上,修正了风和温度的影响,以经批准的一台发动机失效巡航速度备降至计划的延程运行指定备降机场所需的时间,不能超过该飞机延程运行关键系统(不包括货舱抑火系统)最大时限减去 15 min。

(21)构型、维修和程序(CMP,Configuration,Maintenance,and Procedures)文件:是指为了满足延程运行型号设计批准的要求,经局方批准的特定机体/发动机组合的构型、维修和程序文件。该文件包括最低构型、运行和维修相关要求,硬件寿命限制和最低设备清单等。

(22)空中停车(IFSD,In-Flight Shut-Down):是指飞机在空中,发动机因自身原因诱发、飞行机组引起或外部因素导致的失去推力并停车。

B737-NG、B757-200、B737-CL 延程运行关键系统分别如表 4-2-2、表 4-2-3、表 4-2-4 所示。

表 4-2-2　B737-NG 延程运行关键系统

Airframe System	FAA MMEL ETOPS Restriction
Air Conditioning Packs	21-1
Electrical/Electronic Equipment Cooling Blower	21-27
Equipment Cooling Automatic Flow Control Valve/Overboard Exhaust	21-40
Air Distribution Riser Shutoff Valves	21-45
Autopilot Systems	22-1
Autopilot Disengaged Warning System	22-2
STAB out of Trim Light	22-11
Mode Control Panel Switches	22-15
Engine Driven Generator System	24-1
APU Generator System	24-2
Transformer Rectifiers	24-6
Engine Overheat and Fire Detection System	26-2
APU Fire Detection System	26-8
Wing-Body Overheat Detector System (Left)	26-12
Wing-Body Overheat Detector System (Right)	26-13
Lower Cargo Compartment Fire Detection/Suppression System	26-19
Fuel Boost Pumps (Main Tank)	28-1
APU Fuel Valve	28-4
Cross-Feed Valve Open Light	28-5
Flight Deck Fuel Quantity Indicators (Main Tanks)	28-6
Flight Deck Fuel Quantity Indicators (Center Tank)	28-7
System B Pumps	29-2
Wing Anti-Ice Valves	30-1
Engine and Nose Cowl Anti-Ice Valves	30-3
Pitot/Static Probe Heaters	30-5
Vertical Stabilizer Pitot Heaters	30-6
TAT Probe Heater	30-7
AOA Sensor Heaters/Stall Warning System Sensor Heaters/Alpha Vane Heaters	30-8
Pitot, Pitot/Static and Temperature Probe Heater Lights	30-9
Electrically Heated Windshields	30-11
Cowl Anti-Ice Lights	30-17
Inertial Reference System	34-35
Flight Management Computer	34-36
Manifold Isolation Shutoff Valve	36-1
Ground Pneumatic Connector Check Valve	36-2
Pre-Cooler Control Valves	36-3
Engine Bleed Air Shutoff Valves (PRSOV)	36-5
Engine Bleed Trip off Lights	36-8
Auxiliary Power Unit	49-1
APU Auto Shutdown System	49-3
APU EGT Indicator	49-5
APU Surge Control System	49-9
Start Power Unit	49-15
Start Converter Unit	49-16
Ignition System	74-1
Oil Quantity Indication Systems	79-1
Starter Valves	80-3

表 4-2-3　B757-200 延程运行关键系统

Airframe System	FAA MMEL ETOPS Restriction
Air Conditioning Packs	21-51-1
Pack Flow Control and Shutoff Valves	21-51-2
Air Cycle Machine	21-51-7
Equipment Cooling	21-58
Instrument Cooling Monitor System	21-58-12
AutoPilot System	22-10
Engine Driven Generator	24-00-1
APU Driven Generator	24-00-2
Hydraulic Motor Generators	24-25-1
Hydraulic Motor Generator Valves	24-25-2
Engine Fire Detection Systems	26-11-1
Engine Overheat Detection System	26-11-2
Engine Strut Overheat Detection System（Rolls Royce）	26-12-1
APU Fire Detection System	26-15-1
APU Fire Extinguishing System	26-22-1
Main Tank Fuel Pumps	28-22-1
Dual Fuel Crossfeed Valves	28-22-3
Crossfeed Valve Light	28-40-1
Fuel Tank Quantity Indicating Systems	28-41-1
Fuel Quantity Processor	28-41-2
Ice and Rain Protection	30-11-1
Engine Anti-Ice Valves	30-21-1
Pitot Static Probe Heater System	30-31-1
Capt Pitot and F/O Pitot Heat Indicating System	30-31-5
Temperature（TAT）Probe Heater System	30-33-1
Flight Deck Window Heat	30-41-1
Engine Indicating and Crew Alerting System-Display Unit	31-41-1
Cabin Interior Illumination System	33-21-1
Instrument Source Select Switches	34-00-1
Inertial Reference System	34-21-1
Electronic Flight Instrument Symbol Generators	34-22-5
Integrated Standby Flight Display	34-24-2
Flight Management Computer（Including CDU，HMCDU，and MCDU）	34-61-1
Air Supply Control and Test Unit（-400ER）	36-00-1
Engine Bleed Pressure Regulating and Shutoff Valves	36-11-1
Left and Right Bleed Isolation Valves	36-11-6
Engine Pressure Regulating Valve	36-11-8
Engine Firewall Shutoff Valves	36-11-9
Precoolers	36-12-1
Fan Air Control System	36-12-2
Auxiliary Power Unit	49-11-1
Electronic Engine Control Systems	73-21-1
Electronic Engine Control Systems Inop Lights	73-21-2
Fuel Flow Indicating System	73-31-1
Ignition System	74-00-1

表 4-2-4　B737-CL 延程运行关键系统

ATA Chapter	B737-300/400/500 ETOPS Significant Systems	ATA M/M Subsection
21 Air Conditioning	Pressurization Control	-30
	Cooling-Flow Control and Shutoff	-50
	Temperature Control	-60
24 Electrical Power	AC Generator Drive	-10
	AC Generation	-20
	DC Generation	-31
26 Fire Protection	Engine Fire Detection System	-11
	APU Fire Detection	-15
	Lower Cargo Compartment Smoke Detection System	-16
	Lower Cargo Compartment Fire Extinguishing System	-23
28 Fuel	Engine Fuel Feed System	-22
	APU Fuel Feed System	-25
30 Ice/Rain Protection	Wing Thermal Anti-Ice System	-11
	Inlet Cowl Anti-Ice System	-20
	Pitot/Static Probe and Alpha Vane Heaters	-30
	Window Anti-Icing System	-40
34 Navigation	Weather Radar System	-41
36 Pneumatic	Engine Air Distribution System	-11
	APU Bleed System	-14
49 Airborne Auxiliary Power	APU Power Plant	-10
	APU Engine	-20
	APU Fuel System	-30
	APU Ignition and Start System	-40
	APU Controls	-60
	Exhaust Gas Temperature Indicating System	-70
	APU Lubrication System	-90
CFM56-7		
71	Power Plant	ALL
72	Engine	ALL
73	Engine Fuel and Control	ALL
74	Ignition	ALL
75	Air	ALL
76	Engine Controls	ALL
77	Engine Indicating	ALL
79	Oil	ALL
80	Starting	ALL

(23)双重维修(Dual Maintenance):或者称相同维修、多重维修、同时维修,是指对相同的或相似的延程运行关键系统实施的维修。"相同的"延程运行关键系统中进行的双重维修是指在同一次例行维修或非例行维修时,对相同的但是独立的延程运行关键系统的相同部件进行的维修。"相似的"延程运行关键系统的双重维修是指在同一次维修中,对两个发动机中机驱动部件所实施的维修。

(24)加速 ETOPS:FAA 和 EASA 在 1995 年开始使用加速 ETOPS,作为运营商在没有获得特定机身/发动机组合方面实际运营经验时去获得 ETOPS 运营批准的一种方式。加速 ETOPS 方面,监管机构要求航空公司需向其证明它们已具备了所有必需的 ETOPS 流程,并对流程进行了验证。2007 年,FAA ETOPS 规则允许使用加速 ETOPS 的方法,并在 2008 年 FAA ETOPS 的咨询通告 AC120-42B 中第四章 ETOPS 资格认证中对其进行了深入的解释说明且在附录三中给出了 ETOPS 的批准方法。在 EASA AMC.20-6 第二版第三章中提出基于运营人的机身/发动机组合的可用性和之前的运行经验的 ETOPS 获准方法。加速 ETOPS 的审批不要求具有机身/发动机组合的使用经验且审批方法中可减少先前服役的运行经验。虽然不同的监管机构有不同的经验要求,但加速 ETOPS 方法已经成为大多数 ETOPS 操作审批的标准,而不是例外。ICAO 中针对双发运输类涡轮飞机运营的"加速"EDTO 特殊批准主要针对运营人在航空器的直接实际作业经验不足一年的情况下计划开始 EDTO 运营;或运营人积累了航空器的直接实际作业经验,但在航空器 120 min 改航时间 EDTO 运行经验不足一年的情况下计划进行超过 120 min 的 EDTO 运行。

(25)早期 ETOPS:指在未获得 ETOPS 认证的机身/发动机组合的非 ETOPS 服务经验的情况下获得 ETOPS 的型号设计批准。FAA 的 14 CFR 第 25 节的适航标准中给出了运输类飞机使用早期 ETOPS 方法的型号设计批准申请人的要求,第 33 节的适航标准中对涡轮喷气发动机提出了早期 ETOPS 合格性的设计和测试要求。EASA 在 AMC20-6 第二版的第二章型号设计批准中对早期 ETOPS 的批准方法进行了说明。只要当局认可批准计划的所有方面,那么在该机身/发动机组合投入运营时,ETOPS 批准被认为是可行的。但当局也必须确保批准计划达到 AMC 以及飞机和发动机认证机构规定的安全水平。

(26)合适机场(Adequate Airport):对于所考虑或所使用的飞机满足 FAR-139(或与之等效的规定)的要求和性能要求的机场。在预计的最大着陆重量时要能满足性能要求,跑道长度、坡度等要能适合飞机起降。此外,还要求机场有合适的 ATC 设备、足够的导航设施(ILS、VOR、NDB 等)、照明设施、足够的营救设备、消防设施、供油能力、修理设施、容纳旅客的能力以及合适的开放时间。不要求 $ACN \leqslant PCN$,允许超载起降(一般在硬道面上允许 ACN 比 PCN 大 5%,在软道面上允许 ACN 比 PCN 大 10%)。

(27)改航速度(Diversion Speed,或称转场速度):航空公司为确定允许的延程飞行区和飞行计划所选用的改航备降机场的飞行速度,如 $M0.78/290KIAS$、$M0.80/310KIAS$、LRC 等。

(28)改航距离(Diversion Distance,或称转场距离):沿着假定的一发故障时的延程飞行的改航剖面,使用最大连续推力以延程运行的改航速度飘降、在尽可能高的高度改平,然后以延程运行的改航速度巡航,在批准的改航时间内按 ISA、无风计算的飞机飞过的距离。

4.3 批准延程运行的条件

为了使用某种机型实施延程运行,必须在两个方面获得批准。首先,飞机厂家必须对该机型(飞机和发动机的组合)实施延程运行获得型号设计方面的批准。其次,航空公司从主管当局获得使用方面的批准。

4.3.1 型号设计方面的批准

型号设计方面的批准是根据发动机的可靠性及飞机系统的设计和飞机系统的可靠性、有效性验证飞行做出的。

4.3.1.1 型号设计批准

依照 CCAR-121 运行的所有双发飞机和多于两台发动机的载客飞机(客机)应当遵守 CCAR-121 W 章的规定实施延程运行。依据 CCAR-121.711(c)款的要求,只有当机体/发动机组合获得中国民航颁发的延程运行型号设计批准时,合格证持有人才有资格实施延程运行。延程运行飞机的适航要求在 CCAR-121.153 条飞机的审定和设备要求中有规定,在《运输类飞机适航标准》(CCAR-25-R4)附录 K 中也有详细的要求。在咨询通告《延程运行和极地运行》(AC-121-FS-2019-009R2)中也更为详细地规定了对延程运行型号设计的要求。

CCAR-121 第 121.718 条延程运行型号设计批准依据中规定,除 2015 年 2 月 17 日之前制造的安装两台以上发动机的载客飞机和仅用于实施飞行时间少于或等于 75 min 的延程运行的双发飞机之外,合格证持有人只能在下列条件下实施延程运行:飞机获得延程运行型号设计批准且每架用于延程运行的飞机均符合其构型、维修和程序文件的要求。延程运行型号设计批准通常在相应的机体/发动机组合的《飞机飞行手册》(AFM)、型号合格证(TC)或补充型号合格证(STC)中给出。

4.3.1.2 可靠性要求

飞机制造商必须证明:该机型的机体/发动机组合在运行中具有足够的可靠性,并满足安全实施延程运行的要求。

发动机可靠性由空中停车率来衡量。大部分适航审定当局要求该型号发动机的空中停车率由使用该型号发动机的机队的使用情况统计得出,FAA 要求型号发动机累计飞行小时应在 25 万小时以上,EASA 要求型号发动机累计飞行小时在 10 万小时以上。发动机作为机体/发动机组合的一部分,若其空中停车率(基于 12 个月滚动平均值)超过下列数值,合格证持有人应当综合评审其运行,以识别任何共因影响和系统错误。

(1)如果延程运行时间小于或等于 120 min,空中停车率为 0.005 次/1000 发动机工作小时。

(2)如果在北太平洋区域延程运行时间超过 120 min 且小于或等于 207 min,以及在其他区域延程运行时间超过 120 min 且小于或等于 180 min,空中停车率为 0.003 次/1000 发动机工作小时。

(3)如果在北太平洋区域延程运行时间超过 207 min,以及在其他区域延程运行时间超过 180 min,空中停车率为 0.002 次/1000 发动机工作小时。

在空中停车率超过上述数值 30 天之内,合格证持有人应当向局方提交一份调查报告以及已采取的任何必要的纠正措施。

4.3.1.3　时限系统的时间限制要求

包括延程运行在内的远程飞行运行安全依赖于飞机各个关键系统的可靠性,运行也必须遵守时限系统的时间限制。为了达到实施延程运行的系统设计和可靠性标准,飞机可能需要一些附加设备。飞机上至少要有三套独立的交流电源,每套电源都要至少能为通信、导航设备及基本飞行仪表(高度、速度、姿态、航向仪表)和警告系统提供电力,保证足够的信息。APU 也需要达到更高的标准以便提高其可靠性及空中启动性能。在某些型号的飞机上要装液压驱动发电机(HMG)以提高延程运行的签派灵活性。必要的话,用 HMG 提供额外交流电源。在 HMG 是唯一的电源的情况下,驾驶舱中的主要设备应能正常工作。电子设备冷却系统也需要达到更高的标准。货舱灭火能力应加强,抑制火情的时间必须延长到所申请的改航时间再加 15 min 以满足延程运行的要求,确保飞机能安全飞行并在一个适宜机场着陆。由于飞机一发故障后改航备降机场时可能要在较低的高度上飞行较长的时间,飞机必须有足够的防冰和除冰能力。

4.3.1.4　有效性验证飞行

获得型号设计批准的最后一步是制造厂要进行有效性验证飞行(验证飞行剖面见图 4-3-1)。这种验证飞行的目的是检验飞机的飞行品质和在各种系统不工作的情况下的机组的工作负担,检验和延程运行有关的飞机性能,例如某些设备不工作对着陆性能的影响等。

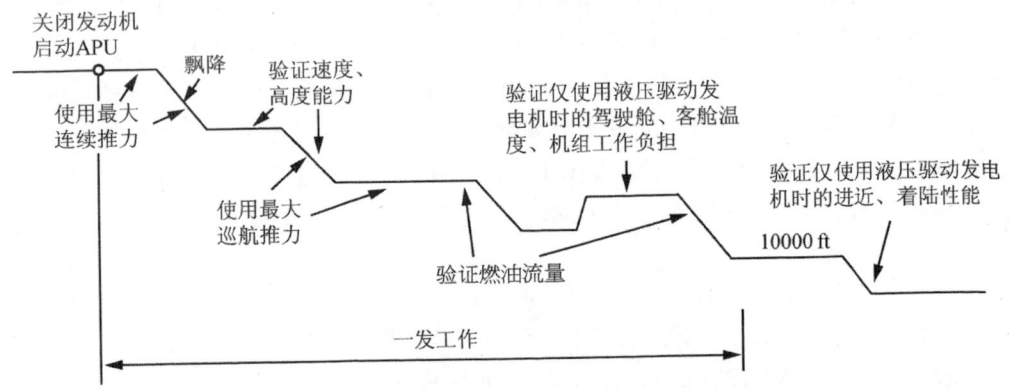

图 4-3-1　飞机制造厂的有效性验证飞行使用的飞行剖面

早期机型如 A300、A310、B757、B767 等双发飞机最初是为较短航线设计的,实施远程越洋飞行必须进行一些特别的验证和需要积累一定的运行经验,以满足航程延长的需要。而且,当时的发动机技术不够先进,无法在飞机交付时就能证明其空中停车率已达到 180 min ETOPS 的要求,这些双发飞机只有在投入运营一段时间之后才会相继获得 180 min ETOPS 要求的能力。目前,新机型如 B777 不存在这个问题,其在技术上更为先进,可靠性进一步提高,并经过了大量的试飞、测试工作才投入市场,在进行商业运营前就已经由适航部门批准具备了 180 min ETOPS 要求的能力。

机体/发动机组合获得 ETOPS 型号设计批准后,例如在 FAA 批准的《飞机飞行手册》中将插入以下说明(下面是波音公司的说明,其他飞机制造厂商的说明与此类似):本飞机/发动机组合的型号设计可靠性和性能已按照 FAA 咨询通告 120-42A 通过审定,如果该飞机的构型符合波音 D6T11604 号文件《延伸航程运行的构型、维修和程序》的要求,则可以实施延伸航程运行,但本结论并不能当作实施延伸航程运行的运行批准。

在 AFM 的"LIMITATION"一节的"Kind of Airplane Operation"款中的"Extended Over-water Operations(可以做延伸到水上的飞行)",并非指飞机可以实施 ETOPS 运行,而是指飞机可以在离最近的海岸线 50 NM 之外的水上飞行,这要求飞机上配备有救生筏之类的特殊设备,否则其航线离最近的海岸线不能超过 50 NM。

下面是对获得 ETOPS 型号设计批准的 B737-800 在其 AFM 中加入的说明:

B737-800 AFM   SECTION 3 EXTENDED RANGE OPERATIONS

The type design reliability and performance of this airplane/engine combination has been evaluated in accordance with FAA Advisory Circular 120-42A and found suitable for Extended Range Operations when configured in accordance with FAA approved Boeing document, D044A007, "737-600/-700/-700C/-800/-900/-900ER ETOPS Configuration, Maintenance & Procedures". This finding does not constitute approval to conduct extended range operations.

The APU must be started and running prior to reaching a point farther than 60 minutes from an adequate alternate airport and kept running for the duration of the ETOPS flight. APU in-flight starts have been successfully demonstrated to 41,000 feet pressure altitude. APU in-flight starts may be attempted at any altitude. [737-NG 的最大审定使用高度就是 *41000 ft*。]

按照 B737 的 AFM、CMP 和 FCOM 中的说明,在进入 ETOPS 航段前就应该启动 APU 并使其保持运转,即使没有发生一发失效(因为实施延程运行必须至少要有 3 套交流电源)。因此,B737-CL/NG 的全发、一发的临界油量中都包括 APU 的耗油。而 B757、B767、B777 的 CMP 和 FCOM 没有这种说明,即全发工作时不需要启动 APU,因此,这些机型仅一发失效的临界油量中包括 APU 的耗油。B747 全发、一发失效的临界油量中都不包括 APU 的耗油,因为 B747 有发动机带动的 4 台发电机,即使一发失效实施 ETOPS 飞行,也不需要 APU 的交流电源,空中不需要 APU 工作,在空中也不能启动 APU。APU 可在地面启动并且起飞时保持运转一直工作到 20000 ft,在飞行中 APU 能为一个空调包提供引气直到 15000 ft,在飞行中 APU 不能提供电力。

空中客车公司对获得 ETOPS 型号设计批准的飞机将在 AFM 第 6 部分"附录和补充"/6.02.02 节"延程运行"中加入以下说明(以 A319-132 为例):

本附录适用于按 AMJ 120-42/IL 20 的规定运行的飞机。

只要是按照空客文件 AI/EA-4000"延程运行的标准"最新修订版来配置、维护和运行,本飞机的型别设计可靠性和性能就符合 AMJ 120-42/IL 20 的标准。单发巡航速度的最大改航时间不可超过 120 分钟。

在其机组使用手册 FCOM 2.04.40 节中也有相关的说明:

当飞机按照适用的空中客车工业公司文件"延程运行标准"的最新版本,即空中客车公司的 CMP(Configuration, Maintenance and Procedure)文件的规定来配置、维护和运行,则本飞机的发动机装置的系统设计和可靠性符合 AMC 20-6(EASA)或 AC120-42A(FAA)或 FAR 25-1535(FAA)中公布的双发飞机延程运行(ETOPS)标准。

飞机上有些系统(Time-Limited-System)和延程飞行时间有关,例如,要求货舱火情抑制系统抑制火情的时间不得少于批准的延程时间再加 15 min。飞机制造厂商会在《飞机飞行手册》(AFM)、CMP 文件或其他文件中给出和延程飞行时间有关的系统的最大能力。例如,波音在 CMP 文件的"关于已证实的能力的声明"部分给出了货舱火情抑制系统的最大能力(抑制火情的最长时间）：

Configuration, Maintenance, and Procedures for

Extended Range（ER）Operation, Model 757

757 机型延伸航程运行的构型、维护和程序

3.6 Part D—Demonstrated Capability Statements

关于已证实的能力的声明

　　这部分的目的是列出已经证实系统能力的项目,这些项目在对批准延程运行的要求进行评估时可能是很重要的。B757 飞机的大部分系统的安全性和性能特点与延伸航程的考虑事项（如到达适宜机场的时间等）无关。其安全性和性能特点可能受到延伸航程影响的那些系统,都已经针对包含一个到达适宜机场以单发速度飞行时间不超过 180 min 的点的运行进行了评估,且结果被认为是令人满意的。

　　在对批准延程运行的要求进行评估时可以考虑表 4-3-1 中的项目。

表 4-3-1　批准延程运行考虑的项目

项目	涉及的系统	已经证实了的能力
1	货舱火情抑制系统	120 min 的延程运行: 对 1 个货舱门的后货舱,抑制火情的能力是 167 min 对 2 个货舱门的后货舱,抑制火情的能力是 170 min 对前货舱,抑制火情的能力是超过 167 min 180 min 的延程运行: 对 1 个货舱门的后货舱,抑制火情的能力是 242 min 对前货舱,抑制火情的能力是超过 242 min
2	电气电源系统	基于飞行时间 10 h 和改航适宜机场 180 min 的分析——不受限制
3	电气/电子冷却系统	基于分析和 4 h 不启用主冷却系统的演示飞行——240 min 的能力
4	气象雷达*	对实施 180 min 的延程运行,气象雷达应该是工作的
5	油量计划	对加入了 Rolls-Royce 服务通告 RB211-72-8773 和/或 RB211-72-8774（高压喷嘴导向叶片和增加了冷却气流的高压涡轮叶片）的飞机,计算油量时应该额外增加 1.5% 的油量[这项是 FAA 180 min ETOPS 要求的,120 min ETOPS 则不需要]
6	放行计划	对 180 min 延程飞行的航班,选择适宜的备降机场时应考虑结冰状况。如预报的航路备降机场的中等或更严重的结冰条件可能影响在该机场按照 AC 120-42A 10.d5（Ⅳ）款确定的时间段内进近和着陆,则该机场不应列为适宜机场[这项是 FAA 180 min ETOPS 要求的,120 min ETOPS 则不需要]
7	APU**	对安装有 ABEX 5 kVA 的液压驱动发电机的 180 min 延程飞行,在到达 ETOPS 进入点前应启动 APU 并在中央油箱的燃油用尽前保持运转

　　注:

　　*:MMEL34-43-1 Weather Radar Systems——May be inoperative provided weather radar is not required by FAR. 在公司的 MEL 中此项应该改为:对实施 180 min 的延程运行,气象雷达应该是工作的。

　　**:这项是 180 min ETOPS 要求的,120 min ETOPS 则不需要。另外,对安装有 VICKERS 10 kVA 的液压驱动发电机的 B757,在 180 min 延程飞行期间不要求 APU 工作,但放行时 APU 应能工作[见 B757 的 DDG（放行偏差指南）]。

4.3.2 使用方面的批准(即运行批准)

由于延程运行的特殊性,航空公司或其他运营者必须获得该国局方对该公司使用该机型实施相关延程运行的批准,这种认证称为延程运行的批准。

4.3.2.1 申请等级和要求

飞机具有延程运行的能力,证明了其可靠性和安全飞行的能力后,为确保安全,适航部门一般还要对运营该机型的航空公司进行一段时间(1~2年)的审核,以确保该航空公司有足够的保障能力、延程运行运营经验,经过这些程序之后,才会批准其实施延程运行航班运营。航空公司申请的改航时间越长,对航空公司的要求就越高。

CCAR-121 的 W 章"延程运行与极地运行"和咨询通告《延程运行和极地运行》(AC-121-FS-2019-009R2)中有获得延程运行批准的详细规定,局方依据规定的要求和限制批准延程运行。按照延程运行申请的要求,合格证持有人向局方递交申请材料,局方对申请材料进行审查和评估,并根据附加指南或审定过程中提出的建议措施,与合格证持有人共同制订验证飞行计划。成功完成验证飞行,合格审定完成后,局方通过向合格证持有人颁发运行规范的相关条款予以批准。目前在我国可以申请的最大改航时间为 75 min、90 min、120 min、180 min、240 min和超过 240 min 的延程运行。在咨询通告《延程运行和极地运行》的附录 1 中给出了合格证持有人申请不同等级的延程运行批准以及每一等级的批准最低要求,不同等级的延程运行规范要求与 FAA 方面的要求基本一致。

4.3.2.2 训练要求

航空公司应对机组和签派员进行和延程运行有关的训练。对于机组和签派员,关于延程运行的训练项目是相似的,所以可使用一个训练提纲。航空公司可以根据自己的情况和要求对提纲进行相应的修改。下面是用于机组和签派员的训练提纲的一个例子。

飞行机组和签派员的训练提纲:

— 熟悉相关规章中的延程运行条例;
— 延程运行飞行计划;
— 一发失效速度;
— 备降机场要求;
— 航路程序;
— 天气预报(如果需要);
— 等时点(临界点)的计算;
— 临界燃油的计算;
— 最低设备清单(和延程运行有关的项目);
— 延程运行特有的系统;
— 签派放行;
— 改航决定;
— 改航剖面;
— 机长的权力和职责。

4.3.2.3 运行和维修要求

要从民航当局获取实施延程运行的批准,航空公司必须向民航当局证明以下两方面的能力：

（1）航空公司能够在延伸航程运行的情况下安全实施运行；

（2）航空公司的维修能力可以使飞机在批准的构型下保持高水平的可靠性。

延程运行的批准，涉及正常维修与规章流程以外的额外规定。飞行员与维修人员必须经过特定的训练，并取得实施延程运行飞行必需的资格。如果航空公司已有丰富的长途飞行经验，可能会实时取得延程运行的批准，否则就可能需要先进行一连串的延程运行认证评估考核，确定其具备实施延程运行飞行的能力，才能得到延程运行的批准。该运行批准并不能超过相关机型的型号认证。航空公司申请的改航时间越长，对其要求就越高。

航空公司为获得延程运行的批准，要：

（1）制订报批计划。

（2）制定延程运行的程序和文件（运行程序和文件、维修程序和文件）。

（3）对机组、签派员等有关人员进行培训。

（4）进行运行方面的验证飞行。

（5）制定延程运行的规范。在运行规范中要包括以下和延程运行有关的内容：

①具有延程运行能力的飞机/发动机组合；

②飞机注册号；

③飞行高度限制；

④最大改航时间（经批准的延程飞行时间）和改航速度；

⑤被授权使用的机场；

⑥被批准飞行的航路和限制。

为获得使用方面的批准，航空公司必须有符合延程飞行要求的操作程序和维护程序（即延程运行手册和维护手册）、推进系统的可靠性记录、飞机维护程序可靠性报告等。在 CCAR-121 的 W 章的第 121.718 条和第 121.719 条有支持延程运行所需的飞行运行方面和维修方面的要求。

航空公司延程运行的飞行运行要求包括：

（1）飞机性能数据。合格证持有人应保证飞行机组人员和签派员或合格证持有人授权实施运行控制的人员可以获得其延程运行所有阶段（包括改航备降阶段）所需的飞机性能数据，否则不能签派或飞行放行飞机实施延程运行。

（2）航路运行中的机场信息。

（3）签派或飞行放行。其主要包括：延程运行备降机场，飞行计划限制，着陆距离，机场救援和消防服务，延程运行指定备降机场最低天气标准，燃油供应，通信，签派或飞行放行，计算机飞行计划，延程运行的例外签派或飞行放行。如果延程运行航路还涉及高原复杂地形改航备降航路的氧气要求，则必须考虑备降机场选择决断点对改航备降计划的影响。

（4）航路运行。其主要包括：机长的权限，起飞后可能使用的备降机场，发动机失效，系统全部或部分失效，其他改航备降情况。

（5）延程运行程序文件。合格证持有人应当根据飞行运行要求制定专门的延程运行飞行机组运行程序，并将这些程序编入相应的手册和文件，且提供给飞行员的这些手册和文件应当同时包括延程运行维修程序中要求飞行机组实施的运行程序。另外，合格证持有人运行手册中延程运行部分的内容及其修订须经局方批准后方可实施。

航空公司实施延程运行的合格证持有人必须满足 CCAR-121.719 条和咨询通告有关的延

程运行维修要求,包括:

(a)维修方案;

(b)延程运行维修文件;

(c)延程运行前维修检查;

(d)双重维修项目的限制;

(e)验证程序;

(f)任务识别;

(g)集中的维修控制程序;

(h)延程运行零部件控制方案;

(i)可靠性方案;

(j)动力系统监控;

(k)发动机状况监控;

(l)滑油消耗量监控;

(m)APU 空中起动监控方案;

(n)构型、维修与程序(CMP)文件;

(o)程序变化。

在延程运行的维护程序中找到对类似任务的处理方法至关重要。如果对两个独立的但是是同类的、和延程运行有关的关键部件进行相同的维护工作,并且无法进行令人满意的地面试验,那么在维修后应该进行验证飞行。其目的是证实故障确实被排除,在这两个部件上不会出现相同的维修错误。在日常维护工作中,在两台发动机的金属屑检测器上安装密封圈以及在左、右两发中更换燃油控制模块便是这种类似任务的例子,应该由两人分别独立完成左、右两发相同的维修工作,以免出现相同的维修错误。这种维修后的验证飞行可以在正常航班的非延程运行段进行。

为了保证安全,所有航空监管机构均会密切监察所有机型的 ETOPS 认证,以及各航空公司的 ETOPS 运行批准,并对其表现进行考核评价。对任何涉及 ETOPS 的技术事件必须加以记录,而经过全球的飞行记录传回来的数据,将会用来评定某一飞机与发动机组合的可靠性,有关的统计资料也会公开。有关的数据必须低于某个型号认证指定的上限,ETOPS-180 的数据限制当然比 ETOPS-120 严格。如果航空公司在运行 ETOPS 航班中保障不力,或经常发生安全隐患,数据记录不良,机型的 ETOPS 认证就会被降级,甚至被临时撤销。

4.3.2.4 验证飞行

当申请在特定区域中使用特定机身发动机组合实施延程运行时,合格证持有人必须在该延程运行区域内申请的航线上,在局方监督下实施真实的验证飞行。验证飞行可以确保要求的延程运行飞行运行和维修方案能够有效地支持其运行。

对于首次申请延程运行批准的合格证持有人,局方应当要求不少于 4 个航段的验证飞行,其中至少前 2 个航段应当是非商业载客运行。对于已经获得延程运行批准的合格证持有人,申请增加新的机身发动机组合、延长批准的最大改航时间或在新的地理区域内实施延程运行,局方应当要求不少于 2 个航段的验证飞行(允许商业取酬)。局方依据合格证持有人所申请延程运行航线上或类似航线上的实际运行经验,来确定验证飞行的次数、验证飞行的方式(载运商业旅客,只载运货物,或者非商业取酬运行)以及需要验证的其他项目。如果批准实施商业取酬的验证飞行,则应当在合格证持有人的运行规范中的相关部分给予临时批准或限制。

验证飞行成功,合格审定完成后,局方修订并颁发无限制条件的运行规范相应条款。

合格证持有人应当在局方监督下使用申请的机体/发动机组合实施验证飞行,该机体/发动机组合应当有足够的能力安全地完成相关运行。在验证飞行中,应当演示下列紧急情况:

(1)一台发动机完全失去推力,并且所有发动机驱动(或正常)电源全部失效(或者是,作为最低要求,在延程运行审定中确定了的临界电气状况);或者

(2)延程运行中,在飞机适航性、机组工作负荷或者飞机性能风险等对安全造成严重威胁的任何其他情况。

4.4　延程运行飞行计划

4.4.1　延程运行飞行计划的内容

延程运行必须包括公司延程运行的所有必要信息,包括以下主要内容:

(1)燃油计划;

(2)计划的飞行高度、飞行航路;

(3)各高度层的气象资料(风向、风速信息及外界温度);

(4)航路各数据(航段预计航迹、航路点之间的距离、预计航段飞行时间);

(5)航线距离和预计航线飞行时间;

(6)起飞机场、目的地机场、航路备降机场和目的地备降机场;

(7)商务业载信息;

(8)等时点(ETP);

(9)从 ETP 到各航路合适备降机场的时间;

(10)ETP 的地理位置数据;

(11)到延程运行合适备降机场的最早到达前一小时、最晚到达后一小时的时间(按预计起飞时刻);

(12)列出航路延程运行合适的备降机场,延程运行的进入点、退出点、燃油关键点信息;

(13)MEL 放行信息;

(14)相关区域电话转接(Phone Patch)通信频率;

(15)目的地备降机场的航路和燃油计划。

4.4.2　延程运行飞行计划的制作方法

4.4.2.1　确定飞行计划航路

以从纽约到马德里的航路为例。假定我们希望沿大圆航线飞行,首先确定在这条航线上的延程运行可选备降机场(Suitable ETOPS Alternate),即对于特定延程运行航线,不考虑当时的临时状况,列入合格证持有人运行规范的可选的航路备降机场。这些机场必须满足 CCAR-121.197 条规定的着陆限制要求。这些机场不仅要满足 CCAR-121 的规章要求、航空公司的性能要求,还要有合适的 ATC 设备,通信导航设备,照明设备和灭火设备及救援、服务设施,开放时间也要合适。

如果要求延程运行可选备降机场成为延程运行指定备降机场（Designated ETOPS Alternate），它必须满足延程运行备降机场的天气标准要求。延程运行指定备降机场的气象条件高于或等于延程运行最低天气标准的延程运行可选备降机场的气象条件。延程运行最低天气标准与延程运行指定备降机场可用的着陆导航设备有关。如果在延程运行指定备降机场的关注时间段内，可用的最新的天气预报表明该机场的气象条件不低于延程运行指定备降机场天气最低标准，则该机场可以在飞行计划和签派或飞行放行单中列为延程运行指定备降机场。运行规范中应当列明延程运行航线的所有延程运行可选备降机场。运行规范中还应当列明这些机场作为延程运行指定备降机场时的最低天气标准。运行规范中延程运行指定备降机场最低天气标准通常可以根据表4-4-1来制定。

表4-4-1　延程运行指定备降机场最低天气标准（AC-121-FS）

进近设施配置	云高	能见度
对于至少有一套可用进近设施的机场，其进近设施能提供直线非精密进近程序、直线类精密进近程序或直线Ⅰ类精密进近程序，或在适用时可以从仪表进近程序改为盘旋机动	最低下降高（MDH）或者决断高（DH）增加120 m（400 ft）	着陆最低能见度增加1600 m（1 mi）
对于至少有两套能够提供不同跑道直线进近的可用进近设施的机场，其进近设施能提供直线非精密进近程序、直线类精密进近程序或直线Ⅰ类精密进近程序（应选择两个服务于不同适用跑道的进近设施）	相应直线进近程序的决断高（DH）或最低下降高（MDH）较大值上增加60 m（200 ft）	着陆最低能见度较高值上增加800 m（1/2 mi）
对于至少有一套Ⅱ类精密进近程序的机场	云高不得低于90 m（300 ft）	着陆最低能见度或跑道着陆最低能见度不得低于1200 m（3/4 mi），或跑道视程（RVR）不得低于1200 m（4000 ft）
对于至少有一套Ⅲ类精密进近程序的机场	云高不得低于60 m（200 ft）	着陆最低能见度不得低于800 m（1/2 mi），或跑道视程（RVR）不得低于550 m（1800 ft）

注：①在确定所使用的仪表进近程序时，应当考虑其他的相关因素。预报的风或阵风值不得超过运行限制值。
②一般不需要考虑条件性天气预报的内容，例如，INTER、TEMPO、PROB ××等后面的内容，但是如果PROB 40或者TEMPO的内容低于适用的运行最低标准，则必须加以考虑。
③当按照MEL的条款实施签派或飞行放行，在确定延程运行指定备降机场最低天气标准的时候，必须考虑MEL项目限制对仪表进近最低标准的影响。
④类精密进近按照非精密进近标准执行。

此外，预报的侧风也必须是可以接受的。一发失效着陆的侧风标准可以减少5 kt。

对于延程运行指定备降机场最低天气标准，CCAR-121的要求与FAA的要求是一致的。但EASA在最低天气标准方面有些差异，EASA的最低天气标准如表4-4-2所示。

表 4-4-2　EASA 延程运行指定备降机场最低天气标准

进近设施配置	仪表飞行规则天气最低云高	仪表飞行规则天气最低能见度
非精密进近	最低下降高(MDH)/最低下降高度(MDA)增加 120 m(400 ft)	着陆最低能见度增加 1500 m
Ⅰ类精密进近	决断高(DH)/决断高度(DA)增加 60 m(200 ft)	着陆最低能见度增加 800 m
Ⅱ类或Ⅲ类精密进近	根据具体情况,由管理局基于以下标准批准: 飞机具有一发不工作Ⅱ类或Ⅲ类着陆的能力; 运营人有正常Ⅱ类或Ⅲ类运行批准	

非精密进近和着陆运行:不使用电子下滑道指引的仪表进近和着陆。

类精密进近和着陆运行:有方位引导和垂直引导,但不满足建立精密进近和着陆运行要求的仪表进近和着陆。

精密进近和着陆运行:使用精确的方位和下滑道指引的仪表进近和着陆,其最低标准由相应的运行类型(分为Ⅰ、Ⅱ、ⅢA、ⅢB、ⅢC 等类型)确定。

非精密进近指只有方向引导,没有垂直引导的(如 NDB、VOR)进近,精密进近指既有方向引导,也有垂直引导的进近,如在 ILS、MLS、精密进近雷达引导下的进近。

决断高度(DA)/决断高(DH):在精密进近中,如果不能建立继续进近所必需的目视参考,则应当开始复飞的特定高度或者特定高。

最低下降高度(MDA)/最低下降高(MDH):在非精密进近或者盘旋进近中,如果不能建立必需的目视参考,则不能继续下降的特定高度或者特定高。

机场运行最低标准:机场用于起飞和着陆时的条件限制。对于起飞,用能见度和/或跑道视程以及云高(需要时)来表示;对于精密进近和着陆运行中的着陆,用与相应运行类型对应的能见度和/或跑道视程,以及决断高度(DA)/决断高(DH)来表示;对于非精密进近和着陆运行中的着陆,用能见度和/或跑道视程、最低下降高度(MDA)/最低下降高(MDH)以及云高(需要时)来表示。

飞行能见度:飞行员在空中从飞机驾驶舱能够在昼间看到或识别前方显著无发光目标物,或在夜间看到或识别前方显著发光目标物的平均水平距离。

目视气象条件:用能见度、离云的距离和云高表示,等于或者高于规定最低标准的气象条件。

仪表气象条件:用能见度、离云的距离和云高表示,低于为目视气象条件所规定的最低标准的气象条件。

超障高度(OCA)/超障高(OCH):为遵循适当的超障准则所确定的相关跑道入口标高或者机场标高之上的特定高度或特定高。

如在放行时某个合适备降机场的气象条件不满足上述最低要求,则它不是延程飞行备降机场,即不能用作延程飞行的备降机场。

4.4.2.2　确定允许的飞行区域

延程运行可选备降机场确定后,我们可以以这些机场为圆心分别画圆,以确定允许的飞行区域。圆的半径是所选定的一发不工作的改航速度和最大改航时间的函数,这个半径也称为改航距离,即在一发故障后以改航速度在最大改航时间内飞过的距离。改航距离是按无风、标准大气、在最佳高度(或与之接近的高度)一发故障、使用最大连续推力、以选定的改航速度飘降(先保持等马赫数下降,当表速增加到选定的表速后保持该表速下降)、在飘降到能以该表速平飞巡航的高度后保持该表速巡航的方式来计算的,见图4-4-1。

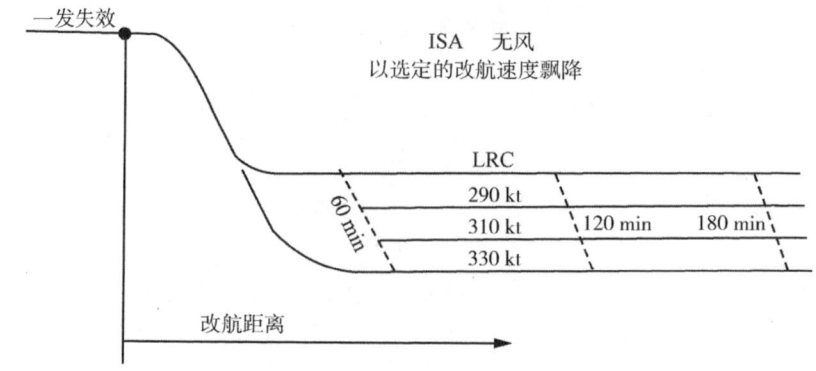

图 4-4-1　延程运行的改航距离

飞机的操作手册或者飞行计划与性能手册中提供了确定改航距离用的图表(见表 4-4-3 和附图 1-16),图表中的速度 0.78/290、0.80/310、0.80/330(M/KIAS)等就是改航速度。

表 4-4-3　B767-200ER 延程运行改航距离

运行区域(改航距离)								
速度 (M/KIAS)	重量 (1000 kg)	改航时间(min)						
		60	70	…	120	130	…	180	190
0.80/330	100	458	532	…	904	979	…	1351	…
	⋮	110	457	531	⋮	902	977	⋮	1348
	120	456	530	…	899	…	…	1343	…
	⋮	⋮	⋮	⋮	⋮	⋮	⋮	⋮	⋮
	180	438	508	…	858	…	…	1379	…
	190	432	501	…	847	917	…	1363	…

影响改航距离的主要因素是改航时间和改航速度,改航时的重量影响不大。选定的改航时间越长、改航速度越大,则改航距离越大。在选择固定速度(M/IAS)改航时,改航重量越大,一般说来改航距离越小(大改航速度、特别小的重量除外);在选择 LRC 速度时,则是改航重量越大,改航距离越大。改航速度由航空公司自己选定,选择大的改航速度,则改航半径大、允许飞行区域大,可使航线缩短、所需要的延程运行指定备降机场减少,但临界油量可能会增加(假设改航备降机场的距离相同)。许多适航当局把改航速度限制在 VMO-20 kt 以下。确定运行区域、计算临界油量、研究越障问题都要使用同一改航速度。

确定延程运行区域(改航距离)用的参考重量早期按照英国民用航空管理局(CAA)的条例 CAP513(CAP,Civil Aviation Publication)参考重量是以结构限制的最大允许起飞重量 MTOW 起飞后 2 h 的重量。后来 JAA、FAA 同意不再给出参考重量的定义,由航空公司确定自己的和 ETOPS 航路结构有关的参考重量。这个重量应尽可能接近真实并要由航空公司所在国的局方审定。

航空公司可以不选择表 4-4-3 中的速度,而选取自己认为更合适的速度,但这时的改航距离及相应的临界油量应该自行计算,波音的 INFLT/REPORT 软件具备这种功能。

现在让我们来看一下按改航时间分别为 60 min、120 min、180 min 确定的圆。由图 4-1-2可见,如按改航时间为 60 min 画圆,即不实施延程运行,那么纽约—马德里就不可能有直达航路。如按 120 min 画圆,可以安排直达航班(见图 4-4-2),可以看出批准实施 120 min 的延程运行可以提供经济上和使用方面的效益(航程和飞行时间缩短、油耗减少)。

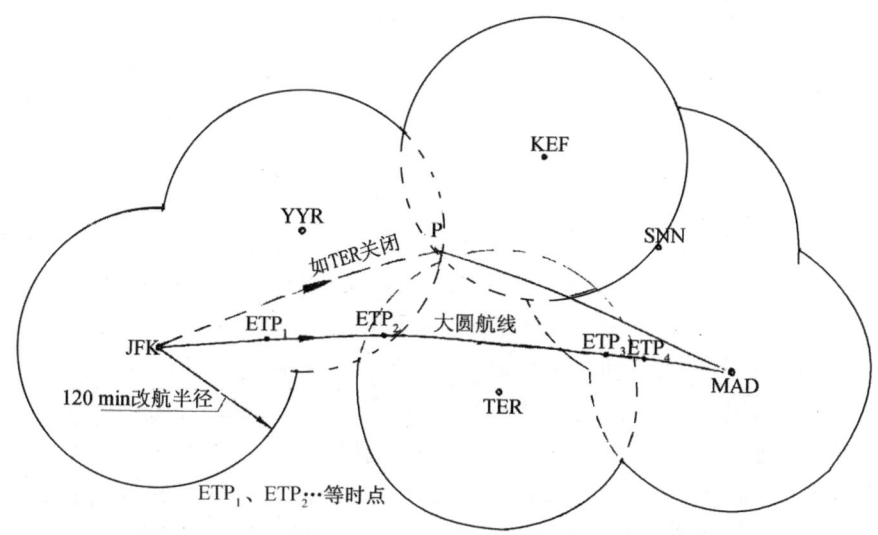

图 4-4-2　按 120 min 规则确定的允许飞行区域和航线

现在来考虑一个备降机场被关闭的情况。假定由于某种原因 TER 被关闭,在这种情况下,图 4-4-2 作出的直达航线超出了 120 min 延程运行的允许飞行区域,因此这条航线是不合法、不允许飞行的,必须沿上面一条较长的航线飞行。如果航空公司被批准实施 180 min 延程运行,则又可以直达飞行(见图 4-4-3)。

此外,还要做的一项检查是在一发故障而另一发使用最大连续推力、以选定的改航速度飞行时的高度能力,亦即看一下是否能越过航路上的障碍物。对于纽约—马德里的飞行,由于是越洋飞行,没有障碍物,高度能力不是限制因素。如果所使用的一个备降机场要求在格陵兰上空飞行,那么一发故障的高度能力可能影响到延程运行的改航速度的选择。一发故障后的净轨迹应以 2000 ft 高度差越过航路上的障碍物,并且在机场标高 1500 ft 以上(见图 4-4-4)。

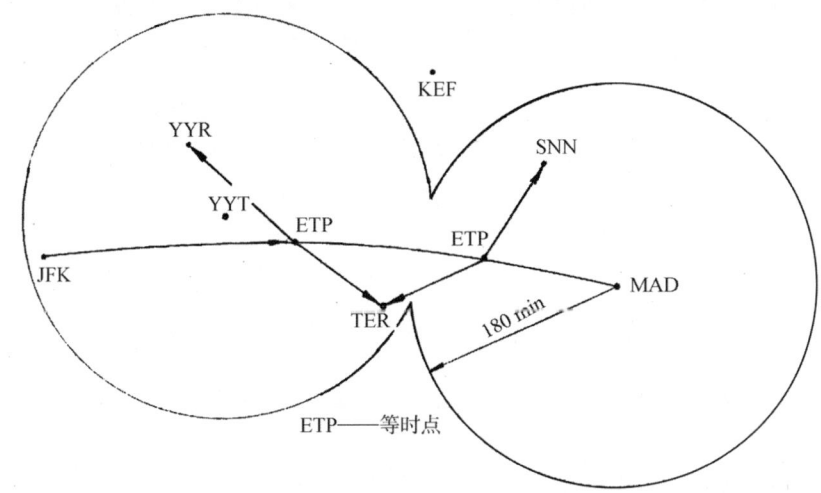

图 4-4-3　按 180 min 规则确定的允许飞行区域和航线

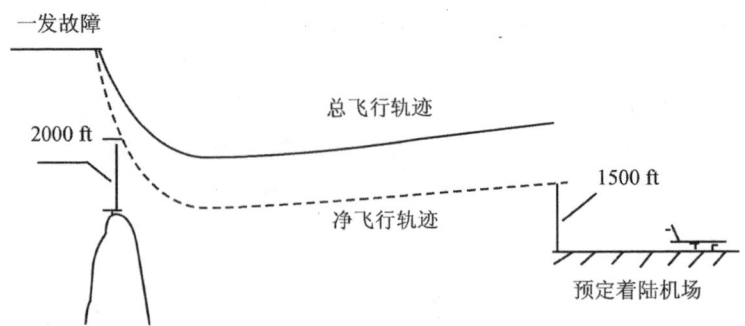

图 4-4-4　飘降轨迹

格陵兰岛的地形在 10000 ft 以上,选择的改航速度大,一发失效后的净轨迹可能不满足图4-4-4 的要求,这就要求重新选择改航速度,确定运行区域和航路,计算 ETP 和油量等。

4.4.2.3　确定进入点(EEP)、等时点(ETP)和退出点(EXP)

在确定了运行区域、航路备降机场和航路之后,要确定延程运行的进入点(EEP)、等时点(ETP)、退出点(EXP),以经、纬度坐标给出这些点。延程运行的进入点即距合适机场以批准的一发不工作的巡航速度飞行、航程为 60 min(双发飞机)或 180 min(三、四发飞机)的那个点。等时点是在预定航路上到两个最近的延程飞行的备降机场改航时间(考虑风的影响)相等的点。飞行计划中应包括等时点,进入点、退出点有时也被包括在计算出的飞行计划中。这些点可以作为附加的航路点输入飞行管理计算机,用于位置监控。延程运行的退出点就是航路上延程运行航段的最后一点。

4.4.2.4　考虑最低设备清单(MEL)的要求

每个航空公司的最低放行标准(即最低设备清单 MEL)必须专门指明和延程运行有关的项目。延程运行的最低放行标准比非延程运行的高(更严格)。

FAA 批准的主最低设备清单(MMEL)上用 ER 标明了和延程运行有关的限制[即对和延程运行有关的项目注上 ER(Extended Range),B777 的 MEL 例外]。例如,在 B767 的主最低设备清单中的第 24-00-2 项列出了对于 APU 驱动的发电机实施延程运行的签派放行要求,见图4-4-5。在 B777 的主最低设备清单中对和延程运行有关的项目不标注 ER,而是指明一发失效后允许的飞行时间,如图 4-4-6 所示。

BOEING

REPAIR INTERVAL		NUMBER INSTALLED		
		NUMBER REQUIRED FOR DISPATCH		
ITEM 24-21-1　APU Driven Generator System(Generator,AGCU, APB)	C	REMARKS OR EXCEPTIONS		
		1	0	(M)(O) May be inoperative provided: 　a) Procedures do not require use of the 　　APU for electrical power, 　b) Auxiliary Power Breaker (APB) remains 　　open, 　c) Both engine driven generator systems 　　operate normally, 　d) Backup AC power system is verified to 　　operate normally before each departure, 　　and 　e) Flight remains within 180 minutes of 　　Landing at a suitable airport.
		NOTE:APU may be used as a pneumatic source.		

图 4-4-5　B767 MMEL

U.S DEPARTMENT OF TRANSPORTATION FEDERAL AVIATION ADMINISTRATION		MASTER MINIMUM EQUIPMENT LIST		
AIRCRAFT: 　　BOEING 767		REVISION NO: 11 DATE: 02/01/91		PAGE: 24-2
SYSTEM & SEQUENCE　ITEM NUMBERS	1.	2. NUMBER INSTALLED		
		3. NUMBER REQUIRED FOR DISPATCH		
24 ELECTRICAL POWER 00-2 APU Driven 　　Generator		4. REMARKS OR EXCEPTIONS		
	C	1	0	*May be inoperative for other than 　ER　operations.
	B	1	0	*(M)Except for ER operations beyond 　120 minutes, may be inoperative 　provided: 　　a) Both engine driven generators, 　　　and either the center OR the 　　　left and right hydraulic Motor 　　　Generators operate normally, 　　　and 　　b) 120 minute ER operations and 　　　less are limited to not more 　　　than three flight days before 　　　repair is made.

图 4-4-6　B777 的 MMEL

注意：主最低设备清单中第 1 列下面的字母 A、B、C、D 表示的是规定的维修间隔(即限定的修复时间,REPAIR INTERVALS),而不是机务维修中的"A 检""B 检""C 检""D 检"。

A 类:列为 A 类的项目必须在 24 h 内[或在经批准的、航空公司制定的 MEL"备注或例外栏(REMARKS OR EXCEPTIONS)"中规定的时间限制内]完成维修。对于列为 A 类的项目没有"持续授权"(即不能续保)。如果由于没有备件或因排班问题使得必须对 24 h 期限进行延长,可以向 FAA(或 CAAC)提出申请。

B 类:此类项目必须在 3 个连续日历日(72 h)内修复,不包括在维修记录本(或记录)上登记该失效项目的那一日。例如,如果一个故障在 26 日上午 10 点被报告、记录在案,则必须在由 26 日 24 时(即 27 日 0 时)开始到 29 日 24 时的时间内修复。带有这类保留故障的航空器在可以进行修复或更换的机场不得起飞。

C 类:本类项目必须在 10 个连续日历日(240 h)内完成修复,不包括登记故障的那日。

D 类:此类项目必须在 120 个连续日历日(2880 h)内完成修复,不包括登记故障的那日。

归为 A、B、C、D 类的原则是该项目对飞行安全影响的程度。归为 A 类的是比较重要、应当尽快修好的项目,归为 B 类的是比 A 类稍微次要的项目。

在安装的项目比正常运行所要求的项目多的情况下,可以把该项目的类别降低一级,例如,如果要求一套高度告警系统且相应的修复间隔为 B,但安装了两套,第一套系统失效可以按照 C 类规定予以保留(10 天)。

合格证持有人应当根据主最低设备清单(MMEL)来制定自己的最低设备清单,其标准不得低于主最低设备清单。延程运行所需的系统冗余水平应当正确体现在延程运行的最低设备清单中。航空公司应该在最低设备清单中标出与实施延程运行可能有关的全部项目(一般在项目左边标上 ER)。在最低设备清单中不是所有项目都明显与延程运行有关。有些项目单独不工作是允许的,但如果同时还有其他设备故障,它必须是正常工作的才可放行实施延程运行。例如,冲压空气驱动的泵(Air Driven Pump),如果不要求安装液压驱动发电机(HMG),则允许该空气驱动泵不工作。该项和延程运行无关,不标注 ER。实施延程运行时条例规定有三套或三套以上的交流电源正常工作才可放行。仅当三个交流电源(两台发动机各带动一台发电机,APU 带动一台发电机)之一不工作时,实施延程运行才需要液压驱动发电机。因此,如果两台发动机驱动的发电机之一或 APU 驱动的发电机不工作,放行实施延程运行就要求液压驱动发电机能正常工作,而如果液压驱动发电机在放行时要求是完好的,那么冲压空气驱动的泵也必须是完好的(要用这台泵产生液压带动 HMG 发电)。可见,如不实施延程运行或实施延程运行但三套正式的交流电源系统是完好的,空气驱动的泵在放行时可以不工作,然而三套正式的交流电源系统之一故障时,为实施延程运行,冲压空气驱动的泵在放行时必须正常,虽然该项并未注明和延程运行有关(未标注 ER),请参考图 4-4-7 所示的原文。

4.4.2.5　延程运行飞行计划

延程运行的飞行计划应该列出延程运行的等时点和延程运行的备降机场。计算所需油量时必须考虑临界燃油方案。

为了对延程运行的飞机做到合法放行,在计划的延程运行的可选备降机场的气象条件必须在最早可能到达时间到最迟可能到达时间的期间内满足作为延程运行指定备降机场的要求。最早可能到达时间按在该机场的第一等时点上改航该机场计算,最迟可能到达时间按在该机场的第二等时点上改航该机场计算(见图 4-4-8)。飞行计划上应列出实施延程运行的备降机场。

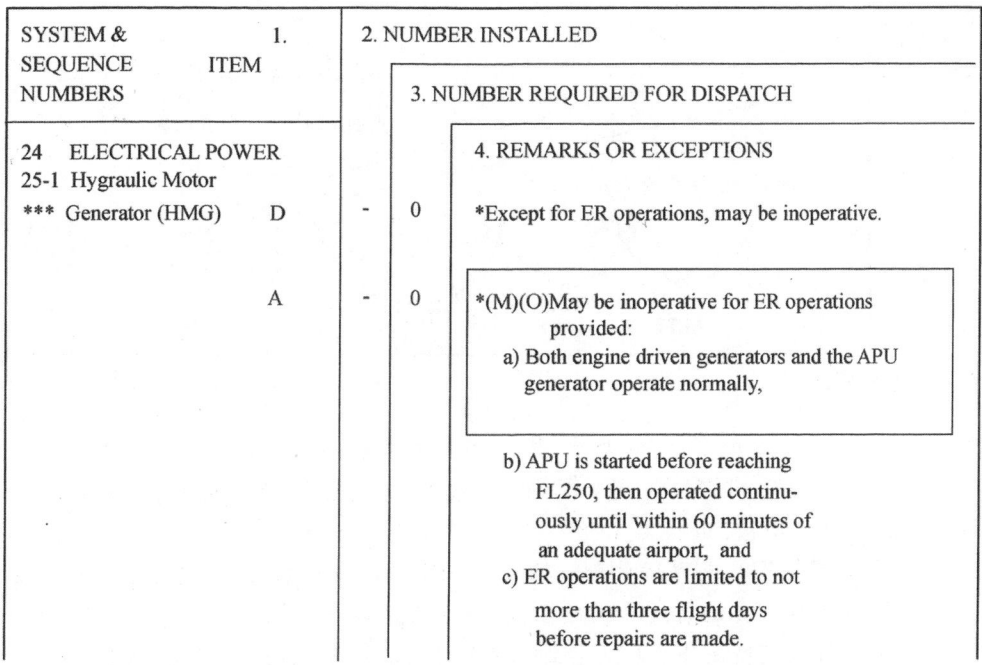

图 4-4-7　液压驱动发电机与冲压空气驱动泵的 MMEL 项

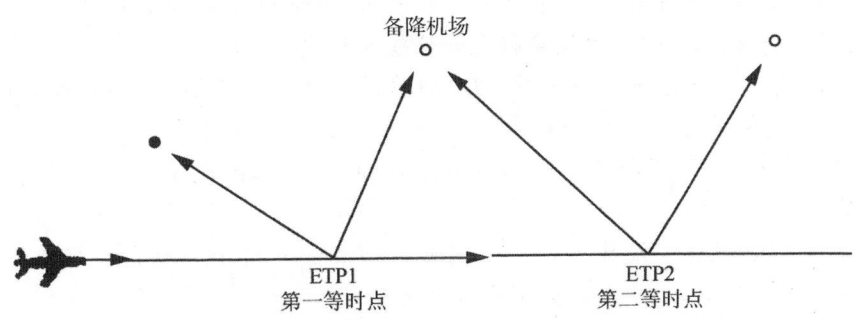

图 4-4-8　确定到达备降机场最早、最晚时刻的示意图

制作延程运行的飞行计划首先要制作出正常飞行计划(可以是利用二次放行或不利用二次放行的飞行计划),算出总油量、允许业载以及在各等时点的剩余油量,然后考虑临界燃油方案。考虑临界油量的目的是要保证在改航时有足够的油量能飞到航路中的各备降机场,对每个等时点都要计算飞到最近的两个指定备降机场所需要的油量,而且计算临界油量时必须

考虑全发工作和一发故障两种情况,因此对每个等时点要计算出四个改航所需油量,以其中最大的一个油量为准。

　　如果在一个等时点上所需要的这个油量大于按正常飞行计划计算出的在该点机上剩余油量,则必须相应增加按正常飞行计划计算出的起飞油量,使得在该点剩余油量等于临界油量(见图4-4-9)。等时点中所需临界燃油量大于根据正常备份油量计算出的飞行计划中在该点的预计剩余燃油量且差值最大,或者所需临界燃油量等于或小于根据正常备份油量计算出的飞行计划中在该点的预计剩余燃油量且差值最小,该点称为燃油关键点(CP)。

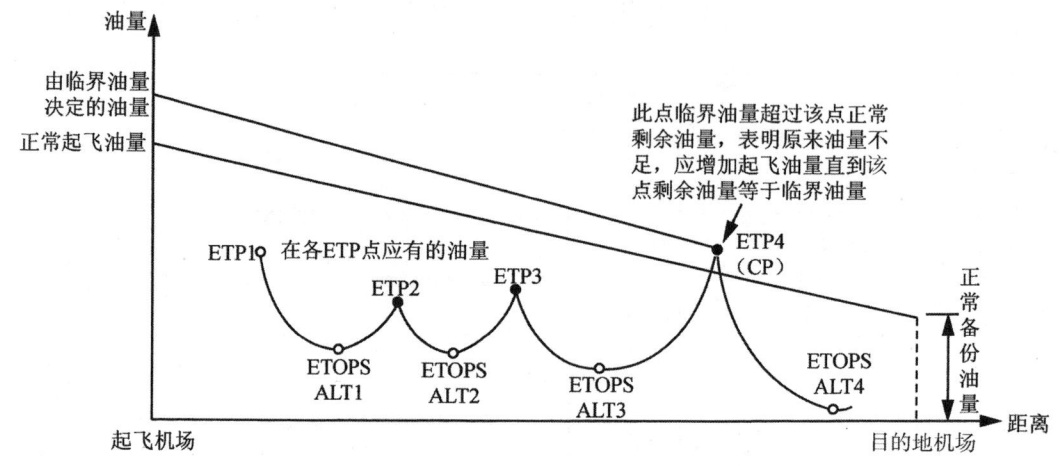

图 4-4-9　临界燃油方案

　　关于延程运行的燃油供应问题,咨询通告《延程运行和极地运行》(AC-121-FS-2019-009R2)中给出了具体要求,如下:

　　(a)任何人都不得签派或飞行放行涡轮发动机飞机实施延程运行,除非,考虑预报的风和其他天气情况,飞机载有其运行所必需的燃油,并且满足以下要求:

　　(i)在以下三种假定情况下,飞机飞往延程运行指定备降机场所需燃油的最大值:

　　假设飞机在临界点迅速释压,然后下降到符合CCAR-121.333条供氧要求的安全高度;

　　假设飞机在临界点迅速释压并且一发同时失效,下降到符合CCAR-121.333条供氧要求的安全高度,然后以经批准的一台发动机失效巡航速度飞行;

　　假设飞机在临界点一台发动机失效,以经批准的一台发动机失效巡航速度飞行,下降到一台发动机失效的巡航高度。

　　(ii)到达备降机场时,保持1500英尺(距离地面高度)的高度等待15分钟,然后,实施仪表进近,着陆。

　　(iii)在使用预报风速来计算上面第(i)段的燃油油量时,应该在实际预报值的基础上加百分之五(逆风分量加百分之五,顺风分量减百分之五)。

　　(iv)按照第(iii)段完成计算后,如果预报有结冰情况的话,根据第(i)段的要求对以下两种情况中的燃油量的最大值进行修正:

　　如果结冰的预报可靠,预报结冰期间的十分之一时间段内,计算机身结冰对燃油消耗的影响(包括无保护表面的结冰以及这个期间发动机和机翼防冰装置的耗油)。如果结冰的预报不是很可靠,根据可用的天气预报,在经批准的一台发动机失效的巡航速度条件下,总温低于+10摄氏度,或者外界大气温度介于−20摄氏度和0摄氏度之间,并且相对湿度高于55%,符合上述条件的飞行时间都认为是预报结冰期。预报结冰期间的十分之一时间段内,计算机身结冰对燃油消耗的影响(包括无保护表面的结冰以及这个期间发动机和机翼防冰装置的燃油消耗)。

预报结冰的整个期间,发动机防冰和机翼防冰(若适用)打开对燃油消耗的影响。

(b)如果合格证持有人没有建立飞机燃油消耗性能衰减的监控程序,那么在计算最终燃油供应量时要增加百分之五,用于补偿巡航燃油消耗性能的衰减。

(c)如果 APU 是必需的电源,那么飞行的相应阶段必须要计算其燃油消耗。

(d)在计算延程运行备降燃油消耗量时,可以假设飞机以经批准的一台发动机失效巡航速度飘降。

上述规章要求计算的燃油供应可以按照图 4-4-10 所示的飞行剖面计算临界油量,计算临界油量时由等时点到指定备降机场一般按大圆距离计算。

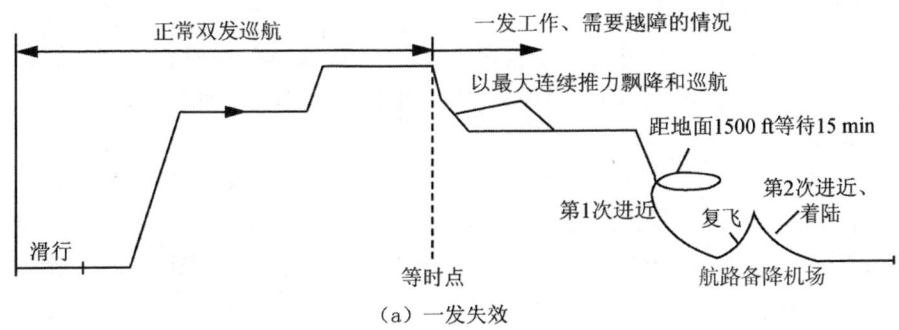

(a)一发失效

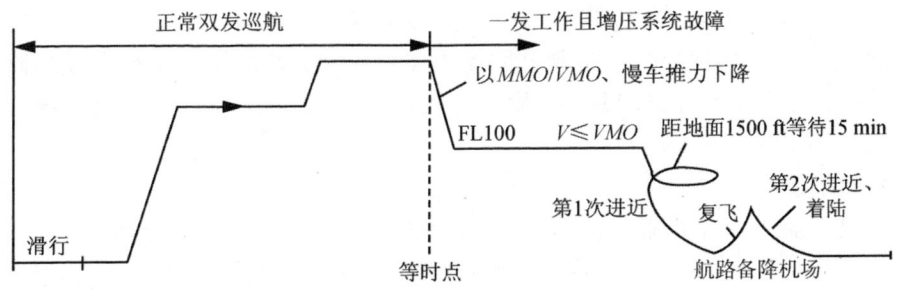

(b)一发失效且增压系统故障

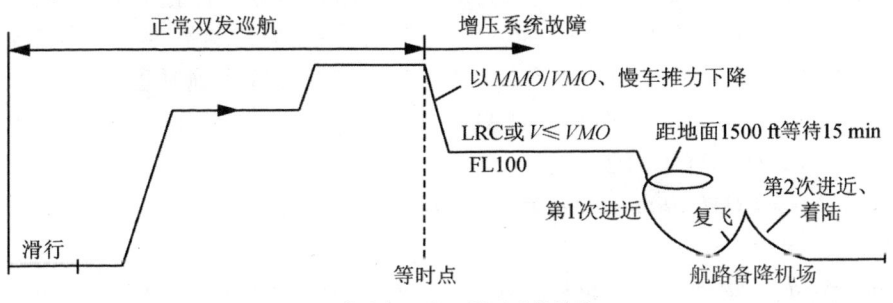

(c)发动机正常、增压系统故障

图 4-4-10　计算临界油量的飞行剖面

对于一发故障的情况,如果航路上有比较高的障碍物,按图 4-4-10(a)飞行剖面计算临界油量;否则,按图 4-4-10(b)飞行剖面计算临界油量,假定在等时点增压失效、紧急下降到

10000 ft，在 10000 ft 以选定的延程运行的改航速度巡航到最近的一个适宜的机场，在一发故障的两种飞行剖面中都要考虑下降到 1500 ft、等待 15 min 后进近并着陆的情况。在受障碍物限制巡航高度高于 10000 ft 时要考虑旅客的补氧问题。使用手册中给出了计算一发故障的临界油量的图表，本书附录 5 中给出了其中部分图表。

计算全发临界油量时假定在等时点增压失效、紧急下降到 10000 ft，然后在 10000 ft 以 LRC 速度巡航（使用手册上只有计算全发 LRC 巡航的临界油量的图表）到最近的一个适宜的机场，下降到 1500 ft、等待 15 min 后进近并着陆，如图 4-4-10（c）所示。

在计算临界燃油量时，还要考虑很多因素对燃油的影响：

（1）温度的影响

在计算临界燃油量时还要考虑温度高于 ISA 的影响及可能的结冰条件，按照图表中的说明进行修正。

（2）飞机性能衰退的影响

计算临界油量时还必须考虑飞机性能衰退的影响，如航空公司没有做性能监控，则必须把计算出的临界油量再多加 5%。如由性能监控得出了实际耗油增加的百分比，则使用这个实际百分数增加油量。

（3）风的影响

如果使用基于局方可接受的风模型（例如 WAFS，World Area Forecast System 模型）的实际预报风，要使用 105% 的顶风、95% 的顺风确定临界油量，否则要把由图表确定的临界油量增加 5%，以考虑气象预报的误差。

（4）MEL 或 CDL 项的影响

如果有 MEL 或 CDL 项，还应该考虑由此造成的耗油增加量。例如，襟翼导轨整流罩丢失，不影响放行，但使阻力增大、航路爬升限制重量减小（或者说使航路爬升或飘降净升限降低）、耗油增加。对于某些系统/部件故障或丢失所产生的影响（修正量）可以从 MEL、CDL、DDG（放行偏差指南）中找到。

（5）APU 的使用

飞机使用手册中给出的临界油量图表是否包括 APU 耗油取决于机型。按照 B737 的 CMP 和 FCOM 中的说明，在进入延程运行航段前就应该启动 APU 并使其保持运转，即使没有发生一发失效。因此，B737-CL/NG 的全发、一发的临界油量中都包括 APU 的耗油。而 B757、B767、B777 的 CMP 和 FCOM 没有这种说明，即全发工作时不需要启动 APU，因此，这些机型仅一发失效的临界油量中包括 APU 的耗油。B747 全发、一发失效的临界油量中都不包括 APU 的耗油，因为 B747 有发动机带动的 4 台发电机，即使一发失效实施延程运行飞行，也不需要 APU 的交流电源，空中不需要也不能启动 APU。

如在 B737-CL/NG 中有以下描述：

B737-CL/NG FCOM/Supplementary Procedures

ETOPS

Operators conducting ETOPS are required to comply with appropriate regulations. An operator must have an ETOPS configured and approved airplane, and approved flight operations and maintenance programs in place to support ETOPS.

APU Operation

Unless otherwise authorized, start the APU before the ETOPS segment.

The APU must be on for the entire ETOPS segment.

在 B737-NG 的 AFM 中也要求在到达 EEP 之前启动 APU 并在整个延程运行航段保持运转。按此说明,全发临界油量应该包括 APU 耗油,正常飞行计划的延程运行航段也应计入 APU 耗油。放行前如 APU 或其发电机不工作,则不能实施延程飞行。在飞行中到达 EEP 前 APU 故障应改航备降机场。

B737 只有 1 个燃油交输阀(交输阀也称交输活门,在右机翼后梁上,可通过右轮舱接近)、头顶板上只有 1 个交输阀开关(Crossfeed Selector)和指示灯。

Fuel Crossfeed Valve Check

During the last hour of cruise, do the following steps:

Crossfeed selector ·· *Open*

Verify that the VALVE OPEN light illuminates bright, then dim.

Crossfeed selector ·· *Close*

Verify that the VALVE OPEN light illuminates bright, then extinguishes.

[FCTM] This verifies that the crossfeed valve is operating so that on the subsequent flight, if an engine fails, fuel is available from both main tanks through the crossfeed valve.

波音使用手册中关于延程运行的部分包含以延程运行的改航速度飞行时一发故障的高度能力和改航油量的图表。其中包括使用最大连续推力和最大巡航推力以不同的延程运行速度巡航的高度能力数值表,这些表反映了飞机的性能能力,这些表没有保守成分在内(即这些表给出的是总高度,而非净高度)。注意,防冰引气对高度的修正是很大的。对于 B767 飞机,使用发动机防冰高度减少 1800 ft,使用发动机和机翼防冰高度减少 3800 ft,在做决断时要考虑防冰的影响。使用手册中也提供了一发故障时在飞行中改航的图表。这些图表可用于快速计算在不同高度上飞行时的改航油量。

在飞行计划中临界油量若检验无问题,则可保证:如果以选定的延程运行改航速度或更低的一个速度飞向在飞行计划中列出的最近的延程运行的备降机场,油量肯定是够的。

使用手册中延程运行部分也提供了一发故障时的巡航数据表格。在决定机组飞一条延程航线以前,航空公司的工程师们应确定所需要的延程运行的改航时间,并选择延程运行速度,确定可用的飞行区域。在放行之前,应制作延程运行计划,在计划中应给出延程运行的等时点和航路中的备降机场,所加油量应能满足临界油量要求,机组应核对飞行计划,证实预定航线处于由所列出的备降机场决定的可用飞行区域内。

4.4.2.6　签派员要做的工作

(1)放行时,延程运行可选备降机场的预报的侧风,包括阵风,在预计最早到达时刻至最晚到达时刻的时段内必须小于等于制造厂证实的最大侧风。

(2)签派员要检查延程运行航线上所有备降机场的气象预报并且已经证实符合公司的关于延程运行的最低设备清单,还应检查航路上的航行通告。如某个(些)合适备降机场的气象预报表明该机场不能作为延程运行备降机场,必须提供另外一些备降机场或一条新航路的新的飞行计划。

(3)在整个飞行阶段,如果延程运行备降机场的条件有重大改变,应及时将情况通知机组。在到达延程进入点之前,应对下述方面进行评估,主要包括:在延程指定备降机场规定的时间段内的天气情况、着陆距离、机场服务以及设备等。例如,在某延程运行备降机场天气预报低于正常着陆最低标准(注意不是延程运行放行的着陆最低标准)。

在起飞机场放行前检查时要满足延程运行的最低天气标准,放行后只要在规定的时间段内航路备降机场天气满足正常的(非延程运行)着陆最低标准即可,则将延程运行备降机场的情况通知机组并选择一个新的延程运行备降机场替代那个不再适宜作为备降机场的机场,如没有可替代的延程运行备降机场,使得原计划航线有一段不在允许飞行区域内,则应备降,或调整航路使之处于允许区域内并重新制作一份飞行计划,新的飞行计划应当包括批准的运行区域内新指定的延程运行指定备降机场的信息。修改后的计划航路上任意一点距离新指定的延程运行指定备降机场的改航时间依然要符合该次签派或飞行放行时确定的最大改航时间限制。如果无法满足上述要求,应当修改计划航路在非延程运行区域完成飞行运行。如果在非延程运行区域内完成飞行运行不可行,则应当选择改航备降。

飞机飞越延程运行进入点之后,如果延程运行指定备降机场的天气条件恶化而低于着陆运行最低标准(或者任何造成机场跑道不再可用的其他情况),合格证持有人不需要因此修改航路或改航备降。机长应当与签派员沟通,对当时的安全运行状况做出合理的评估和判断,确定最安全的运行方案。此时机长有权决定修改航路,备降到延程运行指定的备降机场,或者使用计划航路继续飞行。例如,如果 LPLA/TER(特塞拉)的气象条件不满足要求,则指定 BIKF/KEF[凯夫拉维克(冰岛)]作为新备降机场并调整航路,然后重新制作一份飞行计划航路上传给机组。

4.4.2.7 机组要做的工作

在延程运行期间机组没有特殊的操作程序,但在飞行中机组要监控飞机位置、燃油状态和系统状态,可能要求机组记录发动机数据,以便输入发动机状态监控程序,对发动机状态进行监控。

此外,在飞行进程中机组必须检查延程运行的备降机场的气象条件,且获得沿计划航路预计天气条件满足运行最低标准的所有延程运行可选备降机场的天气、服务和设施设备等相关信息,而不限于飞行计划中的延程运行指定备降机场。

和执行其他飞行任务一样,如果发生了某种事件,机组必须做出是继续飞行还是改航或返航的决断。延程运行的规章条款可保证机组在面对飞行中的事故时有足够的信息做出最好的决断。机组不受延程运行的条例限制,他们必须评价当时的形势并且根据实际情况采取措施。例如,假定发动机在某点失效,特塞拉(TER)是最近的备降机场。机组一直在监听特塞拉和圣约翰斯两处的气象状况。特塞拉的气象条件高于最低标准,但高出最低标准不多,报告的侧风是 25 kt。圣约翰斯有无限的能见度,但距离远 150 NM。机组必须根据获得的信息选择备降哪个机场更好。机组应考虑的因素是:引起改航的失效类型,机场的着陆助航设备和应急设备、跑道长度和气象条件。

在一发故障引起的延程运行的改航情况下,机组必须确定为了一发飞到备降机场,应使用的最好的速度、高度和推力调定值。原定计划是机组应以延程运行的改航速度飞行,但是如果按机组的判断,以另一个速度飞行更有利,那么他可以根据实际情况选择该速度飞行。机组要根据公司的政策和事故发生时的情况选择改航速度、高度和推力调定值。选用最大连续推力,可以得到较高的飞行速度和较高的高度,飞到备降机场的时间短;选用最大巡航推力,则巡航的速度、高度较低,飞到备降机场的时间长。如果时间是最重要的,可以选择以最大连续推力飞行,让飞机在尽可能高的高度上飞行。如果时间不是最重要的,可以选择以最大巡航推力飞行,这将导致较长的改航时间和较低的巡航高度。机组不能修改飞行管理计算机中的一发故障的飘降速度(最大升阻比速度),因此,以延程运行的改航速度(例如 LRC、290 KIAS、310 KI-AS)飞行时,一发故障的高度能力不能通过飞行管理计算机获得。为了得到这些数据,机组必须查阅 QRH(快速参考手册 PI 章)、FPPM(飞行计划和性能手册)。这些手册包括延程运行的航路计划、放行、不同改航速度的高度能力(总高度和净高度)和飞行中改航方面的信息。

4.4.3　延程运行飞行计划的实例

广州(CAN)至温哥华(YVR)备降西雅图(SEA)的延程运行飞行计划(节选)如下。

```
[ OFP ]
------------------------------------------------------------------
              CHINA SOUTHERN AIRLINES FLIGHT RELEASE

CHINA SOUTHERN AIRLINES  -  DISPATCH RELEASE STD/1515   13MAY21
                                              DOC NO    7/0/1

CSN   5053 /13  1515Z        ZGGG - CYVR     1ST ALTN KSEA

AIRCRAFT   B209X  IFR TAKEOFF FUEL 72.6T  TOTAL FUEL 73.0T  .....

(2ND ALTN FOR DESTINATION_____)


(FPL-CSN5053-IS
-B789/H-SADE2E3FGHIJ1J2J4J5LM1OP2RVWXY/LB1D1
-ZGGG1515
-K0946S0950 LMN G471 PLT A599 JTN G327 LAMEN/N0506F310 A593 SADLI
 Y590 BEDAR Y591 FUE Y60 ISAKY Y28 DGC Y14 STOUT/N0494F350 Y14
 SAMON Y142 GTC Y512 BEKEN/M085F350 Y512 ADNAP OTR5 KALNA DCT
 42N160E/M085F370 45N180E 47N170W 50N160W 52N150W
 52N140W/N0487F370 DCT YZT DCT POWOL DCT KEINN V330 FASBO DCT
-CYVR1104 KSEA
-PBN/A1B1C1D1L1O1S2 NAV/ABAS SUR/RSP180 DOF/210513 REG/B209X
 EET/ZSHA0030 RKRR0140 RJJJ0154 BEKEN0331 OATIS0339 ADNAP0351
 KALNA0415 42N160E0506 KZAK0529 43N170E0552 45N180E0641 47N170W0728
 50N160W0814 52N150W0856 52N140W0935 CZVR1000 SEL/DPMQ CODE/781666
 OPR/CSN PER/D RALT/RJBB PASY PANC CYVR RMK/ACASII TCAS)

DISPATCHER ON DUTY SIGNATURE:                   TEL:
                             ..................

I CERTIFY THAT THIS FLIGHT HAS BEEN DISPATCHED IN ACCORDANCE WITH
APPLICABLE REGULATIONS, AND RECEIVED THE FOLLOWING DOCUMENTS:
DISPATCH RELEASE, FLIGHT PLAN, WEATHER AND NOTAM.

PILOT  IN  COMMAND SIGNATURE:
                             ...........................

AMENDMENT LOG (RECORD ANY CHANGES OF INITIAL DISPATCH RELEASE)

TIME.........NEW ITEM........................PIC........DISP........

TIME.........NEW ITEM........................PIC........DISP........

TIME.........NEW ITEM........................PIC........DISP........

REMARK............................................................
```

```
------------------- START OF FLIGHT PLAN -----------------------
CSN5053  13MAY2021   STD/STA 131515/140250   ZGGG 1515 - CYVR 0239
DEST ALTN KSEA
ACFT B209X B787-9, GENX-1B74-75              FUEL BIAS P00.0

DISP
OFP 7/0/1    RLS 1108 13MAY21
WX PROG 1315 1318 1321 1400 1403 1406               OBS 130600
AVG WIND 265/058        AVG W/C P057           AVG ISA P005
AVG TAS/GS 483/539      DEST AVG HLD FF 4416    AVG FF 5658
CLB 310/M.85  CRZ CI85/M0.85/CI85  DSC 310/M.85

ROUTE NO. DEFRTE                      GND/AIR DIST 5970 NM /5350 NM
ZGGG LMN G471 PLT A599 JTN G327 LAMEN A593 SADLI Y590 BEDAR Y591 FUE
Y60 ISAKY Y28 DGC Y14 STOUT Y14 SAMON Y142 GTC Y512 BEKEN Y512 ADNAP
OTR5 KALNA DCT N42E160 DCT N43E170 DCT N45E180 DCT N47W170 DCT
N50W160 DCT N52W150 DCT N52W140 DCT YZT DCT POWOL DCT KEINN V330
FASBO DCT CYVR

FL 311 LAMEN 310 STOUT 350 43N170E 370


                    ALL WEIGHTS IN KILOS
ETOW   244892 ELDW  182262 EZFW  172332 EPLD  045000
MTOW   253530 MLDW  192776 MZFW  181436 APLD  .....

TARGET ARRIVAL FUEL         9930KGS

DEST CYVR       062630  1104

CONT FUEL 5%    003132  0033
ALTN KSEA       002588  0029
HOLD FUEL       002200  0030
EXTR FUEL       002010  0021
TKOF FUEL       072560  1258

TAXI OUTF       000440  0020
LOAD FUEL       073000  1318

CONT FUEL IS CNFMD IN ANY CASE NOT LESS THAN 15 MIN HLD FUEL AT DEST
15 MIN HLD FUEL AT DEST IS 1104 KG

EXTR FUEL INCLUDES: MIN DEP 2010

--------- --------- --------- TANKERING --------- --------- --------
NO TANKERING RECOMMENDED        LOSS FOR EXTRA FUEL: 220 USD/TON

                  FUEL ADJUSTMENTS
ZFW CHG PS 1000  BOF  +333 KGS / MS 1000 BOF -319 KGS
TOW UP 1000 KGS  BOF  +230 KGS TIME +0000 ZGGG/311 LAMEN/310
                                         STOUT/350 43E70/370
TOW DN 1000 KGS  BOF  -226 KGS TIME -0000 ZGGG/311 LAMEN/310
                                         STOUT/350 43E70/370
1FL UP           BOF  -233 KGS TIME +0008 ZGGG/331 LAMEN/330
                                         STOUT/350 HALNA/370
                                         43E70/390
1FL DN           BOF +1471 KGS TIME +0002 ZGGG/291 LAMEN/290
                                         STOUT/300 HALNA/330
                                         43E70/350
```

```
2FL DN              BOF +3489 KGS TIME +0009 ZGGG/266 LAMEN/270
                                              STOUT/280 HALNA/310
                                              43E70/330
CI 150              BOF  +117 KGS TIME -0002 ZGGG/311 LAMEN/310
                                              STOUT/350 43E70/370
CI 0                BOF  -342 KGS TIME +0010 ZGGG/311 LAMEN/310
                                              STOUT/350 43E70/370
                          ALTERNATE SUMMARY
ALTN            MSA    DIST   FL    W/C     TIME    DIFF    FUEL   DIFF
KSEA/SEA/34R    84     137    170   P012    0029            2588
DIV RTE - CYVR YVR JAWBN6 KSEA

KPDX/PDX/28L    54     243    270   P010    0043   +0014    3847  +1259
DIV RTE - CYVR SEA HELNS6 KPDX

KSFO/SFO/28L    100    724    350   P003    0144   +0115    8727  +6139
DIV RTE - CYVR YYJ DCT ELMAA DCT FAMUK Q3 PYE V27 STINS DCT KSFO

KOAK/OAK/30     100    745    350   P003    0147   +0118    8922  +6334
DIV RTE - CYVR YYJ DCT ELMAA DCT EUG J126 OED DCT RBL J3 OAK V301
SUNOL DCT KOAK


RVSM ALTIMETER CHECK (RVSM AIRSPACE ONLY):

AIRPORT ELEVATION  :........FEET      QNH/QFE:........./........HPA

LEFT  ALTIMETER DIS:........FEET      (MAX DIFF<75 FEET)

RIGHT ALTIMETER DIS:........FEET      (MAX DIFF<75 FEET)

PASSING TRANS  ALTITUDE SET 1013.2HPA        (MAX DIFF<150 FEET)
CHECK:   PRIMARY(            )       STANDBY(            )
REACHING CFL   ALTITUDE SET 1013.2HPA        (MAX DIFF<200 FEET)
RECHECK:PRIMARY(            )        STANDBY(            )
ENTRY CLASS II NAVIGATION AREA RECORDED ALTIMETER VALUE:

LEFT    ALTIMETER DIS:........FEET  RIGHT ALTIMETER DIS.......FEET

STANDBY ALTIMETER DIS:........FEET
```

```
                        ETOPS SUMMARY

      CRITICAL POINT FOR FUEL REQUIREMENTS:  N52 01.0  W140 25.1
              CFR AT CP: 12447   FOB AT CP: 15790
        ETOPS ENTRY (RJBB) N37 53.9 E142 55.1 TIME 0336
        ETOPS EXIT  (CYVR) N51 31.5 W133 43.3 TIME 1002

                      DIST     TIME  W/C        ICE    CFR     FOB     COND
ETP1 RJBB/PASY 1027/1047 0253  M007/P005  199    20060   41245   DX
0444 N40 49.8 E155 18.7  - 223NM BEFORE 42E60

ETP2 PASY/PANC  834/901  0225  M011/P016  161    16283   24122   DX
0754 N48 43.0 W164 47.2  - 202NM BEFORE 50N60

ETP3 PANC/CYVR  633/677  0149  M005/P018  123    12447   15790   DX
0934 N52 01.0 W140 25.1  - 16NM BEFORE 52N40

ADDITIONAL ETOPS FUEL WITH ICING COND:  0
NON-ICING CORRECTION:                   0
(IF NO ICING COND ANTICIPATED, NIC MAY BE SUBTRACTED FROM PLNTOF)

ETOPS (180 MIN)
ENRTE ALTNS (WX/NOTAM SUITABILITY PERIOD)
RJBB (1917-0024)    PASY (2144-0310)    PANC (0029-0412)
CYVR (0138-0412)

DEP ATIS:
         ....................................................................
(      )
         ....................................................................

DEP CLNC:
         ....................................................................

         ....................................................................

BLOCK IN   ......   ON    ......   T/O  ......    T/O FUEL  ......

BLOCK OUT  ......   OFF   ......   ETE  ......    LDG FUEL  ......

TOTAL      ......   TOTAL......    ETA  ......    F USED    ......
```

FL	ATA	ZTM/TTME	POSN	ETA	AWY	DST	DTGO	MACH	TAS	WIND	EFR
OAT			MSA		MTK/TTK	TP	WS	IAS	GS		AFR
CLB		003/0000	ZGGG/19		LMN1B	010	5970	CLB		167010	072.6
....			028	....	241/238	562			174	.....	.....
CLB		002/0003	RW01		LMN1B	007	5960	CLB		196023	071.6
....			028	....	212/209	562		310	257	.....	.....
CLB		002/0005	GG512		LMN1B	009	5954	CLB		209023	071.2
....			027	....	111/107	561		310	315	.....	.....
CLB		003/0007	GG516		LMN1B	017	5945	CLB		218023	070.8
....			035	....	068/065	561		310	394	.....	.....

CLB	006/0009	GG519		LMN1B	041	5928	CLB		228026	070.3
....		052		.....	068/065	562	310	446	.....	.....
CLB	006/0015	LMN		G471	048	5887	CLB		238029	069.2
....		055		.....	040/037	557	310	493	.....	.....
CLB	003/0021	XEBUL		G471	029	5839	CLB		244033	068.2
....		055		.....	016/013	555	310	514	.....	.....
CLB	000/0024	VAVSO		G471	002	5809	CLB		239034	067.6
....		045		.....	016/012	559	310	569	.....	.....
CLB	004/0024	TOC		G471	035	5808	CLB	511	238034	067.6
....		045		.....	016/012	559	310	535	.....	.....
311	003/0028	EGEDA		G471	022	5773	.846	511	238036	067.1
M32....		048		.....	339/335	556	01 315	514	.....	.....
	001/0031	-ZSHA			008	5751	.846	511	238037	066.9
		SHANGHAI FIR							.....	.....
311	007/0032	PLT		A599	061	5743	.845	511	238038	066.8
M32....		048		.....	051/047	550	01 315	548	.....	.....
311	002/0038	SAGON		A599	019	5681	.844	511	240038	066.0
M32....		056		.....	051/047	555	01 315	548	.....	.....
311	005/0040	P120		A599	043	5662	.844	509	241039	065.8
M33....		065		.....	051/047	559	01 315	549	.....	.....
311	003/0045	NF		A599	027	5619	.844	509	242040	065.3
M33....		065		.....	050/046	558	01 315	548	.....	.....
311	005/0048	XUVGI		A599	047	5592	.844	509	243041	064.9
M33....		089		.....	050/046	555	02 314	547	.....	.....
311	003/0053	P215		A599	031	5545	.844	509	247040	064.4
M33....		089		.....	051/046	545	02 314	546	.....	.....
311	009/0057	SHR		A599	084	5514	.844	509	250041	064.0
M33....		060		.....	052/047	533	02 314	547	.....	.....
311	003/0106	TUPGA		A599	024	5431	.844	509	257042	063.0
M33....		051		.....	053/048	496	02 314	546	.....	.....
311	001/0108	ELNEX		A599	012	5406	.843	509	258043	062.7
M33....		051		.....	053/048	479	02 314	546	.....	.....
311	006/0110	TOL		A599	050	5394	.844	509	259043	062.6
M33....		051		.....	036/030	470	01 314	535	.....	.....
311	003/0115	KAKIS		A599	029	5344	.844	509	258043	062.0
M34....		025		.....	036/030	450	01 314	537	.....	.....
311	002/0119	NXD		A599	017	5315	.843	507	257043	061.6
M34....		020		.....	080/074	451	01 314	550	.....	.....

```
311        002/0120  LURMA        A599      022  5298  .843 507 258043 061.4
M34....              020     ..... 080/074  451  01    314  550  .....  .....
------------------------------------------------------------------------------
311        001/0123  OLGAP        A599      010  5277  .843 507 260042 061.2
M34....              020     ..... 079/073  452  01    314  548  .....  .....
------------------------------------------------------------------------------
311        005/0124  JTN          G327      047  5266  .842 507 261042 061.0
M34....              020     ..... 105/099  452  00    314  549  .....  .....
------------------------------------------------------------------------------
311        001/0129  NINAS        G327      009  5219  .842 507 264044 060.5
M34....              020     ..... 096/090  452  00    314  550  .....  .....
------------------------------------------------------------------------------
311        001/0130  LASAN        G327      012  5210  .842 507 265044 060.4
M34....              020     ..... 096/090  451  00    314  552  .....  .....
------------------------------------------------------------------------------
311        002/0131  BONGI        G327      018  5198  .842 507 265044 060.2
M34....              020     ..... 096/090  452  00    314  552  .....  .....
------------------------------------------------------------------------------
311        004/0133  BOLEX        G327      034  5180  .842 507 266045 060.0
M34....              020     ..... 061/054  454  00    314  546  .....  .....
------------------------------------------------------------------------------
311        003/0137  TONIX        G327      029  5146  .842 507 267047 059.6
M34....              020     ..... 061/054  449  01    314  546  .....  .....
------------------------------------------------------------------------------
           000/0140  -RKRR                  000  5117  .842 505 269049 059.3
                     INCHEON FIR                           .....  .....
------------------------------------------------------------------------------
311        006/0140  LAMEN        A593      053  5117  .840 505 269049 059.3
M34....              020     ..... 082/075  449  01    314  555  .....  .....
------------------------------------------------------------------------------
310        004/0146  SADLI        Y590      035  5064  .840 505 274053 058.7
M34....              020     ..... 108/101  450  01    314  558  .....  .....
------------------------------------------------------------------------------
310        002/0150  IKEDO        Y590      015  5030  .840 505 276053 058.3
M34....              020     ..... 082/075  450  01    314  556  .....  .....
------------------------------------------------------------------------------
310        003/0151  ELGEP        Y590      029  5015  .840 505 277054 058.1
M34....              020     ..... 082/076  451  01    314  556  .....  .....
------------------------------------------------------------------------------
           000/0154  -RJJJ                  000  4987  .840 505 278055 057.8
                     FUKUOKA FIR                          .....  .....
------------------------------------------------------------------------------
310        004/0154  BEDAR        Y591      033  4986  .839 505 278055 057.8
M34....              020     ..... 082/075  452  02    314  557  .....  .....
------------------------------------------------------------------------------
310        010/0158  BIGIT        Y591      095  4953  .839 505 279056 057.4
M34....              025     ..... 073/066  454  02    314  554  .....  .....
------------------------------------------------------------------------------
310        006/0208  FUE          Y60       053  4858  .839 503 272057 056.3
M35....              036     ..... 059/052  465  04    314  547  .....  .....
------------------------------------------------------------------------------
310        005/0214  ISAKY        Y28       047  4806  .839 503 272058 055.7
M35....              044     ..... 060/052  462  04    314  547  .....  .....
------------------------------------------------------------------------------
310        005/0219  DGC          Y14       044  4759  .840 503 277061 055.2
M36....              042     ..... 053/045  459  03    314  535  .....  .....
------------------------------------------------------------------------------
```

```
310      004/0224  STOUT        Y14    033 4715  .850 494 285061 054.7
M37....            044     ..... 059/051 457 04   314 546 ..... .....
-----------------------------------------------------------------------
CLB      011/0228  HALNA        Y14    097 4682  CLB  494 284079 054.1
 ....             050     ..... 059/052 454 05   310 533 ..... .....
-----------------------------------------------------------------------
350      006/0239  MIHOU        Y14    049 4585  .846 492 293068 053.1
M48....            032     ..... 066/058 439 08   289 522 ..... .....
-----------------------------------------------------------------------
350      004/0244  SUGNO        Y14    032 4536  .846 490 296057 052.5
M51....            020     ..... 067/058 434 07   289 517 ..... .....
-----------------------------------------------------------------------
350      014/0248  SAMON        Y142   119 4504  .846 490 297052 052.2
M52....            026     ..... 072/064 426 05   289 521 ..... .....
-----------------------------------------------------------------------
350      014/0302  NESKO        Y142   125 4385  .846 490 291055 050.8
M53....            040     ..... 073/065 400 03   289 529 ..... .....
-----------------------------------------------------------------------
350      003/0316  GTC          Y512   031 4260  .846 488 283057 049.5
M53....            090     ..... 091/082 404 02   289 542 ..... .....
-----------------------------------------------------------------------
350      006/0319  ELDAK        Y512   055 4229  .846 490 279057 049.2
M53....            090     ..... 091/083 400 02   290 543 ..... .....
-----------------------------------------------------------------------
350      006/0325  SDE          Y512   055 4174  .847 490 272056 048.6
M53....            026     ..... 106/098 405 02   290 544 ..... .....
-----------------------------------------------------------------------
350      005/0331  BEKEN        Y512   041 4119  .850 492 266056 048.0
M53....            020     ..... 107/099 405 02   290 546 ..... .....
-----------------------------------------------------------------------
350      000/0336  *ENTRY       Y512   004 4078  .850 492 261058 047.6
M53....            020     ..... 107/099 403 02   291 546 ..... .....
-----------------------------------------------------------------------
         003/0336  -RJJJ               026 4074  .850 492 261058 047.5
                   FUKUOKA FIR OCEANIC                     ..... .....
-----------------------------------------------------------------------
350      012/0339  OATIS        Y512   108 4048  .850 494 258059 047.3
M53....            020     ..... 117/110 402 02   291 539 ..... .....
-----------------------------------------------------------------------
350      024/0351  ADNAP        OTR5   229 3940  .850 494 244063 046.1
M52....            020     ..... 065/058 425 02   291 569 ..... .....
-----------------------------------------------------------------------
350      028/0415  KALNA        DCT    272 3711  .850 490 258087 043.9
M52....            020     ..... 073/067 419 03   291 576 ..... .....
-----------------------------------------------------------------------
350      023/0444  *ETP1        DCT    223 3439  .850 490 270100 041.2
M53....            020     ..... 072/067 398 03   291 575 ..... .....
-----------------------------------------------------------------------
350      023/0507  42E60        DCT    225 3216  .850 490 280111 039.1
M54....            020     ..... 083/079 381 03   291 588 ..... .....
-----------------------------------------------------------------------
         023/0530  -KZAK               222 2991  .850 490 281098 037.0
                   OAKLAND OCEANIC FIR                     ..... .....
-----------------------------------------------------------------------
350      049/0553  43E70        DCT    448 2769  .850 496 282084 035.0
M55....            020     ..... 071/071 346 00   291 551 ..... .....
-----------------------------------------------------------------------
```

```
370        047/0641  45E80          DCT      434 2321  .850 500  269051 030.5
M49....              020      ..... 067/070  309 01    278  550  .....  .....
------------------------------------------------------------------------------
370        025/0729  47N70          DCT      234 1887  .850 494  225064 026.3
M46....              020      ..... 055/062  299 01    278  566  .....  .....
------------------------------------------------------------------------------
370        021/0754  *ETP2          DCT      202 1653  .850 494  219090 024.1
M48....              020      ..... 053/062  342 02    278  566  .....  .....
------------------------------------------------------------------------------
370        041/0815  50N60          DCT      396 1451  .850 484  213112 022.3
M51....              020      ..... 057/069  379 03    278  578  .....  .....
------------------------------------------------------------------------------
370        038/0856  52N50          DCT      354 1054  .850 484  267085 018.9
M59....              020      ..... 072/086  395 03    278  563  .....  .....
------------------------------------------------------------------------------
370        002/0934  *ETP3          DCT      016 0700  .850 484  309078 015.8
M59....              020      ..... 070/086  367 03    278  563  .....  .....
------------------------------------------------------------------------------
370        025/0936  52N40          DCT      219 0685  .850 488  311077 015.7
M59....              050      ..... 078/095  366 03    278  524  .....  .....
------------------------------------------------------------------------------
           002/1001  -CZVR                   016 0466  .850 488  302052 013.6
                     VANCOUVER FIR                               .....  .....
------------------------------------------------------------------------------
370        028/1002  *EXIT          DCT      245 0450  .850 488  302051 013.5
M58....              050      ..... 077/095  360 02    278  523  .....  .....
------------------------------------------------------------------------------
370        007/1031  YZT            DCT      059 0205  .850 486  292023 011.2
M56....              083      ..... 088/105  353 01    278  508  .....  .....
------------------------------------------------------------------------------
370        006/1037  TOD            DCT      045 0146  .850      298023 010.6
M58....              083      ..... 088/105  353       278  484  .....  .....
------------------------------------------------------------------------------
DES        007/1043  POWOL          DCT      046 0100  DES       311019 010.5
....                 094      ..... 104/121  353       310  411  .....  .....
------------------------------------------------------------------------------
DES        002/1050  KEINN          V330     010 0055  DES       312015 010.4
....                 094      ..... 136/153  354       310  322  .....  .....
------------------------------------------------------------------------------
DES        004/1052  CEESE          V330     020 0045  DES       315013 010.4
....                 093      ..... 136/153  354       310  284  .....  .....
------------------------------------------------------------------------------
DES        008/1056  FASBO          DCT      025 0025  DES       305005 010.3
....                 068      ..... 128/144  355       310  180  .....  .....
------------------------------------------------------------------------------
           1104 CYVR/26L                                                009.9
------------------------------------------------------------------------------


ARR ATIS:
                   .................................................
(          )
                   .................................................


                    WAYPOINT AND WIND DIRECTION/SPEED
          FL100          FL150          FL200          FL310          FL350
   CLIMB 22/023 P13   23/027 P02    24/029 M08    24/033 M32    24/036 M42
```

POS/FRQ	COORD	100	250	310	330	350	370	390
ZGGG	N23 23.5 E113 18.6							
RW01	N23 22.6 E113 17.0							
GG512	N23 16.6 E113 13.4							
GG516	N23 14.0 E113 22.3							
GG519	N23 21.4 E113 39.3							
LMN 116.3	N23 38.9 E114 19.7							
XEBUL	N24 17.0 E114 51.0							
VAVSO	N24 45.6 E114 58.1							
TOC	N24 47.2 E114 58.5	P12 22/024	M18 25/034	M32 24/034	M37 24/035	M42 24/036	M47 24/038	M52 24/041
EGEDA	N25 20.9 E115 06.7	P12 23/025	M18 25/036	M32 24/036	M37 24/037	M42 24/039	M47 24/041	M52 24/043
PLT 111.6	N25 48.4 E114 52.6	P12 23/027	M18 24/037	M32 24/038	M37 24/039	M42 24/041	M47 24/043	M52 24/045
SAGON	N26 30.1 E115 42.5	P12 23/028	M18 25/036	M32 24/038	M37 24/040	M42 24/042	M47 24/044	M52 24/046
P120	N26 43.3 E115 58.3	P12 23/028	M18 25/035	M32 24/038	M37 24/041	M42 24/043	M48 24/045	M52 24/047
NF 292	N27 12.6 E116 33.6	P11 24/028	M18 25/034	M33 24/040	M37 24/042	M43 24/045	M48 24/047	M52 24/048
XUVGI	N27 31.2 E116 55.3	P11 24/029	M18 25/034	M33 24/040	M38 24/043	M43 24/045	M48 24/047	M53 24/050
P215	N28 03.6 E117 33.0	P10 24/031	M18 25/034	M33 25/040	M38 24/043	M43 24/046	M48 24/048	M53 25/051
SHR 114	N28 25.1 E117 58.1	P10 24/033	M18 26/035	M33 25/040	M38 25/044	M43 25/047	M48 25/049	M53 25/051
TUPGA	N29 21.5 E119 08.6	P09 25/035	M18 26/039	M33 26/042	M38 25/045	M43 25/049	M48 25/052	M53 25/054

ELNEX	N29 37.9	P09	M18	M33	M38	M43	M49	M53
	E119 29.4	25/037	26/039	26/043	26/046	26/050	25/053	25/056
TOL 115.9	N29 45.9	P09	M18	M33	M38	M43	M49	M53
	E119 39.5	25/037	26/040	26/042	26/046	26/051	26/054	26/057
KAKIS	N30 29.0	P08	M18	M33	M38	M44	M49	M54
	E120 08.8	24/032	26/040	26/043	26/046	26/050	25/055	25/059
NXD 116.5	N30 53.9	P08	M18	M33	M39	M44	M49	M54
	E120 25.7	24/027	26/041	26/043	26/045	25/049	25/054	25/059
LURMA	N30 58.5	P08	M18	M34	M39	M44	M49	M54
	E120 44.9	24/026	26/042	26/043	26/044	26/047	25/053	25/058
OLGAP	N31 04.4	P08	M18	M34	M39	M44	M50	M54
	E121 09.0	25/024	26/043	26/042	26/042	26/045	26/051	26/057
JTN 109.6	N31 07.4	P07	M18	M34	M39	M44	M50	M54
	E121 20.4	25/022	26/043	26/042	26/042	26/045	26/051	26/058
NINAS	N31 00.0	P08	M18	M34	M39	M45	M50	M54
	E122 15.0	23/021	26/043	26/044	26/045	26/047	26/053	26/060
LASAN	N31 00.0	P08	M18	M34	M39	M45	M50	M54
	E122 25.5	23/021	26/043	26/044	26/045	26/048	26/053	26/061
BONGI	N31 00.0	P08	M18	M34	M39	M45	M50	M54
	E122 38.9	23/020	26/044	27/044	27/045	26/048	26/054	26/061
BOLEX	N31 00.0	P08	M18	M34	M39	M45	M50	M54
	E123 00.0	24/019	26/044	27/045	27/045	27/048	27/053	26/060
TONIX	N31 19.9	P08	M19	M34	M39	M45	M50	M55
	E123 32.6	25/017	25/045	27/047	27/047	27/049	27/053	27/061
LAMEN	N31 36.6	P08	M19	M34	M39	M45	M50	M55
	E124 00.0	25/016	25/043	27/049	27/048	27/050	27/054	27/061
SADLI	N31 49.8	P08	M20	M34	M40	M45	M50	M55
	E125 00.0	26/015	26/037	27/053	27/053	27/053	27/056	27/061
IKEDO	N31 43.2	P08	M20	M34	M39	M45	M50	M55
	E125 39.8	26/016	26/037	28/053	28/053	28/052	27/055	27/060
ELGEP	N31 46.9	P08	M20	M34	M39	M45	M50	M55
	E125 56.3	26/017	27/036	28/053	28/053	28/053	27/055	27/060
BEDAR	N31 54.0	P08	M20	M34	M40	M45	M50	M55
	E126 29.2	26/017	27/034	28/055	28/054	28/054	28/056	27/061
BIGIT	N32 02.3	P08	M20	M34	M40	M45	M50	M55
	E127 07.2	27/016	27/032	28/056	28/056	28/056	28/058	28/064
FUE 115.8	N32 40.1	P09	M21	M35	M40	M46	M51	M55
	E128 49.6	27/011	28/034	27/058	28/063	28/064	28/066	28/070

ISAKY	N33 12.4 E129 39.2	P09 28/013	M22 29/037	M36 27/058	M41 28/064	M46 28/068	M51 28/071	M56 28/076
DGC 114.5	N33 40.6 E130 23.4	P08 28/012	M22 29/039	M36 28/061	M41 28/066	M46 28/071	M51 28/075	M56 29/080
STOUT	N34 11.4 E131 01.0	P07 28/009	M23 30/034	M37 28/060	M42 28/068	M46 28/076	M51 28/082	M57 28/087
HALNA	N34 32.0 E131 32.2	P06 28/007	M23 30/032	M38 29/057	M42 28/068	M47 28/080	M52 29/090	M57 29/095
MIHOU	N35 31.9 E133 05.6	P05 34/003	M24 31/032	M41 30/046	M46 29/053	M50 29/068	M53 29/085	M57 29/101
SUGNO	N35 57.7 E133 56.4	P05 36/007	M25 30/034	M42 30/043	M47 30/047	M52 30/057	M55 30/071	M57 30/087
SAMON	N36 14.6 E134 30.2	P05 36/010	M25 30/035	M42 30/042	M47 30/045	M53 30/052	M56 30/064	M58 30/078
NESKO	N37 06.0 E136 43.2	P04 31/017	M26 29/039	M43 29/048	M48 29/050	M53 29/055	M57 29/061	M59 30/067
GTC 115.5	N37 57.5 E139 06.9	P03 30/024	M27 28/042	M44 28/050	M49 28/053	M53 28/057	M57 28/061	M60 28/064
ELDAK	N38 01.6 E139 45.7	P03 30/023	M27 27/042	M44 28/051	M49 28/053	M53 28/057	M57 28/060	M60 28/063
SDE 116.3	N38 08.3 E140 55.3	P03 30/020	M27 27/043	M44 27/050	M49 27/052	M53 27/056	M57 27/060	M60 27/063
BEKEN	N38 00.4 E142 03.7	P03 30/017	M27 27/040	M43 27/050	M48 27/053	M53 27/056	M57 27/059	M60 27/060
ENTRY1	N37 53.9 E142 55.1							
OATIS	N37 49.0 E143 32.4	P03 31/016	M27 26/036	M43 26/052	M47 26/056	M52 26/059	M57 26/062	M60 25/063
ADNAP	N37 11.8 E145 40.0	P03 28/009	M26 26/035	M41 25/058	M46 25/061	M51 24/063	M56 24/067	M60 24/070
KALNA	N39 09.2 E149 49.5	M00 24/017	M28 27/051	M44 26/071	M48 26/080	M52 26/087	M56 26/093	M60 25/098
ETP1	N40 49.8 E155 18.7							
42E60	N42 00.0 E160 00.0	M07 28/020	M32 28/083	M47 28/103	M51 28/107	M56 28/111	M60 28/111	M60 28/106
43E70	N43 00.0 E170 00.0	M11 29/032	M38 28/064	M51 28/078	M53 28/081	M54 28/084	M53 28/081	M51 28/072

45E80	N45 00.0 E180 00.0	M12 28/036	M42 29/038	M50 28/044	M49 27/049	M47 27/051	M46 27/051	M46 27/049
47N70	N47 00.0 W170 00.0	M11 23/043	M40 22/071	M47 22/077	M47 22/073	M47 22/068	M46 23/064	M47 23/062
ETP2	N48 43.0 W164 47.2							
50N60	N50 00.0 W160 00.0	M01 22/033	M32 22/094	M45 21/112	M49 21/117	M53 21/120	M57 21/112	M58 22/097
52N50	N52 00.0 W150 00.0	M02 24/037	M29 26/059	M45 26/080	M50 26/084	M55 27/085	M60 27/085	M64 27/083
ETP3	N52 01.0 W140 25.1							
52N40	N52 00.0 W140 00.0	M04 27/028	M32 31/062	M47 31/088	M52 31/089	M56 31/088	M58 31/077	M58 31/059
EXIT1	N51 31.5 W133 43.3							
YZT 112	N50 41.0 W127 21.9	M03 29/013	M35 28/013	M51 31/012	M56 31/015	M60 30/018	M58 29/023	M56 29/024
TOD	N50 25.3 W125 52.8	M03 30/013	M35 31/015	M51 33/016	M56 33/017	M59 31/020	M58 30/023	M56 29/024
POWOL	N50 12.3 W124 44.7							
KEINN	N49 49.0 W123 43.9							
CEESE	N49 40.2 W123 36.9							
FASBO	N49 22.4 W123 22.8							
CYVR	N49 11.7 W123 11.0							

```
          FL390          FL350          FL310          FL200          FL100
DESCENT  29/024 M56   31/020 M59    33/018 M51     31/019 M24     32/014 M02


-----------------------  END OF FLIGHT PLAN  ----------------------
```

第5章
装载与配平

航空运输市场竞争激烈,航空公司需要更科学化、规范化和现代化的管理,现代机场也需要提高服务水平,为旅客提供更广泛、更优质的服务。飞机装载配平是地面保障的关键环节,直接影响到飞行安全。荷兰国家航空航天实验室(NLR)曾对 1970—2005 年全球与飞机装载配平有关的不安全事件进行了研究,发现 35 年里共有 82 起有完整记录的飞行事故与装载配平有关,1998—2004 年仅 6 年间有 1200 起航空事件及事故征候是由装载配平问题引发的。飞机装载配平常见安全问题包括:超载起飞;超载落地;错误的业载数据;错误的装载;错误的装载配平计算;错误的装载配平文件;货物移动带来的安全问题;重心接近或超出前极限;重心接近或超出后极限。在我国,此类问题已引起行业管理部门的高度重视,管理当局对飞机装载问题进行了专项整治,取得了明显成效。2019 年 10 月,中国民用航空局下发民航规〔2019〕61 号咨询通告《航空器重量与平衡控制规定》(AC-121-FS-135),为航空运营人重量与平衡控制和管理提供了工作指南,并将此作为日常监管依据。高质量的装载配平能在保证安全的前提下,最大限度地利用飞机的运载能力,是航空公司能够安全、经济营运的重要保证。

5.1 概述

通过制作飞行计划以保证飞机的 $TOW \leqslant MTOW$、$LWD \leqslant MLWD$、$LWA \leqslant MLWA$、$ZFW \leqslant MZFW$,但仅保证重量不超过最大允许值还不够,还必须保证飞机的重心在任意时刻不超出允许的范围。因此,对每种机型都必须根据其操纵性、稳定性的要求及飞机结构限制确定了一个允许的重心范围,如图 5-1-1 所示,正确安排旅客及货物的位置,正确地进行配平,以保证在起降及飞行中任意时刻飞机重心不超出允许范围。

图 5-1-2 为飞机受力示意图,超过重量限制或重心限制可能会危及飞行安全,如起飞时飞机擦尾、结构损伤、气动不稳定、地面不稳定(飞机倾覆)、旅客上下机不安全、耗油增加、疲劳寿命缩短、损伤跑道等。

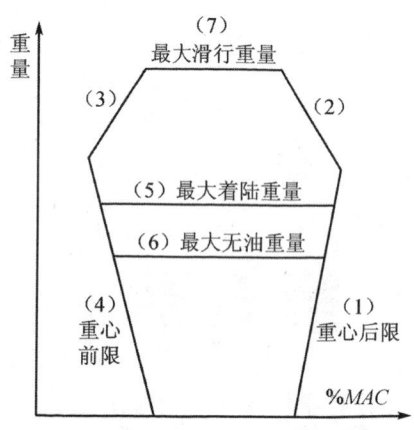

（1）重心后限受飞行稳定性和前轮驾驶效能或松刹车起飞时飞机抬头等因素的限制；

（2）主起落架承载限制、机身中段载荷限制、平尾效能限制；

（3）前起落架承载限制、机身前段和后段载荷限制、机翼动力矩限制；

（4）前起落架承载限制、平尾效能限制（重心太靠前，平尾不能提供足够的平衡力矩）、机身后段载荷限制；

（5）机身前段和后段载荷限制、机翼（根部）承载能力限制；

（6）机身后段载荷限制、起落架和机翼承载能力限制；

（7）机身、起落架和机翼承载能力限制

图 5-1-1　飞机重心范围示意图及有关限制因素的解释

飞机的重心太靠前或者太靠后，都会对飞行造成影响。

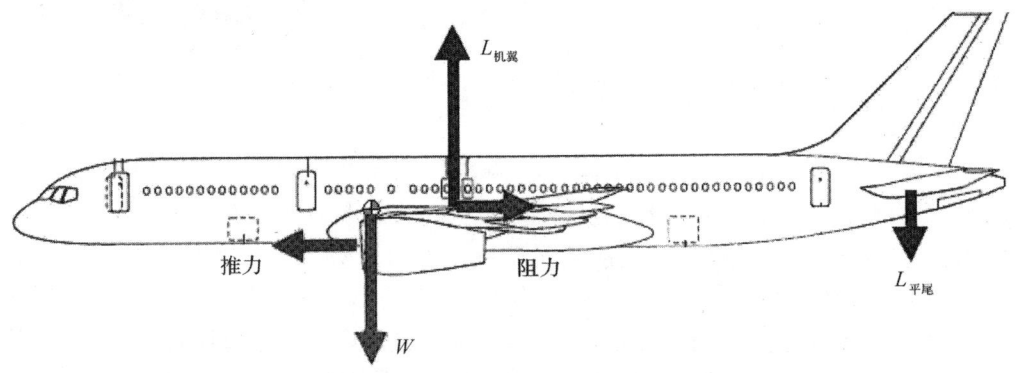

图 5-1-2　飞机受力示意图

飞机重心太靠前的缺点有：重心越靠前，飞机在飞行过程中的操纵性就会越差。这是因为机翼提供的升力大于平尾提供的负升力，重心靠前导致机翼升力的力矩增大，使飞机更难通过升降舵和平尾操纵。所以重心越往后，需要的纵向配平力矩越小，由配平引起的升力损失就越少，相应的油耗也会降低。另外，重心靠前容易导致飞机起飞困难。

飞机重心太靠后的缺点有：重心到平尾、垂尾的力臂短，使侧风时的操纵性降低、最小地面控制速度（V_{MCG}）增大，方向操纵性降低，因此在小重量下重心后限会前移（小重量时的起飞速度、着陆速度小）。例如，对于 B737-800，在小重量时，重心后限受复飞操纵性限制而前移。受操纵性限制的重心后限如图 5-1-3 中曲线①所示。另外，在小重量时，重心如果太靠后，也可能在松刹车起飞时由于起飞推力产生的抬头力矩而使飞机上仰、倾覆，因此重心后限会受此限制而前移，起飞推力的额定值越大，重心后限前移得越多，如图 5-1-3 中虚线所示。

载重平衡问题引起的事故大部分发生在起飞阶段，主要原因包括不正确的配载、超出重心后限以及超重起飞。货运飞行发生载重平衡问题的风险比客运飞行高 8.5 倍。若货机货物固定不牢，还会在飞行中移动，发生危险。

配平不正确也会危及起飞安全，因为安定面配平的结果会影响驾驶杆的杆力，正确的配平抬前轮时会产生可接受的杆力，不正确的配平可能导致起飞时擦尾或过早（如杆力太轻）或过晚（如杆力太重）抬前轮。抬前轮过早可能导致俯仰角过大、爬升梯度减小甚至失速，抬前轮过晚会使起飞距离延长、到障碍物的距离缩短。在受近距障碍物限制时，抬前轮过早或过晚都可能导致可控飞行撞地（CFIT）。

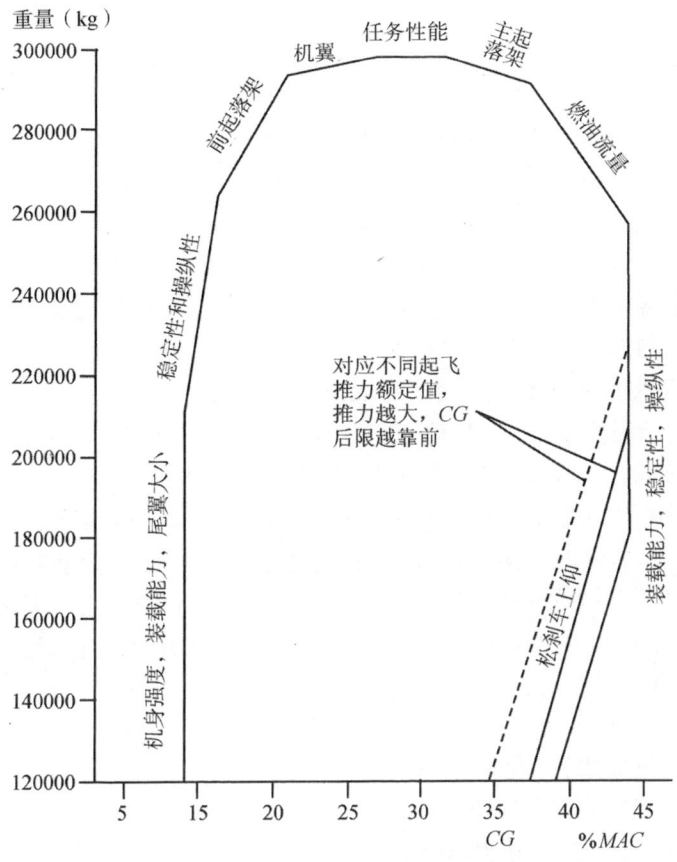

图 5-1-3　B737-800 的重心允许范围

正确配平的前提是对重量、重心的精确计算。旅客、货物、加油量会影响飞机的重量、重心,必须及时通知配载部门旅客、货物和加油量等信息。配平工作由配载人员来完成。配载人员合理安排旅客、货物,确定飞机的重心及配平值(配平调整片的刻度或水平定安面的调定值),并生成装载配平单,飞行员据此单检查旅客、货物的装载情况并在驾驶舱内对配平片或安定面进行调整,使飞机配平并在起飞操作时能有正常的杆力感觉。

因此,应合理地分配旅客座位、货物的舱位以有效控制飞机的重心位置,使飞机的重心处于适当的范围之内,保证飞行安全。科学、有效地控制好飞机的载量,优化装载,可以提高空间利用率,降低运行成本,从而提高运营效益。

5.2　飞机的重心位置

质量是一个恒定值(标量),它不随着物体的空间移动而发生变化。目前重力没有统一的定义。这里不讲定义,只是一般性描述。质量与重量不同,它们之间的大小关系是 $G = m \times g$。在地球的不同地方(纬度),在不同的高度,重力加速度是有微小差别的。所以,重力不是恒定值,它随重力加速度的变化而变化。通常,由于重力加速度在地球上变化很小,其变化可以忽略不计,所以我们认为重力在地球上是一个恒定值。在装载配平中,我们所计算的数值都是物体的质量,而不是重量。但是,我们习惯地称这些数值为重量。

5.2.1 相关术语

5.2.1.1 制造空机重量(*MEW*, Manufacturer's Empty Weight)

制造空机重量是指飞机被制造商生产出来时的重量,是在飞机构型中不可或缺的部分的重量。制造空机重量包括飞机必需的结构和系统、飞机封闭系统内的液体、飞机的座位和座位上的安全带以及灭火器等。制造空机重量所包含的重量项目,不同机型可能不一致,型号相同但编号不同的飞机也可能不一致,因其选择的设备不同,使用和计算时应按各机型的重量项目规定执行。新飞机的制造空机重量由飞机制造厂提供,被记录在飞机的履历册内。

5.2.1.2 基本空机重量(*BEW*, Basic Empty Weight)

基本空机重量是指航空器的结构、动力装置、设施、系统和作为某一特定构型飞机的一个整体所必需的设备的重量,以及航空运营人认为该航空器所配标准项目的重量。基本空机重量是在制造空机重量的基础上加上标准项目(SI, Standard Items)的重量。标准项目是指该型号航空器配备的设备和加载的液体,这些设备和液体不是某一特定构型飞机的作为一个整体所必需的一部分,但对于同型号的航空器是相同的。标准项目主要涉及但不限于以下项目:

(1)不可用燃油和其他不可用液体(如油箱内不能放出的燃油、液压油、散热器降温系统中的液体、使用不到的饮用水);

(2)发动机滑油;

(3)厕所用液体(洗涤水)和化学用品;

(4)灭火器、烟火信号装置和应急氧气设备;

(5)厨房、餐厅、酒吧内的结构、烤箱等;

(6)其他所需的辅助电子设备。

5.2.1.3 运行空机重量(*OEW*, Operational Empty Weight)

运行空机重量也称为使用空机重量,是在基本空机重量或机队空重的基础上加上运行项目(OI, Operational Items)的重量。咨询通告《航空器重量与平衡控制规定》(AC-121-FS-135)对运行项目有明确的说明:

运行项目是执行特定运行所必需的,但未包含在基本空机重量之中的人员、设备和给养。对不同机型,这些项目可能是不同的,包括但不限于以下项目:

(1)机组人员、非机组乘员(不占旅客座位的加机组成员)及其行李;

(2)飞行资料(如飞行手册、航图、放行单等)、飞行必备文件(如航空器国籍登记证、适航证、机组执照、航行记录本、无线电台执照等)和导航设备;

(3)用于服务旅客的物品,包括耳机、枕头、毛毯、拖鞋和杂志等;

(4)供客舱、厨房等使用的可移动设备;

(5)食物和饮料;

(6)可用液体,但不包括可利用载荷中的液体;

(7)所有用于飞行的必需应急设备(如氧气瓶、灭火瓶、防烟罩、医疗箱、急救箱、防疫包、加长安全带、逃生绳、滑梯包、消防斧、手电筒、扩音器等);

(8)救生衣、救生筏和应急定位发射机等;

(9)货箱、货板,以及货物的系留、集装设备等;

(10)饮用水;

（11）可放出的不可用燃油；

（12）放在航空器上又不作为货物计算的备用件；

（13）公司视为标准配置的所有其他设备。

5.2.1.4　干使用重量（DOW，Dry Operating Weight）

运行项目按基本配备（标准机组、乘务组人数和标准配餐等）配置的航空器运行空机重量（OEW）称为基本运行重量（BOW，Basic Operating Weight）。对基本运行重量（机组、乘务组人数和配餐等）做修正，修正后的基本运行重量称为干使用重量（DOW）。当运行项目调整值为 0 时，$DOW=BOW$。干使用重量就是飞机起飞时的起飞重量减去可用燃料和有效载荷以后的重量，这是每天用作确定当天飞机起飞重量的基础的重量。

5.2.2　飞机重心的表示方法

飞机各部分所受重力的合力叫飞机的重力。飞机各部分所受重力的合力的作用点，即重力的作用点为飞机的重心（CG，Center of Gravity）。飞机的重心是一个假设的点，假定飞机的全部重量都集中在这个点上，这个点可以支撑起飞机，使飞机保持平衡，如图 5-2-1 所示。飞机俯仰、偏航和滚转三个方向的转动都是围绕飞机的重心进行的。

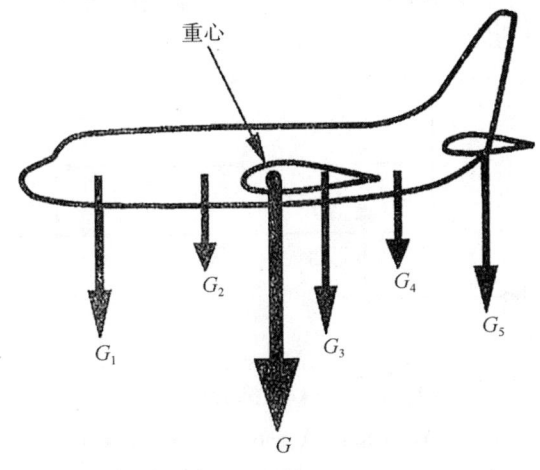

图 5-2-1　飞机的重心

装载配平确定飞机重心的力学原理是合力之矩定理：一个力系的合力对任一点力矩等于各分力对同一点的力矩之和。对空间力系，力矩是矢量。对平面力系，力矩可用标量表示。对确定飞机重心的情况而言，重心是平面力系，而且是平行力系。确定飞机重心的情况，可简化为下例。

假设图 5-2-2 所示的系统由板子及三个重物组成，其重量分别为 W_E、W_F、W_C、W_P，试确定这个系统的重心。

这个例子中的板子重量 W_E 相当于飞机的使用空机重，W_P 相当于旅客重，W_F 相当于油量，W_C 相当于货物重量，整个系统的重心相当于飞机的重心。

可取任一点 O 作为矩心（O 点可在板子以外），规定抬头力矩（顺时针）为正，低头力矩（逆时针）为负，此系统的合力（相当于飞机总重）$W=W_E+W_P+W_F+W_C$。按合力之矩定理有：

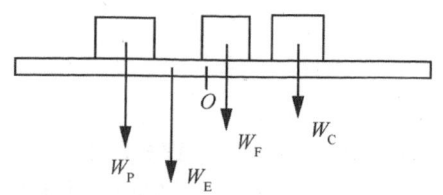

图 5-2-2 确定飞机重心的示例图

$$W \times X = W_E \times L_E + W_P \times L_P + W_F \times L_F + W_C \times L_C = M_E + M_P + M_F + M_C$$

式中：X、L_E、L_P、L_F、L_C 分别为各力对矩心 O 点之力臂，在 O 点之后的力臂为正，由上式算出的 X 即重心到 O 之力臂。这样就确定了力系之合力（即飞机重心）的位置。

飞机具有左右对称性，因此飞机重心一定在其对称面内。飞机构件的重量和装载物品的重量都集中在过机身纵轴并与飞机左右对称面相垂直的平面内。这样简化后，飞机重心位于机身轴线上，当然也可看作在飞机左右对称面和机翼弦平面的交线上。这两者都是工程近似简化。

通常用重心到某一基准位置的距离来表示重心位置。为了方便计算飞机的重心位置，各型飞机可按其构造特点设定基准点和坐标轴。例如，某飞机取飞机构造水平线与中央翼的翼梁中心线的交点为重心来计算坐标的基准点，即坐标原点（如图 5-2-3 所示）。图中，c_A 是飞机平均气动弦长；x_b 为中央翼双梁中心线与飞机构造水平线的交点到平均空气动力弦前缘的距离；x_{CG} 为重心到坐标原点的距离。

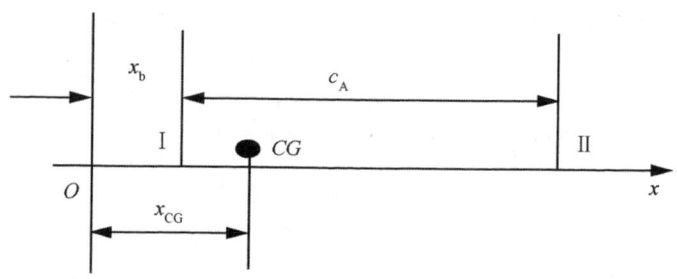

图 5-2-3 计算飞机重心的坐标

现在多以平均空气动力弦（MAC, Mean Aerodynamic Chord）法来表示飞机的重心位置，即用重心投影点与平均空气动力弦的前缘之间的距离占平均气动弦长的百分比来表示重心的位置。$0\% MAC$（LEMAC）表示重心位于 MAC 的最前缘，$100\% MAC$（TEMAC）表示重心位于 MAC 的最后缘。

MAC 是机翼上的一条特定弦线，如图 5-2-4 所示。因为机翼产生的俯仰力矩受机翼的很多因素影响，如机翼的弯曲、弧度、伸展等特性，计算起来较为复杂，因此用与实际机翼气动性能相近的以 MAC 为宽度的简单矩形机翼来替代计算，在实际运用中其精确度和通用性是被认可的。每种机型的 MAC 的长度和所在位置都是固定的，都已经在飞机的技术说明书中写明。飞机的重心和焦点位置都是相对于平均气动弦而言的。

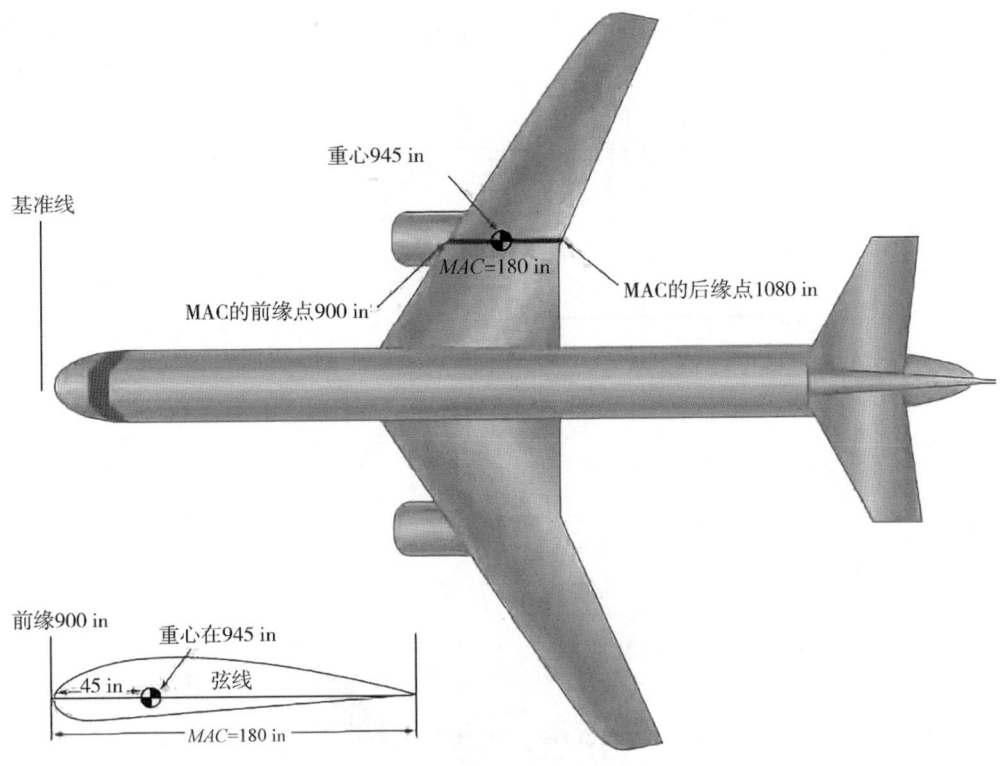

图 5-2-4 平均气动弦(MAC)

假设重心的投影点到平均空气动力弦前缘点的距离为 X_T,平均气动弦长为 b_A,如图 5-2-5 所示,则重心相对位置表示为:

$$X = (X_T/b_A) \times 100\%$$

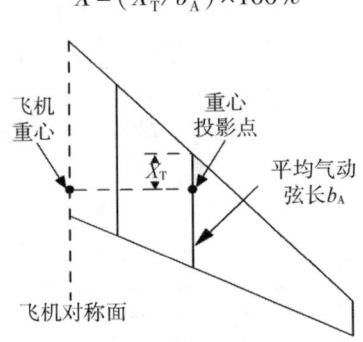

图 5-2-5 飞机重心位置

5.2.3 飞机重心的计算方法

5.2.3.1 代数法

为了便于求算飞机装载后的重心位置,可以根据计算的需要,在飞机机身中轴线上指定一点作为基准点。飞机的每项重量与基准点的水平距离,就是该项重量的力臂。每项重量乘以它的力臂就是该项重量构成的力矩。按照这个规定逐一计算出各项重量(包括空机重量)构成的力矩值(负值为低头力矩,正值为抬头力矩),把所有力矩值加总得到力矩总和,再除以飞机的总重量,就得到飞机重心到基准点的力臂,再利用重心公式即可计算出飞机重心相对平均气动弦的位置,如图 5-2-6、图 5-2-7 所示。

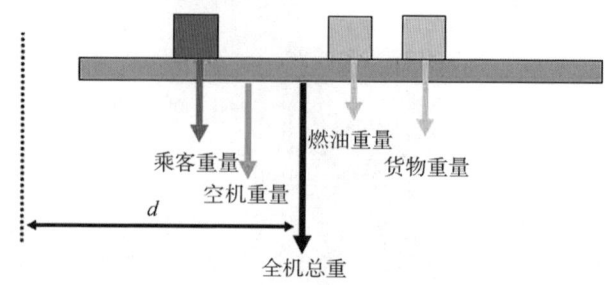

图 5-2-6 飞机重心到基准点的力臂

飞机的重心为:

$$CG = \frac{d - a}{l} \times 100\%$$

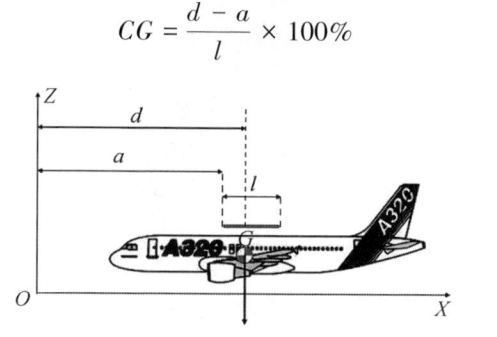

图 5-2-7 飞机的重心位置

5.2.3.2 站位法

站位($B.S.$,Body Station)是用来表示飞机上任何一点的位置的一种度量单位。在设计、制造飞机时,制造厂家选定某一点为站位基准点,该点被定义为 0 站位,其他任何点相对于站位基准点的距离,则称为此任意点的站位,如图 5-2-8 所示。一般取站位基准点右侧各点的站位为正值,左侧各点的站位为负值,这样只要确定了站位基准点的位置,则飞机上任何一点的站位均可确定。

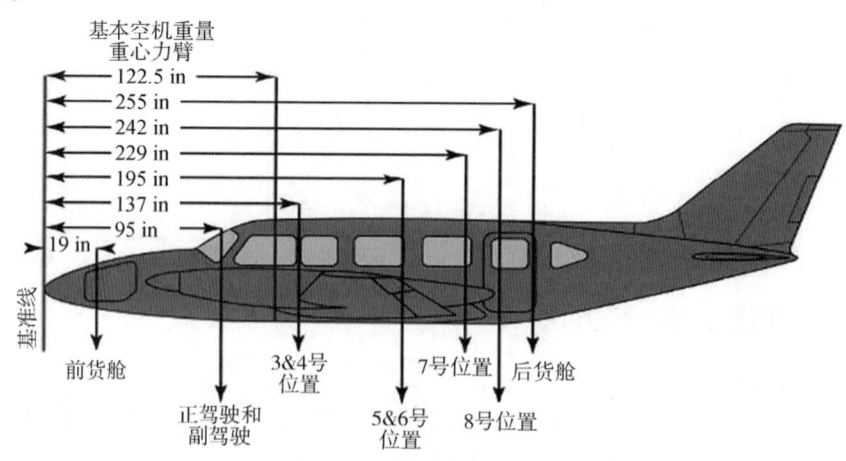

图 5-2-8 飞机的站位

飞机上各个装载项目所在站位数可以直接作为该装载项目的力臂值。可用下式计算出飞

机重心的站位:

$$重心的站位 = \frac{所有力矩之和}{总重量}$$

求出重心的站位后,则可由重心位置公式换算成平均气动弦百分比:

$$重心位置 = \frac{重心的站位 - MAC\ 前缘的站位数}{MAC\ 长度} \times 100\%$$

站位和平衡力臂($B.A.$, Balance Arm)都是沿飞机纵轴的坐标,平衡力臂是到 0 站位的实际距离,如图 5-2-9 所示,而站位一般不是实际距离(仅对原型机,$B.A. = B.S. =$ 到 0 站位的实际距离),如图 5-2-10 所示。例如,对 B737 系列,B737-100 是原型机;对 B757 系列,B757-200 是原型机;对 B767 系列,B767-200 是原型机。因此,在装载配平计算中应该使用平衡力臂而不是站位来计算力矩。

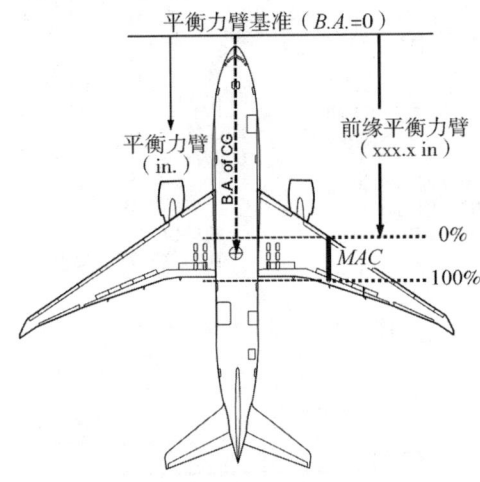

图 5-2-9　平衡力臂

平衡力臂计算飞机重心的公式为:

$$\%MAC = \frac{B.A._{CG} - LEMAC}{MAC} \times 100\%$$

例 5-2-1: B737-300,平均气动弦长 MAC 为 134.5 in,MAC 前缘的 $B.A.$ 为 625.6 in,25%MAC 的 $B.A.$ 为 659.225 in(重心基准一般在 25%MAC 上),如图 5-2-11 所示。

如重心的平衡力臂为 $B.A.$,则用 %MAC 表示的重心位置是:

$$\%MAC = \frac{B.A. - 625.6}{134.5} \times 100\%$$

$$= \frac{659.225 - 625.6}{134.5} \times 100\%$$

$$\approx 25\%$$

5.2.3.3　指数法

在计算时,重量一般以千克(kg)为单位,力臂一般以米(m)为单位,以此计算出的力矩数值很大,计算困难。为了计算方便,通常把单位装载量构成的力矩数再乘以一个适当的缩小系数(如 1/100、1/1000、1/3000 等,因机型不同而不同)作为实际使用的基数,这个基数就称为单

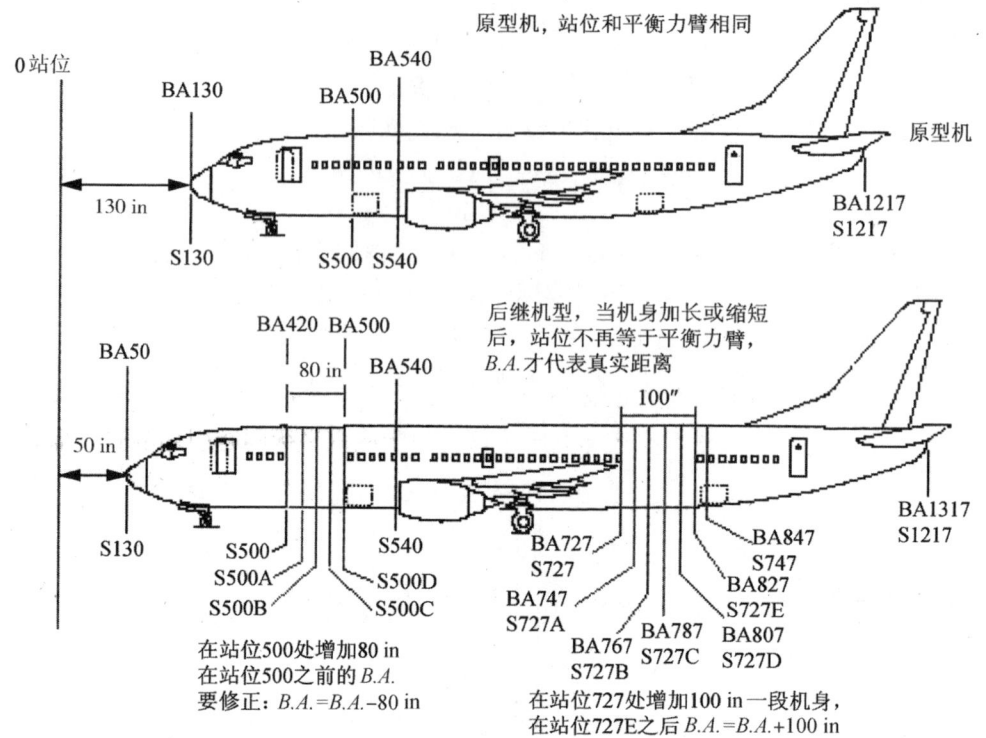

图 5-2-10   站位和平衡力臂

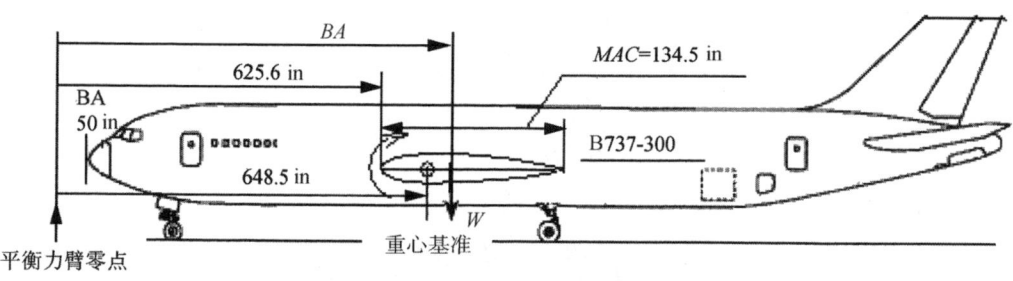

图 5-2-11   确定 B737-300 重心位置的图示

位装载量的指数（INDEX）。指数，即缩小了一定程度的力矩，因此指数的加减即代表了力矩的加减。目前很多种机型在载重平衡的计算中采用指数，不同机型缩小程度不一样，因此指数可能是不同的。

对 B737-300：$INDEX = (B.A. - 648.5) \times W(kg \times in)/30000$

对 B757-200：$INDEX = (B.A. - 1037.8) \times W(kg \times in)/75000$

B757-200 的飞机重心如图 5-2-12 所示，其计算公式为：

$$\%MAC = (B.A. - 991.9)/199.7 \times 100\%$$

5.2.3.4   图解法

在每次飞行前都要准确地计算出飞机的重心位置。采用指数法虽然比采用代数法要简便，但仍需要进行很多的计算，既费时费力，又易出现错误。载重平衡图表是以指数法为基础设计出来的，即指数法的图表化。通过平衡图计算飞机的装载位置要比采用指数法简便得多。详细内容见 5.4 和 5.5 小节。

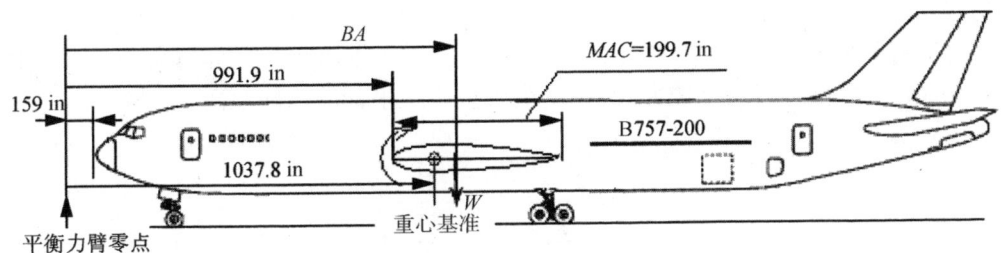

图 5-2-12　B757-200 重心

5.2.4　重心位置的变化与审核

5.2.4.1　重心位置对飞行稳定性的影响

飞机的重心位置一般都投影在飞机的轴线上,以平均空气动力弦线作为 x 轴,向机尾为正,坐标原点取在弦的前端点。这样全机重心位置可用 x 的坐标值来表示,即 x_{CG}。为了排除飞机尺寸的影响,更具可比较性,将 x_{CG} 除以飞机的平均气动弦长 c_A,并用 $\bar{x}_{CG}$ 表示。

现简单说明飞机重心位置对飞机飞行平衡稳定的重要性。设飞机在定常飞行时,作用在飞机上的各力矩取得平衡。若受到向上的小扰动,飞机迎角增大,在机翼、平尾等多处产生附加升力,构成总附加升力 ΔL。这个 ΔL 一定加在全机的焦点上,因焦点就是飞机升力增量的作用点(在平均空气动力弦上)。飞机静稳定性,指飞机的迎角变化后恢复原来姿态的倾向,能自动低头改变迎角,减小升力,静稳定性的大小取决于飞机焦点与飞机重心的相对位置关系。重心在焦点之前,飞机迎角增大会形成下俯力矩,飞机具有纵向静稳定性;反之,重心在焦点之后,飞机迎角增大会形成一个上仰力矩,使迎角进一步增大,飞机是纵向静不稳定的,会导致飞行事故的发生。下面用计算公式进一步说明飞机重心的前后位置对飞行稳定性的影响。设外界干扰使飞机迎角有一增量 $\Delta\alpha$,飞机升力随之有增量 ΔL(或 ΔC_L)。干扰消失瞬间,飞机产生一个力矩增量 ΔM_y,或力矩系数增量 ΔC_m:

$$\Delta M_y = (\bar{x}_{CG} - \bar{x}_A)\Delta C_L S \cdot \frac{1}{2}\rho V^2 c_A \tag{5-2-1}$$

$$\Delta C_m = (\bar{x}_{CG} - \bar{x}_A)\Delta C_L \tag{5-2-2}$$

或表示为

$$\frac{\partial C_m}{\partial C_L} = (\bar{x}_{CG} - \bar{x}_A) \tag{5-2-3}$$

式中:$\partial C_m/\partial C_L$ 可简写为 C_{mC_L},称纵向静稳定度。当 $C_{mC_L}<0$ 时,说明重心位于焦点之前,飞机具有纵向静稳定性。重心越靠后,或焦点越靠前,飞机的纵向静稳定性越差。为保证大型运输机有足够的静稳定性,重心在焦点之前的相对距离 $(\bar{x}_{CG} - \bar{x}_A)$ 不得小于 10%,这就限制了飞机重心的后限。

飞机在起飞和着陆过程中,对飞机重心位置也有一定要求。虽然航班商载不再变化,但起飞和着陆过程中,起落架的收放、襟缝翼的使用以及燃油的消耗等都会导致飞机重心发生变化。

5.2.4.2　重心位置的变化

飞机重心随飞机装载重量的大小和分布位置的不同而变化。

（1）飞机重心位置的计算

设飞机各部分的重力分别为 $W_1, W_2, W_3, \cdots, W_n$，重力作用点与坐标原点的距离 x 相应为 $x_1, x_2, x_3, \cdots, x_n$。根据平行力合成原理，得出全机重心距坐标原点的距离 x_{CG} 为

$$x_{CG} = \frac{\sum_{i=1}^{n} m_i g x_i}{\sum m_i g} = \frac{\sum_{i=1}^{n} m_i x_i}{\sum m_{GR}} \tag{5-2-4}$$

式中：m_{GR} 为飞机运行当时的总重量。由上式可得飞机重心相对于飞机平均空气动力弦的位置

$$\bar{x}_{CG} = \frac{x_{CG} - x_b}{c_A} \times 100\% \tag{5-2-5}$$

式中：c_A 是飞机平均气动弦长；x_b 为图 5-2-3 中中央翼双梁中心线与飞机构造水平线交点到平均空气动力弦前缘的距离；x_{CG} 为重心到坐标原点的距离。

（2）飞机加减物重后其重心的改变量

设飞机重力为 W_0，重心位置为 x_0。现求在其原有重心之后 Δx_i 处加一物重 ΔW_i，该飞机重心位置的变动量 Δx_{CG}。为求得 Δx_{CG}，将图 5-2-13 中各力对 A 点取矩。因合力对 A 点之矩等于各分力对 A 点之矩的代数和，则有

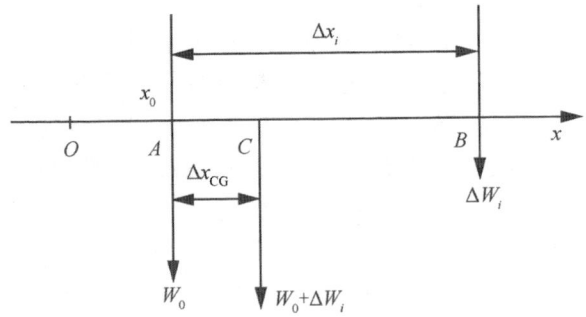

图 5-2-13　加载后飞机重心位置的变化

$$m_0 g \cdot 0 + \Delta m_i g \cdot \Delta x_i = \Delta x_{CG} \cdot (m_0 + \Delta m_i) g \tag{5-2-6}$$

$$\Delta \bar{x}_{CG} = \frac{\Delta m_i \cdot \Delta x_i}{m_0 + \Delta m_i} \tag{5-2-7}$$

如果 $\Delta m_i \leqslant m_0$，则上式可简化为

$$\Delta \bar{x}_{CG} \approx \frac{\Delta m_i \cdot \Delta x_i}{m_0} \tag{5-2-8}$$

将 Δx_{CG} 用平均气动弦长的百分比表示，则有

$$\Delta \bar{x}_{CG} = \left[\frac{\Delta m_i \cdot \Delta x_i}{(m_0 + \Delta m_i) \cdot c_A} \right] \times 100\% \tag{5-2-9}$$

式中：$\Delta \bar{x}_{CG}$ 为代数值，其正负取决于 Δm_i 和 Δx_i。加物重时，Δm_i 为正值；减物重时，Δm_i 为负值。若载重位置位于原重心之后，Δx_i 取正值；载重位置位于原重心之前，Δx_i 取负值。若求出的 $\Delta \bar{x}_{CG}$ 为负值，表示重心位置前移；若 $\Delta \bar{x}_{CG}$ 为正值，表示重心位置后移。

5.2.4.3　重心位置的审核

通常，飞机在不同装载状态下，飞机重心是否在允许范围内，可以从该机型的说明书中查

核。但装载发生较大变化时，就需要计算重心位置的变化量，以确定飞机重心是否在允许范围内。从上面的讨论得知，只需审核飞机重心相对于飞机的平均空气动力弦前后位置。对于飞机重心的上下、左右位置都无须再审核了，飞机设计的构造形式已保证它不会有安全问题。

5.3　最大业务载重量的计算

无论哪种交通运输工具，受其自身结构强度、客货舱容积、运行条件及运行环境等限制，都必须有最大业务载重量的限制。飞机是在空中飞行的运输工具，要求其具有更高的可靠性和安全性以及更好的平衡状态，而装载量和装载位置是直接影响飞行安全和飞机平衡的重要因素。因此，严格限制飞机的最大业务载重量具有更加重要的意义。

飞机的最大业务载重量受到由飞机的设计制造者规定的飞机的最大允许起飞重量（MTOW）、最大着陆重量（MLDW）、最大无油重量（MZFW）的限制，以及飞机的基本重量、起飞油量、航段耗油量、飞机的最大业载限额（由飞机的设计制造者规定）等因素的制约。计算飞机的最大业务载重量应该迅速、准确，因此工作人员应该清楚地了解计算最大业载重量所涉及的几个重量数据的意义，熟练掌握最大业务载重量的计算方法。

5.3.1　实际业务载重量

实际业务载重量（Pay Load）是指飞机上实际装载的旅客、行李、货物和邮件的重量之和。计算实际业务载重量时，行李、邮件、货物的重量按照实际重量计算，旅客的重量则不易很准确地计算。最初是对每个旅客进行称重，得出所有旅客的总重量，但随着乘机旅客人数的增多，这种做法既费时费力，也是对旅客的不尊重，因此改为按一定的标准折合，折合的原则是折合成的旅客总重量不低于又尽量接近于实际旅客体重的总和，使飞机不超载飞行，又尽量减少空载。

目前确定旅客和行李重量的四种方法有：标准平均重量、根据调查数据确定的平均重量、按座位数分级的平均重量、实际重量。

（1）标准平均重量

标准平均重量通常根据人口普查数据或专项调查获取的数据以及旅客各年龄段的出行情况等计算得出，并定期予以更新。例如可以通过《国民体质监测公报》所依据的调查结果或由运营人实施调查所得数据来确定标准平均重量。目前是通过对现阶段各航空运营人在其重量与平衡程序中使用的旅客重量进行普查，然后以普查所获数据为基础确定标准平均重量。

为了便于进行重量与平衡计算，标准平均重量包含了旅客携带进客舱的个人物品与手提行李的重量。因此，在确定标准平均重量时，除了考虑旅客自身的重量外，还应考虑旅客的夏、冬两季服装重量及其携带进客舱的个人物品与手提行李重量。每位旅客的手提行李和个人物品标准平均重量为 7.5 kg（旅客手提行李重量限制为 10 kg；个人物品重量限制为 5 kg）。旅客服装重量按照夏装 2.5 kg、冬装 5 kg 计算。我国按照 CCAR-121 运行的航空运营人，境内运行一般成人标准平均重量为 75 kg；儿童（满 2 周岁但不满 12 周岁）标准平均重量为 38 kg；婴儿（不满 2 周岁）标准平均重量为 10 kg。但目前由于各地的情况不同，采用的折合标准不尽相同，在具体工作中应按当地的规定执行。

为了保证标准平均重量的可信度,在制定标准平均重量时应以人口普查数据或航空运营人在特定条件下调查得到的数据为基础,并定期予以更新。

（2）根据调查数据确定的平均重量

利用局方可接受的调查方法,统计旅客的个人物品和手提行李重量,可为这些包含在旅客重量之内的物品确定一个合适的限额。正确实施的调查可以使运营人得到以相对小规模样本为基础的、适用于大容量总体的可信推论。在设计调查时,运营人需要考虑:获得预期置信度所要求的样本容量;样本选择过程;调查的类型（平均重量或项目数量）。调查得到的平均重量适用于调查航线或区域航线。为了使用调查得出的平均重量,运营人必须每36个月定期验证这些调查数据的有效性。

（3）按座位数分级的平均重量

按座位数分级的平均重量就是把标准偏差部分加到平均重量里,以提高实际重量不会超过平均重量的可信度。

如果运营人采用无手提行李大纲或者不允许携带任何手提行李进入客舱,运营人可将表5-3-1的旅客重量数据各减去5 kg（基于只有一半旅客携带个人物品的假设）。如果运营人确定的男女旅客占比不能与表中给出的数值精确匹配,运营人可以在表格的相应列间进行插值计算。中小客舱航空器运营人在使用表5-3-1所确定的按座位数分级的旅客平均重量标准计算旅客重量时,如果所载运的旅客为非标准重量群体（如运动员团队等）,或存在明显超过正常比例的其他超重旅客时,运营人应采取其他可行的方法（如实际重量）,以确保航空器不会超载,并在安排旅客座位时采取分散安排的方式等,防止出现重心偏出。

例5-3-1:运行30座（旅客座位）航空器的某运营人实施调查,统计出其航班上男女旅客占比各为50%。如果运营人使用的是经批准的有手提行李大纲,该运营人应使用78 kg的旅客重量。如果运营人使用的是无手提行李大纲,则该运营人应使用73 kg的旅客重量,并按照每件9 kg的标准统计所有机旁装载行李。

表5-3-1　成人旅客按座位数分级的平均重量

审定的最大座位数	男女旅客占比										
	0/100	10/90	20/80	30/70	40/60	50/50	60/40	70/30	80/20	90/10	100/0
1~4	使用实验重量（kg）										
5	84	85	86	87	88	88	89	90	91	92	93
6~8	81	82	83	84	85	86	87	87	88	89	90
9~11	79	80	81	81	82	83	84	85	86	87	88
12~16	77	78	79	79	80	81	82	83	84	85	86
17~25	75	74	75	76	77	78	79	80	81	82	82
26~30	74	74	75	76	77	78	79	80	81	82	82
31~53	72	73	74	75	76	77	77	78	79	80	81
54~70+	71	72	73	74	75	75	76	77	78	79	80

（4）实际重量

可以通过以下任一方式确定旅客的实际重量：

①登机前在重量衡器上对每位旅客称重。

②询问每名旅客的体重，运营人在询问（根据自愿原则）到的重量的基础上至少增加 5 kg 的衣服重量。如适用，运营人还可以在某些航线上或在某些季节中增加衣服重量的限额。

为了确定个人物品、手提行李、交运行李、机旁装载行李或重行李的实际重量，运营人应该在重量衡器上对其进行称量。

在这四种方法中，按实际重量的方法得到的统计数据最接近真实，但在运行时可能有操作难度。平均重量是经常选用的方法，这种方法消除了很多潜在的与相对较轻物体重量有关的误差源。前三种平均重量的方法中，标准平均重量更简便和高效；根据调查数据确定的平均重量可能更符合运营人具体的运行情况；按座位数分级的旅客平均重量则比标准平均重量具有更高的安全裕度。

飞机的最大业务载重量限额是飞机制造厂商根据该型飞机的结构强度和各种性能要求所规定的最大业务装载量限额。在对飞机进行装载时，实际装载量也不应超过飞机的最大业务载重量限额。

5.3.2　计算最大业务载重量的意义

飞机的最大业务载重量是指执行航班任务的飞机允许装载的旅客、行李、邮件、货物的最大重量。

计算最大业务载重量的意义是：

（1）确保飞行安全，避免超载飞行。

超载飞机表现出的主要问题有：

①需要较高的起飞速度；

②需要较长的起飞跑道；

③减小了爬升速度和角度，降低了最大爬升高度；

④缩短了航程；

⑤降低了巡航速度；

⑥降低了操纵灵活性；

⑦需要较高的落地速度；

⑧需要较长的落地滑行距离。

这些降低飞机效率的因素在某些情况下可能并不会有严重影响，但如果发生机翼表面结冰或飞机出现故障等情况时，则可能造成极其严重的后果。因此，实际的业务载重量绝对不应超过本次航班的最大业务载重量，否则将造成飞机超载。

（2）充分利用飞机的装载能力，尽量减少空载。

计算出飞机的最大业务载重量和实际业务载重量后，就可以知道航班的剩余业载有多少。此时如果还有旅客要求乘坐本次航班或者还有可由本航班运载的货物的话，则可适量地接收旅客和货物，最大限度地减少航班的空载，提高飞机的客座利用率和载运率，进而提高运输经济效益。

5.3.3 最大业务载重量的计算方法

飞机的起飞重量、着陆重量和无油重量的实际值不应超过各自的最大值,因此应有:

修正后的基本重量+起飞油量+实际业务载重量≤最大起飞重量

修正后的基本重量+备用油量+实际业务载重量≤最大着陆重量

修正后的基本重量+实际业务载重量≤最大无油重量

由上式可以计算出三个最大业务载重量:

最大业务载重量①=最大起飞重量−修正后的基本重量−起飞油量

最大业务载重量②=最大着陆重量−修正后的基本重量−备用油量

最大业务载重量③=最大无油重量−修正后的基本重量

飞机实际可用的最大业务载重量应为这三个最大业务载重量中的最小者,并且不应超过飞机结构限制的最大业务载重量限额,因此应有:

允许的最大业务载重量=

$\min\{$最大业务载重量①、最大业务载重量②、最大业务载重量③、最大结构业载限额$\}$

5.4 载重与平衡的原理

5.4.1 重心包线的力学原理

飞机的操纵性、稳定性及飞机结构限制等要求确定了一个允许的重心范围,如图 5-4-1 为 B737-800 的重心包线。载重平衡计算的最终目的是必须让飞机重心落在这个包线的范围内,否则无法保证飞机安全飞行。

空机的重量和重心位置是已知的,因此相对于某个基准点的力矩数是可计算的。只要计算出燃油、旅客、行李、邮件和货物的力矩数,再与基本重量的力矩数相加,就可得到飞机装载后的总力矩数。利用飞机的总力矩数和飞机的总重量,可计算出飞机的重心位置,如图 5-4-2 所示。

5.4.2 指数法计算重心的原理

如果把基准位置置于飞机前面,计算的飞机的实际力矩数值非常大,飞机的重心包线也难以识别。如果把基准位置置于合理位置,重心包线容易识别,但是力矩数据还是比较大。因此,利用指数法将力矩按一定比例缩小,一方面便于计算,另一方面便于飞机重心包线的识别。另外,为避免指数出现负值,通常还会增加一些数值。不同的机型增加的数值不一样。

$$INDEX = \frac{M}{C} + K$$

式中:M 为力矩数;C 为力矩常数(缩小的程度);K 为基准常数(避免产生负数)。力矩常数(C)和基准常数(K)的值取决于飞机类型和航空公司的政策,如表 5-4-1 所示。

WEIGHT AND BALANCE
CONTROL AND LOADING MANUAL
SAMPLE MANUAL

737-800

CERTIFIED WEIGHT AND CENTER OF GRAVITY LIMITS (Continued)

C.G. LIMITS - MTW 79242 KG, MLW 66360 KG, MZFW 62731 KG

The following diagram represents the certified Center of Gravity Limits in Metric units:

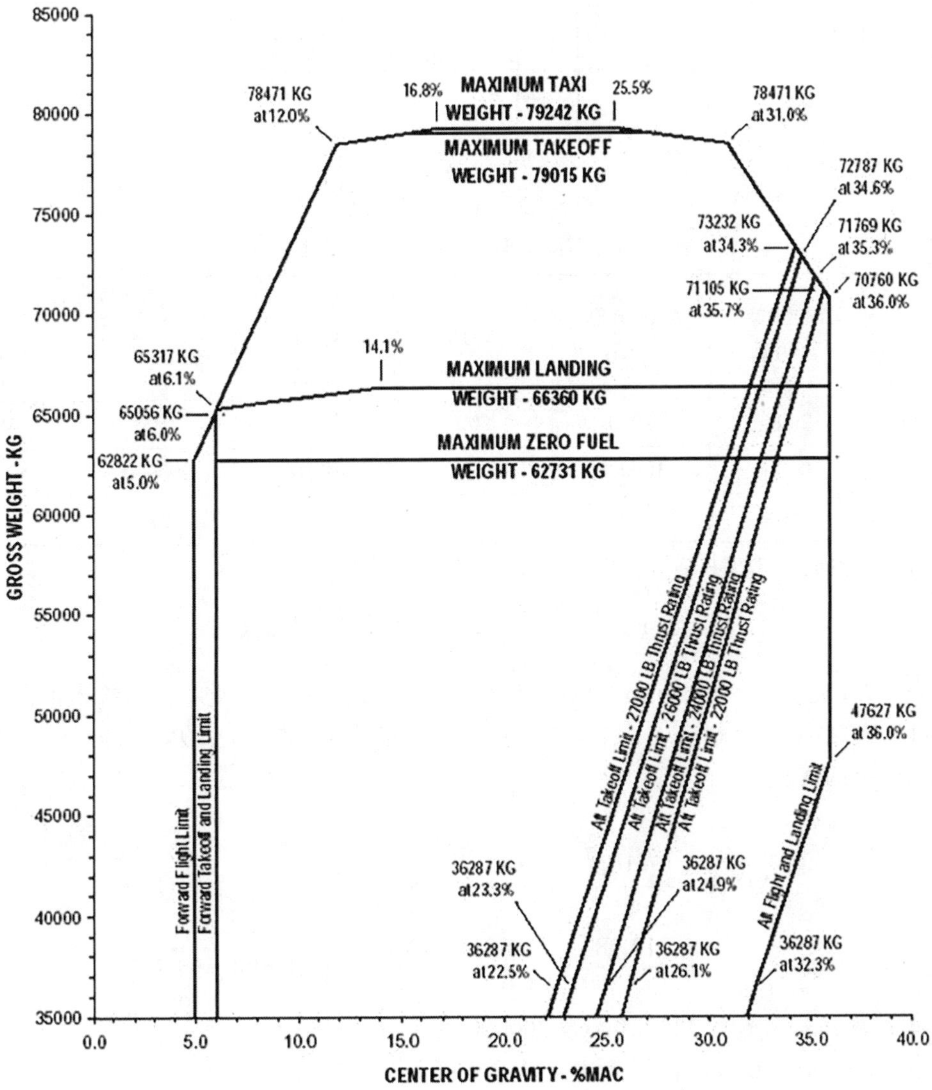

図 5-4-1　B737-800 的重心包线

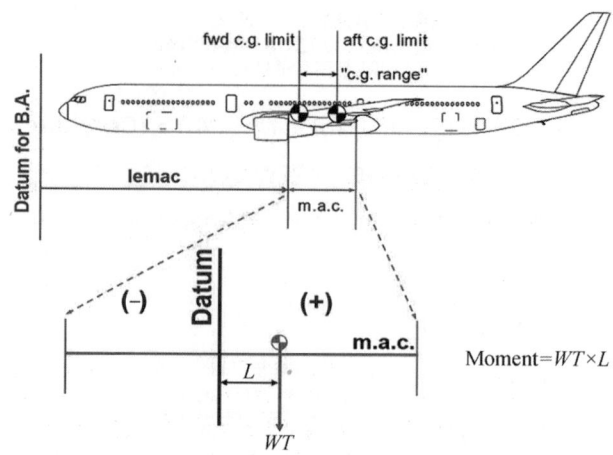

图 5-4-2　飞机的总力矩

表 5-4-1　不同机型的力矩常数和基准常数

机型		A300B2/ A300B4	A300-600/ A310	A318	A319/ A320/ A321	A330/ A340	A340-500/ A340-600
C	（kg·m）	2000	1000	500	1000	2500	5000
	（lb·in）	200000	100000	40000	100000	200000	500000
K		50	40	50	50	100	100

飞机的指数通常用以下公式进行计算：

$$INDEX = \frac{W \times (B.A. - Datum\ B.A.)}{C} + K$$

式中，$Datum\ B.A.$ 是飞机空机重心到基准点的距离；$B.A.$ 是飞机实际重心到基准点的距离。

例 5-4-1：A320：$C = 1000$ kg·m，$K = 50$，基准力臂 $H_{25} = 18.850$ m，$DOW = 44000$ kg，$HARM = 18.982$ m，求 $INDEX$。

解：

$$INDEX = \frac{(HARM - H_{25}) \times DOW}{C} + K$$

$$= \frac{(18.982 - 18.850) \times 44000}{1000} + 50 = 55.808 \approx 55.81$$

例 5-4-2：$Datum\ B.A. = 1323.6$ in，$C = 300000$ kg·in，$K = 50$，

（1）当 $DOW = 194500$ kg，位于 $B.A. = 1342.9$ in 时，计算 DOI；

（2）$\Delta W = 8000$ kg，位于 $B.A. = 1200.0$ in 时，计算 ΔW 产生的指数 $\Delta INDEX$；

（3）计算装载后的指数 $INDEX$。

解：

（1）$DOI = \dfrac{194500 \times (1342.9 - 1323.6)}{300000} + 50 \approx \boxed{+62.5}$

（2）$\Delta INDEX = \dfrac{8000 \times (1200.0 - 1323.6)}{300000} \approx \boxed{-3.3}$

（3）$INDEX \approx 62.5 - 3.3 = \boxed{+59.2}$

把飞机基本重量、燃油重量和业载构成的力矩数分别乘以缩小系数，就可以得到飞机基本重量的指数、燃油的指数和业载的指数，把这三个指数相加，就得出飞机装载后的总指数。根据飞机装载后的总指数和总重量，就可以确定装载后飞机的重心位置。目前绝大多数机型采用这种指数法确定飞机的重心。

5.4.2.1　基本使用指数和干使用指数

（1）基本使用指数（BOI）

运行空机重量（OEW）中包括机组、乘务组、配餐等的重量。对同一架飞机，机组、乘务组的人数或配餐标准可能会发生变化，所以 OEW 也会变化。基本配备（标准机组、乘务组人数和标准配餐等）的 OEW 称为基本使用重量（BOW,Basic Operating Weight）。基本使用重量（BOW）及其重心位置（HARM,或 BA）可通过对飞机称重得到。按运行空机重量计算的指数再加某一基准常数得到的值，称为基本使用指数（BOI,Basic Operating Weight Index）。

例如，

对 A310-200：$BOI = (HARM - 26.67) \times BOW/2000 + 40$；

对 B737-300：$BOI = (HARM - 648.5) \times BOW/30000 + 40$；

对 B757-200：$BOI = (HARM - 1037.8) \times BOW/75000 + 50$；

对 B747-400：$BOI = (HARM - 1323.6) \times BOW/286000 + 50$。

（2）干使用指数（DOI）

如机组、乘务组人数和配餐和基本配备中的标准不同，则要对基本使用重量做修正，修正机组、乘务组人数和配餐等重量后的使用空机重量称为干使用重量（常用 DOW 或 DRY OPTG WT表示），对基本使用指数 BOI 做相应的修正后得到的指数称为干使用指数（DOI,Dry Operating Weight Index）。在载重配平手册上通常给出了对 BOI 进行修正的方法或数据。例如，对 A310-200，驾驶舱中每增加一名机组人员，重量增加 80 kg，指数减 1.2，在前（后）厨房每增加100 kg 配餐，指数减 1.3（加 1.2）等。B737-300 的载重配平手册把基本使用重量称为"BASIC APS WEIGHT"，把对应的基本使用指数称为"BASIC APS INDEX"，干使用重量称为"ADJUSTED APS WEIGHT"，相应的干使用指数仍称为"BASIC APS INDEX"。

飞机在大修之后 OEW 及重心位置可能会变化，DOI（BOI）也就相应变化，大修之后应重新称重确定 DOI（BOI）。因此，同一机型的飞机 DOI（BOI）可能不同。

5.4.2.2　旅客指数和货物指数

旅客及货物产生的力矩（指数）与它们的重量和位置有关。客舱一般分为前客舱、中客舱、后客舱等若干区。货舱一般有前货舱、后货舱，大飞机可能有中货舱和散货舱。同样的重量在不同段或不同的货舱内产生的指数不同。

（1）旅客指数计算

飞机的载重与平衡手册（WBM）人员章节给出了每个乘客座椅重心位置和力臂的信息，通过这些数据可以计算出每个座椅单位重量下的 $\Delta INDEX$。在 AHM 560 中也给出了每千克单位重量下的这些 $\Delta INDEX$ 值。

表 5-4-2　A330 座位位置和对应力臂

Row No.	WINDOW SEATS (lines A,C,H,K)				CENTER SEATS (lines D,E,F,G)		NUMBER PAX
	H-ARM		NUMBER PAX		H-ARM		
	（m）	（in）	LH	RH	（m）	（in）	PAX
1	—	—	—	—	13.885	546.65	2
2	14.520	571.65	2	2	14.901	586.65	2
3	15.536	611.65	2	2	15.917	626.65	2
4	16.552	651.65	2	2	16.933	666.65	2
5	17.568	691.65	2	2	17.949	706.65	2
6	18.584	731.65	2	2	18.965	746.65	2

Note：passenger seat GC is usually located in the middle of the seat.

A330 的基准力臂 $H_{25}=33.1555$ m，力矩常数 $C=2500$ kg·m，基准常数 $K=100$。

$$\Delta INDEX(1\ kg)_{\text{Center seat row 1}}=\frac{1\times(13.885-H_{25})}{C}\approx\frac{13.885-33.1555}{2500}=-0.00771$$

$$\Delta INDEX(1\ kg)_{\text{Window seat row 2 LH}}=\Delta INDEX(1\ kg)_{\text{Window seat row 2 RH}}$$

$$=\frac{1\times(14.520-H_{25})}{C}=\frac{14.520-33.1555}{2500}\approx-0.00745$$

旅客重量一般按照表 5-4-3 的数据进行计算。

表 5-4-3　标准平均重量（境内运行）

标准平均重量	每位旅客重量
成年旅客平均重量	75 kg
儿童旅客平均重量（满 2 周岁但不满 12 周岁）	38 kg
婴儿旅客平均重量（不满 2 周岁）	10 kg

但是，因为飞机舱位座椅较多，尤其是大型飞机，在舱单中不可能考虑每一排座椅的 $\Delta INDEX$。因此，飞机座舱被分为几个部分（通常为 2~4 个部分），只计算这几个区域的 $\Delta INDEX$，并确定每个座舱对指数的整体影响，如图 5-4-3 所示。

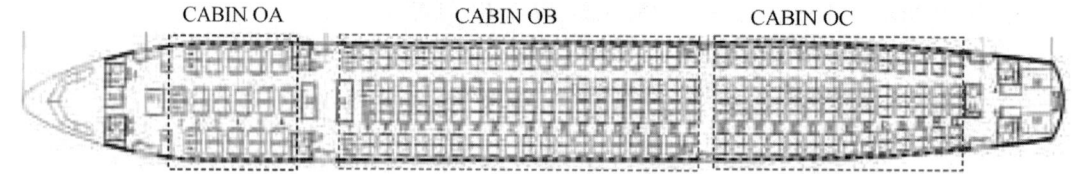

图 5-4-3　客舱舱位布局

同样的重量在不同段或不同的货舱内产生的指数不同，对每个舱段或者说对每个区，要确定它的"平均位置"（假定该区的旅客或货物都集中在这个点上）和可能产生的力矩的最大误差。

以图 5-4-3 中的客舱区为例，设该区共有 $N=24$ 个座位，每个座位（旅客）的平衡力臂为 L_i，设每个旅客重量 $W_i=$ 常数，则"平均位置"的平衡力臂 L 为

$$L=\sum_{i=1}^{N}W_iL_i\Big/\sum_{i=1}^{N}W_i=\sum_{i=1}^{N}L_i/N \tag{5-4-1}$$

当假定旅客(不论几个)的重量始终集中在这个平均位置上时,这个舱的旅客所产生的力矩是人数(重量)的线性函数,如图 5-4-4(b)中的斜直线所示。实际旅客就座时的重心并不一定在这个平均位置上,实际产生的力矩和按上述假定算出的力矩有差别,在计算可能产生的最大力矩误差时,假定旅客按图 5-4-4 中给出的顺序就座,这种顺序是符合人们选择座位的心理的,这样产生的力矩如图 5-4-4(b)中的实曲线所示

$$力矩 M = \sum_{i=1}^{n} W_i L_i \tag{5-4-2}$$

$$力矩 M 的误差 = \sum_{i=1}^{n} W_i L_i - L \times \sum_{i=1}^{n} W_i \tag{5-4-3}$$

曲线与斜直线的水平距离之差就是力矩的误差,这个误差可由上式算出。如果旅客按照图 5-4-4(a)中编号为 8、7、6、5、4、3、2、1、16、15、14、13……的顺序择座,则力矩变化如图 5-4-4(b)中的虚曲线所示,这样就可得到向两个方向(使重心向前、向后移动)的力矩的最大误差。如座位分布如图 5-4-4(c)所示,平均位置就不在区的中间,所得的力矩曲线就不这样对称。

确定了各舱段的"平均位置"后,就可算出每个舱段每××名旅客或每×××千克货物对重心基准产生的力矩(设它们都集中作用在平均位置上)和相应的指数,如平均位置在重心基准点之前,指数为负,反之为正。如果把中间某客舱区的平均位置选作重心基准,该区的旅客就不产生力矩,指数为零。客舱区、货舱区越靠前或靠后,同样多的客、货产生的力矩就越大,指数也就越大。

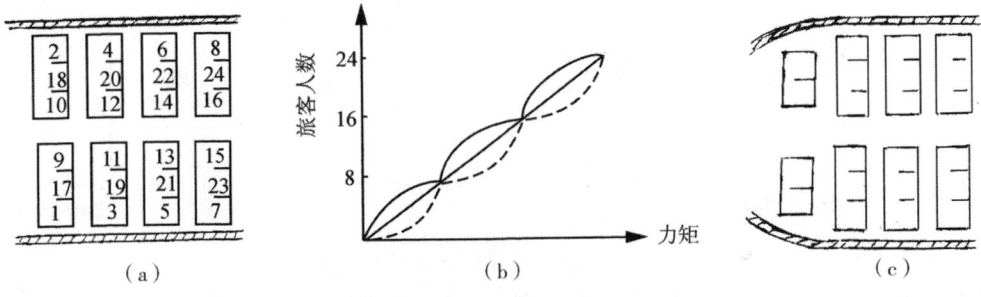

（a）　　　　　　　　（b）　　　　　　　　（c）

图 5-4-4　不同旅客人数可能产生的力矩最大误差

在每一个舱段内,同样的旅客坐在该区靠前或靠后处产生的力矩差别较大,但是在配平时的力矩计算结果相同,即指数都相同,这是因为计算时是假定他们的重量都集中在平均位置上。当然,事实并非如此,因为旅客选取座位时是趋于均匀分布的,旅客不会都集中在该区的后面或前面(见图 5-4-5),在进行装载配平时已针对各区不同旅客数、旅客选取座位的趋势计算了可能造成的力矩偏差的最大值及由此可能造成的重心偏差,对允许的重心范围做了相应的缩小。因此,在一个区内的旅客不论坐在何处都按相同的算法计算即可。当然,乘务员应使旅客均匀就座,不要把旅客都集中在前面或后面。

图 5-4-5　B747-400 的客、货舱布局图

（2）货物指数计算

货物及交运行李重量是按实际重量计算的，与旅客指数相似，需要分舱，确定力臂，计算力矩。货舱一般有前货舱、后货舱，大飞机可能还有中货舱和散货舱。根据货舱的平衡力臂和指数公式计算每个货舱的单位重量所对应的力矩和 $\Delta INDEX$。如果分区舱位的平均位置在重心基准点之前，$\Delta INDEX$ 为负值，反之为正值。另外，不同类型的集装器（ULD，Unit Load Device）计算出的分区货舱重心位置不同，如图 5-4-6 所示。

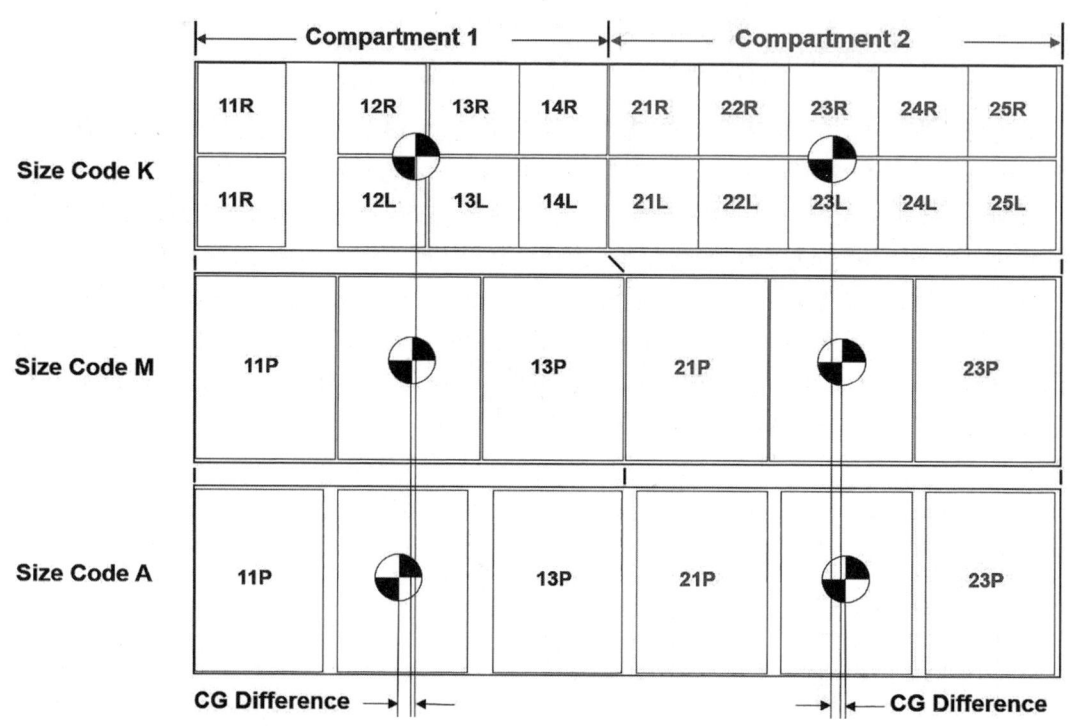

图 5-4-6　不同类型的集装器确定的舱位重心位置

载重平衡手册上一般都有布局图（参见图 5-4-5），给出客舱各段（区）的位置及座位数、各货舱的位置及最大装货量。安排旅客、货物时不得超过这些限制。

5.4.2.3　燃油指数

燃油指数是根据指数计算公式得出的。

对 B737-300：燃油指数 $INDEX=(BA-648.5)\times W(kg\times in)/30000$；对 B747-400：燃油指数 $INDEX=(HARM-1323.6)\times W(kg\times in)/286000$，$HARM$ 是燃油重量所在位置的平衡力臂。

飞机的燃油一般优先储存在机翼主油箱中，有些飞机在机身中央油箱或者尾翼配平油箱还储存燃油。如图 5-4-7 所示为 A340 的油箱布局，翼尖防波（缓冲）油箱（Surge Tank）只用于收集溢流燃油；左机翼防波油箱的溢流燃油回到 1 号主油箱，右机翼防波油箱的溢流燃油回到 2 号主油箱；防波油箱通过通气漏斗与大气相通，通气漏斗和释压活门在每个防波油箱的接近门上，如防波油箱中的油太多，则会从通气漏斗中排出机外。不同机型燃油油箱的布局不完全相同，对飞机重心的影响也不同。

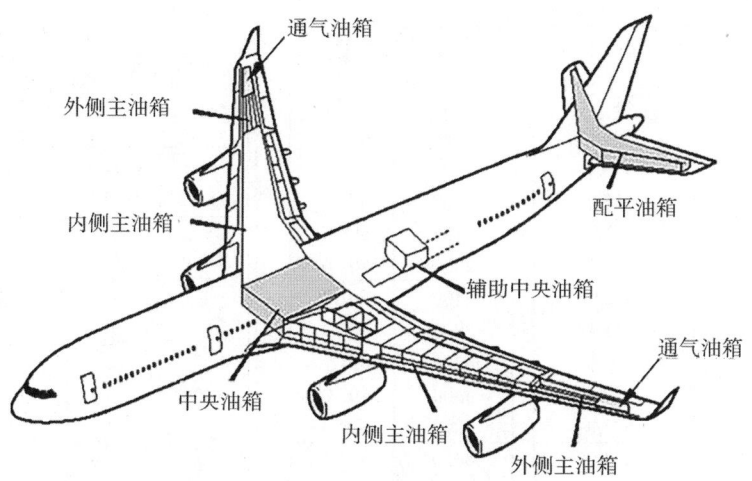

图 5-4-7　A340 的油箱布局

飞机的载重平衡手册燃料章节提供了有关飞机油箱不同油量对应的油箱力臂,并描述了自动加油顺序。根据这些数据,可以确定总油量信息,包括完成加油顺序后每个油箱中的燃料量和相应的总油量的重心位置,计算出单位燃油量所对应的 $\Delta INDEX$。

燃油指数的大小与所加油量多少有关,其变化关系或用一个数值表(如 BAE146-100、B757-200 的燃油指数表),或用一个图(如 A310-200 的燃油指数曲线)给出,这由用户在订货时自行选定。飞机制造厂商在飞机载重平衡手册中给出该机型的油箱装载油量的体积与对应的平衡力臂 BA 的对照表,该表是使用有限元法计算出来的。根据所加燃油的密度,通过该表可计算出燃油重量及其指数的对应关系即燃油指数表(或画出燃油指数曲线)。图 5-4-8 所示为 B737-300 的燃油指数表和对应的燃油指数曲线。

一般加油时油先加到翼尖油箱,然后加到翼根油箱、机身油箱中;用油顺序相反。由于机翼后掠,当油少时,油都在矩心之后,力矩为正,即指数为正,随油量增加,油的重心前移。油量从零开始增加时,力矩先增大,即指数先增大,当油量增加到处于矩心之前时,再加入的油量的力矩为负,故总的燃油力矩开始逐渐减小,甚至小于零,即指数变小,变负。不同机型的油箱位置不同,加油的顺序也可能不同,因此燃油指数的变化规律各不相同。为了使飞机在飞行中随油量的减少不使飞机的重心变化过大,都是设计为使燃油指数围绕零在某一区间内变化。

另外,燃油指数还与其密度有关,因为密度不同,相同重量的燃油的体积不同,其重心位置就不同,对重心基准点的力矩即指数也就不同。如表 5-4-4 左半部分和图 5-4-9 所示,使用时要按照所加的燃油密度来确定燃油指数。

B737-300

COMBINED MAIN TANKS 1&2	
VOLUME L	B.A. IN.
400	606.7
800	609.4
1200	611.8
1600	613.8
2000	615.5
2400	616.9
2800	618.4
3200	619.6
3600	620.8
4000	622.0
4400	623.0
4800	624.0
5200	624.9
5600	625.9
6000	626.7
6400	627.5
6800	628.3
7200	629.2
7600	630.0
8000	630.9
8400	631.7
8800	633.5
9200	635.7
9600	638.2
10000	640.7
10400	643.5
10800	646.4
11200	649.5
11348	650.7

CENTER TANK	
VOLUME L	B.A. IN.
400	600.9
800	601.0
1200	601.2
1600	601.5
2000	601.6
2400	601.7
2800	601.9
3200	601.9
3600	602.0
4000	602.0
4400	602.1
4800	602.1
5200	602.1
5600	602.1
6000	602.1
6400	602.2
6800	602.2
7200	602.2
7600	602.1
8000	601.7
8400	601.1
8755	600.4

利用左面的燃油体积与平衡力臂对照表及燃油密度可以得到下面的燃油指数表或燃油指数曲线

TOTAL FUEL INDEX TABLE			
WT (KG)	INDEX UNITS	WT (KG)	INDEX UNITS
500	-0.7	9000	+0.3
1000	-1.2	* 9111	+0.7
1500	-1.7	9500	+0.1
2000	-2.1	10000	-0.7
2500	-2.5	10500	-1.5
3000	-2.8	11000	-2.3
3500	-3.0	11500	-3.1
4000	-3.3	12000	-3.9
4500	-3.4	12500	-4.7
5000	-3.6	13000	-5.4
5500	-3.7	13500	-6.2
6000	-3.8	14000	-7.0
6500	-3.8	14500	-7.8
7000	-3.7	15000	-8.6
7500	-3.1	15500	-9.5
8000	-2.2	16000	-10.6
8500	-1.1	**16140	-10.8
* TANKS 1+2 FULL			
** TANKS 1+2+C/S FULL			

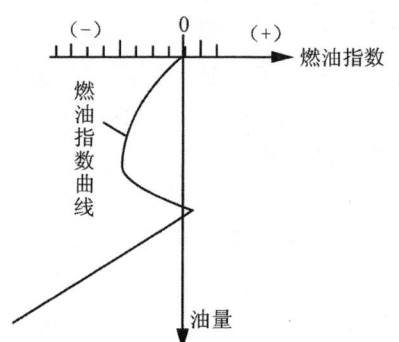

图 5-4-8　B737-300 的燃油指数表与燃油指数曲线

表 5-4-4　燃油指数与货物指数表

FUEL DENSITY（燃油密度）
KG/L　KG/USG
① 0.75　2.84
② 0.77　2.91
③ 0.79　2.99
④ 0.81　3.07
⑤ 0.83　3.14
TOTAL FUEL INDEX
WEIGHT　FUEL DENSITY

KG	①	②	③	④	⑤
42000	1	1	1	1	1
44000	1	1	1	1	1
46000	1	1	1	1	1
48000	2	2	1	1	1
50000	2	2	2	2	1
⋮					
60000	3	3	4	3	3
62000	2	2	3	4	4
64000	1	2	2	3	4
66000	0	1	2	2	3
68000	100	0	1	2	2
70000	99	100	0	1	1
72000	98	99	99	0	1
74000	98	98	99	99	0
76000	97	98	98	99	99
78000	97	97	98	98	99
⋮					

前货舱 重量(KG)	指数	后货舱 重量(KG)	指数
0	0	0	0
	100		1
182	99	171	2
548	98	515	3
913	97	859	4
1279	96	1203	5
1644	95	1547	6
2009	94	1891	7

在采用指数法时，在计算无油重量和起飞重量的指数时，在最后结果中舍去百位以上的数就得到相应的指数。例如：

DOI	40	40	40
后货舱指数	5	5 因	5
后客舱指数	7	7 为	7
前货舱指数	−4	96	100−4
前客舱指数	−3	97	100−3
起飞油量指数	−2	98	100−2
――――――			
起飞重量指数	43	343	300＋43

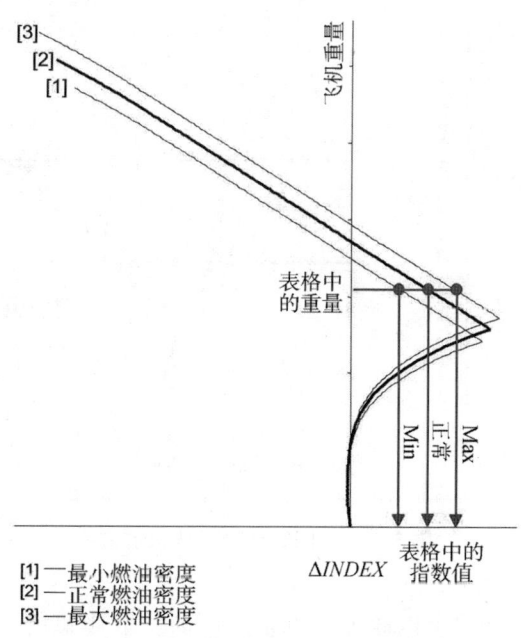

[1]—最小燃油密度
[2]—正常燃油密度
[3]—最大燃油密度

图 5-4-9　不同密度下的燃油指数曲线

　　有的机型前客、货舱的指数采用了一种特殊的表示方法，前客舱、前货舱（在重心基准之前）的指数本应是负数，却以 100 加该负数作为其指数，如表 5-4-4 前货舱、后货舱的指数示例所示，对燃油指数也采取这样的处理方法，如表 5-4-4 左下部油量大于 68000 kg 时的指数所示,其中指数 98 = 100-2,98 实际上表示指数是-2。

5.4.3 图解法计算重心的原理

采用指数法计算飞机的重心需要进行很多的计算,易出现错误。配载平衡图表是以指数法为基础,将指数法图表化。采用图解法确定飞机重心要比采用指数法简便得多,如图 5-4-10 所示。

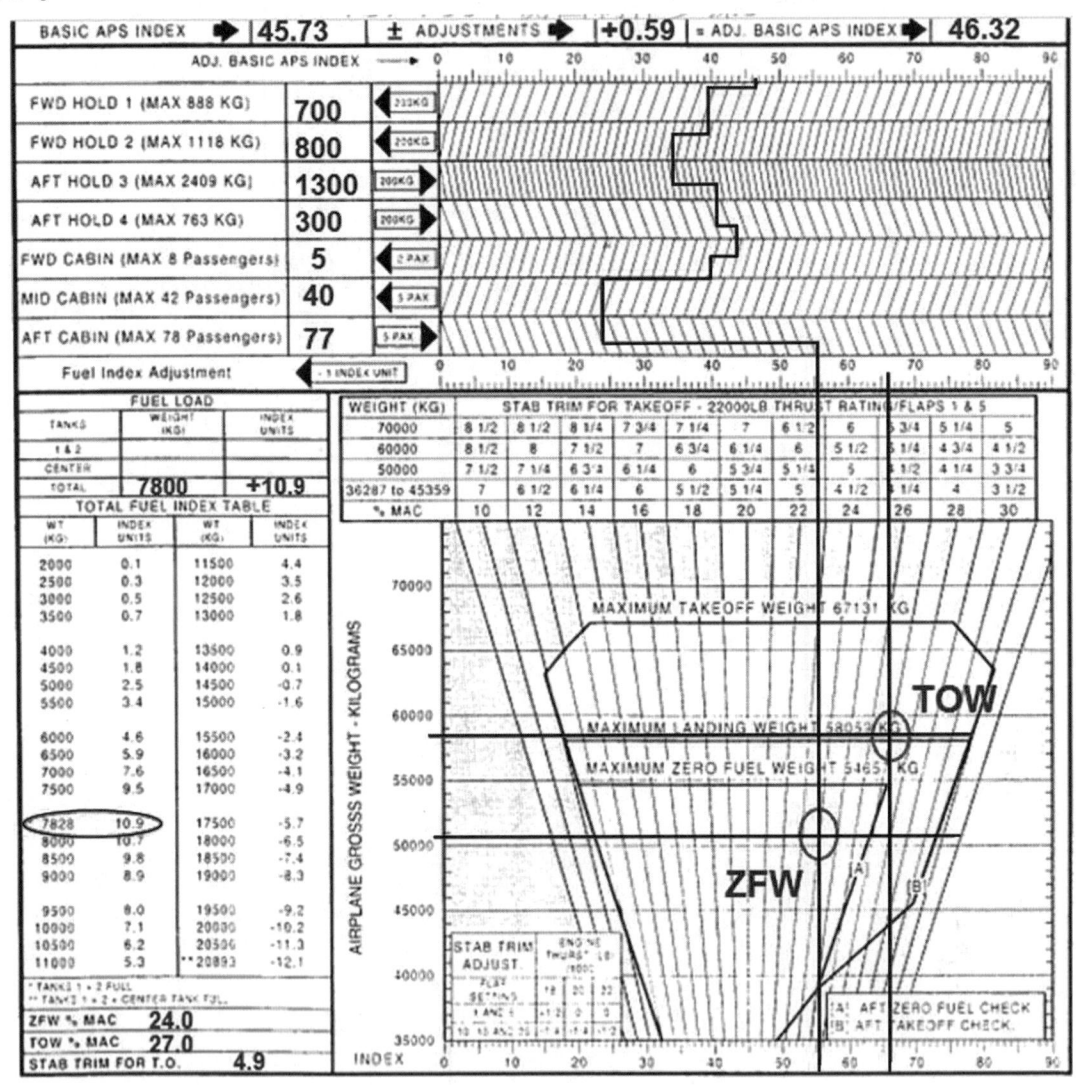

图 5-4-10 图解法确定飞机的重心

5.4.3.1 干操作指数(DOI)

公司在新购置飞机时一般可以通过飞机制造厂商获取飞机的基本使用指数,飞机在大修之后,可以通过重新称重 BOW 计算获得 BOI。如果机组、乘务组人数和配餐和基本配备中的标准不同,则要对 BOW 做修正,获得干使用重量(DOW),通过 DOW 计算获得 DOI。配载平衡图表上给出了对 BOI 进行修正的方法或数据,如图 5-4-11 所示,或者有些直接给出了 DOI 的计算公式。不同机型的指数修正量不同。

ITEMS	WT(kg)	INDEX
BASIC INDEX CORRECTIONS		
BASIC OPERATING WEIGHT INCLUDES COCKPIT CREW AND CABIN ATTENDANTS		
COCKPIT CHANGE + 1 MEMBER 80 kg – 0.98 INDEX		
* ZONE D -1.14		
* ZONE F +1.36		
TOTAL CHANGE		
* INDEX PER 100 kg ADDED		
BI ±	ADJ =	DOI

图 5-4-11　A320 *DOI* 计算

5.4.3.2　旅客指数的图形表示

每个舱位以一定数量的旅客作为一个标准指数增量 $\Delta INDEX$（不同舱位指数增量不同），总的旅客数量就是标准指数增量标准的几倍，按照倍数画出相应长度的指数增量线段，箭头方向为移动方向，不同的客舱舱位箭头的方向不同，如图 5-4-12、图 5-4-13 所示。

$$\Delta INDEX = \frac{旅客人数 \times 标准旅客平均重量 \times (舱位平均平衡力臂 - 基准平衡力臂)}{力矩常数}$$

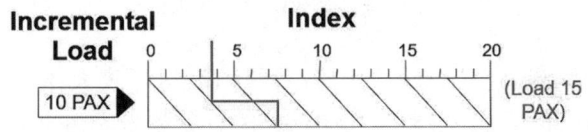

图 5-4-12　旅客指数的图形表示（1.5 倍的标准指数增量）

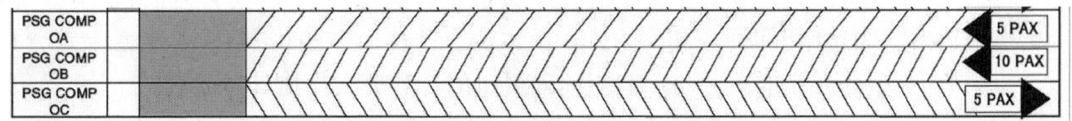

图 5-4-13　旅客指数图形

5.4.3.3　货物指数的图形表示

货舱的每个舱位都以一定重量的货物作为一个标准指数增量 $\Delta INDEX$（不同舱位指数增量不同），总的货物就是标准指数增量标准的几倍，按照倍数画出相应长度的指数增量线段，箭头方向为移动方向，不同的货舱舱位箭头的方向不同。货物指数的图形表示方法与旅客指数方法类似，如图 5-4-14 所示。

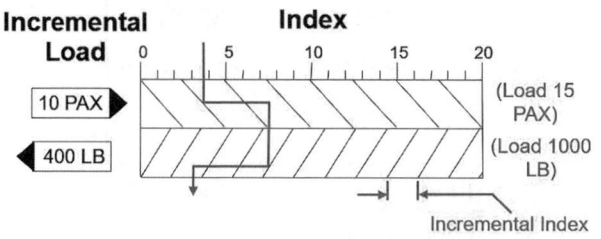

图 5-4-14　货物指数图形(2.5 倍的标准指数增量)

5.4.3.4　无油指数

在配平图上确定重心时,对于以条格图表示客货指数的情况,由干使用指数(DOI)开始,从上至下分别按各区的旅客人数、各货舱的货物重量,按照图中箭头方向相应左右移动,把指数相加减,得出的无油重量所对应的指数,即为无油指数,如图 5-4-15 所示。

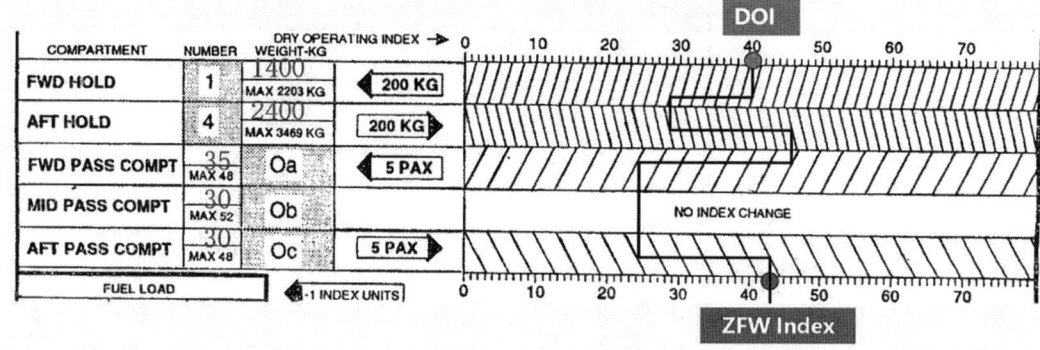

图 5-4-15　无油指数

5.4.3.5　飞机重心的确定

配定平衡是指飞机在不同的飞机重量下,其相应重心位置都满足飞机的操纵稳定和结构强度等要求。当然飞机起飞重量也应满足该机使用手册和起飞机场等限制。图 5-4-16 所示为飞机重心的配定平衡图。

当前一些文件资料和图书手册所用的配载平衡图表达形式大体有两类。第一类,基准点是飞机平均空气动力弦的前缘点,用平均气动弦长的百分数来表达重心的前后限,不同的飞行重量有其各自的重心前后限。第二类,取距平均空气动力弦前缘的 25%(可任意取)弦长处为基准点,用重力在其重心处对基准点力矩的大小表示重心的偏离量;并规定,向机头方向表示低头力矩,取负值,向机尾方向表示抬头力矩,取正值。若该力矩为正,则重心在基准点之后;若该力矩为负,则重心在基准点之前。利用前面讲过的计算公式算出具体的数值。

图 5-4-16(b)给出的垂直式配定平衡图是由水平的平衡式网络图 5-4-16(a)转动 90°,再翻过来的图像。图 5-4-16(c)所表示的各条等百分数的平均气动弦长数值直线是相互平行的。图 5-4-16(b)和图 5-4-16(c)都是配定平衡的状态图,但表达手段有质的区别。

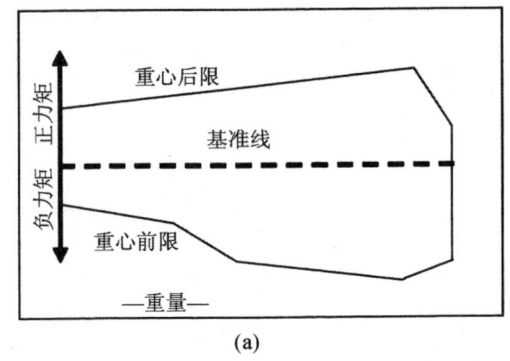

(a)

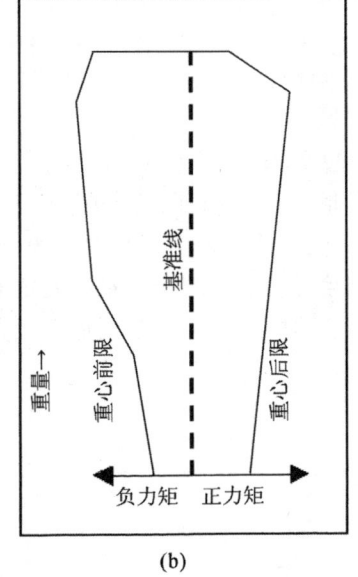

(b)

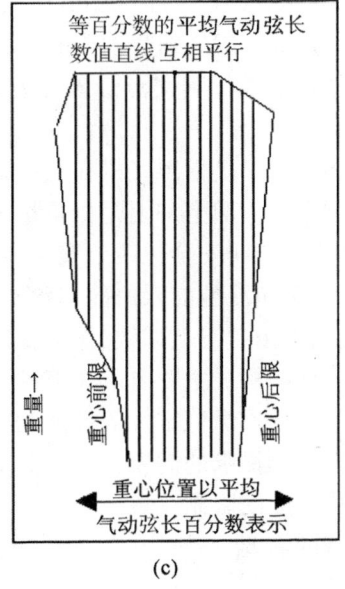

(c)

图 5-4-16 飞机重心的配定平衡图

下面做具体说明。

设飞机机体纵轴 x 的原点在平均空气动力弦的前端点,向后(机尾)为正。全机重量分类分段各自集中,各自重量用 m_1、m_2、$\cdots$、m_i、$\cdots$、m_n 来表示,它们各自的坐标位置依次表示为 x_1、x_2、$\cdots$、x_i、$\cdots$、x_n。这样全机的重量和重心可算得:

$$m_{GR} = \sum_{i=1}^{n} m_i \tag{5-4-4}$$

$$x_{CG} = \frac{\sum_{i=1}^{n} m_i x_i}{\sum m_{GR}} , \quad \bar{x}_{CG} = \frac{x_{CG}}{c_A} \tag{5-4-5}$$

将 m_{GR} 值填写在图 5-4-16(c)上,$\bar{x}_{CG}$ 填到图 5-4-16(c)中某一相应平均气动弦长百分数垂线上,审核是否超出前后限,是否超重。

对于图 5-4-16(b),则是:

①选定 x 轴的坐标原点,在概念上可随便选定,但为实际运用方便,坐标原点一般取在飞机重心前后限的中间,如距平均空气动力弦前缘的 25% 弦长处。

②图 5-4-16(b)不要求算出重心位置,但要求算出对选定坐标原点的总力矩。

③对图 5-4-16(b)的具体做法是,首先将全机的重量分类分段各自集中,各重量为 m_1、m_2、$\cdots$、m_i、$\cdots$、m_n;用图 5-4-16(c)中的新坐标原点的坐标轴来表达各自集中重量的 x 坐标值 x_1、x_2、$\cdots$、x_i、$\cdots$、x_n,然后计算全机重量

$$m_{GR} = \sum_{i=1}^{n} m_i \tag{5-4-6}$$

以及对新坐标原点的总力矩,即

$$M_{y\sum} = \sum_{i=1}^{n} m_i g x_i \tag{5-4-7}$$

式中:$M_{y\sum}$ 为正值则为抬头力矩,为负值则为低头力矩;m_i 是装载取正值,是卸载取负值;x_i 的坐标值在原点前取负值,在原点后取正值。最后将上述 m_{GR} 和 $M_{y\sum}$ 按各自的比例缩小在图 5-4-16(b)中,根据 m_{GR} 值找到竖轴上的点,根据 $M_{y\sum}$ 值找到横轴上的点,然后过这两点分别作各自坐标轴的垂直线得到交点。该点若在网络点限制区内,就满足飞行的各项要求,可以起飞。

上述配定平衡图只是指出飞机在某机场的起飞总重量及其重心位置,没有体现出审核的内容。应当在配定平衡图上加上全机重量大小的限制、飞机重心前后的限制。这种限制线构成一个区域,其边界线叫安全包线。安全包线是在飞机设计过程中积累形成的,由飞机设计部门提供。通常,安全包线的前限由前起落架恒等载荷线、燃油使用线和水平安定面配平线等构成;安全包线的后限由起飞上仰线、操纵稳定性所要求的后限和主起落架恒等载荷线等构成。

整个安全包线的构成主要有下述五种因素:

①飞行中飞机的操纵稳定性等性能的需要;

②飞机地面滑跑操纵的需要;

③飞机结构强度的限制;

④燃油转载的需要;

⑤阵风载荷的限制。

图 5-4-16(b)既包含了代表重力对坐标原点力矩大小的横坐标轴,还包含了代表飞机重量大小的竖坐标轴以及安全包线,在一定程度上可以说明飞机装载的特性;但飞行人员仍觉得有不足之处,因此还在图上作出了关键重量的等值线,如使用空机重量、无油重量、起飞重量和着陆重量的等值线,以及平均气动弦长的 25% 处,它的飞机重心公式表达式为:

$$\bar{x}_{CG} = \left[\frac{\sum m_i x_i}{\sum m_i} \frac{1}{c_A} + 0.25 \right] \times 100\% \tag{5-4-8}$$

上式与前面讲的计算公式唯一的差别是括号内多了一个 0.25,因为坐标轴原点不在平均空气动力弦的前端点,而是在离端点的 $0.25c_A$ 处,实际上是一样的。只要

$$\frac{\sum m_i x_i}{\sum m_i} \frac{1}{c_A} = \Delta \bar{x}_{CG} \tag{5-4-9}$$

计算结果相同,得出的平均气动弦长改变值是一样的,即在同一条平均气动弦长百分比的等值线上。所以在以飞机重力为竖轴和以飞机重力矩为横轴的坐标图上,其重心位置的平均气动弦长的百分比等值线是一条斜直线。

因此,飞机的重心包线图可以转换成用飞机的力矩和重量确定的图形包线,如图 5-4-17 所示。

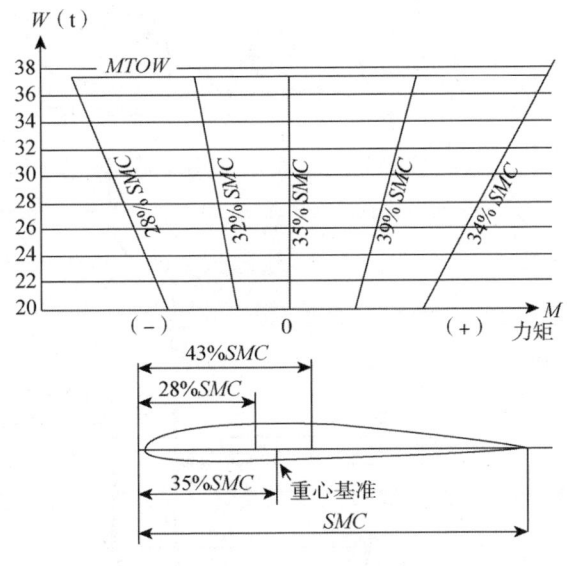

图 5-4-17　飞机的重心包线

如图 5-4-18 所示，M_E、M_P为负（低头力矩），M_F、M_C为正（抬头力矩）。以重心基准线 O 为起点，先把其中之一（如 M_E）标在图上，再依次加上其余的几个力矩，最终得到的 AB 线就反映了合力矩的大小，代表重量 $W = \sum W_i$ 之线与 AB 线的交点即反映了合力之作用点，即飞机最终的重心位置。

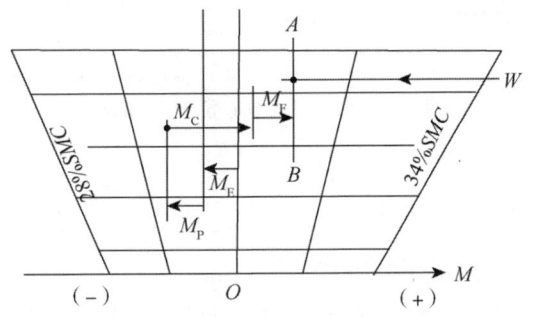

图 5-4-18　飞机的重心变化

通过无油重量对应的无油指数和无油重量直线相交，可以确定无油重量对应的飞机重心，即无油重心；根据起飞油量查出对应的燃油指数，加到无油指数上，得到起飞重量对应的指数（起飞指数），通过起飞指数和起飞重量直线相交，可以得到起飞重量对应的重心位置，即起飞重心；根据总油量查出对应的燃油指数，加到无油指数上，得到滑行重量对应的指数（滑行指数），通过滑行指数和滑行重量直线相交，可以得到滑行重量对应的重心位置，即滑行重心；根据着陆时剩余油量查出相应的燃油指数，加到无油重量的指数上得到着陆重量时的指数（着陆指数），通过着陆指数和着陆重量直线相交，可以确定着陆重量对应的重心位置，即着陆重心。

因此，通过干操作指数，再经过旅客指数修正、货物指数修正、燃油指数修正等就可以获得无油重心、起飞重心、着陆重心、滑行重心等，如图 5-4-19 所示。

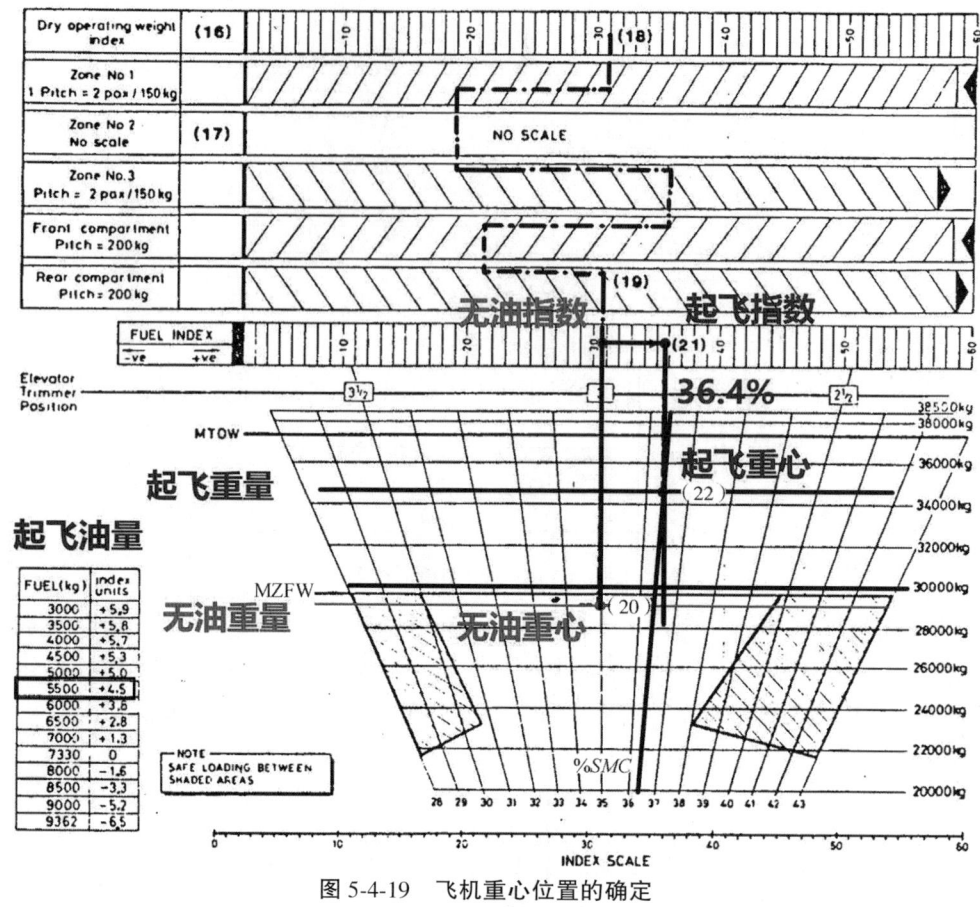

图 5-4-19　飞机重心位置的确定

5.4.4　飞机的配平原理

5.4.4.1　水平安定面的配平

水平安定面是飞机水平尾翼的一部分,其作用是使飞机具有适当的俯仰稳定性。当飞机水平飞行时,水平安定面不会对飞机产生额外的力矩;当飞机受到扰动而抬头时,此时作用在水平安定面上的气动力就会产生一个使飞机低头的力矩,使飞机恢复到水平飞行姿态;当飞机受到扰动而低头时,就会产生一个抬头力矩。飞机在起飞之前要对水平安定面进行配平,也就是我们所说的起飞配平。起飞配平是指在飞机起飞前,根据飞机的重心位置和重量分布,通过调整水平安定面的角度,使得飞机在起飞过程中保持平稳飞行姿态的过程,如图5-4-20所示。正确的起飞配平设置对于保证飞机的安全起飞和飞行至关重要。

起飞配平的目的是保持飞机在起飞过程中的平稳飞行姿态。如果飞机的重心位置过于靠前或者靠后,会导致飞机在起飞时出现过于倾斜或者仰角过大的情况,从而影响飞机的稳定性和操控性。因此,通过调整水平安定面的角度,可以使飞机在起飞过程中保持合适的仰角和侧倾角,从而保证飞机的稳定性。起飞配平还可以影响飞机的起飞性能。根据飞机的重心位置和重量分布,合理的起飞配平设置可以减小飞机的阻力,增大飞机的升力,从而缩短起飞滑跑距离和增大起飞速度。在起飞过程中,飞机需要克服地面摩擦力和重力,通过调整水平安定面的角度,可以减小飞机受到的阻力,使飞机更容易脱离地面,提高起飞性能。

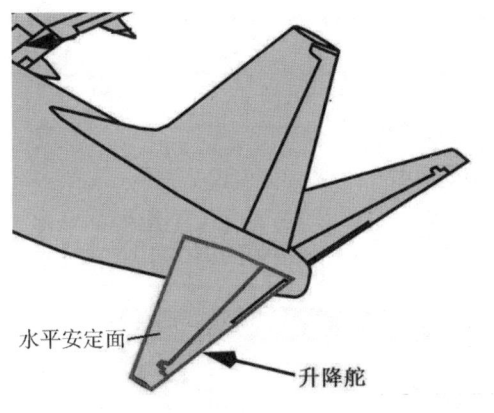
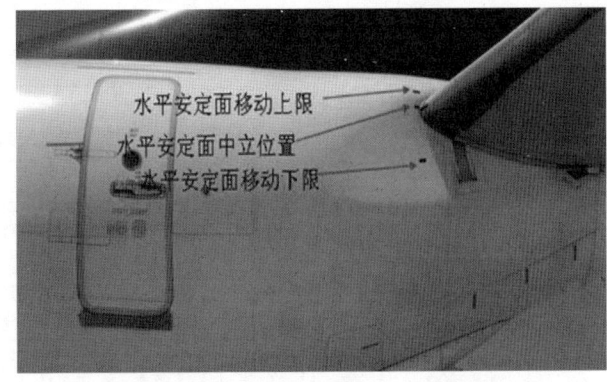

图 5-4-20　水平安定面

　　水平安定面配平用来平衡飞机的纵向俯仰力矩和驾驶杆的杆力。如果飞机的重心靠后，则驾驶杆杆力太小，可能造成操纵飞机起飞时抬头过快、过猛，为此需要把水平安定面前缘向上偏转，让飞机产生一个低头力矩，使得拉杆时有正常的杆力感觉。反之，如果飞机的重心靠前，则需要把水平安定面的前缘向下偏转。在调节水平安定面角度时，升降舵一般保持在中立位置，这样可以保证有足够的俯仰操纵性。因此，飞机在起飞时，起飞配平值必须在指定的位置，处于"起飞绿区"，如图 5-4-21 所示。

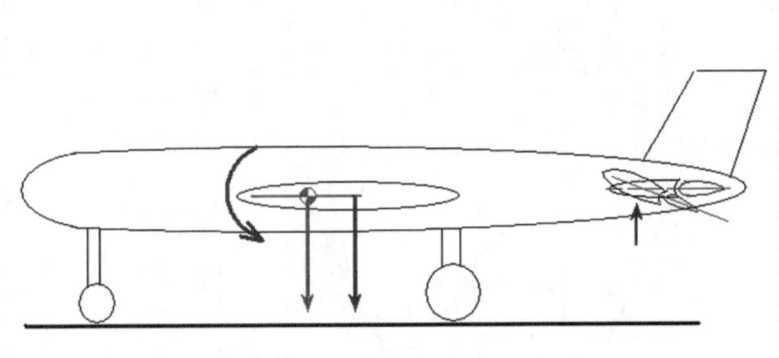
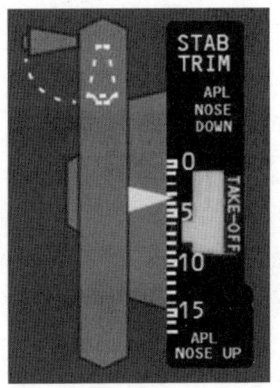

图 5-4-21　水平安定面的配平

5.4.4.2　配平值的确定

　　在实际操作中，起飞配平设置通常由飞行员根据飞机的性能手册和起飞配平图进行调整。飞机的性能手册会提供飞机的重心范围和配置限制等信息，起飞配平图则会根据飞机的重心位置和重量分布情况，给出相应的水平安定面配平值。飞行员根据这些信息，可以通过调整水平安定面的角度，使飞机在起飞过程中保持平稳飞行姿态。

　　在配载平衡图上，当确定了起飞重心后，可以根据相应的坐标或者表格确定起飞配平值，如图 5-4-22、图 5-4-23 所示。机组根据这个配平值在起飞前通过配平手轮或俯仰配平电门、多功能控制显示单元（MCDU）进行配平，调节水平安定面的角度。起飞配平设置应该根据飞机的实际情况进行调整。不同的飞机型号和不同的起飞条件会有不同要求。因此，飞行员需要根据飞机的性能特点和起飞条件，进行合理的起飞配平设置。

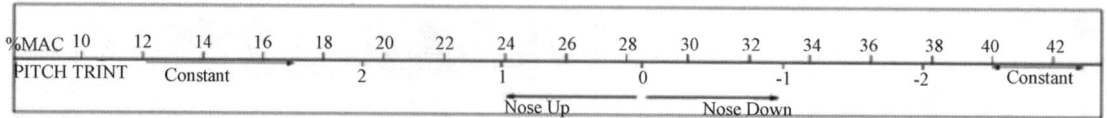

图 5-4-22　坐标形式的配平值确定

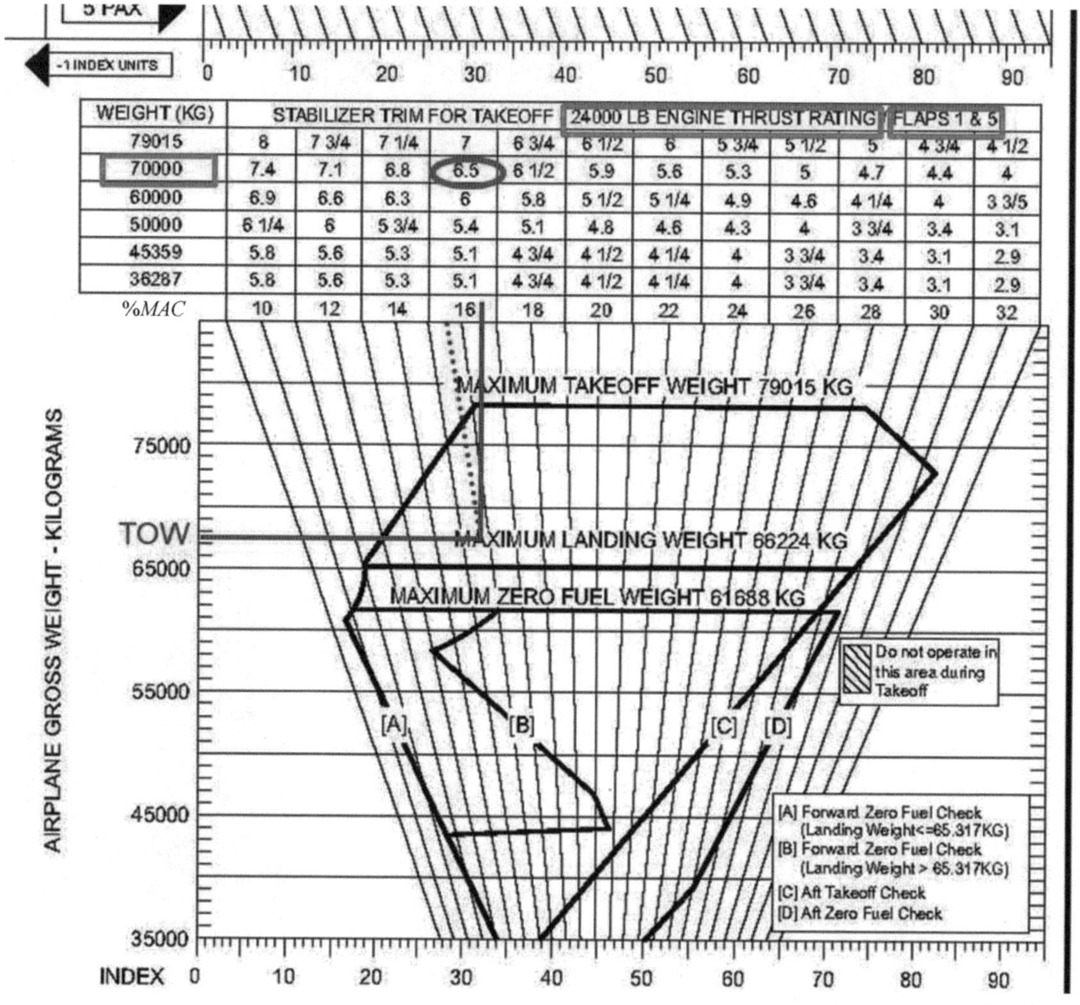

WEIGHT (KG)	STABILIZER TRIM FOR TAKEOFF				24000 LB ENGINE THRUST RATING					FLAPS 1 & 5	
79015	8	7 3/4	7 1/4	7	6 3/4	6 1/2	6	5 3/4	5 1/2	5	4 3/4	4 1/2
70000	7.4	7.1	6.8	6.5	6 1/2	5.9	5.6	5.3	5	4.7	4.4	4.2
60000	6.9	6.6	6.3	6	5.8	5 1/2	5 1/4	4.9	4.6	4 1/4	4	3 3/5
50000	6 1/4	6	5 3/4	5.4	5.1	4.8	4.6	4.3	4	3 3/4	3.4	3.1
45359	5.8	5.6	5.3	5.1	4 3/4	4 1/2	4 1/4	4	3 3/4	3.4	3.1	2.9
36287	5.8	5.6	5.3	5.1	4 3/4	4 1/2	4 1/4	4	3 3/4	3.4	3.1	2.9
%MAC	10	12	14	16	18	20	22	24	26	28	30	32

图 5-4-23　表格形式的配平值确定

水平安定面配平值的大小与飞机重心、起飞重量（松刹车重量）、襟翼位置和发动机推力等因素有关。

（1）起飞重量、重心和襟翼对配平值的影响

不同的起飞重量有不同的配平值。较小的机型，往往划分为 2～3 个重量范围，用 2～3 个重量来代表，图 5-4-24 所示为 B737-400 装载配平手册中确定安定面配平值的图表。在 QRH 或载重平衡图表中有时只给出一个重量的配平调定值，再给出一个重量修正值，如表 5-4-5 所示，表中的配平值按图 5-4-24 中的中间曲线读出。

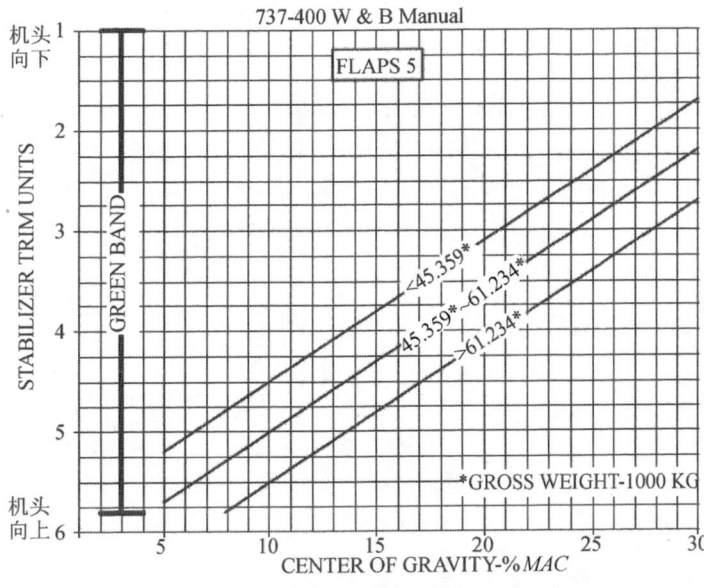

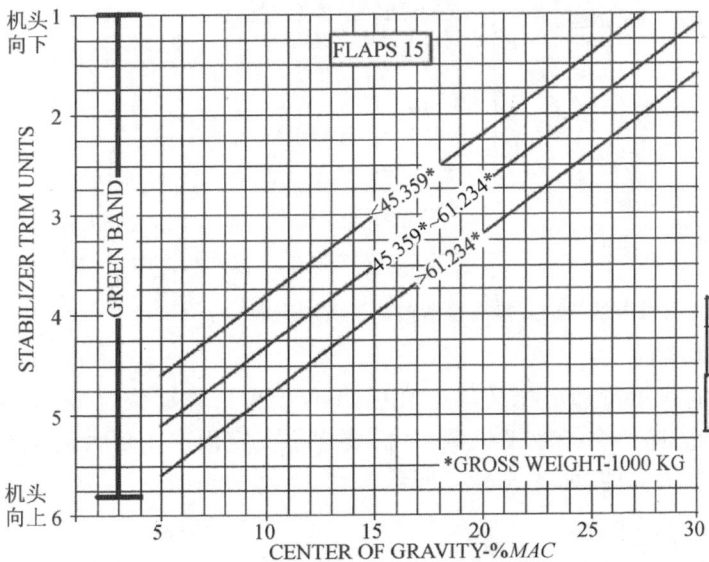

图 5-4-24　B737-400 确定安定面配平值的图表

表 5-4-5　B737-400 QRH 中确定配平值的图表

Stab Trim Setting　Max Takeoff Thrust　　　　　　737-400/CFM56-3_23.5K　**QRH**

Flaps 5

C.G.%MAC	6	10	14	18	22	26	30
STAB TRIM	$5\frac{1}{2}$	5	$4\frac{1}{2}$	$3\frac{3}{4}$	$3\frac{1}{4}$	$2\frac{3}{4}$	$2\frac{1}{4}$

Flaps 15

C.G.%MAC	6	10	14	18	22	26	30
STAB TRIM	5	$4\frac{1}{4}$	$3\frac{3}{4}$	3	$2\frac{1}{2}$	$1\frac{3}{4}$	1

For weights at or below 45360 kg subtract 1/2 unit from above value.

For weights at or above 61235 kg add 1/2 unit from above value.

水平安定面的配平值与飞机重量、重心、襟翼位置有关。重量增大,配平值增大;重心向后,配平值减小;襟翼角度增加,配平值减小,见图 5-4-25。

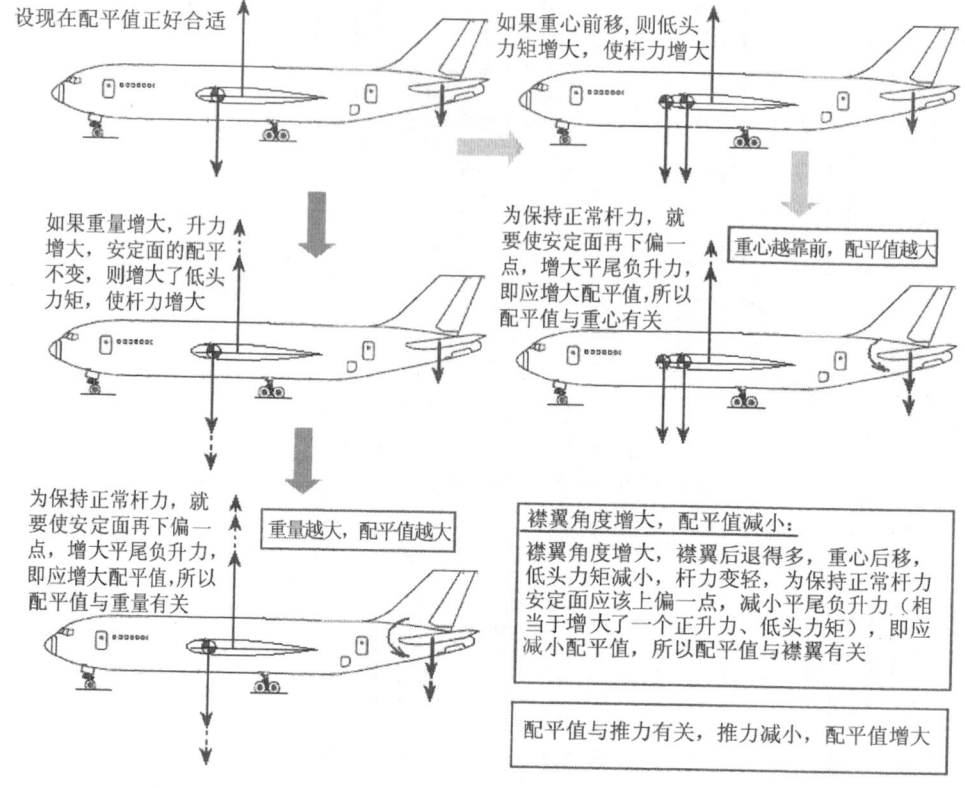

图 5-4-25　飞机重量、重心、襟翼位置对配平值的影响

(2)发动机推力对配平值的影响

有些机型在根据起飞重心、起飞重量确定配平调定值时,不仅要考虑襟翼位置,还要考虑推力的大小。这是因为发动机推力减小会导致飞机低头,需要水平安定面产生更大的抬头力矩,则水平安定面的配平值需要更大一些,如图 5-4-26 所示。

图 5-4-26　发动机推力对水平安定面的影响

因此,采用减推力起飞时的配平值与采用最大起飞推力起飞时的配平值是不一样的。如果不知道减推力的程度,可按使用全推力到推力减少 15% 的情况查图表。例如,有的机型规定在使用全推力到推力减少 15% 时查一张图表,减推力超过 15% 查另一张图表或对由上一张图表查出的配平值做一些修正。推力减小 15%,配平值增大 0.3~0.5(即安定面向下偏得更多些,增大些抬头力矩),如图 5-4-27(a)、(b)所示。

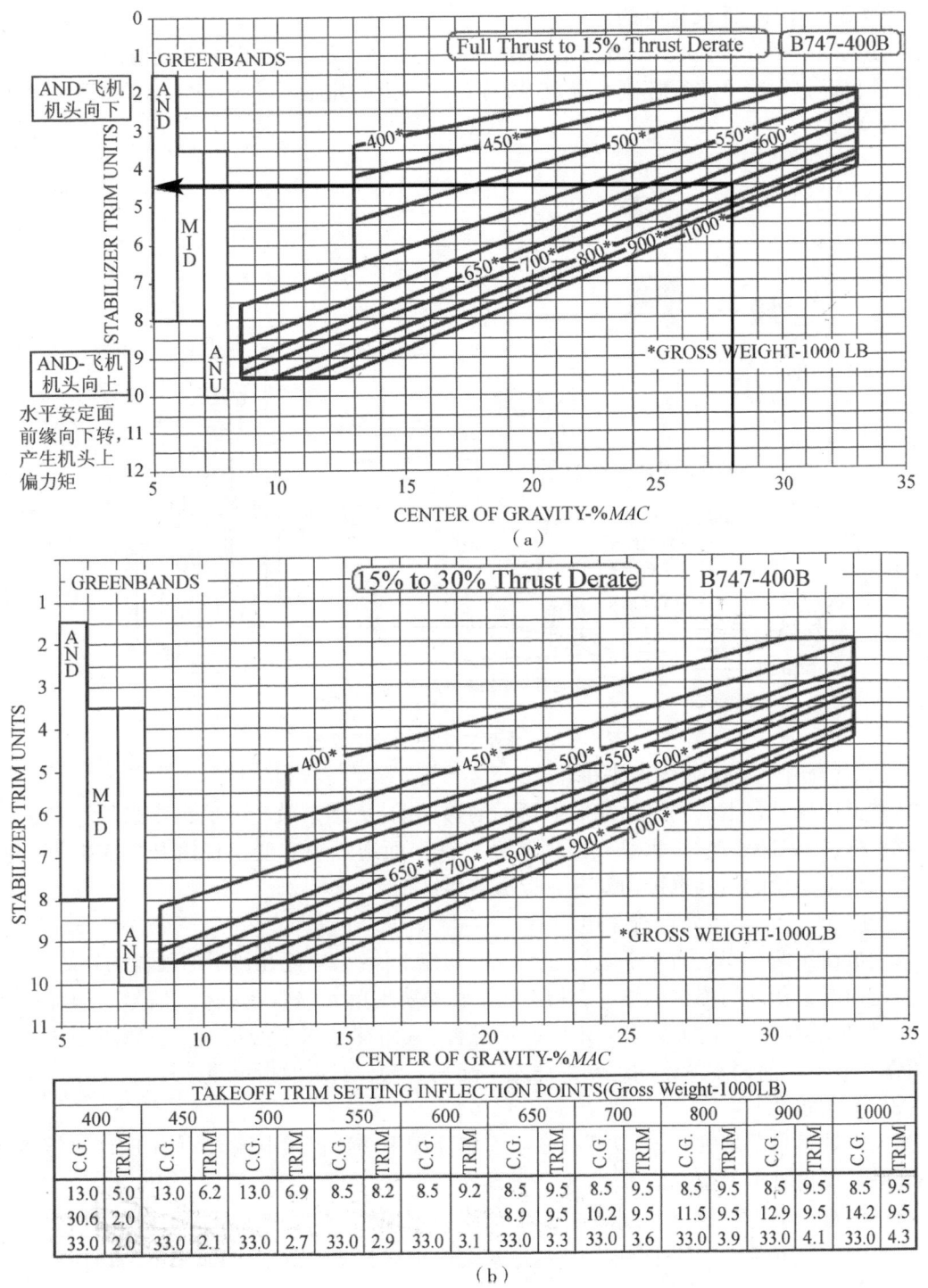

表格：

TAKEOFF TRIM SETTING INFLECTION POINTS(Gross Weight-1000LB)																			
400		450		500		550		600		650		700		800		900		1000	
C.G.	TRIM	C.G.	TRIM	C.G.	TRIM	C.G.	TRIM	C.G.	TRIM	C.G.	TRIM	C.G.	TRIM	C.G.	TRIM	C.G.	TRIM	C.G.	TRIM
13.0	5.0	13.0	6.2	13.0	6.9	8.5	8.2	8.5	9.2	8.5	9.5	8.5	9.5	8.5	9.5	8.5	9.5	8.5	9.5
30.6	2.0									8.9	9.5	10.2	9.5	11.5	9.5	12.9	9.5	14.2	9.5
33.0	2.0	33.0	2.1	33.0	2.7	33.0	2.9	33.0	3.1	33.0	3.3	33.0	3.6	33.0	3.9	33.0	4.1	33.0	4.3

（b）

图 5-4-27　B747-400 推力对配平值的影响

　　配平值增大,水平安定面前缘向下偏转,平尾产生附加的向下的力,使机头上仰;配平值减小,水平安定面前缘向上偏转,平尾产生附加的向上的力,使机头下俯。因此,应通过调节水平安定面配平值的角度来控制飞机的俯仰稳定性,以确保飞机飞行的安全。

5.4.4.3　水平安定面配平的方式

驾驶员用水平安定面配平手轮人工操纵水平安定面,或用安定面配平电门进行电动操纵;自动驾驶仪通过数字式飞行控制系统(DFCS)自动操纵安定面。水平安定面的配平操作如图5-4-28所示。

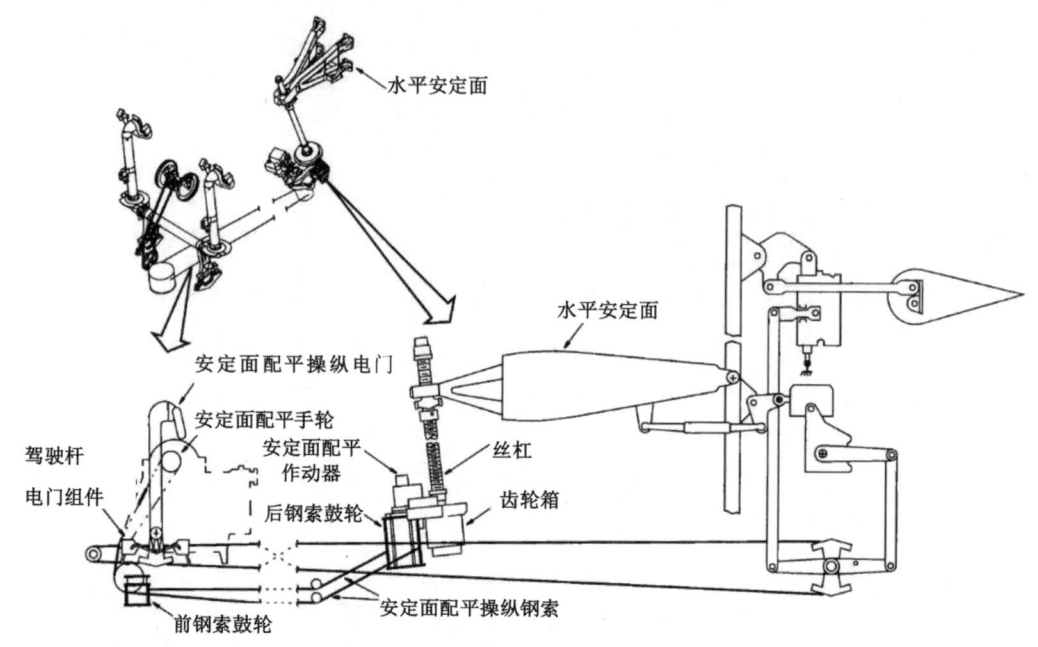

图 5-4-28　水平安定面的配平操作

（1）人工配平——安定面配平手轮

驾驶员用操纵台上的安定面配平手轮手动移动前、后钢索鼓轮。后钢索鼓轮移动齿轮箱和起重螺杆,当螺杆移动时,水平安定面移动。安定面通过中立变换杆给升降舵提供机械输入。人工配平手轮的转动也带动安定面指示器指针指示。

（2）电动配平——安定面配平电门

驾驶员使用两个安定面配平电门进行主电动配平操纵,电门在每个驾驶盘的外侧,电门控制给安定面配平作动筒提供电输入,并给飞行数据采集单元(FDAU)传送信号。当配平作动器工作时,带动齿轮箱,进而带动安定面起重螺杆,从而移动水平安定面。齿轮箱回传安定面前、后钢索鼓轮。前钢索鼓轮的运动带动人工配平手轮和安定面指示器指针。

（3）自动配平

当飞机处于自动驾驶时,自动驾驶仪给安定面配平马达提供电输入,安定面位置传感器给自动驾驶仪提供安定面位置信号,实现了自动配平。

水平安定面的配平值增大,水平安定面前缘向下偏转,平尾产生附加的向下的力,使机头上仰;反之,水平安定面的配平值减小,水平安定面前缘向上偏转,平尾产生附加的向上的力,使机头下俯。

5.4.5　对飞机无油重心的限制

按前述的办法可确定飞机起飞时的重心,是否只要起飞重心在允许的范围内就可以保证安全?回答是否定的。因为如果起飞重心在允许范围内,但比较靠近边界,那么在飞行中由于燃油消耗、重心移动、人员在飞机内走动(比如乘务员推餐车前后送餐盒)等,都会使飞机重心变化,有可能超出允许范围而危及飞行安全。另外,旅客重量的误差,计算旅客、货物力矩(指数)时的误差(旅客和货物的实际重心不在各舱段的平均位置产生的误差),都有可能使重心偏出允许范围。解决此问题的方法是缩减重心包线,确定无油重心的允许范围。

首先必须考虑燃油消耗对重心的影响,把无油时的重心范围缩小以应对燃油消耗对重心的影响,如图 5-4-29 所示。例如,如燃油消耗可能使重心前移 2%SMC、后移 3%SMC,则必须把重心前限后移 2%SMC、重心后限前移 3%SMC。

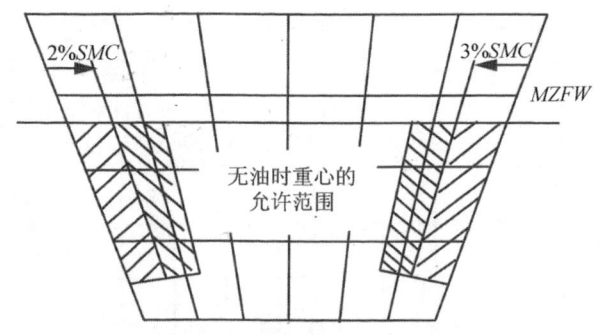

图 5-4-29　确定无油重心允许范围的示意图

此外,还必须计算下述各种因素造成的最大可能的力矩误差:

(1)旅客重量与规定的平均重量不同所造成的力矩误差;

(2)旅客分布造成的力矩误差(即上节提到的最大力矩误差);

(3)人员及服务车移动造成的力矩误差;

(4)货物重心与货盘、集装箱几何中心不重合造成的力矩误差,因为在配平计算各货舱指定重量为×××千克的货物产生的力矩时假定它们在货盘、集装箱的中心;

(5)飞机起飞、着陆加减速时油箱内燃油的涌动(Slosh)造成的燃油重心的变化;

(6)收放起落架造成的力矩的变化;

(7)收放襟翼造成的力矩的变化;

(8)燃油密度的差别造成的力矩的变化。

对造成重心向前、向后移动的各种随机误差分别求均方根值,分别加上其他向前、向后的固定性的误差,计算出这些力矩误差使无油重量的重心可能向前、向后移动的距离,再把图 5-4-29 中的左、右边界线分别向右、左移动这一距离以抵消这种误差的影响,确保飞机重心在任何情况下都不会超出允许范围。因此,无油重心被限制在一个较小的范围内,安排旅客、货物时必须使无油重心落在此范围内。图 5-4-30 是 B757-200 缩减重心包线后的重心限制。

5.5　配载平衡图表的填制方法

飞机制造厂商可以提供多种形式的配载平衡图表,用户可自行选择,也可要求飞机制造厂商提供做一些修改后满足要求的配载平衡图表。配载平衡图表彼此可能有很大不同,但原理是相同的。常用的有 Alignment Bar 型和 Universal Index 型配载平衡图表,下面分别介绍这两种配载平衡图表的填制方法。

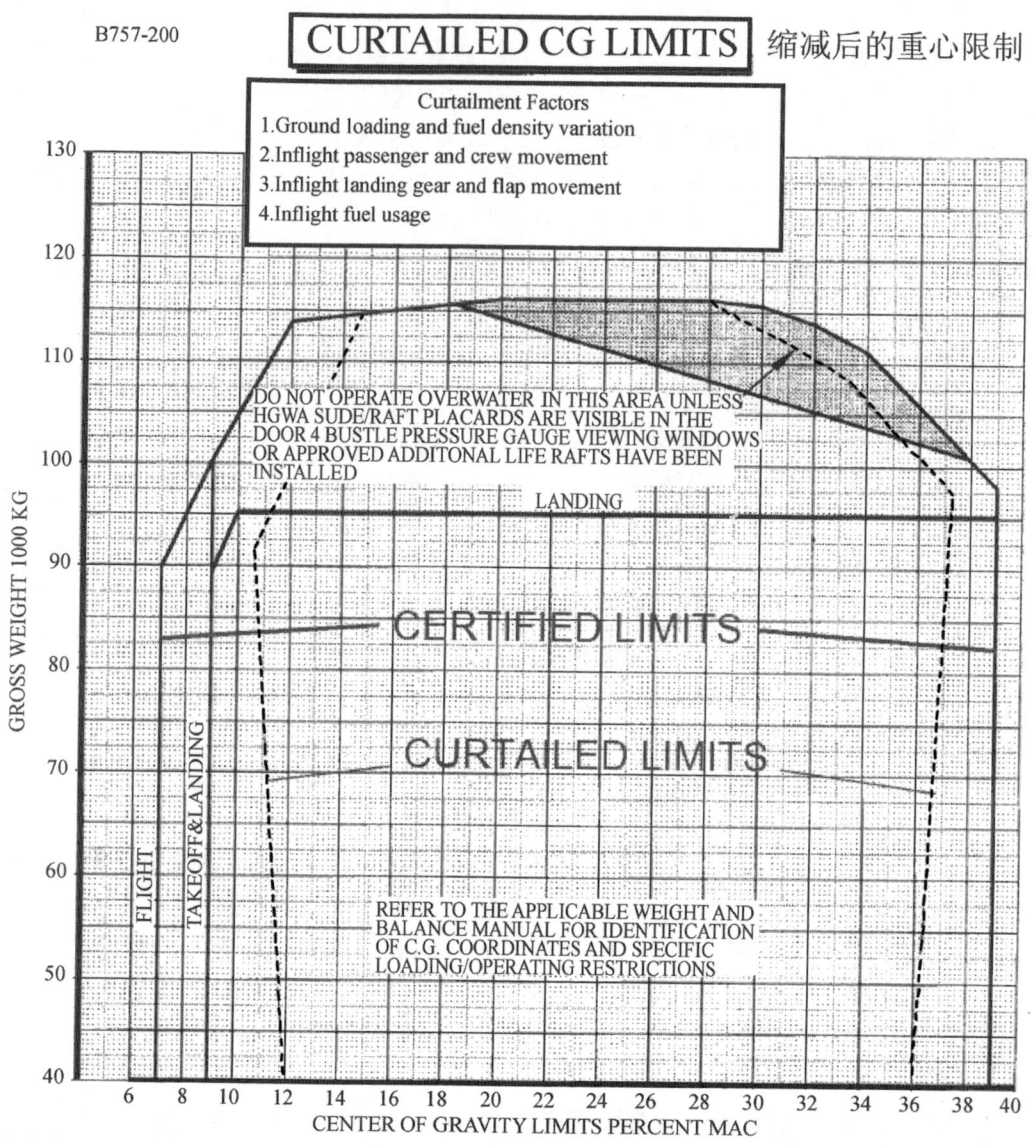

图 5-4-30　B757-200 缩减重心包线后的重心限制

5.5.1　Alignment Bar 型载重平衡图

5.5.1.1　BAe146-100 飞机实例

以 BAe146-100 飞机为例,其载重平衡图见图 5-5-1,该图(1a)、(2a)、…、(23)表示工作步骤。假设:$OEW = 22600$ kg,$DOI = 31$,旅客 62 人,每人重 75 kg,行李及货物共重 2000 kg,1 区安排 12 人,2 区 36 人,3 区 14 人,前货舱装货 1200 kg,后货舱装货 800 kg,起飞油量 5350 kg,飞到目的地机场时耗油 2350 kg,$MTOW = 37308$ kg,$MLW = 33270$ kg,$MZFW = 29700$ kg,要求完成配载平衡图、确定无油重心、起飞重心及升降舵配平调整片位置。

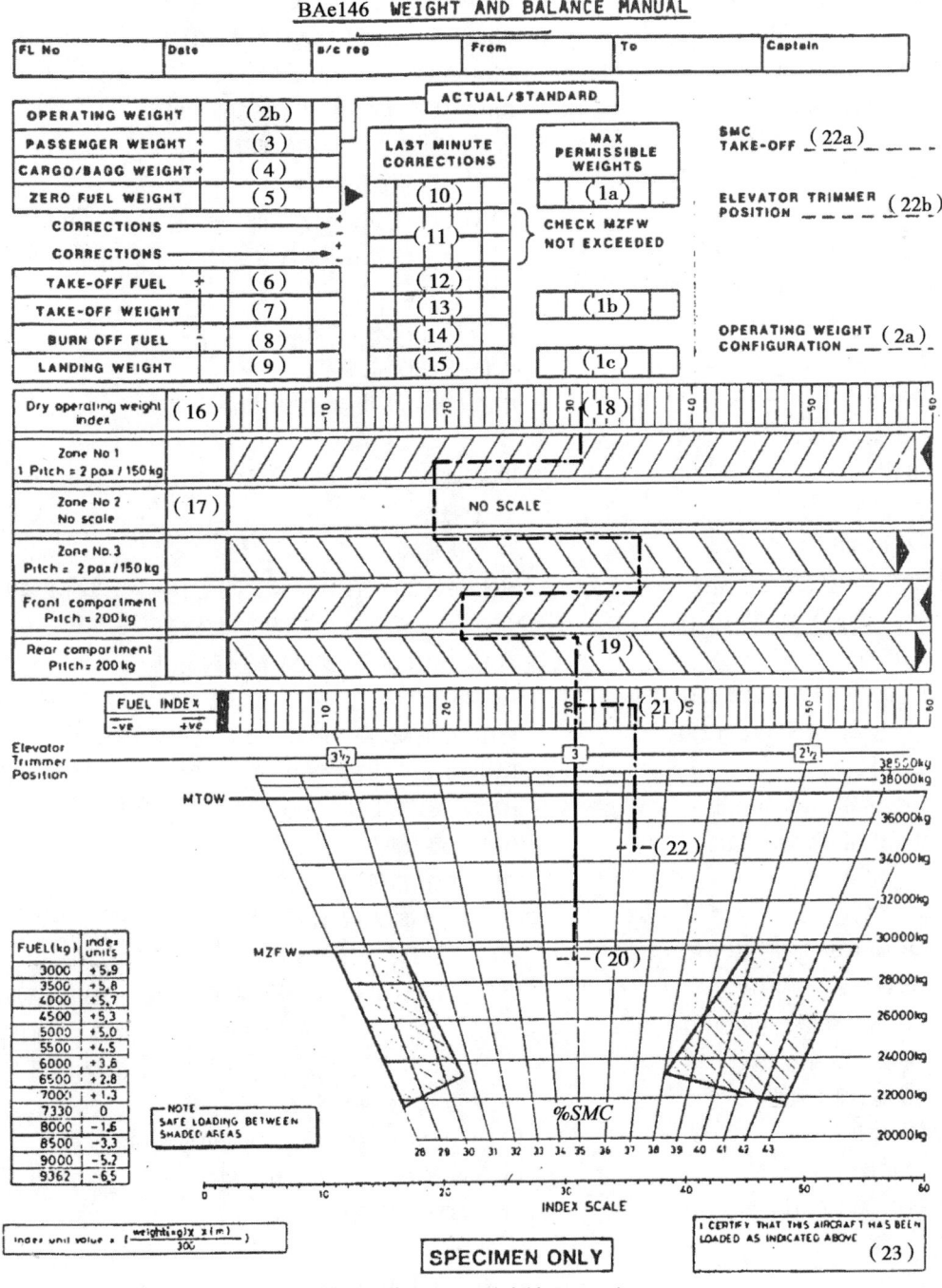

图 5-5-1　BAe146-100 载重平衡图使用说明

下面给出载重平衡图上有关缩写和术语的说明。

A—乘务员座位

A/C REG—飞机注册号

ADDRESS—收电人地址

LMC—最后一分钟变化

M—邮件；男性旅客

MID COMPT—中舱

ADLT、ADT、AD—成年人

AFT COMPT(AFT HOLD)—后舱

ALLOWED TRAFFIC LOAD—允许业载

A/F—成年女性旅客

APS—已准备好、可以使用的飞机

B—行李;公务舱

BAG—行李

BASIC OPTG WT—基本使用重量

BASIC APS WT—基本使用重量

BOW—基本使用重量

BULK—散货舱

C—货物

CA—中国民航航班号的头两个字母

CAB BAG—装在客舱中的行李

CAPT—机长

CARGO HOLD—货舱

CH/CHD—小孩(单独占座位)

CL/CPT—客舱等级/货舱位置

CONTAINER—集装箱

CONF—布局,指头等舱和普通舱的
座数,如写:F18/Y210

COMP—计算

CPM—集装箱、板分布信息

CREW—机组/乘务组人数,
如:3/7

DEST—目的地

DLI—Dead Load Index,即干使
指数与各货舱中货物、行
李、邮件指数之和(即不能
移动的载荷的指数)

E—经济舱,即普通舱

FL NO—航班号

FLT(FLIGHT)—航班号

VERSION—客舱布局,同 CONF

VIP—要客

WT—重量

Y—普通舱

ZFCG—无油重心

ZFW—无油重量

MTOW—最大允许起飞重量,指在跑道
头松刹车起飞的最大允许重量

ORIGINATOR—发电人地址

PAD—需要减载时,可优先减下去的
旅客,如本公司的职工、持免
票或优惠票的旅客

PANTRY—与正常配餐标准相比,增
加或减少的配餐重量

PAX 或 PASS 或 PSGR—旅客

PALLET—货板,货盘

P/L—Pay Load,业载

RECHARGE—指向谁收取加油费,该
项可填客户的代号

REV—修改,修正,修订

S—贮藏室

SI—补充信息

SEATING CONDITION—就座状态,各客舱段的旅客人数,
如 F14/Y80

STAB TRIM SET—安定面配平调定值

T—总计

T/C—旅游舱(经济舱)

TAIL STAND—尾撑杆,尾支架

TOCG—起飞重心

TOW—起飞重量

TR—过站(指机上在经停站中转的客
货、邮的人数、件数重量)

UD—上层舱(UP DECK)用

ULD—Unit Load Device,指集装箱(Container)或货盘(Pallet),
如配平图表上有 ULD 项,则表示要填写它们的重
量,如它们计入货物、行李重量中,则无此项

F—头等舱(或用 F/C 表示)

UNDERLOAD—空余吨位,即尚可再增加的业载

FWD COMPT(FWD HOLD)—前舱

FWD LINE—代表重心前限的线

G—厨房

I/INF—不占座位的婴儿

L—厕所

LDM—装载信息

不同机型的配载平衡图表上可能使用的英语词汇不同,但含义相同。

载重平衡图的填制方法:

在(1a)、(1b)、(1c)处分别填上 *MZFW*、*MTOW*、*MLW*,每格填一个相应的数字。

在(2a)处填上由前面的表得到的代号如(2a)或填上使用运行空机重量 *OEW* 和干使用指

数 DOI,如 22600 kg 和 31,在(2b)处填上 22600。

在(3)处填上旅客重 62×75＝4650 kg。

在(4)处填上货物和行李重 2000 kg。

在(5)处填上无油重量 29250 kg。

在(6)处填上起飞油量 5350 kg。

在(7)处填上起飞重量 34600 kg。

在(8)处填上消耗的油量 2350 kg。

在(9)处填上着陆重量 32250 kg。

在(10)~(15)处填上由于最后增减(在填完表之后又有增减)旅客、货物得到的无油重量、起飞重量、着陆重量等。在填表时注意检查无油重量、起飞重量、着陆重量是否超过最大允许值。

在(16)处填上干使用指数 31。

在(17)的五个格中分别填上客舱各区安排的旅客人数、各货舱装的货物重量。

然后找到干使用指数 31 的位置(18),按图中箭头所示画出各区、各货舱的指数变化,最后得到无油重量的指数(19)。

向下画垂线找出与 $ZFW＝29250$ kg 的交点(20),读得 ZFW 的重心为 35%。

查出与起飞油量 5350 kg 对应的燃油指数 4.65,加到无油重量的指数上得到起飞重量的指数(21)。

向下作垂线,找出与 $TOW＝34600$ kg 的交点(22),读得起飞重心为 36.4%及升降舵配平值 2.8,分别填到(22a)和(22b)处。

最后由机长在(23)处签名。

着陆时剩余油量 3000 kg 对应的指数为 5.9,着陆时重心为 36.8% SMC。

在完成配平图后即将起飞前,可能又有旅客、货物的增减,此时应在"最后一分钟变化"下的(10)~(15)处填上改变后的各项重量,注意 ZFW 不要大于 $MZFW$。由于"最后一分钟变化"造成的指数变化可由各区(各货舱)的斜格宽度来判读,对代表无油重量 ZFW、起飞重量 TOW 指数的垂线修正得出新的重心位置。载重平衡图的填制结果如图 5-4-2 所示。

关于填写配载平衡图表的几点说明:

(1)收电人不止一个时,按其优先顺序写入。

(2)关于旅客重量:一般使用平均重量,例如,可以规定一名旅客加 5 kg 手提行李按 77 kg 或 80 kg 计,也可以对男、女旅客规定不同的平均重量,对冬、夏季规定不同的平均重量(因为穿衣服多少不同),不同国家的这个平均重量可能是不同的。2~12 岁的儿童可按成人重量的一半来计算,2 岁以下的婴儿按 5~10 kg 计算。如是团体包机,整架飞机上都是特殊旅客,如携带武器装备的军人或大块头的运动员等,与规定的平均重量明显差别较大,可以根据实际情况采用不同的平均重量,对这种情况也可使用实际旅客重量。如飞机上有一大部分旅客明显高于或低于规定的平均重量,那么对这部分人应使用实际重量,对其余旅客仍用平均重量。另外,对于一些很小(例如座位数在 10 个左右)的飞机,也应使用实际旅客重量。实际旅客重量可以通过称量得到或通过口头询问旅客得到。国际航空运输协会(IATA)的机场运行手册第 531 款 1.2.2.2 节规定:婴儿的重量算在成人的标准体重内。有些载重平衡图表上按此规定执行,不再另算婴儿的重量。

(3)填写旅客数时,如航空公司对男、女旅客规定的体重不同,则应填入男、女旅客各多

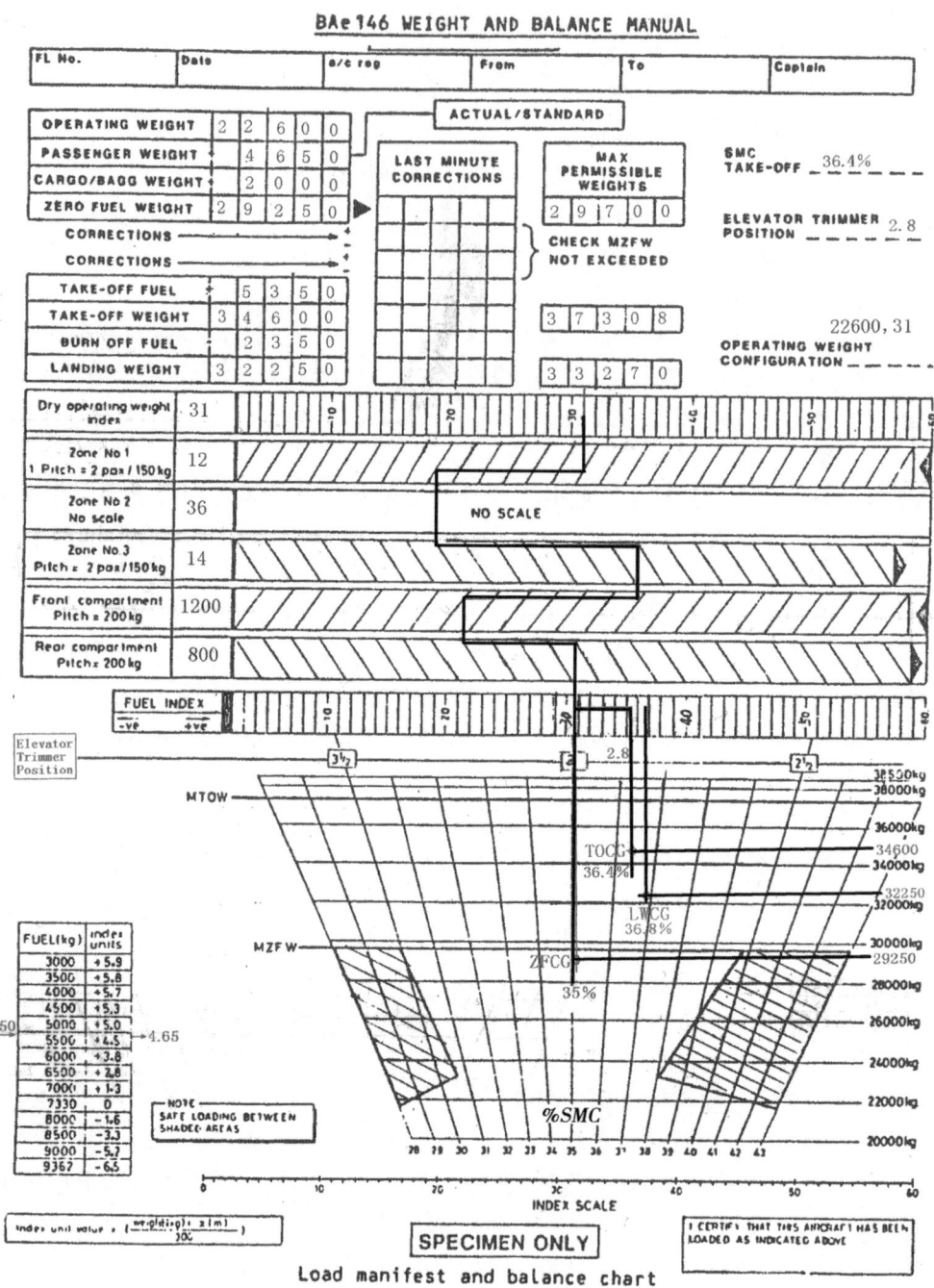

图 5-5-2　BAe146-100 载重平衡图的填制结果

少,否则,即使配平图上分 M、A/F 两项,也不必管它。

(4)在 REMARK 项下可以注明 PAD 旅客的人数、坐的位置,如 F1/Y4 等,需要减载时,可先把他们减下去。

(5)一个航班当有经停站时,就将每个经停站都填写在 DEST 项中,如西安(SIA)至东京(TYO)的航班经停上海(SHA)、福冈(FUK),则应把上海(SHA)、福冈(FUK)、东京(TYO)都填写在 DEST 项中,并应写明在这每个地方中转的旅客数和货物重量及以此站为终点的旅客数和货物重量(分别填入上、下两栏),在阴影区中填入小计。对于这种有经停站的情况,起飞油量可按到第一站计算(为该站选一备降机场),如在该经停站不能加油或油价太高或为缩短航班时间不中间加油,则起飞油量是整个航班所需的油量,但航程油量(Trip Fuel)和着陆重量是到达第一经停站的值。

(6)注意配载平衡图表上给出的各舱的座位数、最大允许的装货量,分配旅客、货物时不得超过这些限制。货物对货舱地板产生的压强不得超过该货舱地板能承受的最大压强(这个最大压强与机型、货舱有关),否则应加垫板以分散压力、减小压强。

(7)"最后一分钟变化"的修改中如重量变化不大,可只改动无油重量、业载、起飞重量、着陆重量等项,重心和配平值可以不重新确定。

(8)配载平衡图表中的起飞油量(Take-off Fuel)是滑行到跑道头准备起飞时的油量,等于停机坪油量(Ramp Fuel,即总加油量)减去滑行到跑道头所用油量。

(9)燃油指数与其密度有关,有些配载平衡图表中的燃油指数表格中针对不同的密度给出不同的数值,要按照所加的燃油的密度来确定燃油指数。

(10)向各货舱安排货物时,除要遵守各货舱的最大载量限制外,还要注意"累计载重"限制,即两个或几个货舱的总载量也不能超过一定的值,此限制值小于这几个货舱的最大载量之和。

(11)飞机上积累的灰尘、水汽、重新喷涂的油漆、遗失在飞机上的物品都会增加飞机的重量,改变其重心,通过对飞机进行称重可以确定其精确的重量、重心及 BOI。一般每 4 年称重一次。

(12)应注意下述可能导致飞机倾覆(机尾触地)的操作和情况:旅客由后客舱门下飞机;前货舱卸空而后货舱是满的;大风对飞机产生的力矩;下雪时飞机上的积雪产生的力矩;有较大斜坡的停机坪;以及飞机重量轻且重心靠后的情况。

(13)货舱中的货物必须牢牢地固定,以免飞行中货物移动造成重心变化。

(14)基本使用重量(BOW)一般不包括集装箱、货盘(板)的重量,箱、板的重量被计入货物、行李的重量中,即配载部门报的货物、行李重量包含集装箱、货盘的毛重。如货物、行李重量不含箱、板的重量,在配载平衡图表上的 ULD 项中应填写箱、板的重量。

另外,不同的机型其载重平衡图不同,其相应的表格内容和填制方法不尽相同。

例如:在 B757-200 的配载平衡图表上给出了对基本使用指数做修正用的表格"ADDITION AND DEDUCTION INDEX TABLE",当机组、乘务组人数和配餐与标准配备不同时,则使用此表计算增减的人员或配餐重量造成的指数增减量,然后加到基本使用指数上。表中"FLIGHT DECK OCCUPANT"指驾驶舱中的人员,"FWD CABIN ATTENDANT DOOR1"指前舱门 1 处乘务员,表中给出每增减 1 人(按 80 kg)指数的增减值。对表中各项,若是增加,则对应的指数取表中左面的正负号,反之,取表中右面的正负号。若增减的重量不是表中规定值,对应的指数=实际增减重量×表中的指数/表中的重量。例如,在前舱门 1 处增加 1 名乘务员,体重按72 kg 计,则指数减少量=72×0.7/80。该表中关于各客舱、货舱中旅客、货物的增减造成的指数的增减量是用于计算"最后一分钟变化"的。

再如:B737-300 的配载平衡图表关于燃油指数的处理有所不同,该图给出的例子表明了画图的方法。其道理是一样的,只不过把单独查出的燃油指数再加到无油重量的指数上的过程用连续的画图过程代替了,把 A310 和 B737-300 的两张图对比分析就不难明白其中的道理。

5.5.1.2　A310-200 飞机实例

A310-200 执行上海至成都的航班,已知:基本使用重量(机组 2 人,乘务员 7 人)BOW = 80000 kg, BOI = 39.9, $MZFW$ = 111500 kg, $MTOW$ = 131500 kg, $MLWD$ = 120000 kg,扣除滑出油量后的起飞油量为 15000 kg,航程油量(TRF)为 9670 kg,本航班机组增加 2 人,在前厨房(D 区)增加配餐 100 kg,在后厨房(E 区)增加配餐 200 kg,头等舱成人旅客 15 名,普通舱 OB 区成人旅客 90 名、OC 区成人旅客 100 名,成人旅客按每人 80 kg 计。交运行李 200 件,共重 4000 kg,放于货舱 C4 区;邮件 2 包,重 500 kg,放于货舱 C5 区;另有货物 19 件,共重 7000 kg,分别放于货舱 C1 区 3000 kg、C2 区 2000 kg、C5 区 2000 kg。填写载重配平图表,求:无油重量的重心、起飞时的重心及配平值、着陆时的重心。假设填写配载平衡图表后,又增加了 2 名去成都的旅客,被安排在普通舱内,这 2 名旅客没有交运行李或货物。

大部分空客机型在后部至少有一个散货舱——货舱 5。如货物、行李重量不含集装箱、货板的重量,在配载平衡图表上的 ULD 项中应填写所用的箱、板的重量,并把它们加到货物、行李的重量上,即箱、板的重量被计入业载中。

A310-200 载重表和平衡图的填制方法与 BAe146 是类似的,其结果见图 5-5-3、图 5-5-4。

在配载平衡图表中(或其背面)有对 BOI 做修正的说明或表格,见 A310 配载平衡图,如图 5-5-4 所示。

5.5.1.3　A320-200 飞机实例

A320-200 飞机实例如图 5-5-5 所示。

5.5.1.4　A310 飞机有经停站实例

A310 飞机有经停站实例如图 5-5-6 所示。

5.5.1.5　B737-300 飞机实例

B737-300 飞机实例如图 5-5-7 和图 5-5-8 所示。

5.5.2　Universal Index 型配载平衡图

Universal Index 型配载平衡图表的载重表与 Alignment Bar 型配载平衡图表中的载重表填制方法是一样的。下面重点介绍配载平衡图的填制,如图 5-5-9 所示。

例 5-5-1:B737-800 干使用重量 90600 lb,干使用指数 42,客舱 OA 段 10 名旅客,客舱 OB 段 70 名旅客,客舱 OC 段 60 名旅客,货舱 CPT1 装货物 500 lb,CPT2 装货物 500 lb,CPT3 装行李 500 lb,CPT4 无行李,最大允许起飞重量 172500 lb,最大允许着陆重量 144000 lb,最大无油重量 136000 lb,最大机坪重量 170000 lb,停机坪油量 28380 lb,滑行油量 225 lb,航程油量 19650 lb,成人重量每人 75 kg,完成指数形式的载重配平单,如图 5-5-10~图 5-5-12 所示。

5.5.3　电子舱单

现在一些航空公司已经使用计算机离港系统完成载重平衡工作,把计算机打印的载重平衡单[EDP(Electronic Data Process)Loadsheet]作为飞行计划的一部分提供给机组。不同公司的载重平衡单可能稍有差别。这种载重平衡单可以在最后一分钟快速打印或发出,在最后方案打印或发送到飞机上之前都可以随时调整旅客或货物的数量。因此,这种方法可以免除载重平衡单上的"最后一分钟变化"。

图 5-5-3 是离港系统(DCS)制作的中国国际航空公司厦门(XMN)至天津(TSN)CA1530/12 航班的 AHM 517 格式的载重平衡单,机号 B2536,全经济舱 140 座,机组、乘务组各 4 人。

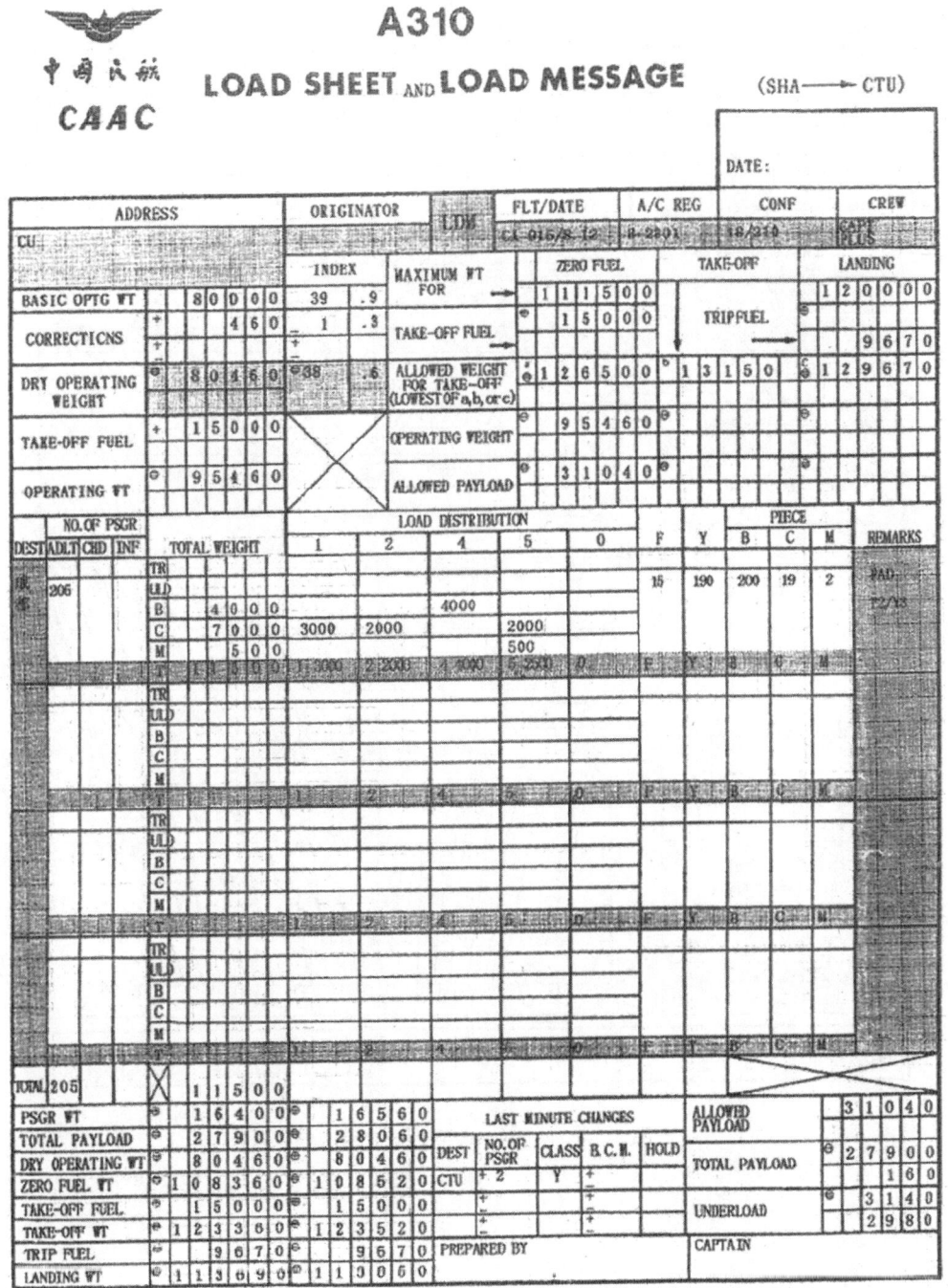

图 5-5-3　A310-200 载重表使用示例

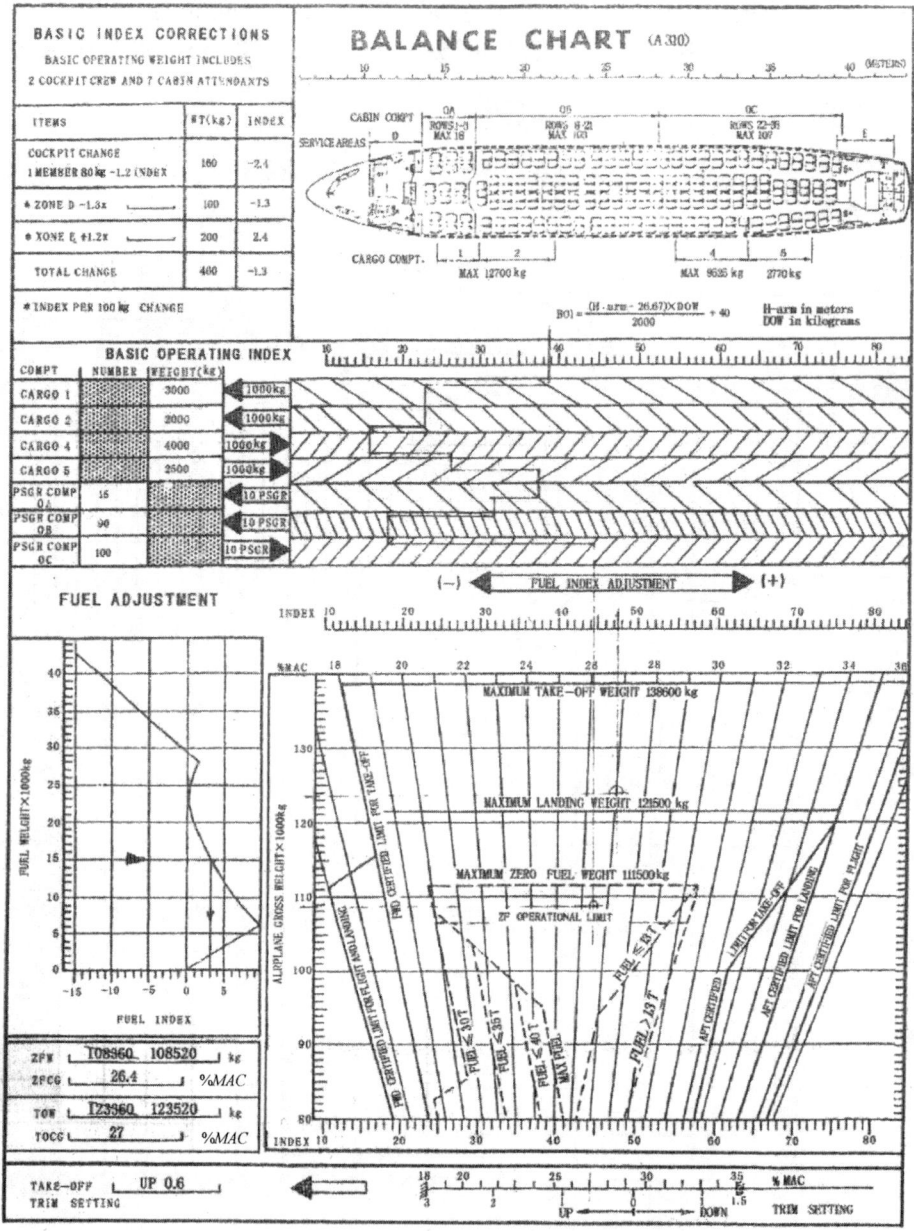

图 5-5-4　A310-200 平衡图使用示例

注:因只增加了 2 名旅客,对重心的影响很小,所以只修改了 *ZFW* 和 *TOW*,没有修正 *ZFCG* 和 *TOCG*。

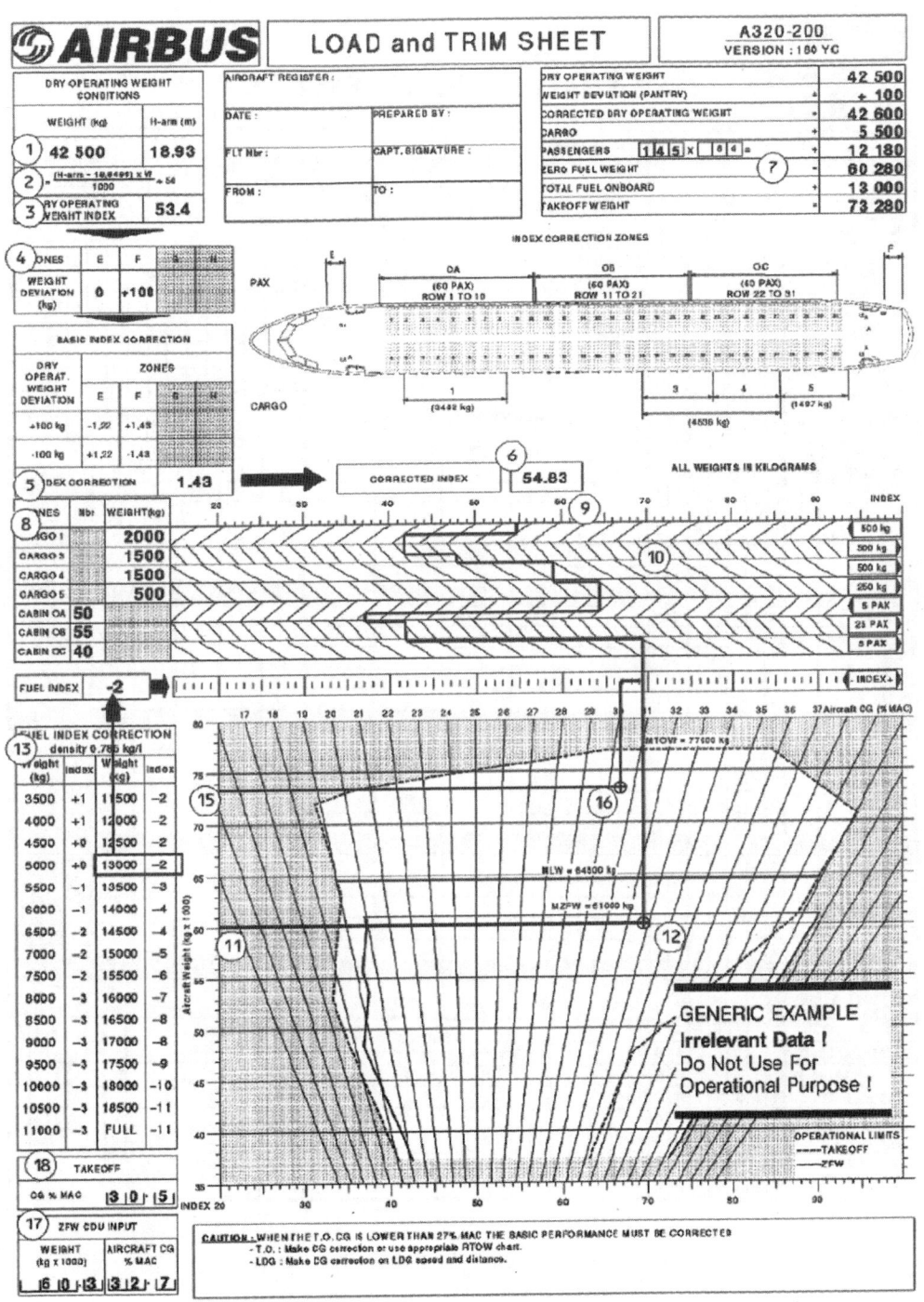

图 5-5-5　A320-200 载重平衡图使用示例

A310
LOAD SHEET AND LOAD MESSAGE

(SHA → TYO)

DATE:

ADDRESS	ORIGINATOR	LDM	FLT/DATE	A/C REG	CONF	CREW
CU			CA 925	B-230	18/210	CAT PLUS 7

			INDEX	MAXIMUM WT FOR →	ZERO FUEL	TAKE-OFF	LANDING
BASIC OPTG WT		80000	39.9		111500		120000
CORRECTIONS	+	460	1.3	TAKE-OFF FUEL →	14500	TRIP FUEL	
	−		− 38.6				8250
DRY OPERATING WEIGHT	⊕	80460	⊕	ALLOWED WEIGHT FOR TAKE-OFF (LOWEST OF a,b, or c)	a 126000 b 13760		c 128250
TAKE-OFF FUEL		14500		OPERATING WEIGHT	94960		
OPERATING WT	⊕	94960		ALLOWED PAYLOAD	31040		

DEST	NO. OF PSGR ADLT / CHD / INF				TOTAL WEIGHT	LOAD DISTRIBUTION 1	2	4	5	0	F	Y	PIECE B	C	M	REMARKS
上海	2	1	1	TR	1050	1050					1	2	6	2	1	PAD P2A3
	18	2		ULD							1	19	17	2	1	
				B	300	300										
				C	2000		2000									
				M	100	100										
				T		1450	2000			F Y B 22 C M						
福冈	3		1	TR	1000		1000				1	2	6	1	1	
				ULD							1	9	8	4	2	
	8	2		B	200		200									
				C	2000	2000										
				M	300		200									
				T		2000	1500			F Y B 14 C M						
东京	20	1	1	TR	1200	1200					3	18	8	2	1	
				ULD							10	133	130	3	2	
	140	3	2	B	2800		1500	1300								
				C	1500				1500							
				M	150				150							
				T		1700	1500	4300	1650	F 13 Y 151 B 138 C M						
				TR												
				ULD												
				B												
				C												
				M												
TOTAL	191	9	5		12600	1200	4950	4800	1650		17	183				

					LAST MINUTE CHANGES					ALLOWED PAYLOAD		31040
PSGR WT	⊕	15640	⊕									
TOTAL PAYLOAD	⊕	28240	⊕	DEST	NO. OF PSGR	CLASS	B. C. M.	HOLD	TOTAL PAYLOAD	⊕	28600
DRY OPERATING WT	⊕	80460	⊕		+		+					
ZERO FUEL WT	⊕	108700	⊕		+		+		UNDERLOAD	⊕	2440
TAKE-OFF FUEL		14500	⊕		+		+					
TAKE-OFF WT	⊕	123200	⊕		+		+					
TRIP FUEL	⊖	8250	⊖	PREPARED BY			CAPTAIN				
LANDING WT	⊕	114950	⊕								

图 5-5-6 有经停站时 A310 载重表使用示例

B737-3YD

BASIC APS WEIGHT		3 0 2 2 0		
APS ADDITIONS	+	2 0 0		
APS DELETIONS	−			
ADJUSTED APS WEIGHT	=	3 0 4 2 0		
RAMP FUEL	+	1 3 7 0 0		
RAMP WEIGHT	=	4 4 1 2 0		
TAXI FUEL	−	2 0 0		
OPERATING WEIGHT	=	4 3 9 2 0		

Priority Address(es)

LOADMESSAGE
ALL WEIGHTS IN KILOGRAMS

Originator Recharge/Date/Time Initials
LDM
Flight A/C Reg Version Crew Date

MAXIMUM WEIGHTS FOR	ZERO FUEL	TAKE-OFF	LANDING
	4 8 3 0 7		5 1 7 0 9
TAKE-OFF FUEL +	1 3 5 0 0	Trip Fuel +	9 0 0 0
ALLOWED WEIGHT FOR TAKE-OFF (Lowest of a,b,c) =	6 1 8 0 7	6 0 9 0 0	6 0 7 0 9
OPERATING WEIGHT			4 3 9 2 0
ALLOWED TRAFFIC LOAD =			1 6 7 8 9

DEST	Passengers			Cab Bag		Total	Distribution - Weight					Remarks	
		No	Weight				1	2	3	4	0	PAX	PAD
	M	120	9240	310	Tr								
	A/F	15	1080		B	2 8 0 0	1000	1600	800				
	Ch	5	150		C	1 4 0 0			800	600		PAX / / PAD / /	
	Inf				M	3 0 0				300			
	T	140	10470	310	T	4 5 0 0	1/1000	2/1600	3/1400	4/900	0/		
	M				Tr								
	A/F				B								
	Ch				C							PAX / / PAD / /	
	Inf				M								
	T				T		1/	2/	3/	4/	0/		
	M				Tr								
	A/F				B								
	Ch				C							PAX / / PAD / /	
	Inf				M								
	T				T		1/	2/	3/	4/	0/		

TOTAL	140	10470	T	4 5 0 0
				3 1 0
PASSENGER				1 0 4 7 0
TOTAL TRAFFIC LOAD	=			1 5 2 8 0
ADJUSTED APS WEIGHT	+			3 0 4 2 0
ZERO FUEL WEIGHT	=			4 5 7 0 0
4 8 3 0 7				
TAKE-OFF FUEL	+			1 3 5 0 0
TAKE-OFF WEIGHT	=			5 9 2 0 0
6 0 9 0 0				
TRIP FUEL	−			9 0 0 0
LANDING WEIGHT	=			5 0 2 0 0
5 1 7 0 9				

ALLOWED TRAFFIC LOAD =		1 6 7 8 9	BALANCE CONDITIONS
TRAFFIC LOAD		1 5 2 8 0	ZERO FUEL
			19 %MAC
UNDERLOAD =		1 5 0 9	TAKE-OFF
			16.3 %MAC

LAST MINUTE CHANGES

DEST	ITEM	CABIN	+/−	WEIGHT	INDEX UNITS

ZFW 45700 KG	MAX CERT ZERO FUEL WEIGHT 48307 KG	LAST MINUTE CHANGES	+	
TW 59400 KG	MAX CERT TAXI WEIGHT 81500 KG		−	
TOW 59200 KG	MAX CERT TAKE-OFF WEIGHT 61235 KG			
LW 50200 KG	MAX CERT LANDING WEIGHT 51709 KG	TOTAL LAST MINUTE CHANGES	=	

INDEX UNITS = WT (KG) X [ARM (IN) -648.5] / 30,000
FOR APS INDEX, ADD 40 UNITS

AUG '89 REV C

图 5-5-7　B737-300 载重表使用示例

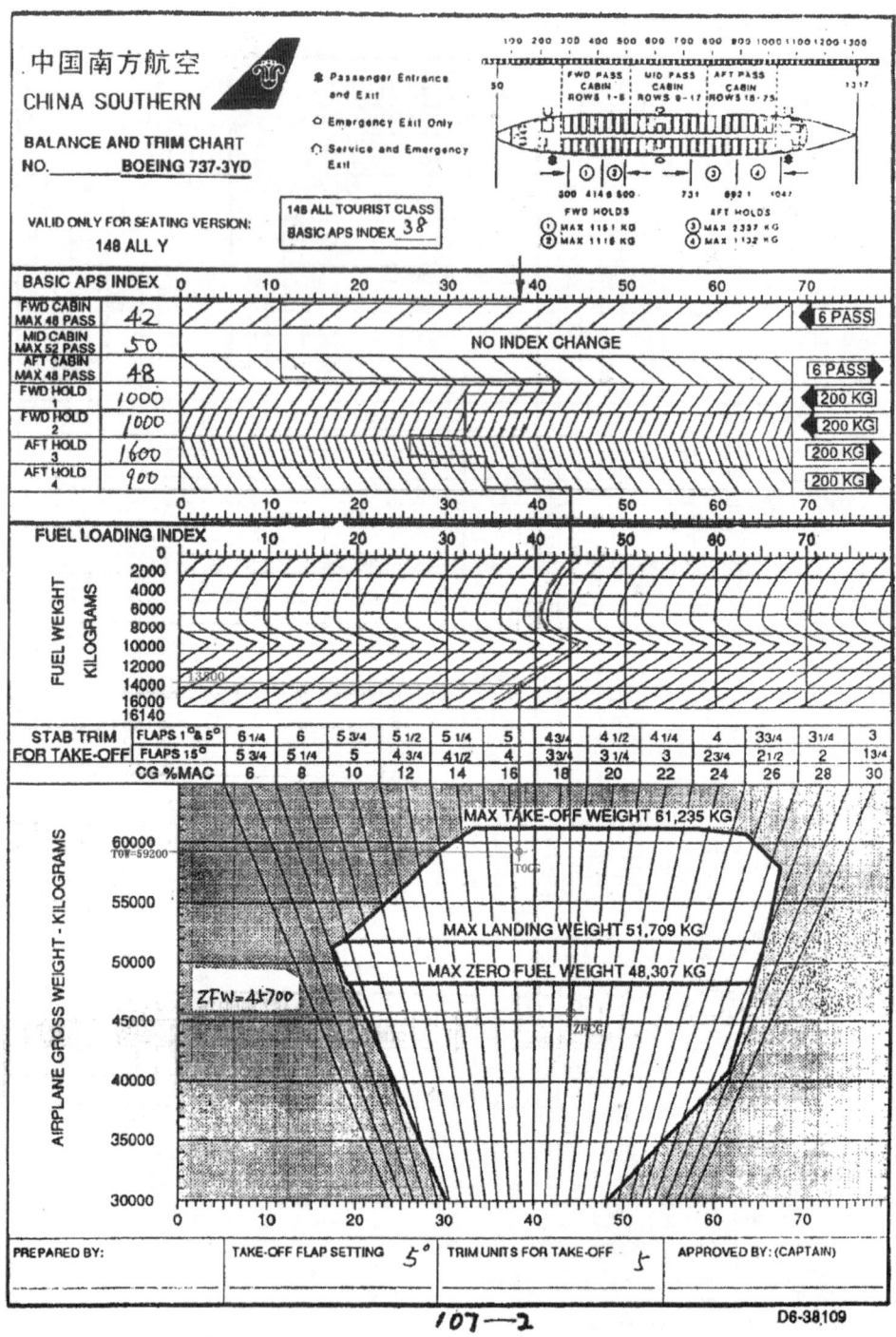

图 5-5-8　B737-300 平衡图使用示例

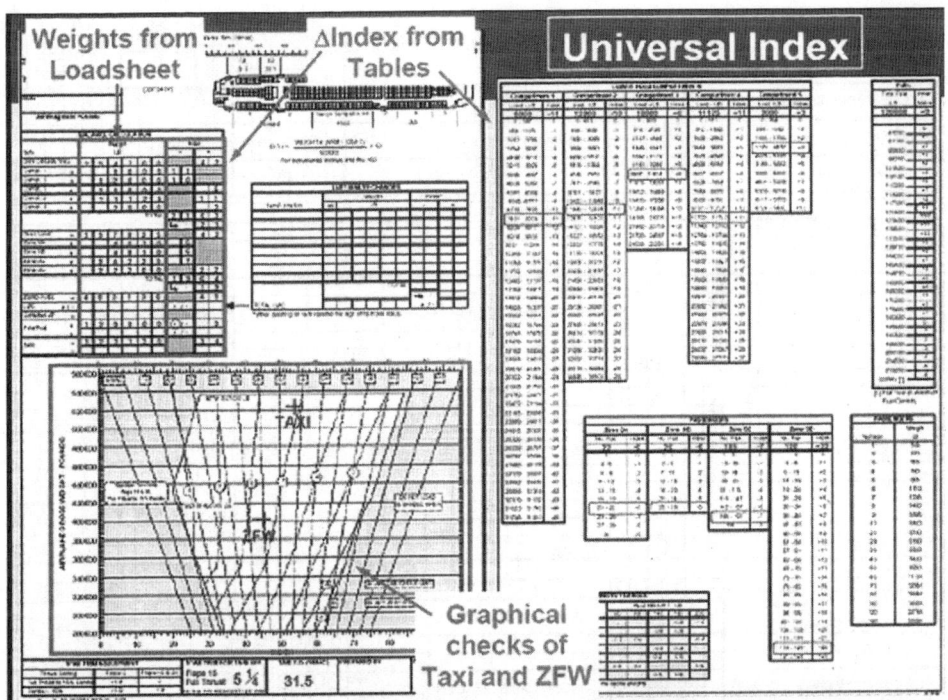

图 5-5-9　Universal Index 型配载平衡图表

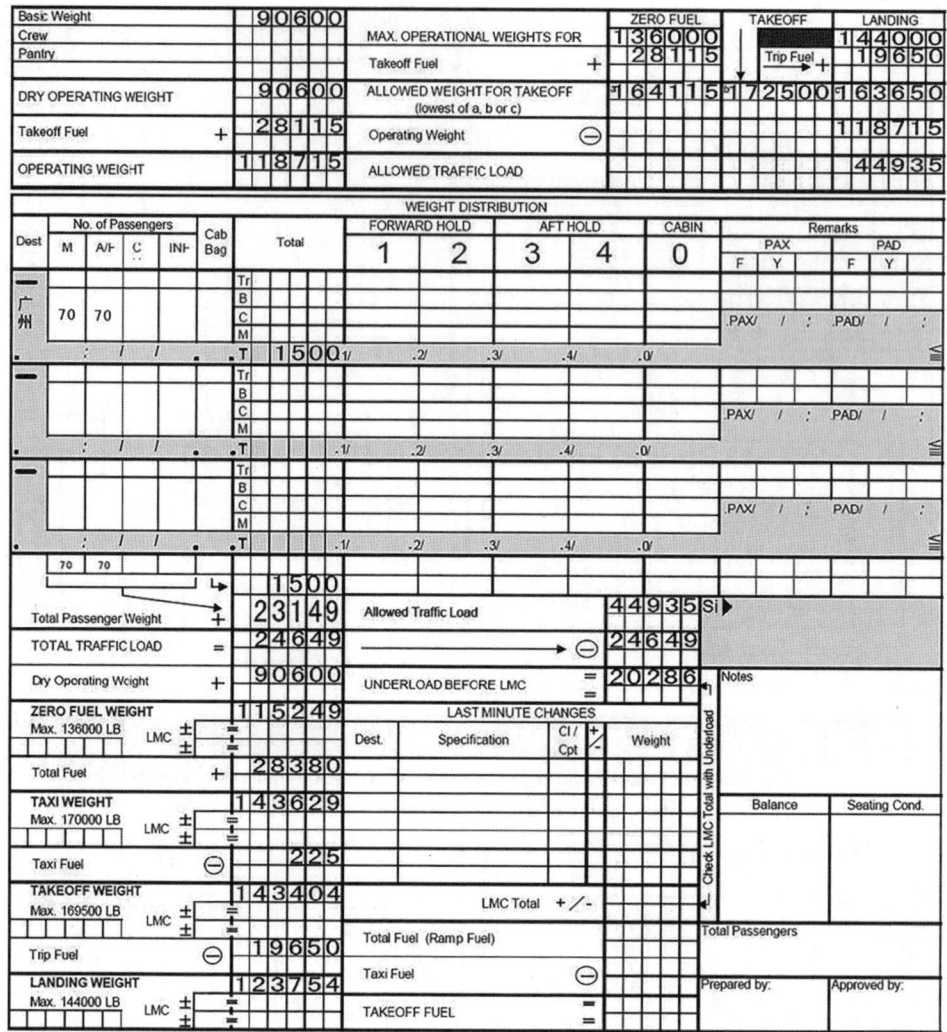

图 5-5-10　B737-800 指数形式的载重表

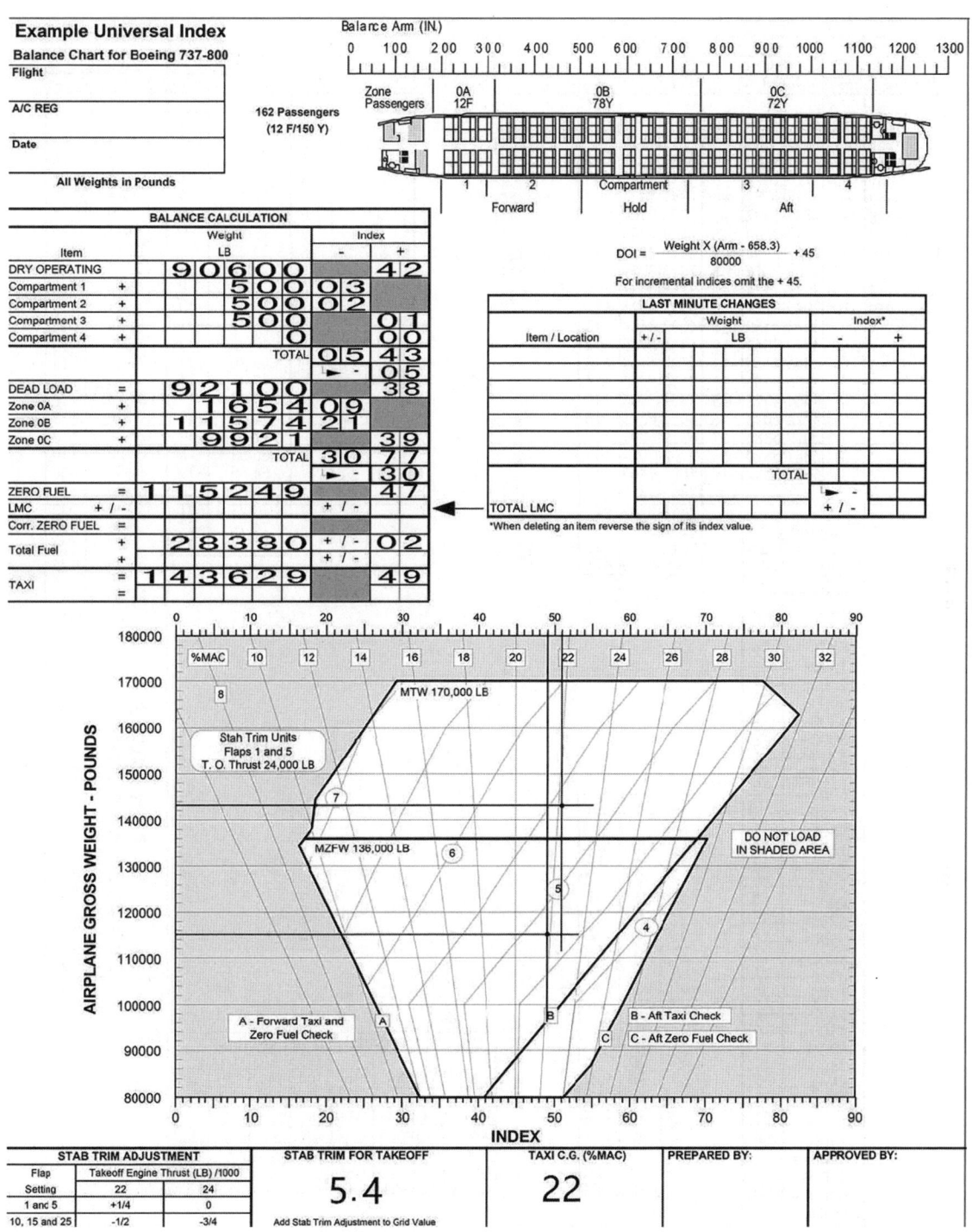

图 5-5-11　B737-800 指数形式的平衡图

LOWER HOLDS							
Compartment1		Compartment2		Compartment3		Compartment4	
Weight-LB	Index	Weight-LB	Index	Weight-LB	Index	Weight-LB	Index
0-94	0	0-153	0	0-191	0	0-94	0
95-283	-1	154-461	-1	192-574	+1	95-284	+1
284-472	-2	462-769	-2	575-957	+2	285-474	+2
473-661	-3	770-1077	-3	958-1340	+3	475-664	+3
662-850	-4	1078-1385	-4	1341-1723	+4	665-854	+4
851-1038	-5	1386-1692	-5	1724-2106	+5	855-1044	+5
1039-1227	-6	1693-2000	-6	2107-2489	+6	1045-1234	+6
1228-1416	-7	2001-2308	-7	2490-2872	+7	1235-1424	+7
1417-1605	-8	2309-2616	-8	2873-3255	+8	1425-1614	+8
1606-1794	-9	2617-2924	-9	3256-3638	+9	1615-1685	+9
1795-1959	-10	2925-3232	-10	3639-4021	+10		
		3233-3539	-11	4022-4404	+11		
		3540-3847	-12	4405-4786	+12		
		3848-4155	-13	4787-5169	+13		
		4156-4463	-14	5170-5552	+14		
		4464-4771	-15	5553-5935	+15		
		4772-5078	-16	5936-6318	+16		
		5079-5386	-17	6319-6701	+17		
		5387-5694	-18	6702-7084	+18		
		5695-5887	-19	7085-7467	+19		
				7468-7850	+20		
				7851-8233	+21		
				8234-8616	+22		
				8617-8999	+23		
				9000-9009	+24		

PASSENGERS					
Zone 0A		Zone 0B		Zone 0C	
No.Pax	Index	No.Pax	Index	No.Pax	Index
1	-1	1	0	1-2	+1
2	-2	2-5	-1	3	+2
3	-3	6-8	-2	4-5	+3
4	-4	9-11	-3	6	+4
5	-5	12-15	-4	7-8	+5
6	-6	16-18	-5	9	+6
7	-7	19-22	-6	10-11	+7
8-9	-8	23-25	-7	12-13	+8
10	-9	26-28	-8	14	+9
11	-10	29-32	-9	15-16	+10
12	-11	33-35	-10	17	+11
		36-39	-11	18-19	+12
		40-42	-12	20	+13
		43-46	-13	21-22	+14
		47-49	-14	23	+15
		50-52	-15	24-25	+16
		53-56	-16	26	+17
		57-59	-17	27-28	+18
		60-63	-18	29	+19
		64-66	-19	30-31	+20
		67-69	-20	32-33	+21
		70-73	-21	34	+22
		74-76	-22	35-36	+23
		77-78	-23	37	+24
				38-39	+25
				40	+26
				41-42	+27
				43	+28
				44-45	+29
				46	+30
				47-48	+31
				49	+32
				50-51	+33
				52	+34
				53-54	+35
				55-56	+36
				57	+37
				58-59	+38
				60	+39
				61-62	+40
				63	+41
				64-65	+42
				66	+43
				67-68	+44
				69	+45
				70-71	+46
				72	+47

FUEL INDEX TABLE	
Total Fuel LB	Index Value
7000	0
9700	+1
11300	+2
12400	+3
13400	+4
14300	+5
15000	+6
15700	+7
16300	+8
16900	+9
18500	+10
19800	+9
21100	+8
22300	+7
23500	+6
24900	+5
26200	+4
27500	+3
28900	+2
30300	+1
31700	0
33100	-1
34500	-2
35900	-3
37400	-4
38500	-5
40200	-6
41500	-7
42700	-8
43900	-9
45100	-10
46300	-11
48652[1]	-12

[1]Full fuel at maximum fuel density.

PASSENGERS	
No.	Weight LB
1	185
2	370
3	555
4	740
5	925
6	1110
7	1295
8	1460
9	1665
10	1850
12	2220
20	3700
30	5550
40	7400
50	9250
60	11100
70	12950
72	13320
78	14430

LAST MINUTE CHANGES					
	ADD WEIGHT(LB)				
LOCATION	50	100	140	180	200
Flight Deck			-1.4		
Fwd Galley Complex	-0.4	-0.7			-1.4
Door 1 Attendant Sta.			-0.9	-1.2	
Door 4 Attendant Sta.			+0.9	+1.2	
Aft Galley Complex	+0.3	+0.7			+1.4

Reverse the index sign when deleting the above weights.

图 5-5-12　B737-800 指数形式的平衡图的指数表

```
AIR CHINA
LOADSHEET                    CHECKED                APPROVED         EDNO
ALL WEIGHTS IN KG                                                    01


FROM/TO FLIGHT              A/C REG VERSION          CREW   DATE      TIME
XMN TSN CA1530/12           B2536   Y140             4/4/0  12JUL97   1029  (0)
                           WEIGHT                   DISTRIBUTION
LOAD IN COMPARTMENTS         3569   1/1017           4/2552  0/0            (1)
PASSENGER/CABIN BAG          5768   77/0/6/1                 TTL  84  CAB 0 (2)
MAX TRAFFIC PAYLOAD         14223   PAX 83                                  (3)
TOTAL TRAFFIC LOAD           9337   BLKD 3                                  (4)
DRY OPERATING WEIGHT        33404
ZERO FUEL WEIGHT ACTUAL     42741   MAX              47627   L    ADJ       (5)


TAKE OFF FUEL              10200   （起飞油量，不含滑出油量）
TAKE OFF WEIGHT   ACTUAL   52941   MAX               58968        ADJ       (6)


TRIP FUAL                  6200    （航程油量）
LANDING WEIGHT   ACTUAL    46741   MAX               51709        ADJ       (7)


BALANCE AND SEATING CONDITIONS              LAST MINUTE CHANGES
DOI      34.46     DLI      44.34     DEST        SPEC  CL/CPT  + -  WEIGHT (8)
LIZFW    37.53     MACZFW   15.74                                          (9)
LITOW    36.58     MACTOW   15.58                                          (10)
LILAW    34.33     MACLAW   14.32                                          (11)
                  DLMAC    19.64                                           (12)
STAB TO   5.1 UP                      （起飞时配平片调整到 5.1 UP 位置）
SEATING                               （就座状态）
OA/2   OB/59   OC/22                  （旅客在客舱 OA、OB、OC 各段就座人数）
UNDERLOAD BEFORE LMC       4886              LMC   TOTAL   + -              (13)
LOADMESSAGE AND CAPTAINS INFORMATION BEFORE LMC
BW   33244 KGS        BI    34.46     （基本重量和基本指数）
TZFW/TSN   33404 KGS                  （到天津旅客下机、货物卸载后的无油重量）
LDM                                   （装载信息，以下每行中各项用 "." 分隔）
CA1530/12. B-2536. Y140. 04/04                                            (14)
- TSN. 77/0/6/1. 0. T3569. 1/1017. 4/2552                                 (15)
. PAX/83. PAD/0                                                           (16)
SI （此处写补充信息，如机上有担架、残疾人的轮椅等，无补充信息则空白）

   BW   33244  BI 34.46
   TSN FRE   2883  POS 0   BAG 686   TRA 0 （到天津的货物、邮件、行李及过站重量）
   =
   07121029/00000000   XMN   006
```

图 5-5-13　电子舱单

图 5-5-13 中:(0)~(16)行的符号及数字的含义是:

(0)CREW 4/4/0 表示机组人数/客舱中的男乘务员人数/女乘务员人数。

(1)货舱中的总货载及在各舱位中的分配数字。

(2)旅客及客舱中的行李总重 5768 kg,旅客总人数 84,其中 77 名成人、6 名儿童、1 名婴儿,客舱中的行李重量为 0,婴儿不占座位("77/0/6/1"的写法表示成年男性旅客/成年女性旅客/儿童/婴儿)。

(3)最大允许业载 14223 kg,占座位的旅客 83 人,此行可能出现:SOC nn/nn(表示各舱中被货物、行李、邮件占据的座位数)。

(4)实际总业载为 9337 kg(=3569+5768),BLKD 3 表示不是为旅客或死重(行货邮)配备的座位数。

(5)实际无油重量、最大无油重量及由于最后一分钟变化(LMC)修正的无油重量,注意:本行中的字母 L 是 Limitation 的首字母,表示本航班的最大允许业载受最大无油重量限制,如果受最大允许起飞重量或最大允许着陆重量限制,L 将出现在标号为(6)或(7)的行中相应位置上。实际上 L 标示的是最大无油重量、最大允许着陆重量限制的起飞重量及本场性能限制的最大起飞重量中最小的一个。

(6)实际起飞重量、最大允许起飞重量及由于最后一分钟变化(LMC)修正的起飞重量。

(7)实际着陆重量、最大允许着陆重量及由于最后一分钟变化(LMC)修正的着陆无油重量。

(8)DOI 是干使用指数,DLI(Dead Load Index)是装完货物、行李、邮件尚未装客时的指数;DEST、SPEC、CL/CPT、+、-、WEIGHT 表示在其下面分别填写最后增(+)减(-)的旅客或货物品名(SPEC)所到的目的地(DEST)、所在的客舱(CL)/货舱(CPT)位置及重量。

(9)无油重量的指数及对应的平均气动弦长(MAC)的百分数。

(10)起飞重量的指数及对应的平均气动弦长(MAC)的百分数。

(11)着陆重量的指数及对应的平均气动弦长(MAC)的百分数。

(12)装完货物、行李、邮件尚未装客时的重量对应的平均气动弦长(MAC)的百分数,即 DLI 对应的平均气动弦长(MAC)的百分数。(注:DLMAC 与 MACDLW 含义相同。)

(13)最后一分钟变化(LMC)前的空余吨位及最后增减的总量。

(14)航班号 CA1530/12,飞机注册号 B2536,经济舱 140 座,机组 4 人/乘务组 4 人。

(15)到天津的成年旅客 77 名/儿童 6 名/婴儿 1 名/客舱中行李 0 kg,总货载 3569 kg,货舱 1 为 1017 kg,货舱 4 为 2552 kg。

(16)占座的旅客数 83 人,需减载时可优先减下去的人数为 0。

由离港系统(DCS)制的计算机装载配平单上用到的缩写符号的英文解释如下:

BI—Basic Index

BLKD—Blocked

BLKD—Blocked-Fitted Seats not Available for Passengers or Deadload

BW—Basic Weight

DCS—Departure Control System

DLI—Dead Load Index

EDNO—Edition Number

EDP—Electronic Data Process

LDM—Loadmessage

LILAW—Loaded Index at Landing Weight

LIR—Loading Instruction/Report

LITOW—Loaded Index at Take-off Weight

LIZFW—Loaded Index at Zero Fuel Weight

MACDLW—Dead Load *MAC*

MACLAW—*MAC* at Landing Weight

MACTOW—*MAC* at Take-off Weight

MACZFW—*MAC* at Zero Fuel Weight

PAD—Passengers Available for Disembarkation：no firm Booking，Low Priority

SI—Supplementary Information

SOC—Seats Occupied by Cargo，Baggage and/or Mail per Class

SPEC—Specification

STABTO—Stabilizer Trim Setting at Take-off

UNDLD—Underload

当公司某种机型有很多架时,如果一架飞机的重量与机队的平均重量相差不超过最大着陆重量的±0.5%,并且基本使用重量的重心与机队的平均重心相差不超过±0.5%*MAC*,可以对该机使用机队平均重量和平均重心进行装载配平计算,见图 5-5-14。

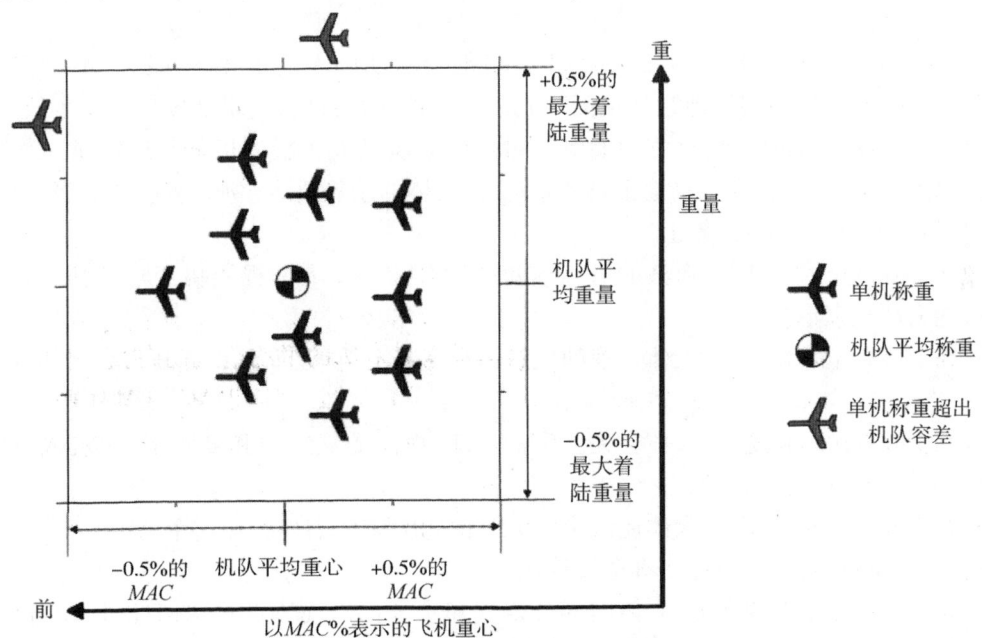

图 5-5-14　机队平均重量和平均重心位置的确定

第6章

供氧分析

随着科学技术的发展,现代民航客机的升限已经超过人类呼吸系统所能承受的生理极限。随着气压高度的增加,空气中的氧气量会逐渐减少,人体的呼吸系统所吸入的氧气量不足以维持正常生理功能。

①在 5000 ft 的高度上,人的眼睛的夜视功能会有功能上的退化。在这个高度夜间飞行时,仪表和航图容易被看错,地貌和地面灯光也容易被误判,这将会增加发生飞行事故的概率。

②在 10000 ft 的高度上,空气中的氧气量仅仅能满足人大脑的最低氧气需求。如果在没有任何保护的情况下,在这个高度上待 4 h 左右,身体的缺氧反应(例如刺痛和头痛)可能不会很明显,但是人将丧失判断能力。

③在 14000 ft 的高度上,血液的氧气饱和度仅有 85%。人的视线将变暗,判断力、记忆力和思考能力严重衰退。

④在 16000 ft 的高度上,人体血液的氧气饱和度降至 79%,而且将逐渐丧失大部分的生理功能。

⑤在 18000 ft 的高度上,人体会失去生命功能,血液的氧气饱和度降至 73%,人可能感觉到头疼,大概 30 min 后就会死亡。

⑥在 20000 ft 的高度上,人体血液中的氧气饱和度降至 71%。在这个高度上,人的意识只能维持 5~15 min,超过这个时间将会导致死亡。

⑦在 25000 ft 以上的高度,人体内部分氮气和水分都以气体形式逸出体外,使身体浮肿。同时人的意识只能维持 3~6 min,死亡紧随其后。

现在的飞机制造厂商在飞行高度超过 10000 ft 的飞机上都采用了密闭增压舱来保障旅客和机组人员的生命安全,并且不论实际的巡航高度是多少,座舱增压维持的座舱高度在 8000~15000 ft。在这种情况下,不需要用供氧设备也能使乘客和机组人员有一个舒适的环境。为了防止在高于 10000 ft 的高度飞行时发生座舱失压,飞机上安装了氧气系统,以保证意外情况下机组和旅客的生命安全。

我国西部地区多高原和山区,航路最低安全高度比较高,尤其是涉及飞越青藏高原的航线,航路最低安全高度超过 6000 m。航空公司在运行高原航线时,必须评估飞机在最不利的时刻发生客舱失压是否可以下降到 3000 m 以下;然而,为满足安全越障的要求,通常无法下降到不需要为旅客供氧的高度,这就要求航空公司必须给出客舱失压后的机组操作程序,确保在满足供氧要求的条件下能够飞至适宜的机场着陆。航行中发生座舱失压的概率很小,客舱失压的潜在危险却很大。因此,飞机在高原航线运行时,必须考虑座舱失压后的供氧问题。

6.1 机载氧气系统介绍

目前国内外民航飞机氧气系统中的氧气源根据储存方式的不同分为两种:一种是氧气瓶氧源,另一种是机载产氧源。其中,氧气瓶氧源又分为气态氧源和液态氧源。气态氧源制作方便,维护和使用比较简单,但是单位体积内贮存的氧气量比较少;而液态氧源体积小,重量轻,但贮存困难,维护工作较为复杂。机载产氧作为一种新型氧源,成为未来发展的趋势,根据制氧的方式不同分为分子筛制氧、化学制氧等多种方式。不过目前绝大多数民航客机采用氧气瓶储存方式的气态氧源以及化学制氧方式的氧气源。

6.1.1 旅客氧气瓶供氧

氧气贮存在高压氧气瓶或者低压氧气瓶里。高压氧气瓶是由热处理合金或在外表面包着金属丝制成,以抗破损。高压氧气瓶外表颜色为绿色,可承受的最大压力为 140 kg/cm²;低压氧气瓶外表颜色为淡黄色,可承受的最大压力为 31.5 kg/cm²。

以 B757-200 为例,该机型飞机上一般装有 11 个氧气瓶。在 21.1 ℃时每个瓶中装有瓶压为 1850 psi、体积为 114 ft³ 的标准氧气量。这些氧气瓶放置在后货舱地板的架子上。每个氧气瓶组件包括一个慢开的截止阀、一个压力表、一个用于瓶压高于 2475～2775 psig(表压力值单位,合 17065～19133 kPa)时排出瓶内气体的过压保护装置。这个过压保护装置连接到飞机蒙皮上一个机身外通风口的管内(见图 6-1-1)。当氧气瓶压力达到限制值时,氧气瓶可以通过管路自动向机身外排出多余的氧气。氧气组件位于旅客和乘员座位的上方。飞机上的每个厕所的夹层上也有一个氧气组件。每个组件都有几个供氧面罩组。打开组件的活门后,氧气面罩会自动脱落。这个组件的活门由电动装置控制,开关一般位于驾驶舱右侧的混合电子设备面板上,或者由位于旅客头顶面板上的一个手动操作开关控制。此外,飞行员也可以通过手动开关打开活门。

旅客氧气系统由 2 个连续流量控制组件来打开和控制进入旅客氧气系统的流量(如图 6-1-2 所示)。一个是电控气动式持续流量控制组件,另一个是气动持续流量控制组件,两者平行相连,每个控制组件可以独立地支持旅客氧气系统,这就增加了系统的可靠性。这些控制组件安装在氧气瓶架前面底端的一个支架上。持续流量控制组件由一个减压装置、一个驱动阀门、一个主流量控制阀门三部分组成。其中,驱动阀门通过一个真空膜盒或者电控的驱动器来控制触发供氧,一般情况下,当座舱高度达到 4250 m 时,真空膜盒机构使驱动阀门打开,开始供氧。

空客机型的旅客氧气瓶供氧与波音机型的供氧方式类似,其氧气瓶供氧系统如图 6-1-3 所示。旅客氧气面罩可通过图 6-1-4 所示的驾驶舱内面板的"PASSENGER"键控制。氧气瓶内的压力可通过图 6-1-5 所示位置显示给驾驶舱内人员。

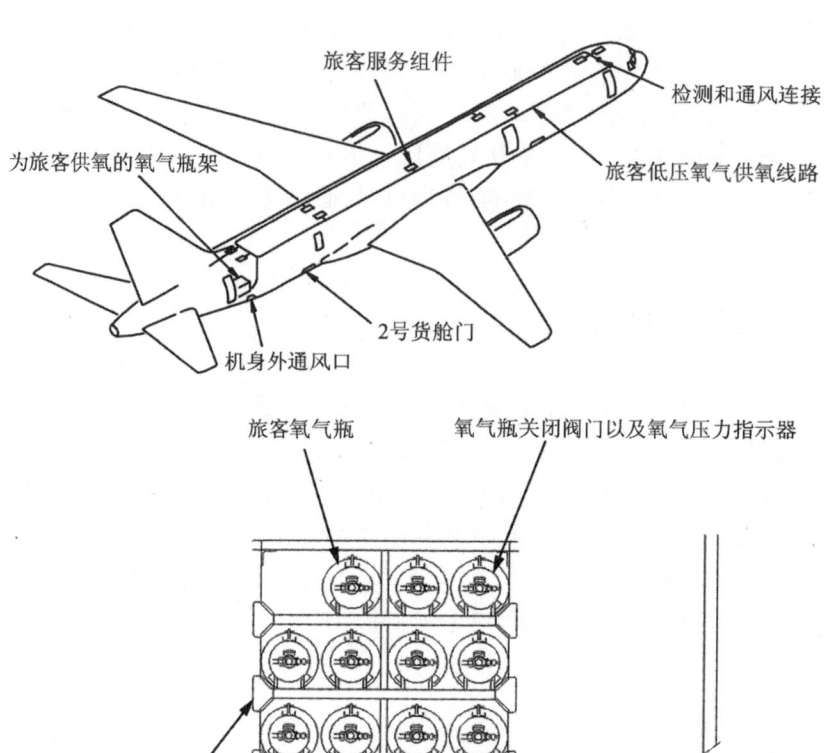

图 6-1-1　旅客氧气瓶架

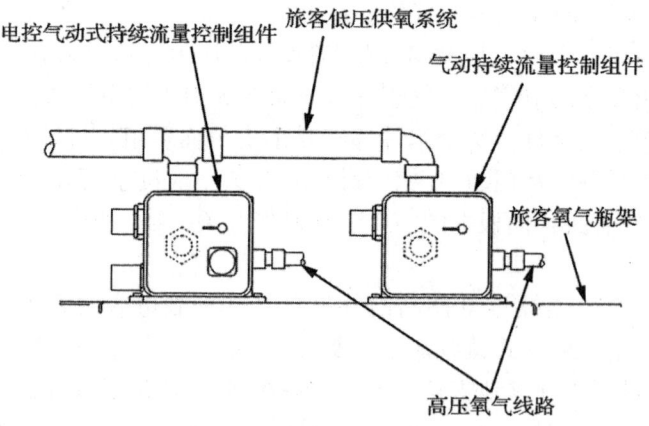

图 6-1-2　连续流量控制组件

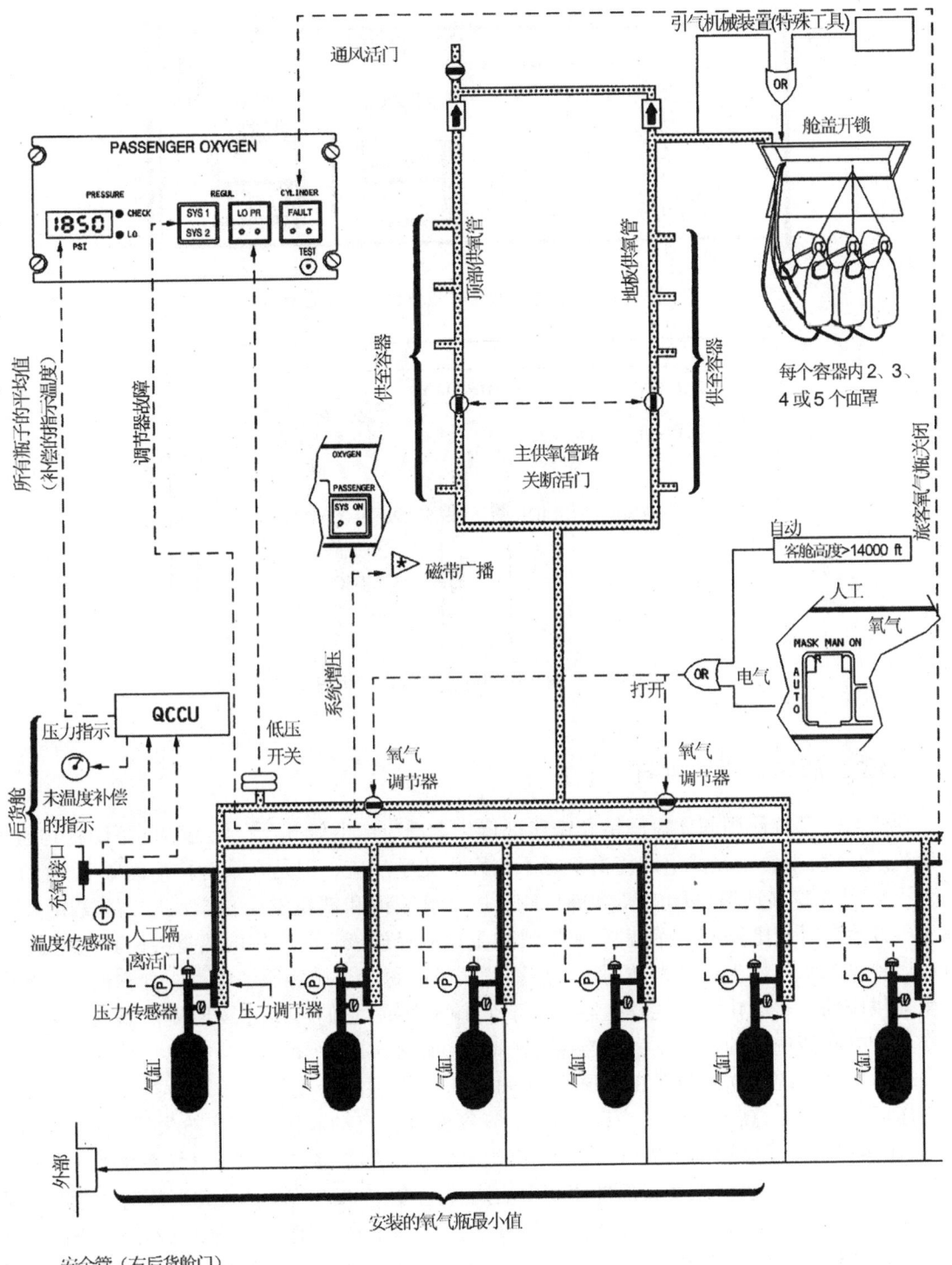

图 6-1-3 空客机型旅客氧气瓶供氧系统

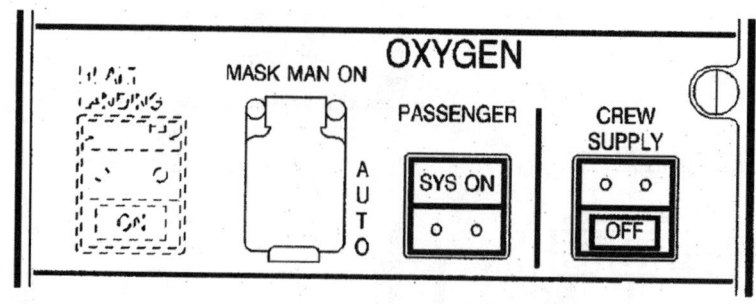

图 6-1-4　空客机型旅客氧气面罩控制面板

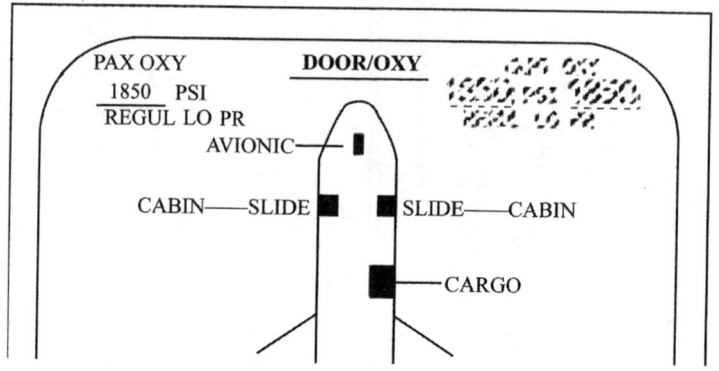

图 6-1-5　空客机型旅客氧气瓶压力显示

6.1.2　旅客化学式供氧

化学氧气系统使用固体氯酸钠和铁粉进行化学反应产生氧气。在低温情况下两种物质不发生化学反应,即使是在严重的撞击下也不会发生化学反应。但是在乘客拉下面罩后,撞针撞击发火帽产生高温环境(2040 ℃/4000 ℉以上),氯酸钠和铁粉发生化学反应,释放出氧气。在氧气发生器外壳上贴有一圈特制的带子,发生反应释放氧气时产生的热量会使其变色,这样就可以检查该氧气发生器是否被用过。旅客氧气是由每一个旅客服务组件中独立的化学氧气发生器供应的。每个化学氧气发生器连接 4 个氧气面罩,它们可以连续地提供氧气。乘务员工作位上方和盥洗室的氧气发生器连接 2 个面罩,如图 6-1-6 所示。

化学氧气系统独立安装在座椅顶部,为每个旅客提供一个化学氧气发生器。飞机发生客舱失压后,客舱压力高度达到某一设定值时,氧气面罩会自动放出。当客舱失压,氧气面罩放出后,旅客拉下面罩就启动了氧气发生器,产生氧气。使用氯酸盐的化学氧气系统的工作压力很低,可以避免高压氧气瓶及其部件使用中的高压危险,在发生坠机、火灾事故时化学氧气系统比氧气瓶供氧系统更加安全。同时,化学氧气系统的维护费用比氧气瓶供氧系统低很多,使用氯酸盐的化学氧气发生器的储存寿命至少是 12 年,使用过氧化物的化学氧气发生器的储存寿命是 5 年。

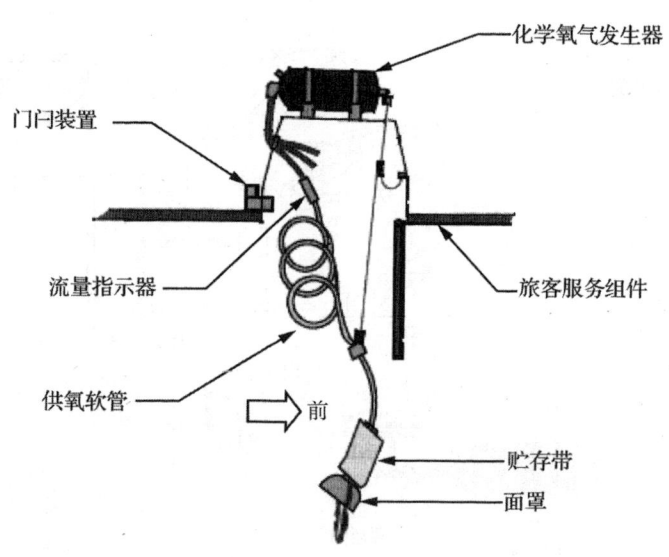

图 6-1-6　化学氧气系统示意图

化学氧气发生器有一个压力电门,当客舱压力高度接近 14000 ft 或舱尾的旅客氧气开关打开时,压力电门会自动启动氧气系统。当化学氧气发生器启动时,PASS OXY ON 灯就会亮起。氧气系统启动后,氧气面罩会自动从存放箱中脱落。任何一个氧气面罩被拉下,都会启动氧气系统。拉下一个氧气面罩会促使一个组件的全部氧气面罩放下,并向所有氧气面罩提供 100% 的氧气。如果氧气发生器发生反应,而氧气面罩箱没有打开,则可以人工将面罩拉出。

化学氧气发生器产生的氧气流量与时间有关,氧气流量随时间(随化学反应的进行)而减小,和供氧压力与客舱高度无关,即无论在什么高度座舱失压、化学氧气发生器开始工作,所产生的氧气流量随时间的变化都是一样的。

6.1.3　机组氧气系统

飞行员的氧气是由圆柱形的高压氧气瓶提供的,如图 6-1-7 所示,机组氧气系统可以通过图 6-1-8 所示的机组氧气系统进行控制。

氧气的压力可以从后顶板的压力指示器中读出。机组氧气系统压力可高达 1850 psi,在氧气瓶和氧气面罩之间的管道上有一个减压器,以便给调节者一个较小压力的氧气流量,如图 6-1-9 所示。氧气流量可以用固定在氧气面罩上的调节器进行调节,当推动 NORMAL/100% 的操纵杆时,调节器将由供应空气和氧气的混合气调节到供应 100% 的氧气。氧气面罩/调节器存放在紧靠每个机组工作位的箱内。使用时,用拇指和食指捏紧释放手柄并取出面罩,捏住释放手柄,面罩头罩自动充气;将头罩套在头上并释放手柄,头罩收紧,面罩将与头部和脸紧贴。

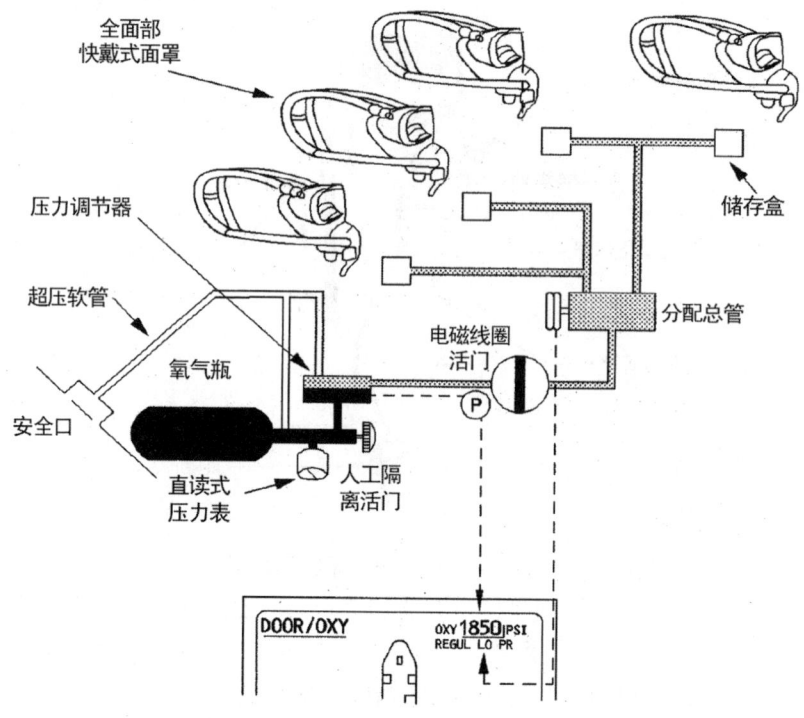

图 6-1-7　空客机型机组氧气系统

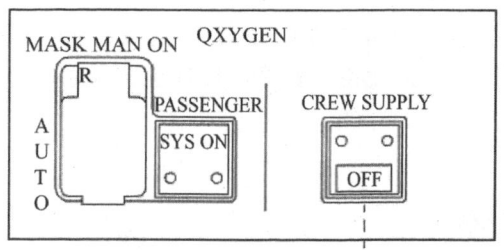

图 6-1-8　空客机型机组氧气系统控制

6.1.4　机组便携式氧气设备

便携式防护性呼吸保护装置(PBE)可供机组人员灭火或进入浓烟区域时使用,如图 6-1-9 所示。它们通常装载在旅客客舱的舱前段和舱尾段,放置在座椅上方,在使用时其可以提供 15～20 min 的氧气供应。急救箱和备用的便携式氧气设备安装在旅客座舱的适当位置。圆柱形氧气瓶上有压力表、压力调节器和开/关阀门。圆柱形氧气瓶被增压到 1800 psi,在这个压力和 70 ℉(21 ℃)的情况下,圆柱形氧气瓶可以容纳 4.25 ft³(120 L)的自由氧。一个氧气瓶有两个连续流量的供氧口,机上人员根据具体需要进行选择。一个出口的氧气流量是 2 L/min,供人员在飞机上行走时使用。另一个出口的氧气流量是 4 L/min,用于急救。各出口的供氧持续时间可以用容量除以流量来计算。

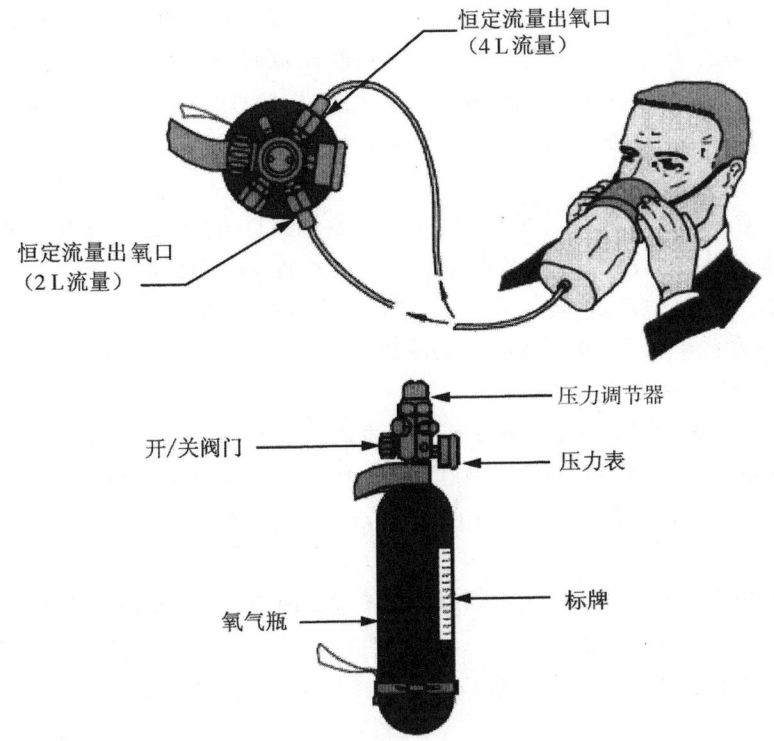

图 6-1-9 便携式防护性呼吸保护装置

6.2 供氧相关规章要求

6.2.1 CCAR-121 有关供氧的要求

第 121.329 条 涡轮发动机飞机用于生命保障的补充供氧要求

（a）在运行涡轮发动机驱动的飞机时，每个合格证持有人应当根据本条的规定，在飞机上配备生命保障氧气和分配设备以供使用：

（1）所提供的氧气量应当至少是为遵守本条（b）和（c）款所必需的量。

（2）每一特定运行所需要的生命保障和急救用氧气量，应当根据座舱气压高度和飞行持续时间来确定，这些座舱气压高度和飞行持续时间要与为该运行和航路制定的运行程序一致；

（3）对具有增压座舱的飞机，氧气量应当根据座舱气压高度和下列假设来确定：座舱增压故障发生在供氧需求临界的飞行高度或者飞行中某点，然后飞机按照飞机飞行手册中规定的应急程序，在不超过其使用限制的情况下，下降到一个允许顺利结束本次飞行的飞行高度；

（4）发生了这种故障之后，座舱气压高度被视为与飞行高度相同，除非能证明，座舱或者增压设备任何可能的故障均不会导致座舱气压高度等于飞行高度。在这种情况下，可将达到的最大座舱气压高度作为审定或者确定供氧量的基础。

（b）机组成员。每个合格证持有人应当按照下列要求为机组成员提供氧气：

（1）在座舱气压高度 3000 米（10000 英尺）以上至 3600 米（12000 英尺）（含）时，应当对在驾驶舱内值勤的每一飞行机组成员提供氧气,并且他们也应当用氧,如果在这些高度上超过30分钟,则对于30分钟后的那段飞行应当对其他机组成员提供氧气;

(2)在座舱气压高度3600米(12000英尺)以上,应当对在驾驶舱内值勤的每一飞行机组成员提供氧气,并且他们也应当用氧,在此高度上整个飞行时间内,应当对其他机组成员提供氧气。

(3)当要求某一飞行机组成员用氧时,他应当连续用氧,除非为执行其正常任务需要除去氧气面罩或者其他氧气分配器。对那些处于待命状态的或者在完成此次飞行前肯定要在驾驶舱内值勤的后备飞行机组成员,视为本款第(1)、(2)项所述的其他机组成员。如果某一后备飞行机组成员不在待命状态,并且在剩下的一段飞行中将不在驾驶舱内值勤,则就补充氧气要求而言,可以将其视为一名旅客。

(c)旅客。除经局方批准外,每个合格证持有人应当按照下列要求为旅客提供氧气:

(1)对于座舱气压高度3000米(10000英尺)以上至4000米(13000英尺)(含)的飞行,如果在这些高度上超过30分钟,则对于30分钟后的那段飞行应当为10%的旅客提供足够的氧气;

(2)对于座舱气压高度4000米(13000英尺)以上的飞行,在此高度上整个飞行时间内为机上每一旅客提供足够的氧气。

(d)对于在特定区域运行符合本条(c)款存在困难的,经局方批准,可以按照以下要求实施运行:

(1)对于座舱气压高度3000米(10000英尺)以上至4300米(14000英尺)(含)的飞行,如果在这些高度上超过30分钟,则对于30分钟后的那段飞行应当为10%的旅客提供足够的氧气;

(2)对于座舱气压高度4300米(14000英尺)以上至4600米(15000英尺)(含)的飞行,足以为30%的旅客在这些高度的飞行中提供氧气;

(3)对于座舱气压高度4600米(15000英尺)以上的飞行,在此高度上整个飞行时间内为机上每一旅客提供足够的氧气。

第121.333条　具有增压座舱的涡轮发动机飞机应急下降和急救用的补充氧气要求

(a)当运行具有增压座舱的涡轮发动机飞机时,合格证持有人应当提供氧气和分配设备,以在座舱增压失效时符合本条(b)款至(e)款的要求。

(b)机组成员。当在飞行高度3000米(10000英尺)以上运行时,合格证持有人应当提供足够的氧气以符合本规则第121.329条要求,但对在驾驶舱内值勤的每一飞行机组成员应提供不少于2小时供氧量。这里的2小时供氧量是指飞机从其最大审定运行高度以恒定下降率用10分钟下降至3000米(10000英尺),并随后在3000米(10000英尺)高度上保持110分钟所必需的供氧量。在确定座舱增压失效情况下驾驶舱内值勤飞行机组成员所需要的供氧量时,可以包含符合本规则第121.337条规定的氧气。

(c)飞行机组人员对氧气面罩的使用。

(1)当在飞行高度7600米(25000英尺)以上运行时,在驾驶舱内值勤的每一飞行机组成员均应当配备有一个氧气面罩,其设计应保证能将其迅速取下戴在脸上,适当固定并密封,在需要时能立即供氧,并且不妨碍该飞行机组成员与其他机组成员之间用飞机内话系统立即通话。当在飞行高度7600米(25000英尺)以上未使用氧气面罩时,它应当保持在备用状态,且位于飞行机组人员在其值勤位置上可以立即取用的范围内。

(2)当在飞行高度7600米(25000英尺)以上运行时,应当有一名在飞机操纵位置上的驾驶员始终使用一个固定在脸上、密封并供氧的氧气面罩。但如果在驾驶舱值勤的每一个飞行机组成员均有一个速戴型氧气面罩,合格证持有人已经证明用一只手在5秒钟内即可以取下并戴到脸上,适当固定、密封并在需要时能立即供氧,并且不妨碍戴眼镜、不会延误飞行机组成员执行其指定的紧急任务、不妨碍该飞行机组成员与其他机组成员之间用飞机内话系统立即通话,则在下列情况下这一名驾驶员不需要戴上和使用氧气面罩:

(i)客座数在30人以上(不包括任何必需的机组成员座位),或者商载大于3400千克(7500磅)的飞机,低于飞行高度层12500米(41000英尺)(含);

(ii)客座数在31人以下(不包括任何必需的机组成员座位),或者商载不大于3400千克(7500磅)的飞机,低于飞行高度层10500米(35000英尺)(含);

(3)在每次飞行的起飞之前,每个飞行机组成员应当对其所使用的氧气设备进行飞行前检查,以确保氧气面罩功能正常、固定合适并连接到适当的供氧接头上,且供氧量及其压力适于使用。

(d)客舱乘务员对便携式氧气设备的使用。在飞行高度7600米(25000英尺)以上飞行期间,每一客舱乘务员应当携带至少可以供氧15分钟的便携式氧气设备,除非经证明,在整个客舱内分布有足够的带有面罩或者备用接口的便携式氧气装置,或者在整个客舱内分布有足够的备用接口和面罩,可以确保在座舱释压时,无论客舱乘务员在何处,每一客舱乘务员均可以立即使用氧气。

(e)旅客。当飞机在飞行高度3000米(10000英尺)以上运行时,除经局方批准外,应当对旅客提供满足下列要求的氧气源:

(1)经审定在飞行高度7600米(25000英尺)以下(含)运行的飞机能在所飞航路的任一点上4分钟之内安全下降到飞行高度4000米(13000英尺)(含)以下时,如果座舱气压高度3000米(10000英尺)以上至4000米(13000英尺)(含)高度上飞行超过30分钟,则对于30分钟后的那段飞行应当以符合本规则第121.335条规定的流量供氧率为至少10%的旅客提供氧气;

(2)当飞机运行在飞行高度7600米(25000英尺)(含)以下且不能在4分钟之内安全下降到飞行高度4000米(13000英尺)时,或者当飞机运行在飞行高度7600米(25000英尺)以上时,在座舱释压后座舱气压高度3000米(10000英尺)以上至4000米(13000英尺)(含)的飞行,如果在这些高度上超过30分钟,则对于30分钟后的那段飞行应当能以符合本规则第121.335条规定的流量供氧率为至少10%的旅客供氧,并且按照适用情况,能够符合本规则第121.329条(c)款第(2)项的要求,但对旅客的供氧时间应当不少于10分钟;

(3)为了对那些由于生理上的原因,在从飞行高度7600米(25000英尺)以上的座舱气压高度下降后可能需要纯氧的机上乘员进行急救护理,在座舱释压后座舱气压高度2400米(8000英尺)以上的整个飞行时间内,应当为2%的乘员(但在任何情况下不得少于1人)提供符合《运输类飞机适航标准》(CCAR-25)第25.1443条(d)款的供氧。应当有适当数量(但在任何情况下不得少于2个)的经认可的氧气分配装置,并为客舱乘务员使用这一供氧源提供办法。

(f)对于在特定区域运行符合本条(e)款第(1)项和第(2)项存在困难的,可以选择符合下列条件并经局方批准:

(1)经审定在飞行高度7600米(25000英尺)以下(含)运行的飞机能在所飞航路的任一

点上4分钟之内安全下降到飞行高度4300米(14000英尺)(含)以下时,如果座舱气压高度3000米(10000英尺)以上至4300米(14000英尺)(含)高度上飞行超过30分钟,则对于30分钟后的那段飞行应当以符合按照本规则第121.335条规定的流量供氧率为至少10%的旅客提供氧气;

(2)当飞机运行在飞行高度7600米(25000英尺)(含)以下且不能在4分钟之内安全下降到飞行高度4300米(14000英尺)时,或者当飞机运行在飞行高度7600米(25000英尺)以上时,在座舱释压后座舱气压高度3000米(10000英尺)以上至4300米(14000英尺)(含)的飞行,如果在这些高度上超过30分钟,则对于30分钟后的那段飞行应当能以符合本规则第121.335条规定的流量供氧率为至少10%的旅客供氧,并且按照适用情况,能够符合本规则第121.329条(d)款第(2)项、第(3)项的要求,但对旅客的供氧时间应当不少于10分钟。

6.2.2　国外有关供氧的条令

JAR/FAR 25.1443(c)规定,对旅客和客舱乘务员,在各种座舱气压高度当使用所提供的氧气设备(包括面罩)吸气时,每个成员补氧的最小氧气流量不得小于为保证下述平均气管氧气分压所需的流量:

(1)在10000 ft<座舱气压高度≤18500 ft时,平均气管氧气分压为100 mmHg,假设每分钟呼吸15 L BTPS(Body Temperature Pressure Saturated,体温与大气压力饱和度)状态的空气,每次最大吸气量为700 cm^3,每两次呼吸之间的时间间隔相同。

(2)在18500 ft<座舱气压高度≤40000 ft时,平均气管氧气分压为83.8 mmHg,假设每分钟呼吸30 L BTPS状态的空气,每次最大吸气量为1100 cm^3,每两次呼吸之间的时间间隔相同。

FAR25.1447　氧气分配装置的设备标准

(c)如果审定的运行高度超过25000 ft,则必须有满足下述要求的氧气分配设施:

(1)对每一个乘员,无论在何处就座,都必须有一个可立即使用的连接到氧气源的氧气分配设备,在每个厕所内至少要有两个连接到氧气源的氧气分配设备。在客舱内的氧气分配设备和氧气出口至少要比座位数多10%。这些额外的分配设备在客舱内尽可能地均匀分布。如果要在30000 ft以上的高度运行,在座舱气压高度超过15000 ft之前,提供所需要的氧气流量的分配设备(氧气面罩)必须能自动出现在乘员面前。必须给机组提供人工释放氧气分配设备并使之能立即可用的手段,以备在自动系统失效时使用。

6.3　座舱失压后的处置程序

一旦发生座舱失压,飞行员应参考飞行中使用的快速检查单,尽快下降高度。飞机发生座舱失压后飞行员的操作见表6-3-1、表6-3-2。

表 6-3-1 急剧失压程序

机长	副驾驶
条件:可能由于增压系统故障或机身损坏造成座舱失压,首先判明情况,检查座舱压差是否减小,座舱高度是否上升,判明确为座舱失压,在飞行高度高于平均海平面高度 14000 ft 时,按以下程序处置	
摘下耳机,戴好氧气面罩	
话筒电门选择 MASK 位,调大麦克风音量,证实通信建立	
旅客信号牌接通	增压方式选择 MAN/MAN-AC(如安装),全关排气活门
	如座舱压差继续下降,报告"压差继续下降"
宣布:"紧急下降"(如需要)	旅客氧气电门接通(如需要)

表 6-3-2 紧急下降程序

机长	副驾驶
紧急下降——宣布 向旅客广播即将开始紧急下降	报告 ATC 并获得该区域高度表设定值
启动开关连续位 概略调低 MCP(模式控制板)高度窗 接通航向选择,右转航向 30°(或根据 ATC 的指挥) 接通高度层改变方式 双发推力手柄——慢车 减速板(手柄置于)——飞行卡位(放减速板) 准确调定改平高度和目标速度	应答机 7700 询问飞机结构有无损坏 (紧急下降一般不放起落架)
"急剧失压,紧急下降检查单"	读急剧失压,紧急下降检查单
证实记忆项目处置内容	
调所需速度	"2000 ft 处改平"(距改平高度还有 2000 ft)
减速板——下卡位(收回减速板)	"1000 ft 处改平"(距改平高度还有 1000 ft)
机组氧气调节器——正常(如需使用氧气时), 启动电门——按需 "完成紧急下降检查单"	"紧急下降检查单完成"
根据天气、氧气、剩余燃油、着陆机场及飞机情况重新设计飞行计划	

6.4 供氧越障分析

6.4.1 障碍物数据的获取

为了提高高原航线运营的可行性和经济性,确保飞机发生发动机失效飘降和客舱失压紧急下降时能够安全超越障碍物,在进行航路性能分析时,地形图及航路障碍物选取的准确性非常重要。为了确保地图作业结果的可靠性,首先必须通过国家测绘地理信息局等权威部门获取准确有效的地形图,通常国家基本比例尺地形图包括 1∶500、1∶1000、1∶2000、1∶5000、

1：1万、1：2.5万、1：5万、1：10万、1：25万、1：50万、1：100万地形图。对于航路安全性研究,1：100万地形图可以满足精度要求。确定待分析航段及逃离航段所涉及的地形图,将相邻的地形图按照经纬度网格准确拼接在一起,形成一张能够覆盖待分析航段及逃离航段的地形图。

根据公布的航路数据中提供的航路点经纬度,在地形图上标出航路点位置,并以航路点为圆心,25 km 为半径作圆。将相邻航路点连接绘制出航线,以 25 km 的垂直距离在航线两侧作平行线,平行线与航路点的圆相切,从而绘制出整个航线的障碍物保护区,如图 6-4-1 所示。然后,筛选在保护区内的障碍物,加德满都到拉萨的航路关键障碍物见图 6-4-2。

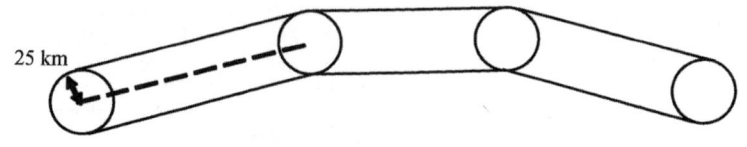

图 6-4-1   画保护区的方法

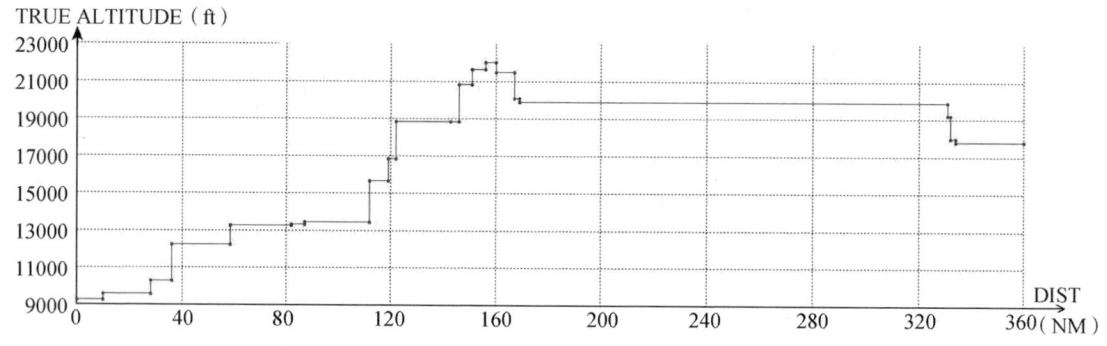

图 6-4-2   加德满都到拉萨的航路关键障碍物

6.4.2   旅客化学式供氧分析

对于安装了化学式供氧的旅客氧气系统,根据前述化学氧气系统的供氧特点,一般在该机型的飞行计划与性能手册(FPPM)中会给出该机型加装的化学氧系统的最大高度包线,图 6-4-3 是 B737-700 的 12 min 化学氧最大高度包线,图 6-4-4 是 B737-700 的 22 min 化学氧最大高度包线。图 6-4-5 是空客机型的 12 min 化学氧最大高度包线,图 6-4-6 是空客机型的 21 min 化学氧最大高度包线。

在考虑化学氧供氧越障时,实际飞机的飞行剖面应在该机型的化学氧包线以下。根据航路关键障碍物的状况,选择管制规定的高度层,并考虑实际飞行过程中不宜频繁下降以减少飞行员的负担,一般取 4 个以下的高度改平。高度确定后,飞机按图 6-4-3 与图 6-4-4 的时间各高度以 V_{MO}/M_{MO} 速度计算(若有结构破损应使用穿越颠簸气流速度)距离,对于波音机型 V_{MO}/M_{MO} 速度可以在该机型的高速性能数据库中得到,如图 6-4-7 所示。

对于空客机型,其 V_{MO}/M_{MO} 速度可在该机型的 FCOM 手册中获得,如图 6-4-8 所示。

利用波音公司提供的性能计算软件 INFLT/REPORT 计算紧急下降的结果如图 6-4-9 所示。利用空客公司提供的性能计算软件 WINPEP 计算紧急下降的结果如图 6-4-10 所示。

巡航过程中一般认为扰流片收起,考虑到有可能发生座舱破损,采用穿越颠簸气流速度。空客机型的穿越颠簸气流速度可以在该机型的 FCOM 手册中获得,如图 6-4-11 所示。

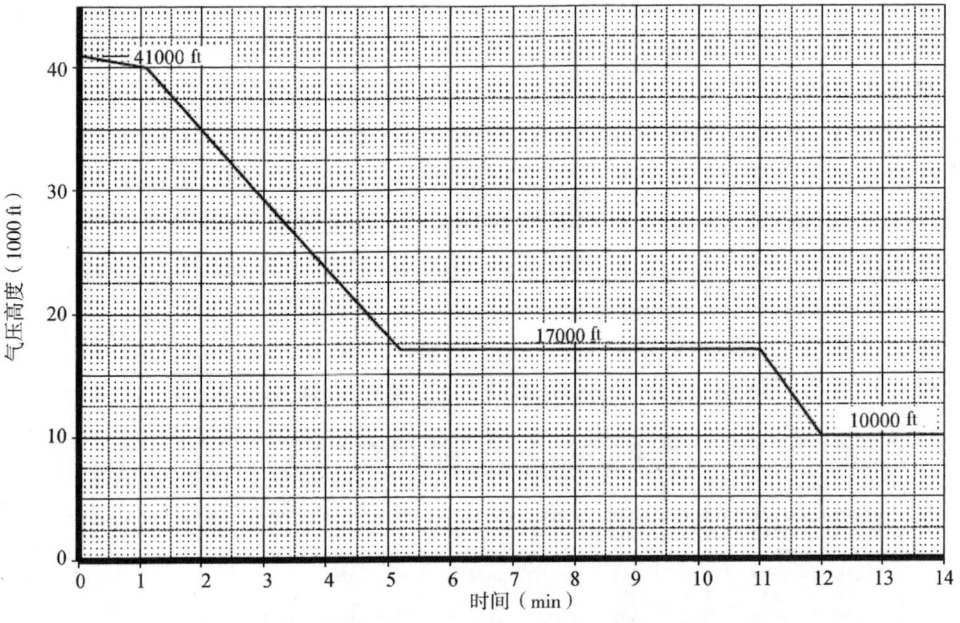

图 6-4-3　B737-700 的 12 min 化学氧最大高度包线

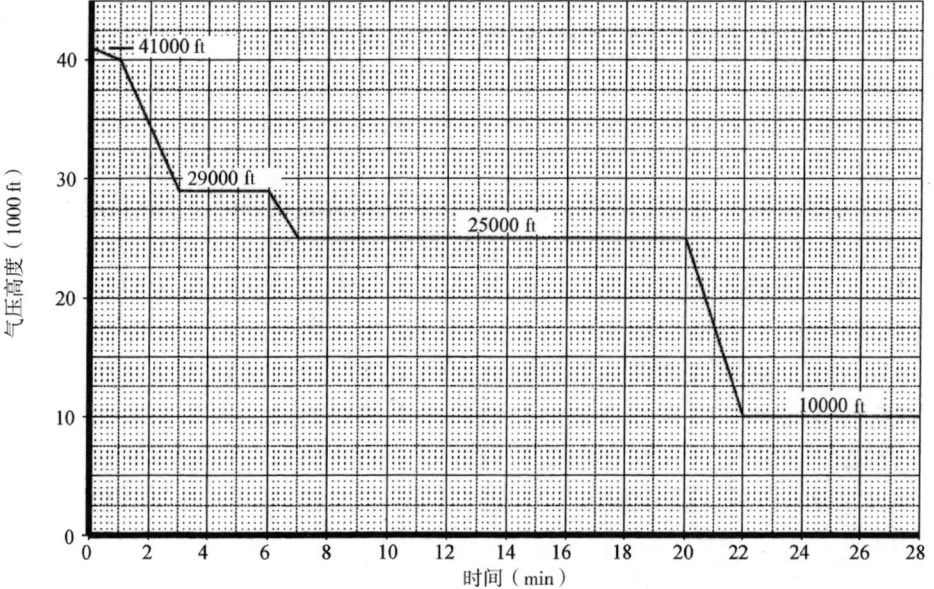

图 6-4-4　B737-700 的 22 min 化学氧最大高度包线

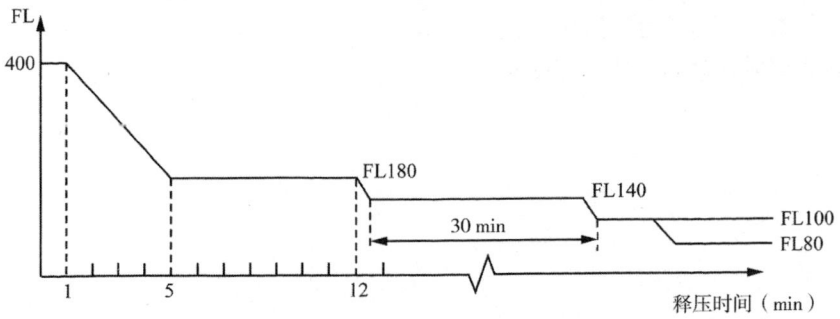

图 6-4-5　空客机型的 12 min 化学氧最大高度包线

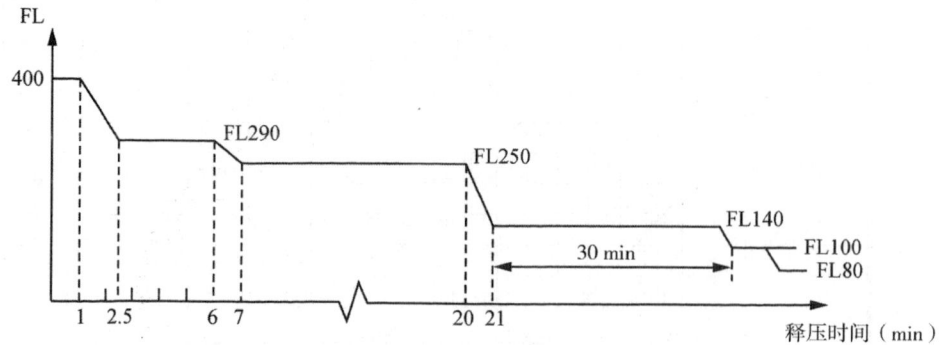

图 6-4-6　空客机型的 21 min 化学氧最大高度包线

```
🖹 37767B.DAT — 记事本                                                    _  □  X
文件(F)  编辑(E)  格式(O)  查看(V)  帮助(H)
.070000.100000.150000.200000.250000.300000.350000.400000.450000.500000
.550000.600000.650000.6775272.00000                                     1
0.     45000.                                                            1
   .235    .280    .344    .397    .443    .486    .525    .561    .595    .627
   .658    .687    .715    .730    .730
   .235    .280    .344    .397    .443    .486    .525    .561    .595    .627
   .658    .687    .715    .730    .730
H
H
H **************** MAXIMUM OPERATING MACH ****************************
H   GOOD FOR GEAR UP OPERATION
/FLOE
/TBLU
MMO 01
      0. 26000. 41000.                                                  11
    .82     .82     .82
H
H   GOOD FOR GEAR DOWN OPERATION
/FLOE
/TBLU
MMO 02
      0. 26000. 41000.                                                  11
    .73     .73     .73
H
H
H   ****************MAXIMUM OPERATING SPEEDS ***************************
H
H
/FLOE
/TBLU
VMO 01   737-700 VMO
      0. 26000. 41000.                                                  11
  340.00 340.00 340.00
H
H
```

图 6-4-7　波音机型的高速性能数据库中 V_{MO}/M_{MO} 速度

　　得到有关距离和高度关系剖面(如需返回起飞机场,考虑转弯应扣除 2.5 min 的距离)后与关键障碍物数据做比较,确定航路决策点。

SPEED LIMITATIONS

MAXIMUM OPERATING SPEED VMO/MMO

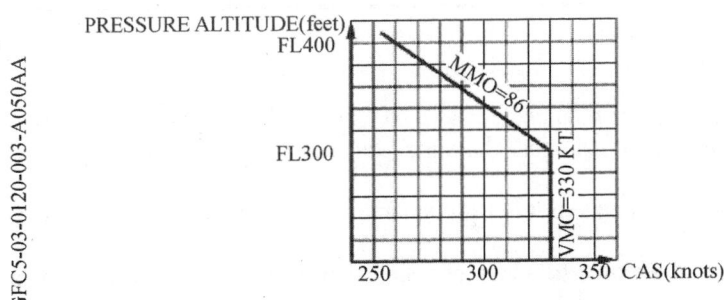

GFC5-03-0120-003-A050AA

The maximum operating limit speed VMO/MMO may not be deliberately exceeded in any regime of flight.

图 6-4-8　空客机型 FCOM 手册中的 V_{MO}/M_{MO} 速度

TABLE NO MODEL CONFIG16 737-700			(2)ENGINES CFM56-7B24/26		DESCENT SCHEDULE 0.82/340		GEAR UP	WIND 0 KTS

ISA C	WEIGHT KG	ALT FT	TIME MIN	FUEL KG	DIST NM	R/C FPM		CARD
0.0	60607.	10000.	6.159	132.202	22.152	-4101.429		1
0.0	60607.	12800.	6.833	140.033	26.602	-4205.461		2
0.0	60607.	13800.	7.070	142.685	28.207	-4243.559		3
0.0	60607.	23600.	9.281	164.437	44.418	-4664.042		4
0.0	60607.	24600.	9.494	166.166	46.105	-4747.930		5
0.0	60607.	26600.	9.871	169.093	49.147	-6940.603		6
0.0	60607.	27600.	10.018	170.186	50.334	-6670.118		7
0.0	60607.	40000.	12.355	184.744	68.751	0.000		8

图 6-4-9　波音机型计算紧急下降的结果

```
C:\PROGRAM FILES\PEP\SESSIONS\IFP\SESSION 2.PRN - 14 101 bytes

☐ Print Preview Mode

                        WEIGHT  :  63000. KG   ISA + 0.0 DG.C

                             100.0 % OF IDL  POWER

                    EMERGENCY DESCENT :  0.82 MN / 350.00 KT


ALT.    ALTG    UGHT    MACH    CAS     TAS    WIND    TIME    FUEL    DIST    RATE     GRDT    ALPH    CL       CD
( FT )  ( FT )  ( KG )  ( )     ( KT )  ( KT ) ( KT )  ( MN )  ( KG )  ( NM )  (FTMN)   (DEG.)  (DEG.)  ( )      (

41000.  41000.  63000.  0.820   243.3   470.3   0.0    0.00    0.    0.0    -4656.9   -5.61    3.72  0.59363  0.058
40000.  40000.  62999.  0.820   249.0   470.3   0.0    0.21    1.    1.7    -4704.8   -5.67    3.42  0.56579  0.055
39000.  39000.  62998.  0.820   254.9   470.3   0.0    0.43    2.    3.3    -4748.5   -5.72    3.23  0.53920  0.053
38000.  38000.  62996.  0.820   260.8   470.3   0.0    0.63    4.    5.0    -4800.6   -5.78    3.06  0.51384  0.051
37000.  37000.  62995.  0.820   266.9   470.3   0.0    0.84    5.    6.6    -4863.6   -5.86    2.90  0.48965  0.049
36089.  36089.  62994.  0.820   272.6   470.3   0.0    1.03    6.    8.0    -4932.4   -5.94    2.76  0.46861  0.047
36089.  36089.  62994.  0.820   272.6   470.3   0.0    1.03    6.    8.0    -5413.5   -6.53    2.76  0.46808  0.047
36000.  36000.  62994.  0.820   273.2   470.5   0.0    1.04    6.    8.1    -5424.7   -6.54    2.75  0.46607  0.047
35000.  35000.  62993.  0.820   279.5   472.7   0.0    1.23    7.    9.6    -5557.3   -6.67    2.60  0.44419  0.045
34000.  34000.  62992.  0.820   285.9   474.8   0.0    1.40    8.    11.0   -5702.0   -6.81    2.46  0.42350  0.044
33000.  33000.  62991.  0.820   292.4   476.9   0.0    1.58    9.    12.3   -5852.3   -6.96    2.33  0.40394  0.042
32000.  32000.  62990.  0.820   298.9   479.0   0.0    1.75    10.   13.7   -6006.5   -7.11    2.21  0.38543  0.041
31000.  31000.  62989.  0.820   305.6   481.1   0.0    1.91    11.   15.0   -6182.5   -7.29    2.09  0.36789  0.040
30000.  30000.  62988.  0.820   312.3   483.2   0.0    2.07    12.   16.2   -6370.1   -7.48    1.98  0.35128  0.039

                                                                                            Print    Close
```

图 6-4-10　空客机型计算紧急下降的结果

THRUST AND AIRSPEED

Set the thrust to give the recommended speed (see table on next page). This thrust setting aims to obtain, in stabilized conditions, the speed for turbulence penetration given in the graph below.

Change thrust only in case of an extreme variation in airspeed, and do not chase your Mach or airspeed.

R A transient increase is preferable to a loss of speed that decreases buffet margins and is difficult to recover.

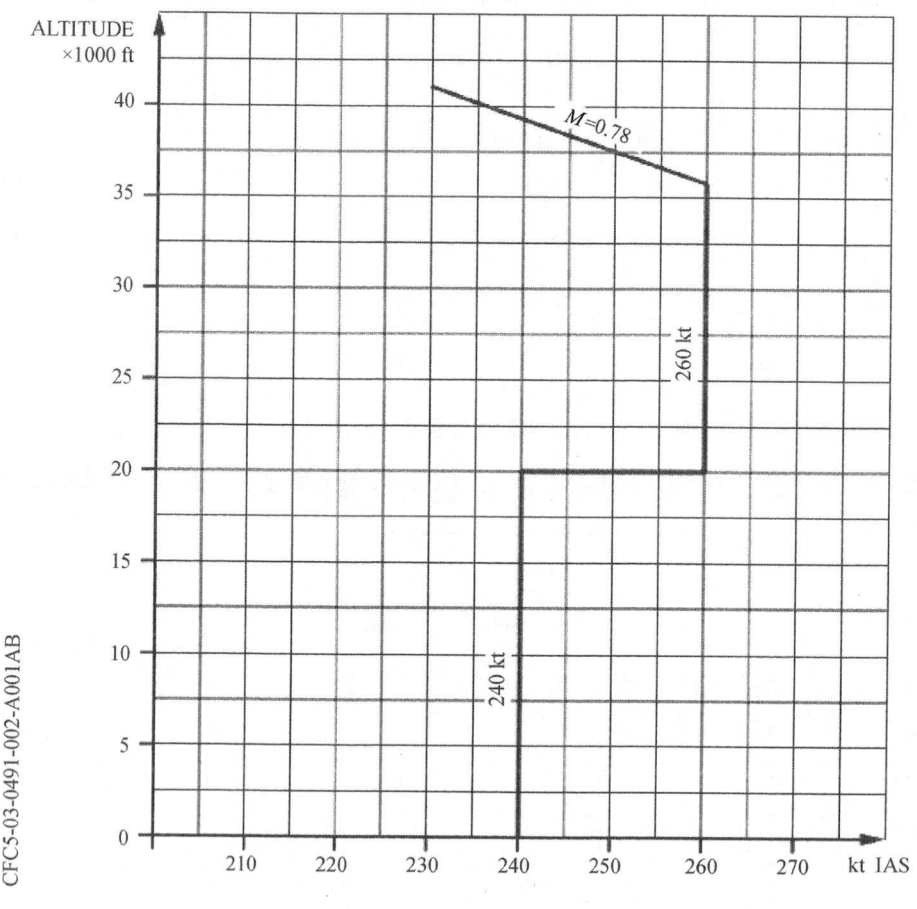

CFC5-03-0491-002-A001AB

图 6-4-11   空客机型 FCOM 手册中的穿越颠簸气流速度

例如,某航线按照该机型的最大审定高度、航线障碍物的分布情况以及该机型的 22 min 化学氧剖面,确定去程选取的高度为 39800 ft、23600 ft、13800 ft 和 10000 ft,根据该机型计算的紧急下降剖面、V_{MO}/M_{MO} 速度和去程风速,确定各段飞行的时间和平均真空速后输入 TIP 软件中,如图 6-4-12 所示。

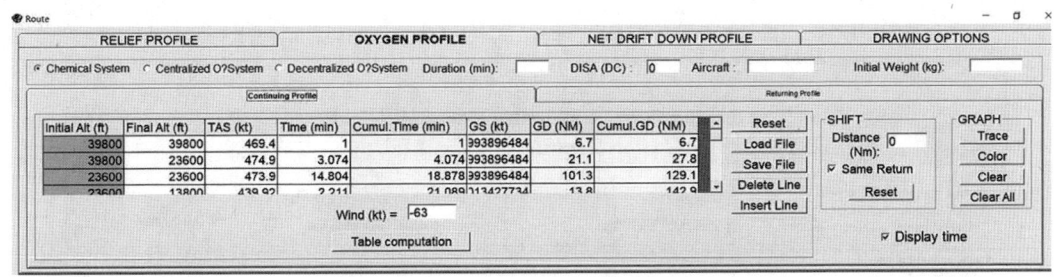

图 6-4-12　TIP 软件中去程参数设定

根据 TIP 软件中的去程参数,绘制去程化学氧越障飞行剖面,如图 6-4-13 所示。

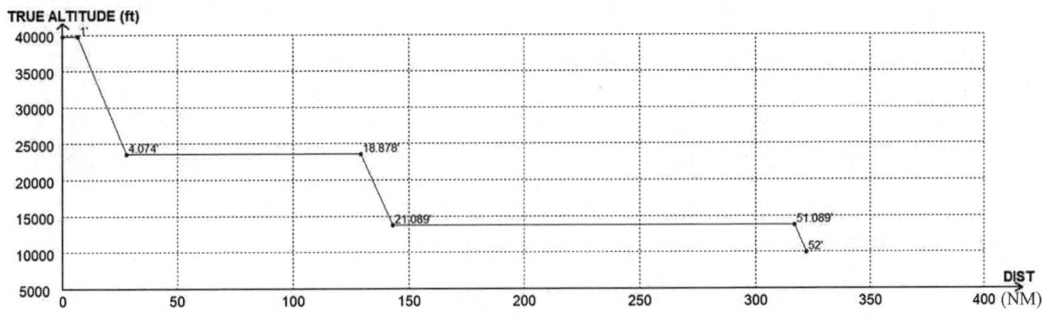

图 6-4-13　TIP 软件中去程化学氧越障飞行剖面

确定回程选取的高度为 39800 ft、24600 ft、12800 ft 和 10000 ft,根据该机型计算的紧急下降剖面、V_{MO}/M_{MO} 速度和回程风速,确定各段飞行的时间和平均真空速后输入 TIP 软件,如图 6-4-14 所示。

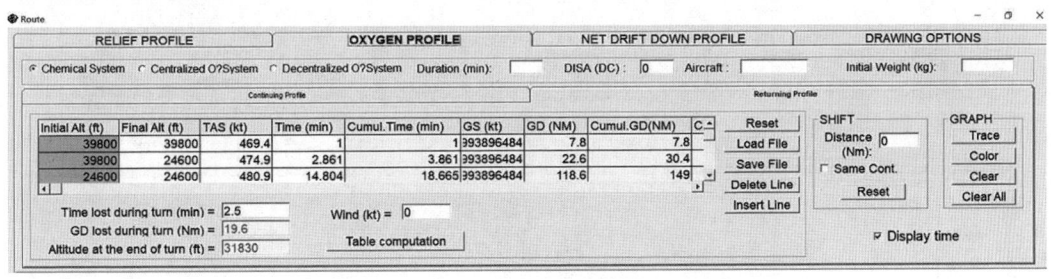

图 6-4-14　TIP 软件中回程参数设定

根据 TIP 软件中的回程参数,绘制回程化学氧越障飞行剖面,如图 6-4-15 所示。

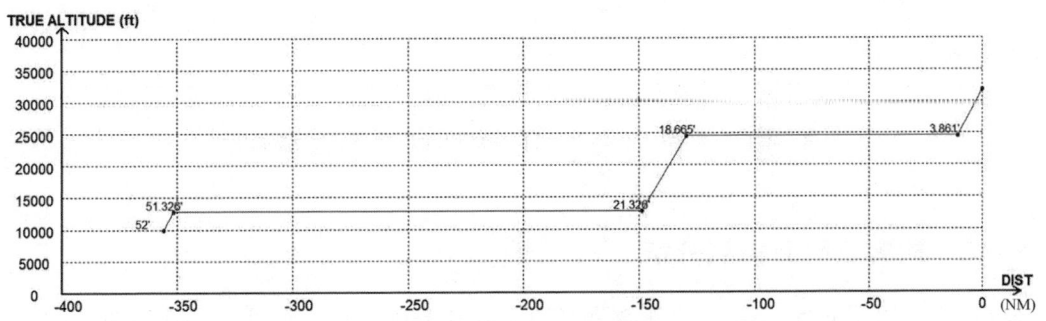

图 6-4-15　TIP 软件中回程化学氧越障飞行剖面

当考虑该航线在 350~850 NM(该航线在 350 NM 和 850 NM 位置附近分别有可供使用的航路备降场)发生座舱失压的方案时,将障碍物高度统一抬升 2000 ft,以考虑失压超障余度的要求。如将去程和回程的关键点位置均取在 565 NM 位置,则发现无法满足去程的越障要求,如图 6-4-16 所示。

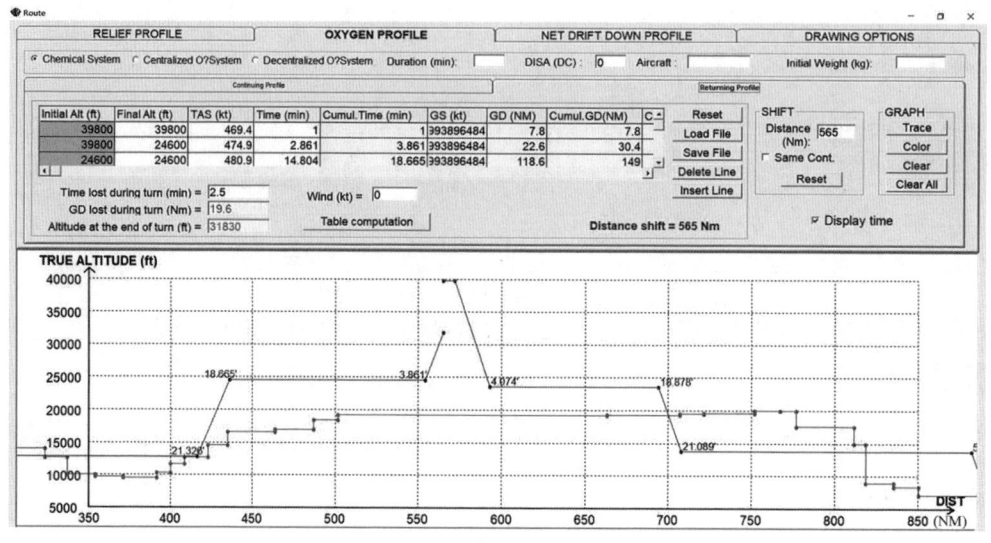

图 6-4-16　航线 350~850 NM 区间决策点在 565 NM 处的情况

通过调整去程和回程的决策点位置,得到能够越障的方案为:对于回程的决策点位置应在 565 NM 处,对于去程的决策点位置应在 680 NM 处,如图 6-4-17 所示。当失压发生在 565~680 NM 处时,需要考虑在航路两侧是否有可用备降机场,以及在改航过程中仍需评估失压后的越障问题。具体越障评估方法同上,这里不再赘述。

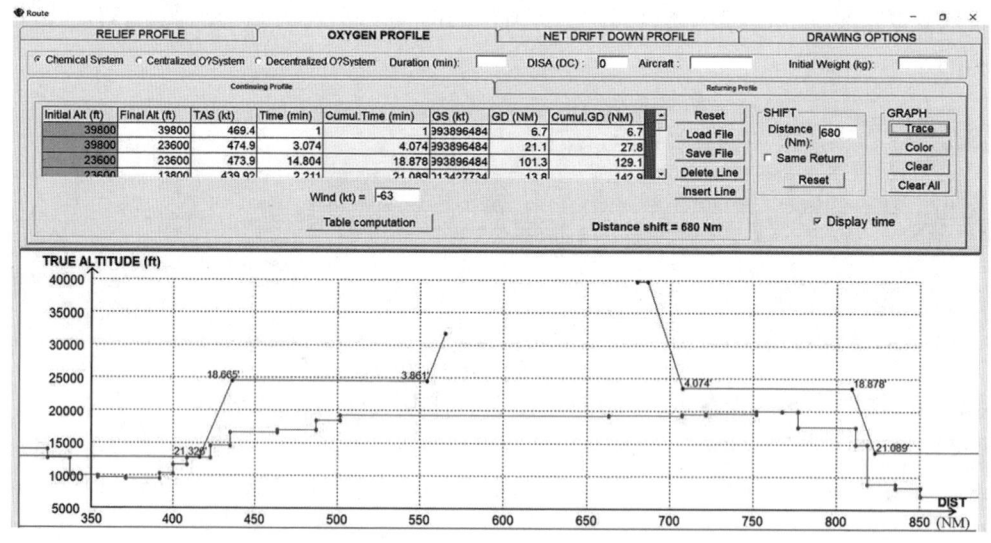

图 6-4-17　航线 350~850 NM 去程和回程决策点位置确定

6.4.3　旅客氧气瓶供氧分析

表 6-4-3 给出了 B757-200 从失压直到紧急下降到 14000 ft 改平、平飞再下降到 10000 ft 所

需要的氧气量,在 14000 ft 气压高度平飞的时间就是失压后的总时间减去紧急下降时间后剩余的那部分,换言之,紧急下降时间加平飞时间就是失压后的总时间。表 6-4-3 对紧急下降到14000 ft 以下再改平的情况也适用,表中给出的氧气量此时是稍有富余的。

表 6-4-3　B757-200 失压后总时间计算需氧量

No.OF OCCUPANTS IN PASSENGER CABIN	TOTAL POST DECOMPRESSION TIME(HOURS)	PRESSURE ALTITUDE AT DECOMPRESSION(ft)				
		27000	31000	35000	39000	42000
		LITERS REQUIRED				
100	0.17	1091	1224	1370	1585	1724
	1	1355	1524	1723	1986	2159
	2	1880	2049	2248	2511	2684
	3	2404	2574	2773	3036	3209
150	0.17	1621	1813	2038	2337	2543
	1	2017	2263	2254	2938	3194
	2	2804	3051	3342	3726	3982
	3	3590	3838	4129	4513	4769
200	0.17	2166	2426	2728	3129	3404
	1	2694	3026	3416	3931	4274
	2	3744	4076	4466	4981	5324
	3	4792	5126	5516	6031	6374
250	0.17	2712	3037	3417	3922	4266
	1	3372	3787	4277	4924	5353
	2	4684	5100	5590	6237	6666
	3	5994	6412	6902	7549	7978
300	0.17	3257	3649	4107	4714	5128
	1	4049	4549	5139	5917	6433
	2	5624	6124	6714	7492	8008
	3	7196	7699	8289	9067	9583

根据客舱就座人数(包括乘务员)、计划巡航高度(即失压时的高度)和预计的失压后在10000 ft 以上高度飞行的总时间由表 6-4-3 查出所需的氧气量。如果改平、巡航高度高于14000 ft,所需的氧气要多一些,表 6-4-4 给出了 B757-200 失压后的改平、巡航高度高于 14000 ft时每巡航一分钟所需的额外氧气量。

表 6-4-4　B757-200 在 14000 ft 以上高度改平每分钟额外需氧量

No.OF OCCUPANTS IN PASSENGER CABIN	ADDITIONAL OXYGEN REQUIRED (LITERS PER MINUTE ABOVE 14000 FT PRESSURE ALTITUDE)				
	INTERMEDIATE PRESSURE ALTITUDE				
	15000	17000	21000	25000	29000
100	18	123	179	254	309
150	27	185	261	371	451
200	36	246	350	498	606
250	45	308	440	625	760
300	54	369	529	752	915

用在14000 ft以上高度上巡航的时间(min),即图6-4-18中的t',乘以查询表6-4-4中的每分钟所需额外氧气量,即可得出总的所需的额外氧气量,把该数值加到由表6-4-3查出的该数值上就是所需的总氧气量。根据氧气总量和氧气瓶压力(见表6-4-5、表6-4-6),确定所需的氧气瓶的数目。注意:查表6-4-3时用的失压后的总时间要包括t'在内,因为表6-4-4给出的仅是额外氧气量;按旅客实际人数加乘务员人数查表,不要另加10%旅客人数。

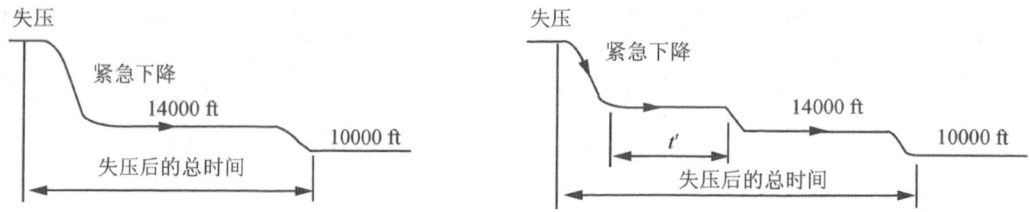

图6-4-18   失压后时间计算

表6-4-5   B757-200氧气瓶压力与氧气量关系表

CYLINDER PRESSURE AT 21 ℃ PSI	OXYGEN VOLUME(1000 LITERS)										
	NUMBER OF 114 CUBIC FOOT CYLINDERS INSTALLED										
	1	2	4	5	6	7	8	9	10	11	12
100			0.1	0.1	0.1	0.1	0.2	0.2	0.2	0.2	0.3
200	0.1	0.3	0.7	0.9	1.1	1.3	1.5	1.7	1.9	2.1	2.3
300	0.3	0.7	1.4	1.8	2.2	2.6	2.9	3.3	3.7	4.0	4.4
400	0.5	1.0	2.1	2.7	3.2	3.8	4.3	4.9	5.4	6.0	6.5
500	0.7	1.4	2.8	3.5	4.3	5.0	5.7	6.4	7.1	7.9	8.6
600	0.8	1.7	3.5	4.4	5.3	6.2	7.1	8.0	8.9	9.8	10.7
700	1.0	2.1	4.2	5.3	6.3	7.4	8.5	9.5	10.6	11.7	12.7
800	1.2	2.4	4.9	6.1	7.4	8.6	5.5	11.1	12.3	13.6	14.8
900	1.4	2.8	5.6	7.0	8.4	9.8	11.3	12.7	14.1	15.5	16.9
1000	1.5	3.1	6.3	7.9	9.5	11.1	12.6	14.2	15.8	17.4	19.0
1100	1.7	3.5	7.0	8.7	10.5	12.3	14.0	15.8	17.5	19.3	21.1
1200	1.9	3.8	7.7	9.6	11.5	13.5	15.4	17.3	19.3	21.2	23.1
1300	2.1	4.2	8.4	10.5	12.6	14.7	16.8	18.9	21.0	23.1	25.2
1400	2.2	4.5	9.1	11.3	13.6	15.9	18.2	20.5	22.7	25.0	27.3
1500	2.4	4.9	9.8	12.2	14.7	17.1	19.6	22.0	24.5	26.9	29.4
1600	2.6	5.2	10.5	13.1	15.7	18.3	21.0	23.6	26.2	28.8	31.5
1700	2.7	5.5	11.1	13.9	16.7	19.5	22.3	25.1	27.9	30.7	33.5
1800	2.9	5.9	11.8	14.8	17.8	20.8	23.7	26.7	29.7	32.6	35.6
1900	3.1	6.2	12.5	15.7	18.8	22.0	25.1	28.3	31.4	34.6	37.7
2000	3.3	6.6	13.2	16.5	19.9	23.2	26.5	29.8	33.1	36.5	39.8
	CREW SYSTEM		PASSENGER SYSTEM								

表 6-4-6 B757-200 氧气瓶压力温度修正表

CYLINDER PRESSURE AT 21 ℃ (psi)	PRESSURE ADJUSTMENT FOR EACH 5 ℃ (psi)
400	7
600	11
800	14
1000	17
1200	21
1400	24
1600	28
1800	31
2000	34

表 6-4-5 中的氧气瓶的数量可按以下公式算出：

假设 1 个氧气瓶（体积为 114 ft³、76 ft³ 或 39 ft³）在 21 ℃ 时瓶压为 P_{21}，此时瓶中的可用氧气量为：

$$Q = 28.32 \times V \times (P_{21} - 100) / (1840 + 14.7)$$

式中：Q——1 个氧气瓶温度为 21 ℃ 时可用的氧气量，单位为 L；

28.32——单位换算系数，1 ft³ = 28.3168 L ≈ 28.32 L；

V——氧气瓶的体积，114 ft³ 或 76 ft³ 或 39 ft³；

P_{21}——21 ℃ 时的瓶压，单位为 psi；

100——考虑氧气瓶低压时不可用的氧气量，在瓶压为 100 psi 时可用氧气量为 0；

14.7——海平面标准大气压强 P_0，$P_0 = 1013.25$ hPa = 760 mmHg = 14.7 psi；

1 850——21 ℃ 时氧气瓶容许的最大压强，单位为 psi。分母用 1850（氧气系统压力）代替 1854.7 的（1840+14.7）计算结果与表 6-4-5 中数据更相符。

一个氧气瓶在温度 T(K) 时的压力 P 与 T_{21}（21 ℃ = 294.15 K）时的压力 P_{21} 满足：

$$P_{21} = T_{21} \times P / T$$

N 个 114 ft³ 或 76 ft³ 或 39 ft³ 的氧气瓶在温度为 T、压强为 P 时的可用氧气量在 760 mmHg、21 ℃ 时的体积 Q 是：

$$Q = N \times 28.32 \times V \times (T_{21} \times P / T - 100) / 1854.7$$

式中：Q——可用氧气量在 NTPD 状态（760 mmHg、21 ℃）时的体积，单位为 L；

P——114 ft³ 或 76 ft³ 或 39 ft³ 的氧气瓶在温度 T 时的压力，单位为 psi。

对于 N 个体积为 11 ft³ 或 7.15 ft³ 或 4.25 ft³ 的手提氧气瓶：

$$Q = N \times 28.32 \times V \times (T_{21} \times P / T - 50) / 1854.7$$

表 6-4-6 中氧气瓶压力温度修正量是由气体状态方程 $PV/T = Const$ 计算出来的，式中 P、V、T 分别表示气体的压强、体积、绝对温度。现在氧气瓶中气体质量、体积 V 不变，所以 $P/T = Const$，即

$$P/T = P_1/T_1$$

式中：P_1、T_1 分别表示在某种给定状态下的压强和温度。

例如，表示 70 °F（21.29 ℃）时的压强和温度：$T_1 = 273.15 + 21.29 = 294.44$ K，于是：

$$P = P_1 \times T/T_1 = P_1(1 + \Delta T/T_1) = P_1 + P_1 \times \Delta T/T_1$$

式中：$\Delta T = T - T_1$，$P_1 \times \Delta T/T_1$，即压强修正量 ΔP。

由此式就可算出表 6-4-6 给出的压强修正值 ΔP，例如：

21 ℃，瓶压 = 1850 psi 时，每高（低）5 ℃ 的 $\Delta P = \pm 1850 \times 5 / 294.44 = \pm 31.42$ psi

$21\ ^{\circ}\text{C}$,瓶压 $=1000$ psi 时,每高(低) $5\ ^{\circ}\text{C}$ 的 $\Delta P=\pm 1000\times 5/294.44=\pm 16.98$ psi

$21\ ^{\circ}\text{C}$,瓶压 $=2000$ psi 时,每高(低) $5\ ^{\circ}\text{C}$ 的 $\Delta P=\pm 2000\times 5/294.44=\pm 33.96$ psi

按上述方法,根据航线上的障碍物的分布情况,选择巡航高度层,确定需要的供氧剖面,根据剖面确定所需的氧气量,计算所需的氧气瓶的数目。

空客机型的旅客氧气瓶氧气系统的计算方法相比波音机型的计算方法较为简单。空客公司给出了不同高度上的旅客氧气面罩的氧气流量,如表 6-4-7 所示。以加德满都—拉萨航线为例,航空公司性能人员可以使用空客公司提供的 TIP 软件中的 ROUTE 模块的 GASEOUS 功能进行供氧计算,如图 6-4-19 所示。

表 6-4-7　空客机型的旅客氧气瓶供氧的面罩氧气流量

Flight Altitude(ft)	Flow Consumption per Mask (L/min) NTPD	
	A319/320/321 Central Flow Regulator	A340-500/-600/A330-200/-300/A340-300 Enhanced Central Flow Regulator
10000	0.370	0.645
11000	0.422	0.669
12000	0.463	0.691
13000	0.496	0.712
14000	0.523	0.732
15000	0.718	0.810
16000	0.849	0.984
17000	0.962	1.153
18000	1.071	1.317
19000	1.178	1.476
20000	1.281	1.630
21000	1.456	1.783
22000	1.620	1.932
23000	1.780	2.076
24000	1.934	2.215
25000	2.074	2.350
26000	2.227	2.479
27000	2.367	2.603
28000	2.502	2.723
29000	2.632	2.840
30000	2.785	2.952
31000	2.879	3.060
32000	2.997	3.165
33000	3.110	3.266
34000	3.219	3.363
35000	3.322	3.457
36000	3.427	3.548
37000	3.524	3.636
38000	3.618	3.719
39000	3.707	3.799
40000	3.781	3.876

需要注意的部分是,在向 ROUTE 模块输入氧气面罩的流量时应按照局方的规定根据数值所在的高度乘以按规定在该高度上需要提供氧气的客舱人数百分比。输入氧气面罩流量后可以在如图 6-4-20 所示的空白位置输入时间与计算好的真空速,通过调整各高度上的时间在软件上选取能够刚好越过下方地形+2000 ft 的剖面,TIP 软件可以根据用户的输入值自动计算所需的氧气瓶的数目,当然用户需要不断调整决策点的位置以获得最小的放行氧气瓶数目。

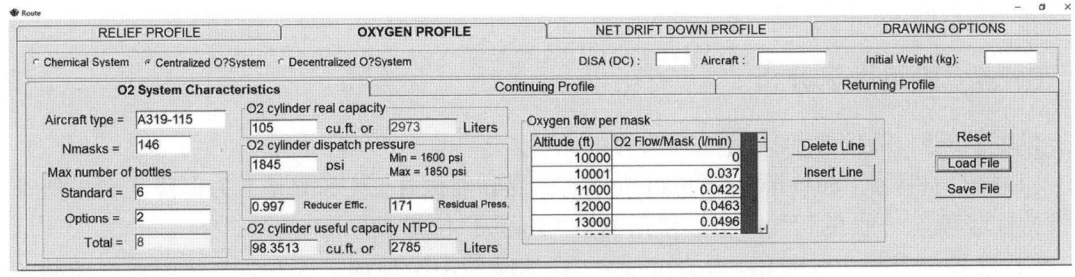

图 6-4-19　空客机型的 TIP 软件的 ROUTE 模块

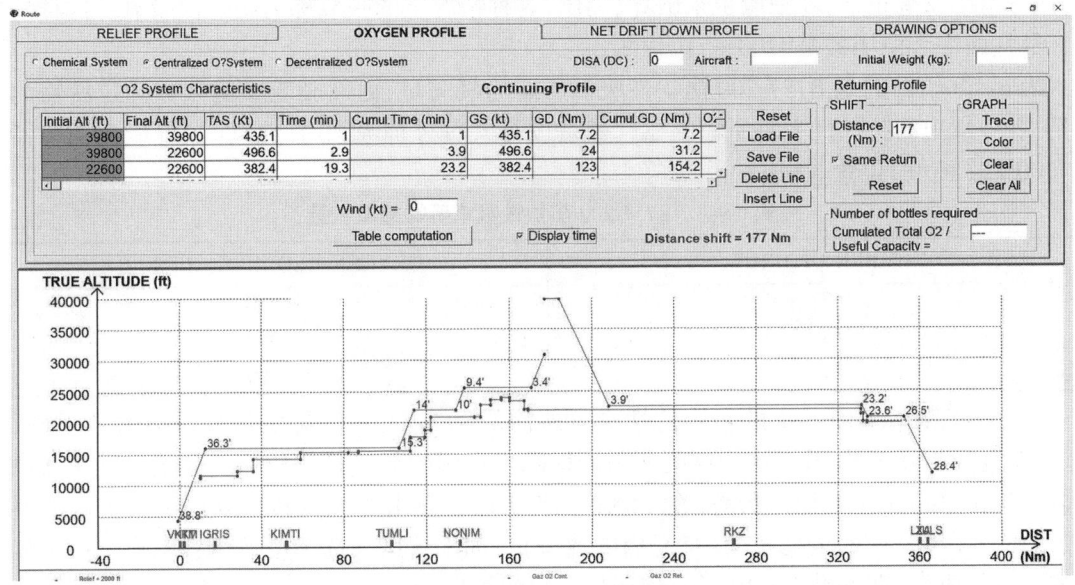

图 6-4-20　空客机型的 TIP 软件的 ROUTE 模块氧气瓶供氧分析

安装 6 个氧气瓶可满足航路上一旦出现座舱增压失效后的旅客供氧需要,最低放行压为 1600 psi,具体程序如下:

决策点 DP(N2816.9,E08758.0),若座舱失压发生在 DP 点之前,则向加德满都返航;若座舱失压发生在 DP 点之后,则向拉萨继续飞行。

向加德满都飞行剖面见图 6-4-21,具体运行程序为:客舱失压后使用紧急下降模式,下降到 25600 ft,以 275 kt 巡航,保持该高度飞行,飞过 NONIM 点之后,下降到 22000 ft,保持该高度飞行,飞过 TUMLI 点之后,下降到 16000 ft,随后飞行至加德满都着陆。

向拉萨飞行剖面见图 6-4-22,具体运行程序为:客舱失压后使用紧急下降模式,下降到 22600 ft,以 275 kt 巡航,保持该高度飞行至拉萨着陆。

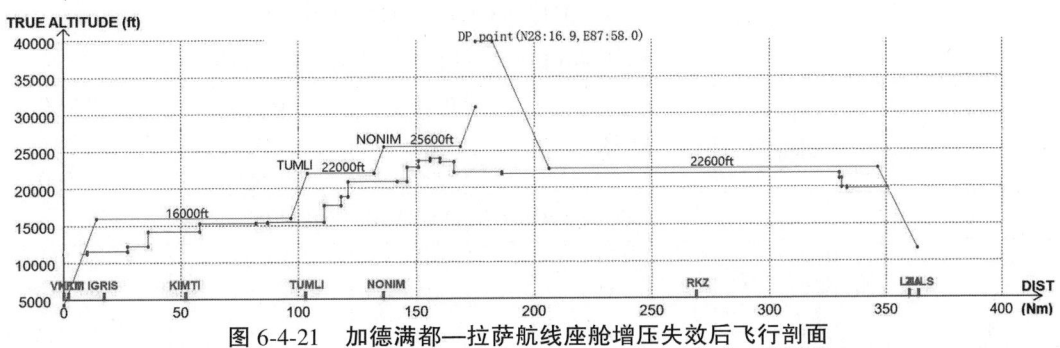

图 6-4-21　加德满都—拉萨航线座舱增压失效后飞行剖面

6.4.4　机组供氧分析

机组供氧时,需要吸氧的情况有两类:

(1)维持性吸氧(Sustenance Breathing)。在驾驶舱破坏、失压情况下为保障生命的吸氧称为维持性吸氧。一般在氧气面罩上使用的是"Normal"位供氧。

(2)防护性吸氧(Protective Breathing)。在驾驶舱中出现烟雾或有害气体时需要给机组供氧,此时增压系统正常,驾驶舱也未破坏、失压,这种情况下的吸氧称为防护性吸氧。这种情况下在氧气面罩上使用的是"100%"位供氧。

表 6-4-8 给出的是 B757-200 防护性吸氧所需的氧气量,由驾驶舱中的人数确定所需的氧气量,表中给出的防护性吸氧所需的氧气量是按每人每分钟吸 20 L 纯氧[氧气压力调节器放在"100%"(纯氧)位置]算出的。

表 6-4-8　B757-200 防护性吸氧所需的氧气量

NUMBER OF CREW	OXYGEN REQUIRED(LITERS)
2	660
3	990
4	1320

表 6-4-9、表 6-4-10 给出的是维持性吸氧所需的氧气量,用于确定从失压后紧急下降、改平、巡航直到下降至 8000 ft 这一期间所需的保障生命用的氧气量,表 6-4-9 是按氧气压力调节器氧气放在"Normal(正常)"位置、在 14000 ft 改平巡航算出的。表 6-4-10 还给出了调节器设定为"100%(纯氧)"时的修正值(括号内的数据)。如果营运人选择按吸纯氧来确定机组氧气系统的放行压力,则查表 6-4-10 时应使用括号中的数字。

表 6-4-9　B757-200 维持性吸氧所需的氧气量表(一)

NUMBER OF CREW	OXYGEN REQUIRED FOR LEVEL OFF AT 14000 FT(LITERS)			
	TOTAL POST DEPRESSURIZATION TIME（HR）			
	2	3	4	5
2	660	960	1270	1570
3	980	1440	1900	2360
4	1310	1920	2530	3150

表 6-4-10　B757-200 维持性吸氧所需的氧气量表(二)

NUMBER OF CREW	ADDITIONAL LITERS REQUIRED FOR EACH MINUTE HELD AT INTERMEDIATE ALTITUDE OTHER THAN 14000 FT				
	INTERMEDIATE PRESSURE ALTITUDE FT				
	UP TO 13999	14000	14001 TO 17999	18000 TO 21999	22000 TO 25000
	REGULATOR ON "NORMAL" OR（100%）				
2	0(22)	0(17)	1(16)	3(12)	6(11)
3	0(33)	0(25)	2(24)	5(18)	8(16)
4	0(44)	0(34)	2(32)	6(25)	11(21)

表 6-4-8、表 6-4-9 和表 6-4-10 中的氧气量都包括正常使用时的氧气泄漏及检查氧气面罩时消耗的氧气量。按驾驶舱中的人数及失压后在 8000 ft 以上高度飞行的总时间由表 6-4-9 确定维持性吸氧所需的氧气量。当改平高度不等于 14000 ft 时，还应由表 6-4-10 确定需要额外增加的氧气量，表 6-4-10 给出的只是在改平高度上氧气流量的增量，用在改平高度上的巡航时间乘以这个增量就得到所需额外增加的氧气量，把该数值加到由表 6-4-9 查出的氧气量上就得到维持性吸氧所需的总氧气量。

实际供氧量应该取防护性吸氧和维持性吸氧中所需的氧气量中大的一个作为机组氧气系统所需的总的氧气量，然后查表 6-4-5、表 6-4-6 确定放行时机组氧气系统中的氧气瓶应达到的压力。

除上述防护性吸氧和失压后补氧的情况外，在 41000 ft（12500 m）（不含）以上飞行时，规章要求一名执飞的飞行员必须戴上氧气面罩并吸稀释氧，因此，对计划的在 41000 ft（12500 m）以上的那段飞行时间应该计算所需的额外氧气量。在 25000 ft（7600 m）（不含）以上飞行时，规章要求当驾驶舱只剩下一名执飞的飞行员时，他也必须戴上氧气面罩并吸氧，如有这种情况也应该额外增加氧气。对这种座舱正常情况下所需的额外氧气，可以按下述方法修正：

（1）在氧气面罩调节器置于"NORMAL"位时，按 2.05 L/人/min 增加氧气量；或者，只有一个 114 ft³ 氧气瓶的机组氧气系统按 1.2 psi/人/min 增加放行压力，有 2 个 114 ft³ 氧气瓶的机组氧气系统按 0.6 psi/人/min 增加放行压力。

（2）如正常使用期间氧气面罩调节器置于"100%"位，按 13 L/人/min 增加氧气量；或者，只有一个 114 ft³ 氧气瓶的机组氧气系统按 8 psi/人/min 增加放行压力，有 2 个 114 ft³ 氧气瓶的机组氧气系统按 4 psi/人/min 增加放行压力。

这步确定的氧气量修正值要分别加到由表 6-4-8、表 6-4-9 和表 6-4-10 确定的氧气量上。如果直接修正压力，则按上述原则修正由表 6-4-5 确定的放行压力，然后按表 6-4-6 做温度修正。

对只有 1 个 76 ft³（或 39 ft³）氧气瓶的系统，按上述各 "ΔP/人/min" 的 1.5（或 3）倍增加放行压力。这里 2.05 L/人/min 是座舱未失压时所需的额外氧气量，与表 6-4-10 座舱失压时的量不同。

在确定机组氧气系统所需的氧气量时，在 8000 ft 以上的高度失压后的总飞行时间不能少于 2 h，即使实际飞行时间不到 2 h。

对配备 12 min 或 22 min 化学氧旅客氧气系统的飞机，维持性吸氧按失压后总飞行时间 2 h 计算就可以了，此时防护性吸氧和维持性吸氧所需的氧气量近似相同。对配备化学氧系统的飞机，其手册中一般给出一个简单的数值表（如表 6-4-11、表 6-4-12 所示），可以很方便地确定机组氧气系统的最小放行压力。

表 6-4-11　B737-700 型号为 76 ft³氧气瓶所需压力 psi（22 min）

BOTTLE TEMPERATURE		NUMBER OF CREW USING OXYGEN		
℃	℉	2	3	4
50	122	735	1055	1360
45	113	725	1040	1340
40	104	715	1020	1320
35	95	700	1005	1300
30	86	690	990	1280
25	77	680	975	1255
20	68	670	960	1240
15	59	655	940	1215
10	50	645	925	1195
5	41	635	910	1175
0	32	620	890	1150
−5	23	610	875	1130
−10	14	600	860	1110

表 6-4-12　B737-700 型号为 76 ft³氧气瓶所需压力 psi（12 min）

BOTTLE TEMPERATURE		NUMBER OF CREW USING OXYGEN		
℃	℉	2	3	4
50	122	530	735	945
45	113	520	725	930
40	104	510	715	915
35	95	505	700	900
30	86	495	690	885
25	77	485	680	870
20	68	480	670	860
15	59	470	655	840
10	50	460	645	830
5	41	455	635	815
0	32	445	620	800
−5	23	440	610	785
−10	14	430	600	770

　　空客机型的机组氧气计算方法分为两种,在 FCOM 手册中均有详细的介绍,如图 6-4-22 与图 6-4-23 所示。

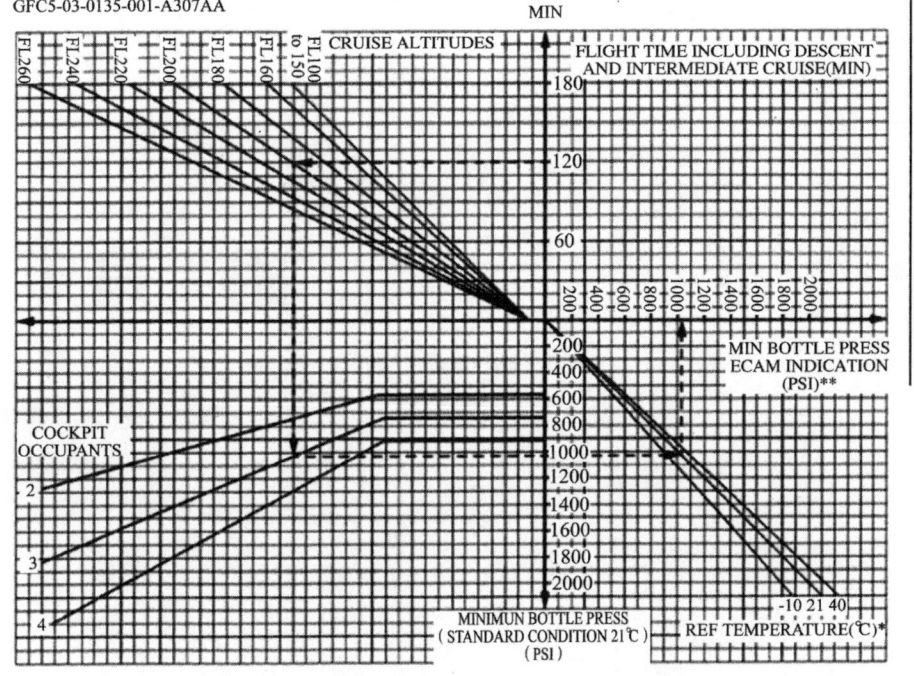

图 6-4-22 空客机型 FCOM 手册计算机组氧气方法(一)

COCKPIT FIXED OXYGEN SYSTEM

MINIMUM FLIGHT CREW OXYGEN PRESSURE

REF TEMPERATURE *		°C	− 10	0	10	20	30	40	50
		°F	14	32	50	68	86	104	122
MIN ** BOTTLE PRESSURE (PSI)	2 CREW MEMBERS		520	540	560	580	600	620	640
	2 CREW MEMBERS +1 OBS		660	690	710	740	760	790	810
	2 CREW MEMBERS +2 OBS		810	850	880	910	940	970	1000

* REF TEMPERATURE :
. On ground : REF TEMPERATURE = (OAT + COCKPIT TEMP) / 2
. In flight : REF TEMPERATURE (°C) = CAB TEMP (°C) − 10
        or
        REF TEMPERATURE (°F) = CAB TEMP (°F) − 18
** MINIMUM BOTTLE PRESSURE TO COVER :
− Preflight checks
− Usage of oxygen when only one pilot is in the cockpit
− Unusable quantity (to ensure regulator functioning with minimum pressure)
− Normal system leakage
− and
  · Protection after loss of cabin pressure with mask regulator on NORMAL (diluted oxygen) :
    − During emergency descent for all cockpit members for 15 minutes (refer to cabin fixed oxygen system)
    − During cruise at FL 100 for 2 crew members for 105 minutes
or
  · Protection against smoke with 100 % oxygen for all cockpit members during 15 minutes at 8000 feet cabin altitude.

Note : The above times are based on the use of a sealed mask, may be shorter for bearded crew (in terms of performance, pressure or duration).

图 6-4-23 空客机型 FCOM 手册计算机组氧气方法(二)

附　录

附录1　制作飞行计划相关图表(波音飞机)

737-800/CFM56-7B26
FAA
Category C Brakes

BOEING

Flight Planning and Performance Manual

FLIGHT PLANNING
Text

Introduction

This chapter contains flight planning data to determine trip fuel/time, reserve fuel and enroute terrain clearance capability. The data includes engine bleed effects for normal air conditioning operation, i.e. two packs at normal flow, one pack bleeding each engine.

Simplified Flight Planning

Flight Planning Allowances

Simplified Flight Planning charts enable rapid determination of estimated trip time and fuel from brake release to landing. Additional flight planning information, including maneuver allowances is summarized below.

Ground Operations

Fuel can be saved by minimizing APU utilization. Average APU fuel flow rate under normal operation is 225 lb per hour on the ground.

Taxi fuel allowance is approximately 25 lb per minute.

APU Operation During Flight

For APU operation during flight, increase fuel flow according to the following table. These increments include the APU fuel flow and the effect of increased drag from the APU door.

PRESSURE ALTITUDE (1000 FT)	APU FUEL FLOW (LB/HR)
39	100
35	100
31	110
25	130
20	150
15	160
10	180
5	200

Climb

Trip Fuel and Time charts are based on an initial climb speed of 280 KIAS. Local ATC may require that 250 KIAS not be exceeded below 10000 ft. Approximately 50 lb of additional fuel is burned when this restriction is imposed.

Altitude Selection

Best fuel mileage for a given speed schedule is achieved at optimum altitude. Fuel mileage penalties for operation at off-optimum altitudes are shown in the following table.

OFF-OPTIMUM CONDITION	FUEL MILEAGE PENALTY %		
	LRC	.79M	CI30
2000 FT ABOVE OPTIMUM ALTITUDE	1	1	1
OPTIMUM ALTITUDE	0	0	0
2000 FT BELOW	1	2	2
4000 FT BELOW	4	6	5
6000 FT BELOW	7	10	8
8000 FT BELOW	9	16	12
10000 FT BELOW	12	20	15
12000 FT BELOW	16	23	20

Cruise

Long Range Cruise is the speed which gives 99% of the maximum fuel mileage at zero wind. For cruise within 2000 ft of optimum altitude Long Range Cruise may be approximated by a constant .79M or Cost Index 30.

Increase total fuel flow during cruise approximately 100 lb/hr for engine anti-ice on or 300 lb/hr for engine and wing anti-ice on.

Every 1000 lb reduction in landing weight decreases trip fuel at optimum altitude by approximately 0.5%.

Descent

Optimum descent speeds for minimum trip fuel are a function of weight. Trip Fuel and Time charts are based on .78/280/250 descent speed, and include straight in approach allowances only. Each additional minute of flaps down maneuvering consumes approximately 110 lb of additional fuel with the gear retracted and 130 lb with the gear extended.

Missed Approach

Approximately 280 lb of additional fuel is burned during the missed approach maneuver, based on applying go-around power from the final approach configuration, retracting flaps and gear while climbing to 1500 ft and accelerating to 250 KIAS.

Optimum Altitude

Optimum altitude for best fuel mileage is presented for LRC/.79M speed schedule. Fuel mileage penalties for flying off-optimum altitude are tabulated under Flight Planning Allowances.

Short Trip Cruise Altitude

Short range operation may be limited by the distance required to perform the climb and descent. The Short Trip Cruise Altitude chart shows the maximum altitude at which it is possible to cruise for various times in level flight.

附图 1-1　制作飞行计划的相关图表概述(1)

FLIGHT PLANNING
Text

BOEING
Flight Planning and Performance Manual

737-800/CFM56-7B26
FAA
Category C Brakes

The chart also shows the altitude profile for minimum trip fuel as used to determine Short Trip Fuel and Time.

Trip Fuel and Time

Trip Fuel and Time charts are provided for Long Range Cruise and .79M at constant altitude to determine trip fuel and time from brake release to touchdown. A similar chart for a cost index of 30 (which approximates Long Range Cruise) is also shown. APU usage, taxi, inflight flaps down maneuvering (other than straight in approach) and reserve fuel should be added to the trip fuel obtained from the trip fuel and time charts to obtain the total fuel required. These fuel allowances are presented under Flight Planning Allowances. Additional fuel for extended inflight traffic delays should be determined from the holding table.

To determine trip fuel and time enter the appropriate chart with trip ground distance, adjust for the anticipated wind condition, and proceed vertically to cruise pressure altitude. Trip fuel is read to the right from the lower set of altitude lines by going to the reference landing weight and adjusting to the planned landing weight using the guidelines. Trip time is read to the left from the upper altitude lines, adjusting for ISA deviation as necessary.

For winds greater than those shown, enter chart directly with adjusted ground distance and ignore wind adjustment guidelines. Adjusted ground distance is ground distance multiplied by the ratio Average TAS/ (Average TAS + Wind Component), where headwind is negative and tailwind is positive.

Step Climb Trip Fuel and Time

Best fuel mileage is obtained if the flight is planned at optimum altitude for the cruise weight and speed schedule. Since optimum altitude increases as the airplane weight decreases, a step climb procedure bracketing optimum altitude is necessary.

The Step Climb Planning charts provide trip fuel and time for LRC, .79M and cost index 30 from brake release to touchdown, similar to the constant altitude Trip Fuel and Time charts. The charts are based on 4000 ft step climbs to 2000 ft above optimum altitude.

Short Trip Fuel and Time

Short Trip Fuel and Time charts include fuel and time to climb to cruise altitude, cruise, descent and straight in approach. The charts are based on the cruise altitude for minimum trip fuel for the planned landing weight. For distances greater than 500 nm, or other altitudes, use the Trip Fuel and Time charts.

To determine fuel and time, enter the appropriate chart with ground distance, adjust for wind and proceed vertically to the planned landing weight. Trip fuel is read to the right from the lower set of landing weight curves. Trip time is read to the left from the upper curve and is valid for all landing weights.

Holding Planning

This table provides total fuel flow information necessary for planning holding and reserve fuel requirements. The chart is based on the higher of the maximum endurance speed and the maneuvering speed.

The fuel flow is based upon flight in a racetrack holding pattern. For holding in straight and level flight reduce the table values by 5%.

Fuel Tankering

The Fuel Tankering table below is provided for LRC/ .79M cruise schedule based on a step climb procedure to bracket optimum altitude. When a fuel price differential exists between two stations, this table may be used to determine if fuel may be economically transported for a subsequent flight sector.

Enter the table with the trip distance to read the break-even fuel price ratio. As noted, the break-even fuel price is the fuel price at departure multiplied by the break-even fuel price ratio. To justify tanker operation economically, the fuel price at destination must be greater than the break-even fuel price.

TRIP DISTANCE (NM)	BREAK-EVEN PRICE RATIO
200	1.015
400	1.032
600	1.046
800	1.062
1000	1.078
2000	1.179
3000	1.313
4000	1.500

Passenger Oxygen Requirements

This airplane is equipped with a 12 minute or optional 22 minute chemical passenger oxygen system. The altitude envelope provided shows the maximum altitude that the airplane may be flown during a cabin depressurization event and still support the physiological requirements of the passengers using the oxygen system installed. The envelope is intended to assist in terrain clearance planning.

The maximum altitude envelope assumes an immediate descent is made to 10000 ft or the lowest safe altitude whichever is higher as stipulated in the Emergency Descent procedure of the Quick Reference Handbook. Should the presence of terrain necessitate a level-off at an altitude higher than 10000 ft, the required terrain clearance descent profile should at no time exceed the

附图 1-1　制作飞行计划的相关图表概述(2)

737-800/CFM56-7B26
FAA
Category C Brakes

◊ _BOEING_

Flight Planning and Performance Manual

FLIGHT PLANNING
Text

maximum altitude capability of the passenger oxygen system shown by the envelope. Once terrain is cleared, the descent to 10000 ft should be completed.

Flight Crew Oxygen Requirements

Regulations require that sufficient oxygen be provided to the flight crew to account for the greater of supplemental breathing oxygen in the event of a cabin depressurization or protective breathing in the event of smoke or harmful fumes in the flight deck.

This airplane model is equipped with a 12 minute or optional 22 minute chemical passenger oxygen system which limits the extent to which the airplane can be operated at altitudes above 10000 ft following depressurization. As a result the flight crew supplemental oxygen quantity requirement is determined using the fixed maximum cabin altitude profile allowed by the passenger oxygen system. Consideration must also be given to protective breathing oxygen, which is the quantity of oxygen necessary for 15 minutes of pressurized 100% oxygen flow (Emergency setting) for each flight crew member in a fully pressurized cabin (8000 ft cabin pressure altitude). The greater quantity of either the supplemental or protective breathing requirements is used to determine the minimum dispatch cylinder pressures provided. An additional 10% of oxygen quantity is supplied by the values provided in the tables to account for normal system leakage and for use by one active duty pilot at flight altitudes above 25000 feet should the other active duty pilot leave the flight deck.

To determine the minimum dispatch oxygen cylinder pressure enter the flight crew oxygen table with the number of crew plus observers using oxygen and read the minimum cylinder pressure required for the appropriate cylinder temperature.

Driftdown

Net Level Off Weight

This chart is used to determine the maximum weight for terrain clearance based on the approved Airplane Flight Manual net engine inoperative performance at the speed for optimum climb gradient. Regulations require terrain clearance flight planning based on net performance which is the gross (or actual) gradient performance degraded by 1.1%.

The net level off pressure altitude must clear the terrain by 1000 ft. To determine the maximum weight for terrain clearance, enter with net level off pressure altitude, proceed to the ISA deviation and read weight. Adjust weight for anti-ice operation with the weight adjustment shown on the chart.

Driftdown Profiles Net Flight Path

These charts are provided to determine the time, fuel and distance from engine failure at any point during driftdown for legal enroute terrain clearance. Data is based on net performance, Max Continuous Thrust following engine failure and level flight deceleration to optimum driftdown speed (speed for optimum climb gradient).

To determine driftdown time, fuel and distance, enter the appropriate chart with pressure altitude for terrain clearance (terrain height plus 2000 ft), move horizontally to equivalent weight at engine failure (adjusted for ISA deviation and anti-ice operation as indicated) and read fuel burned from the point of engine failure by interpolating between dashed lines. Read time and ground distance from point of engine failure by moving vertically downward from equivalent weight, adjusting distance for wind.

Driftdown/LRC Cruise Range Capability

This chart shows the range capability from the start of driftdown/cruise based on level off following driftdown and targeting the one engine inoperative Long Range Cruise speed.

To determine the fuel required, enter the chart with the desired ground distance, adjust for the anticipated wind, and proceed vertically to the start of driftdown weight and read the fuel required to the right. The time required is read to the left from the single weight line on top. The chart may be used in reverse to determine the range capability knowing the fuel remaining from the start of driftdown.

ETOPS

Regulations require that flights conducted over a route that contains a point further than one hour's time at "normal one engine inoperative speed" from an adequate diversion airport comply with rules set up specifically for extended range operation with two engine airplanes. If the airplane meets these requirements, the approved Airplane Flight Manual (AFM) will contain a statement to this effect.

This section provides the planning and inflight performance information necessary to conduct operations under these regulations. Information is presented for several different driftdown/cruise speed schedules. The information is based on drifting down using Max Continuous Thrust at the Mach number shown until reaching the corresponding airspeed and maintaining that airspeed throughout the remainder of the driftdown and level cruise. Based on route requirements, operators may select any one of these

附图 1-1　制作飞行计划的相关图表概述(3)

FLIGHT PLANNING
Text

BOEING
Flight Planning and Performance Manual

737-800/CFM56-7B26
FAA
Category C Brakes

speeds for their operation, or may elect to use an entirely different speed which is not contained in this section.

Flight Planning

The following information is shown for flight planning purposes and is intended for use mainly by ground personnel.

Data is included for five basic schedules (LRC, .76/280, .79/290, .79/310 and .79/330) throughout this section to allow the operator to evaluate the effect of using speeds up to 330 KIAS (VMO -10).

Area of Operation (Diversion Distance)

The maximum diversion distance used to establish the Area of Operation may be obtained from this chart. The Area of Operation is defined as the region within which the operator is authorized to conduct extended range operation. The distance to the diversion airport from any point along the intended route must be covered within the approved time using the single engine cruise speed (assuming still air and ISA conditions) selected by the operator.

Enter the chart for the appropriate speed with the weight at the point of diversion using the speed schedule and time selected and read the maximum diversion distance.

Critical Fuel Reserves

ETOPS Regulations require reserve planning to include a "Critical Fuel Scenario" calculation. The information shown is the fuel required to satisfy the flight profile as described on the charts. This information is shown for two engine operation at Long Range Cruise, as well as single engine operation at LRC, 280, 290, 310 and 330 KIAS. There are two separate Engine-Inop scenarios; one assumes a loss of pressurization and subsequent cruise at 10000 ft, the other assumes only an engine failure with the airplane drifting down to an applicable level-off altitude for the remainder of the diversion. The ETOPS Critical Fuel required is the greater of All-Engine fuel at Long Range Cruise or Engine-Inop fuel for the speed schedule selected. This fuel is compared to the amount of fuel normally onboard the airplane at that point in the route. If the fuel required by critical fuel reserves exceeds the amount of fuel normally expected, fuel load must be adjusted accordingly.

To determine the fuel required, enter the appropriate chart with desired diversion distance, adjust for forecast wind (factored if applicable), proceed vertically to the weight at critical point and read the fuel required to the right. If the operator is using a wind forecasting model acceptable to the FAA (such as the World Area Forecast System, WAFS), regulations allow the wind factor

applied in this step to be 5% of the forecast wind (increase headwinds, decrease tailwinds), as indicated in the note at the bottom of the chart. If the operator is not using an acceptable wind forecasting model, the calculated diversion fuel is to be increased by 5% instead of factoring the forecast wind values. Adjustments for non-standard conditions are shown at the bottom of the page.

The data does not include an allowance for performance deterioration. However, regulations require a 5% allowance for performance deterioration unless a value has been established by the operator for in-service deterioration.

Net Level Off Weight

These charts are used to determine the maximum weight for terrain clearance based on net engine inoperative performance for the speed schedules shown and is for planning purposes only. Regulations require terrain clearance flight planning based on net performance which is the gross (or actual) gradient performance degraded by 1.1%.

Net level off weight is determined by entering the appropriate chart at the left with the desired level off pressure altitude, adjusting for anti-ice as shown, proceeding across to the correct ISA deviation line and reading the allowable weight at the bottom. The net level off altitude may be obtained by entering the chart with weight and proceeding up to the correct ISA deviation line and reading across to the altitude on the left.

Driftdown/Cruise Range Capability

These charts show range capability from the start of driftdown/cruise when the driftdown/cruise procedure outlined in this section is followed. Charts are shown for each of the speed schedules discussed previously.

To determine the fuel required, enter the appropriate chart with the desired ground distance, correct for the anticipated wind, and proceed vertically to the start of driftdown weight and read the fuel required to the right. The time required is read to the left from the single weight line on top. The chart may be used in reverse to determine the range capability knowing the fuel remaining from the start of driftdown.

附图 1-1　制作飞行计划的相关图表概述(4)

737-800/CFM56-7B26
FAA
Category C Brakes

BOEING

Flight Planning and Performance Manual

ENROUTE
Text

Introduction

This chapter contains inflight data for use as general reference in airplane performance monitoring, flight planning studies and as a supplement to information provided in the Operations Manual.

All Engines

Long Range Cruise Maximum Operating Altitude

These tables provide the maximum operating altitude in the same manner as the FMC. Maximum altitudes are shown for a given cruise weight and maneuver capability. Note that these tables consider both thrust and buffet limits, providing the more limiting of the two. Any data that is thrust limited is denoted by an asterisk and represents only a thrust limited condition in level flight with 100 ft/min residual rate of climb. Flying above these altitudes with sustained banks in excess of approximately 15° may cause the airplane to lose speed and/or altitude. The altitudes shown in the tables are limited to the maximum certified altitude of 41000 ft.

Enroute Climb

The Enroute Climb tables are based on 280 /.78 climb speed. Local ATC may require that 250 KIAS not be exceeded below 10000 ft. See Flight Planning Allowances in Chapter 2 for additional fuel burn when this restriction is imposed.

Long Range Cruise Enroute Fuel and Time

These charts are provided to determine if the fuel remaining is sufficient to complete the trip at Long Range Cruise and to approximate the time remaining. The charts also enable rapid determination of fuel and time required to proceed to an alternate airfield.

Fuel and time are read in a similar manner to the Long Range Cruise trip fuel and time charts in Chapter 2 with distance to destination replacing total trip distance; i.e., climb phase is excluded.

Cruise Table

Cruise tables are provided for Long Range Cruise and .79M speed schedules with both engines operating assuming normal engine bleed for air conditioning (2 bleeds/2 packs on). These tables are similar to the cruise control tables presented in the Performance Inflight chapter of the QRH and are expanded to include true airspeed and Max Cruise Thrust setting information.

To account for APU fuel burn, refer to the APU Operation During Flight table in the chapter 2 text.

Wind-Altitude Trade

Wind is a factor which may justify operations considerably above or below optimum altitude. For example, a favorable wind component may have an effect on ground speed which more than compensates for the loss in air range.

Using the applicable table, it is possible to determine the break-even wind (advantage necessary or disadvantage that can be tolerated) to maintain the same range at another altitude. Tables are provided for Long Range Cruise and .79M. The tables make no allowance for climb or descent time, fuel or distance, and are based on comparing ground fuel mileage.

Descent

Time, fuel, and distance for descent are shown for two descent speed schedules. This data includes the effect of a 250 KIAS speed restriction below 10000 ft and includes a straight in approach with flaps down at the outer marker.

Holding

%N1 required, indicated airspeed and fuel flow information are tabulated for holding at flaps up based on the recommended holding speeds. Small variations in airspeed will not appreciably affect the overall endurance time. The fuel flow is based on a racetrack holding pattern. For holding in straight and level flight, these values may be reduced by 5%.

Engine Inoperative

Altitude Capability

The Engine Inoperative Altitude Capability tables are based on the thrust limited capability with 100 ft/min residual rate of climb. For the engine inoperative case, buffet capability is not considered limiting. Tables are shown for Long Range Cruise and constant speeds of 280, 290, 310 and 330 KIAS for both Max Continuous and Max Cruise Thrust. Adjust the table values for thermal anti-ice operation as shown below the tables.

Cruise Table

Cruise tabulations for engine inoperative are based on use of Max Continuous Thrust. A heavy line is shown to denote the approximate Max Cruise Thrust limits. Tables are provided for Long Range Cruise and constant speeds of 280, 290, 310 and 330 KIAS for both Max Continuous and Max Cruise Thrust. The fuel flow values in these tables reflect only the engine fuel burn. To conservatively account for APU fuel burn, refer to the APU Operation During Flight table in the chapter 2 text.

附图 1-1　制作飞行计划的相关图表概述(5)

ENROUTE
Text

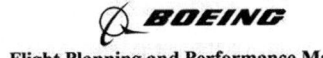
Flight Planning and Performance Manual

737-800/CFM56-7B26
FAA
Category C Brakes

Diversion Fuel and Time

These charts enable rapid determination of fuel and time required to proceed to an alternate airfield (or continue to destination) with one engine inoperative. Data are shown for Long Range Cruise and constant speeds of 280, 290, 310 and 330 KIAS. The fuel required is based on cruise plus descent with straight in approach to touchdown. Diversion fuel and time are determined in a manner similar to the All Engine Long Range Cruise Enroute Fuel and Time chart.

Holding

%N1 required, indicated airspeed and fuel flow are shown for one engine inoperative holding based on the recommended speeds described earlier in this chapter. The fuel flow is based on a racetrack holding pattern and may be reduced by 5% for holding in straight and level flight.

附图 1-1　制作飞行计划的相关图表概述（6）

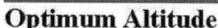

飞行计划

737-800/CFM56-7B26
FAA
Category C Brakes

BOEING
Flight Planning and Performance Manual

FLIGHT PLANNING
Simplified Flight Planning

Optimum Altitude

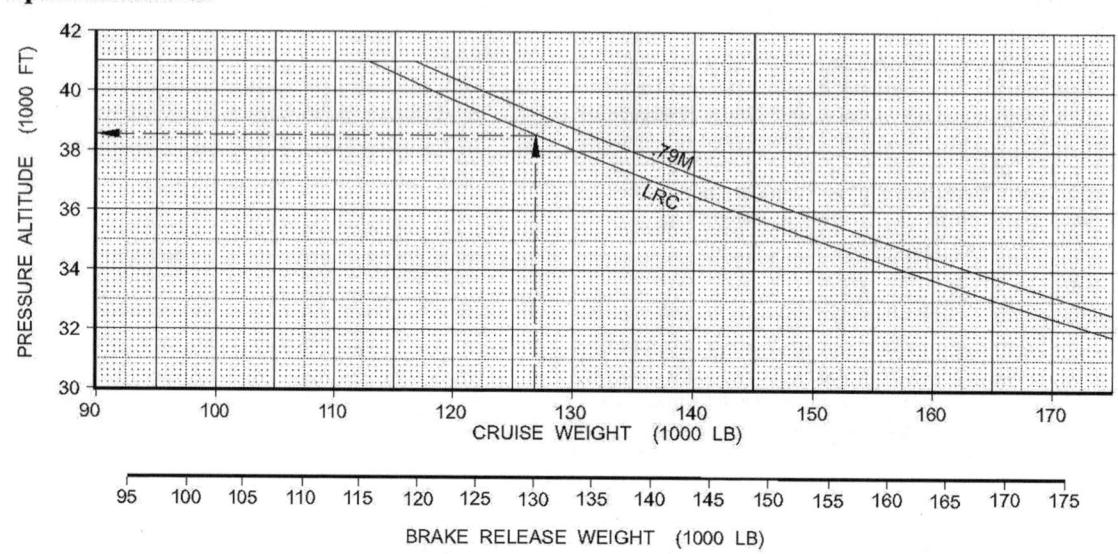

Short Trip Cruise Altitude

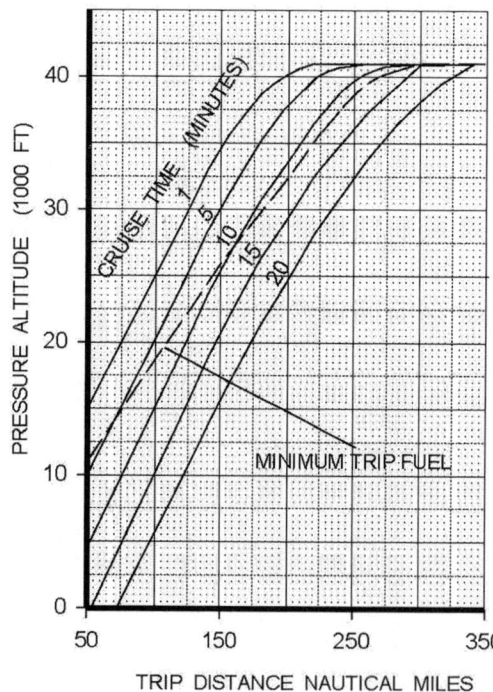

Altitude Adjustments

PARAMETER	ADJUSTMENT
WEIGHT	-200 FT/1000 LB ABOVE 100000 LB
TEMPERATURE	-1300 FT/10°C ABOVE STD +700 FT/10°C BELOW STD
WIND	HEAD: -200 FT/10 KTS TAIL: -1500 FT/10 KTS
AIRPORT ELEVATION	ORIG: +100 FT/1000 FT ELEVATION DEST: +200 FT/1000 FT ELEVATION

附图 1-2　最佳高度图表

FLIGHT PLANNING
Simplified Flight Planning

Flight Planning and Performance Manual

737-800/CFM56-7B26
FAA
Category C Brakes

Long Range Cruise Trip Fuel and Time
200 to 1000 NM
Based on 280/.78 climb and .78/280/250 descent

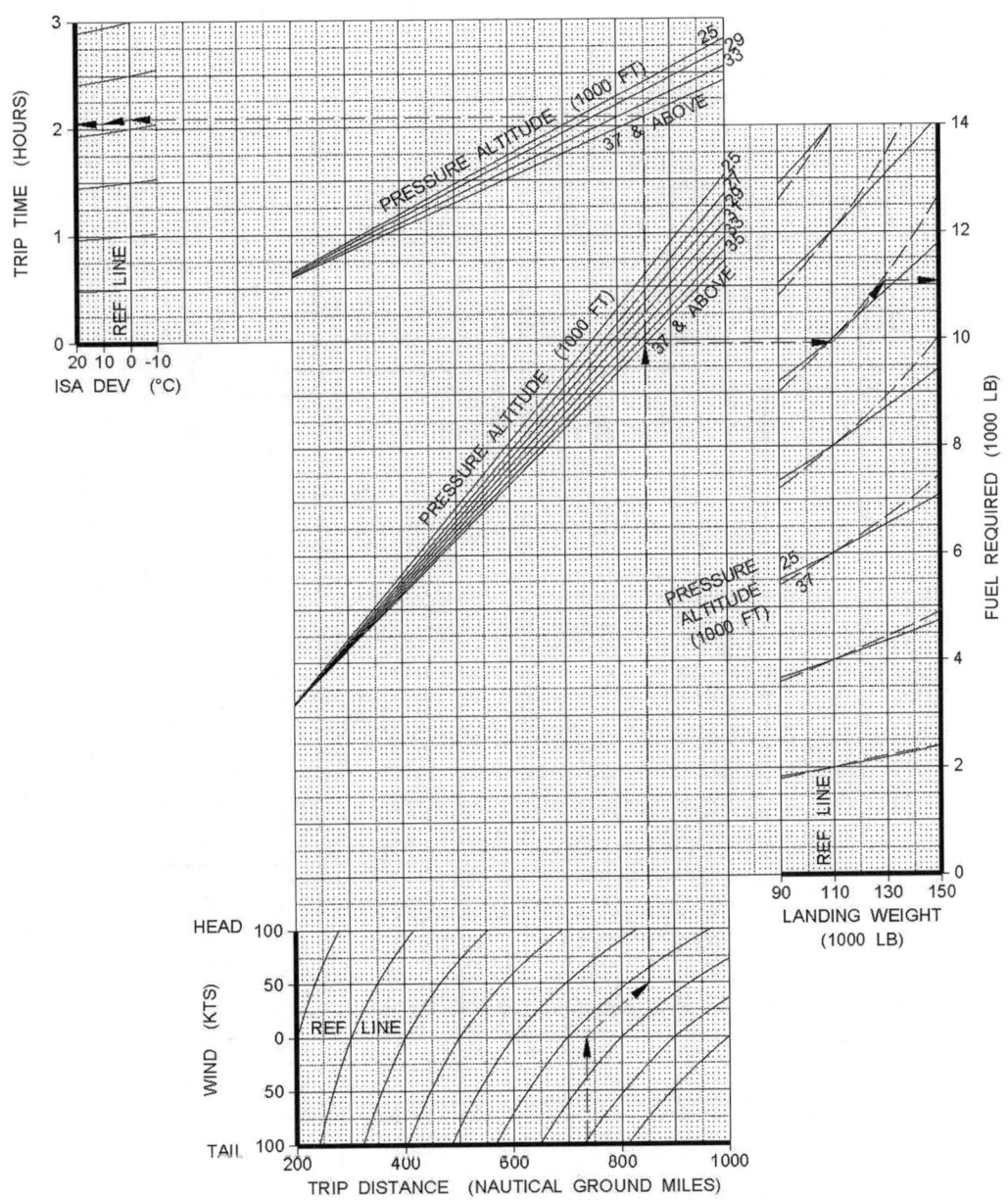

附图 1-3　LRC 巡航模式下的航程油量和航程时间图表

飞行计划

FLIGHT PLANNING
Simplified Flight Planning

BOEING

Flight Planning and Performance Manual

737-800/CFM56-7B26
FAA
Category C Brakes

.79M Trip Fuel and Time
200 to 1000 NM Trip Distance
Based on 280/.78 climb and .78/280/250 descent

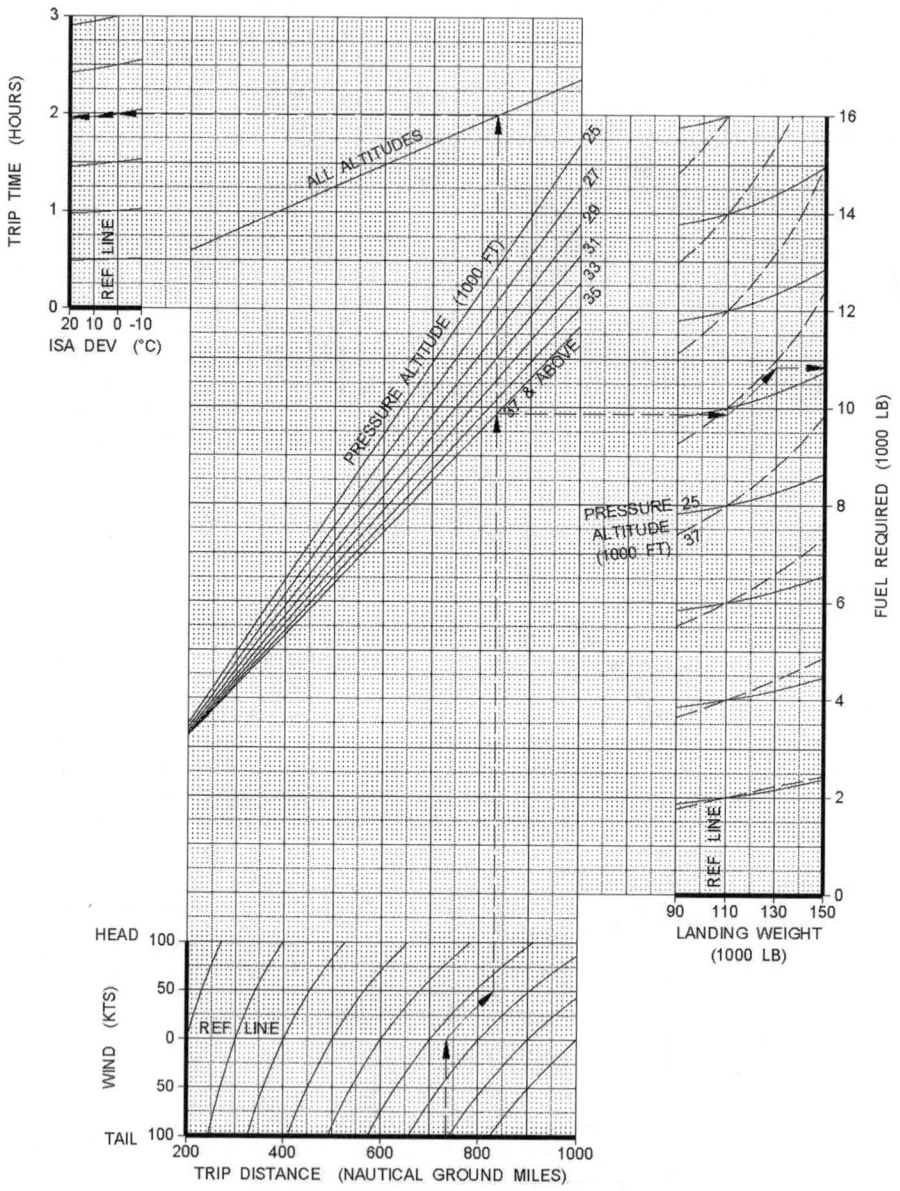

附图 1-4　*M*0.79 巡航模式下的航程油量和航程时间图表

737-800/CFM56-7B26
FAA
Category C Brakes

FLIGHT PLANNING
Simplified Flight Planning

Flight Planning and Performance Manual

Cost Index 30 Trip Fuel and Time
1000 to 5000 NM Trip Distance

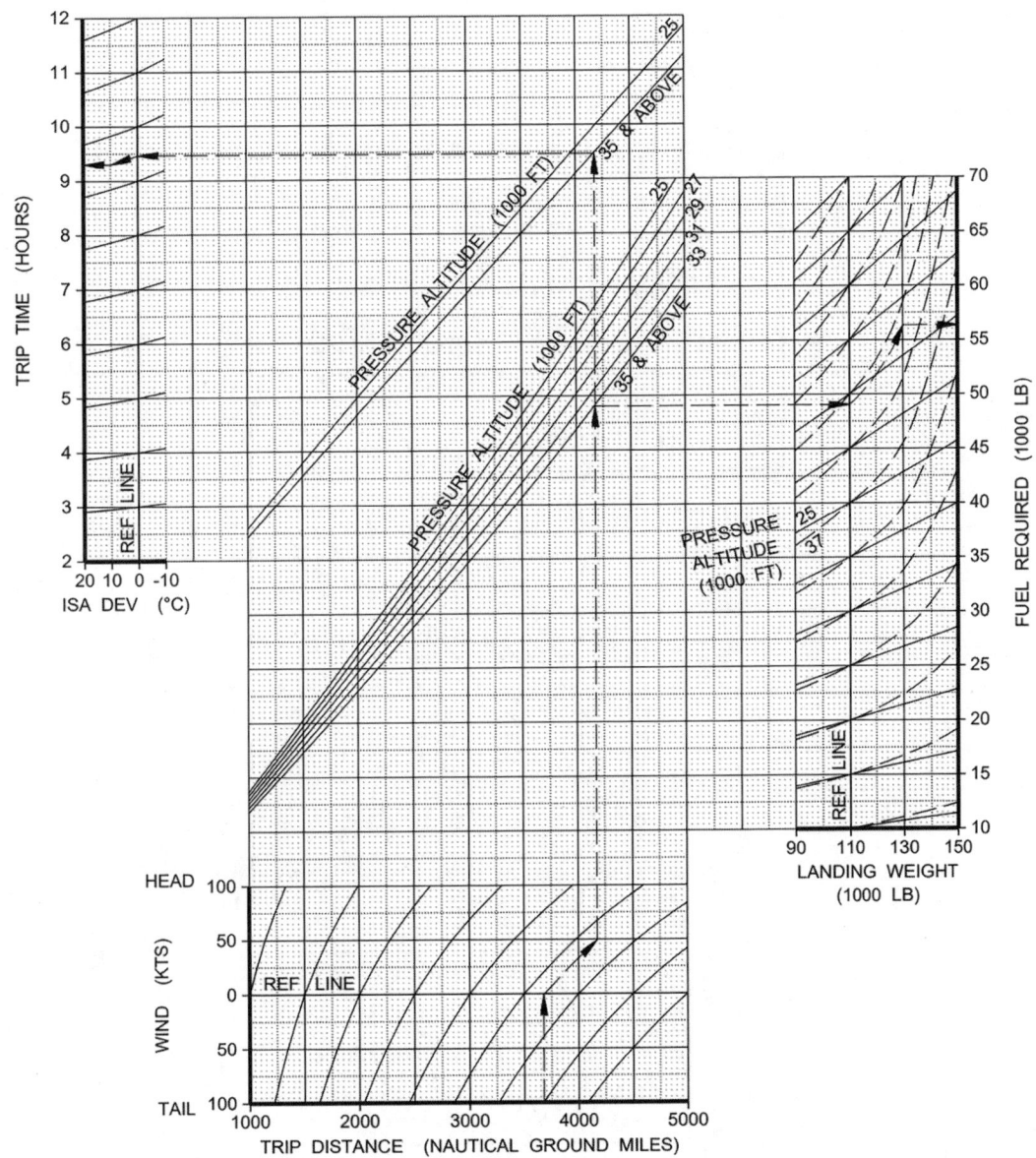

附图 1-5　成本指数 30 巡航模式下的航程油量和航程时间图表

飞行计划

FLIGHT PLANNING
Simplified Flight Planning

BOEING

Flight Planning and Performance Manual

737-800/CFM56-7B26
FAA
Category C Brakes

Long Range Cruise Step Climb Trip Fuel and Time
Based on 280/.78 climb, long range cruise and .78/280/250 descent

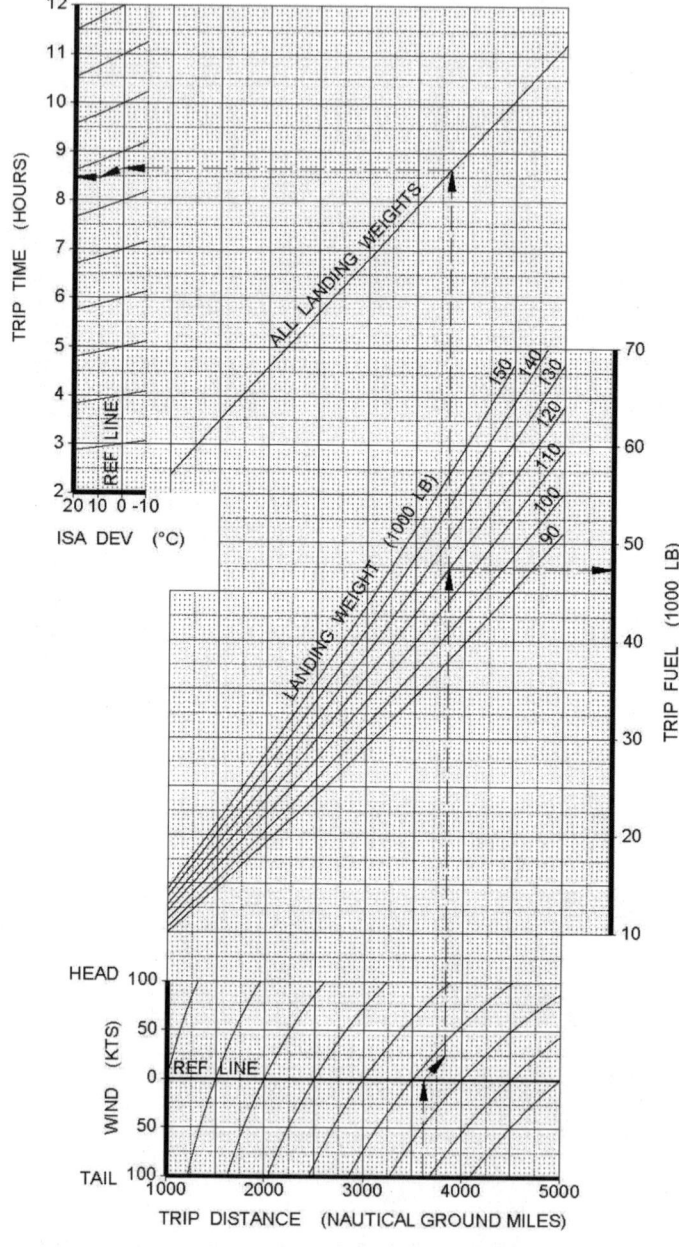

附图 1-6　LRC 阶梯巡航下的航程油量和航程时间图表

737-800/CFM56-7B26
FAA
Category C Brakes

BOEING
Flight Planning and Performance Manual

FLIGHT PLANNING
Simplified Flight Planning

Long Range Cruise Short Trip Fuel and Time
Based on 280/.78 climb and .78/280/250 descent at short trip cruise altitude

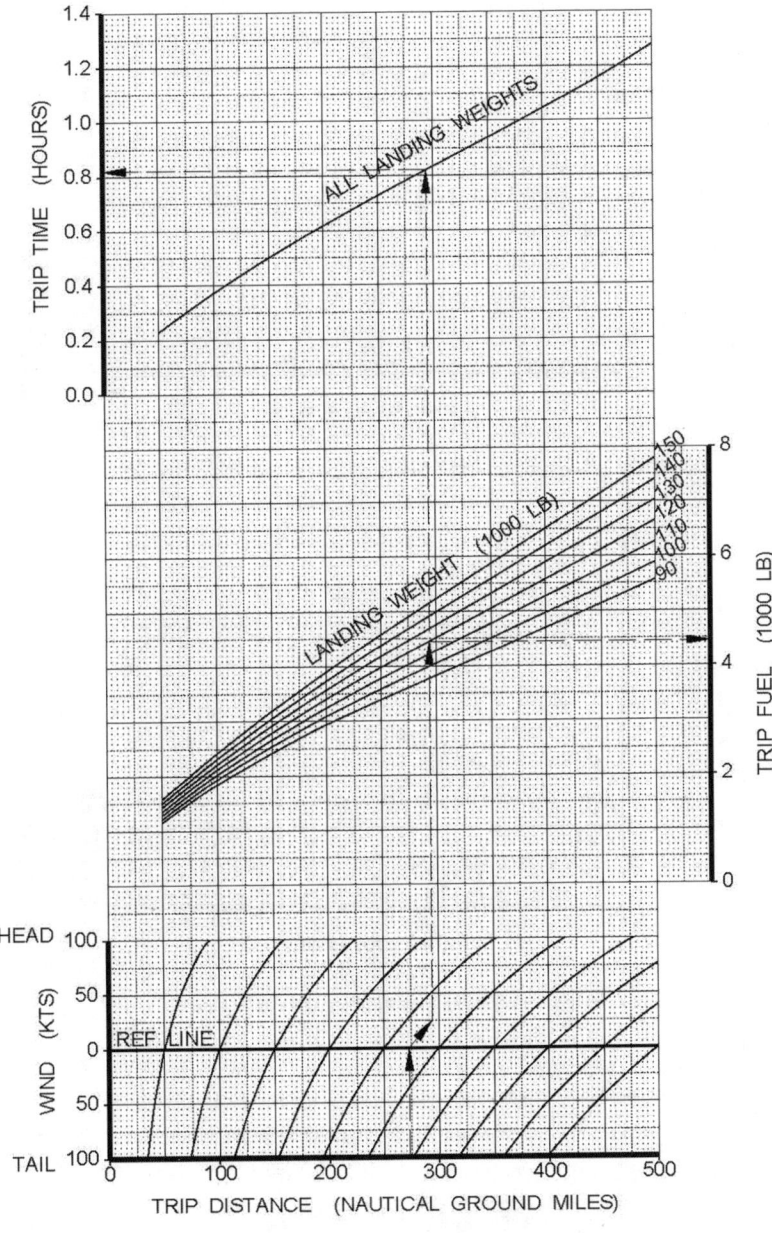

附图 1-7　LRC 改航油量和航程时间图表

FLIGHT PLANNING
Simplified Flight Planning

BOEING
Flight Planning and Performance Manual

757-200
RB211-535E4

Fuel Tankering
LRC/.80M

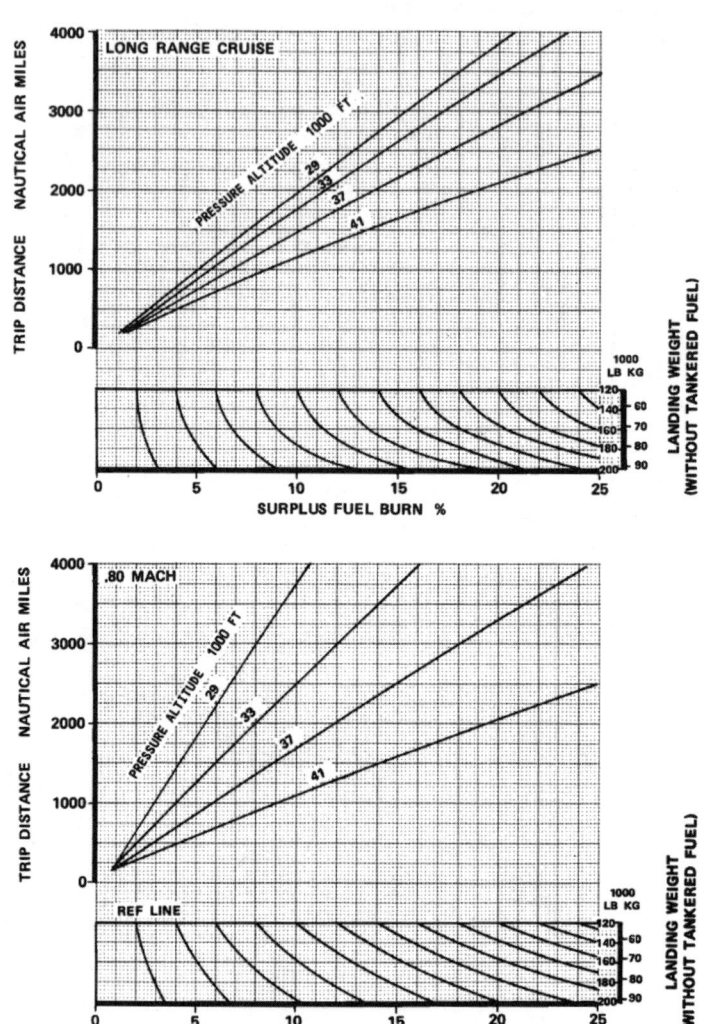

（a）载运燃油分析曲线（B757-200）

Fuel Tankering

The Fuel Tankering table below is provided for LRC/.79M cruise schedule based on a step climb procedure to bracket optimum altitude. When a fuel price differential exists between two stations, this table may be used to determine if fuel may be economically transported for a subsequent flight sector.

Enter the table with the trip distance to read the break-even fuel price ratio. As noted, the break-even fuel price is the fuel price at departure multiplied by the break-even fuel price ratio. To justify tanker operation economically, the fuel price at destination must be greater than the break-even fuel price.

TRIP DISTANCE (NM)	BREAK-EVEN PRICE RATIO
200	1.015
400	1.032
600	1.046
800	1.062
1000	1.078
2000	1.179
3000	1.313
4000	1.500

（b）保本价格率（B737-800）

附图 1-8　保本油价计算图表

ENROUTE
All Engine

Ɓ **BOEING**
Flight Planning and Performance Manual

737-800/CFM56-7B26
FAA
Category C Brakes

Wind-Altitude Trade
LRC

PRESSURE ALTITUDE (1000FT)	CRUISE WEIGHT (1000 LB)									
	180	170	160	150	140	130	120	110	100	90
41					24	10	2	0	5	15
39				18	7	1	1	5	14	29
37		24	12	4	0	1	6	14	27	44
35	16	7	2	0	2	7	15	27	42	60
33	4	1	1	3	9	17	28	42	59	78
31	0	2	5	11	20	30	43	58	75	95
29	3	8	14	23	33	45	59	74	92	111
27	11	18	26	36	47	60	75	91	108	127
25	22	30	40	51	63	76	91	107	124	143

.79M

PRESSURE ALTITUDE (1000 FT)	CRUISE WEIGHT (1000 LB)									
	180	170	160	150	140	130	120	110	100	90
41					40	15	2	1	12	35
39				29	10	1	1	12	32	63
37		41	19	6	0	3	13	32	60	96
35	26	11	2	0	5	16	35	60	93	133
33	5	0	1	8	20	39	63	92	128	170
31	0	4	12	26	44	67	94	127	165	
29	8	18	32	50	72	98	129	163		
27	25	40	58	79	104	133	164			
25	49	67	88	111	138	168				

Above wind factor tables are for calculation of wind required to maintain present range capability at new pressure altitude, i.e. break-even wind.

Method:

1. Read wind factors for present and new altitudes from table.
2. Determine difference (new altitude wind factor minus present altitude wind factor); this difference may be positive or negative.
3. Break-even wind at new altitude is present altitude wind plus difference from step 2.

Example: LRC at 33000 ft, 130000 lb, present wind -20 knots (headwind)

To find break-even wind for
A) Climb to 37000 ft
B) Descent to 29000 ft

A) Wind factors from LRC table at 130000 lb are 17 (33000 ft) and 1 (37000 ft). Difference is -16 so break-even wind is -20 plus -16, i.e. -36 knots (headwind).
B) Similarly, wind factors are 17 (33000 ft) and 45 (29000 ft). Difference is +28 so break-even wind is -20 plus (+28), i.e. 8 knots (tailwind).

附图 1-9　保本风速表

上述风速因子表用于确定在新高度上的保本风速（Break-even Wind，或称得失相当风速），即在新高度上和当前高度有相同航程能力的风速。航程能力相同是指每消耗一磅燃油飞过的地面距离相同。没有风速因子表时可按这一原则确定保本风速。

本表用法如下：

（1）按照巡航速度从相应的表中根据飞机重量读出当前高度和新高度上的风因子；

（2）计算差值：ΔS＝新高度上的风因子－当前高度上的风因子，此差值可正可负；

（3）新高度上的保本风速＝当前高度上的风速＋ΔS（如算出的保本风速为负，表示是顶风）。

注意：风速（指沿航路的风分量）必须以节（kt）为单位，顶风取负值，顺风取正值。

例：飞机重量 160 klb，在 31000 ft 高度以 LRC 巡航，风速－20 kt（顶风），试确定在 35000 ft 和 27000 ft 高度上的保本风速。

（1）在 31000 ft 的风因子＝5　　在 35000 ft 的风因子＝2　　在 29000 ft 的风因子＝26

（2）计算差值 ΔS　　　　$\Delta S＝2-5＝-3$　　　　$\Delta S＝26-5＝21$

（3）计算保本风速　　　　$-20+(-3)＝-23(\mathrm{kt})$　　　$-20+21＝1(\mathrm{kt})$

这就是说：如改飞 35000 ft，该高度上的顶风小于 23 kt 才合算（才可省油），如改飞 27000 ft，该高度上的顺风大于 1 kt 才合算（才可省油）。

飞行计划

ENROUTE
All Engine

Flight Planning and Performance Manual

737-800/CFM56-7B26
FAA
Category C Brakes

280/.78 Enroute Climb
ISA & Below

PRESSURE ALTITUDE (FT)	UNITS MIN/LB NM/KTAS	BRAKE RELEASE WEIGHT (1000 LB)									
		180	170	160	150	140	130	120	110	100	90
41000	TIME/FUEL					28/4500	22/3800	19/3300	17/2900	15/2600	13/2300
	DIST/SPD					180/413	139/408	118/405	102/403	89/402	77/400
40000	TIME/FUEL					24/4100	20/3600	18/3200	16/2800	14/2500	12/2200
	DIST/SPD					149/407	125/404	108/402	95/400	83/399	73/397
39000	TIME/FUEL				25/4400	22/3900	19/3400	17/3100	15/2700	13/2400	12/2100
	DIST/SPD				159/406	133/402	115/400	100/398	88/397	78/396	68/395
38000	TIME/FUEL			27/4800	23/4200	20/3700	18/3300	16/2900	14/2600	12/2400	11/2100
	DIST/SPD			169/405	140/401	121/399	106/397	94/395	83/394	73/393	64/392
37000	TIME/FUEL		29/5200	24/4500	21/4000	19/3600	17/3200	15/2900	13/2600	12/2300	11/2000
	DIST/SPD		182/405	147/400	127/397	112/395	99/393	88/392	78/391	69/390	61/389
36000	TIME/FUEL	31/5600	26/4800	22/4200	20/3800	18/3400	16/3100	14/2800	13/2500	11/2200	10/2000
	DIST/SPD	195/404	155/398	133/395	117/393	104/392	92/390	82/389	73/388	65/387	57/386
35000	TIME/FUEL	27/5100	24/4500	21/4100	19/3700	17/3300	15/3000	14/2700	12/2400	11/2200	10/1900
	DIST/SPD	165/397	141/394	123/391	109/390	98/388	87/387	78/386	69/385	62/384	55/383
34000	TIME/FUEL	25/4800	22/4300	20/3900	18/3600	16/3200	15/2900	13/2600	12/2400	10/2100	9/1900
	DIST/SPD	149/392	130/390	115/388	103/386	92/385	82/384	74/383	66/382	58/381	52/380
33000	TIME/FUEL	24/4600	21/4200	19/3800	17/3400	15/3100	14/2800	13/2500	11/2300	10/2100	9/1800
	DIST/SPD	137/388	121/385	108/384	96/383	87/382	78/381	70/380	62/379	55/378	49/377
32000	TIME/FUEL	22/4400	20/4000	18/3600	16/3300	15/3000	13/2700	12/2500	11/2200	10/2000	9/1800
	DIST/SPD	125/382	111/381	100/379	90/378	81/377	73/376	65/376	58/375	52/374	46/373
31000	TIME/FUEL	20/4200	18/3800	17/3500	15/3200	14/2900	13/2600	11/2400	10/2100	9/1900	8/1700
	DIST/SPD	114/377	102/375	91/374	82/373	74/372	67/372	60/371	54/370	48/369	43/369
30000	TIME/FUEL	19/4000	17/3600	16/3300	14/3100	13/2800	12/2500	11/2300	10/2100	9/1900	8/1700
	DIST/SPD	103/371	93/370	84/369	76/368	69/368	62/367	56/366	50/365	45/365	40/364
29000	TIME/FUEL	18/3800	16/3500	15/3200	14/2900	12/2700	11/2400	10/2200	9/2000	8/1800	7/1600
	DIST/SPD	95/366	85/365	77/365	70/364	64/363	57/362	52/362	47/361	42/360	37/360
28000	TIME/FUEL	17/3600	15/3300	14/3100	13/2800	12/2600	11/2300	10/2100	9/1900	8/1700	7/1500
	DIST/SPD	87/362	79/361	71/360	65/359	59/359	53/358	48/358	43/357	39/356	34/356
27000	TIME/FUEL	16/3500	14/3200	13/2900	12/2700	11/2500	10/2200	9/2000	8/1800	7/1700	7/1500
	DIST/SPD	80/357	72/356	66/356	60/355	55/355	49/354	45/354	40/353	36/352	32/351
26000	TIME/FUEL	15/3300	14/3000	12/2800	12/2600	11/2400	10/2200	9/2000	8/1800	7/1600	6/1400
	DIST/SPD	73/353	67/352	61/352	55/351	51/351	46/350	41/350	37/349	33/348	30/348
25000	TIME/FUEL	14/3200	13/2900	12/2700	11/2500	10/2300	9/2100	8/1900	7/1700	7/1500	6/1400
	DIST/SPD	67/349	62/348	56/348	51/347	47/347	43/346	38/346	35/345	31/345	28/344
24000	TIME/FUEL	13/3000	12/2800	11/2600	10/2400	9/2200	9/2000	8/1800	7/1600	6/1500	6/1300
	DIST/SPD	62/345	57/345	52/344	47/344	43/343	39/343	36/342	32/342	29/341	26/340
23000	TIME/FUEL	12/2900	11/2700	11/2500	10/2300	9/2100	8/1900	7/1700	7/1600	6/1400	5/1300
	DIST/SPD	57/341	52/341	48/340	44/340	40/340	37/339	33/339	30/338	27/337	24/337
22000	TIME/FUEL	12/2800	11/2600	10/2400	9/2200	8/2000	8/1800	7/1700	6/1500	6/1400	5/1200
	DIST/SPD	53/338	48/337	44/337	41/337	37/336	34/336	31/335	28/335	25/334	22/333
21000	TIME/FUEL	11/2600	10/2500	9/2300	9/2100	8/1900	7/1800	7/1600	6/1500	5/1300	5/1200
	DIST/SPD	48/334	44/334	41/334	37/333	34/333	31/332	28/332	26/331	23/331	20/330
20000	TIME/FUEL	10/2500	10/2300	9/2200	8/2000	8/1900	7/1700	6/1500	6/1400	5/1300	5/1100
	DIST/SPD	45/331	41/331	38/330	35/330	32/330	29/329	26/329	24/328	21/327	19/327
18000	TIME/FUEL	9/2300	9/2100	8/2000	7/1900	7/1700	6/1600	6/1400	5/1300	5/1200	4/1000
	DIST/SPD	37/325	35/324	32/324	29/324	27/324	24/323	22/323	20/322	18/321	16/320
16000	TIME/FUEL	8/2100	8/1900	7/1800	7/1700	6/1600	6/1400	5/1300	5/1200	4/1100	4/1000
	DIST/SPD	31/319	29/319	27/318	24/318	22/318	20/317	19/317	17/316	15/315	13/314
14000	TIME/FUEL	7/1900	7/1800	6/1600	6/1500	5/1400	5/1300	5/1200	4/1100	4/1000	3/900
	DIST/SPD	26/313	24/313	22/313	20/313	18/312	17/312	15/311	14/311	13/310	11/309
10000	TIME/FUEL	5/1500	5/1400	5/1400	5/1200	4/1100	4/1000	3/900	3/800	3/800	3/700
	DIST/SPD	16/303	15/303	14/303	13/302	12/302	11/302	10/301	9/300	8/300	7/298
1500	TIME/FUEL	2/600	2/600	2/600	2/600	2/500	2/500	2/400	1/400	1/400	1/300

FUEL ADJUSTMENT FOR HIGH ELEVATION AIRPORTS		AIRPORT ELEVATION	2000	4000	6000	8000	10000	12000
EFFECT ON TIME AND DISTANCE IS NEGLIGIBLE		FUEL ADJUSTMENT	-100	-300	-400	-600	-700	-800

Shaded area approximates optimum altitude at LRC.

附图 1-10　爬升性能图表(1)

737-800/CFM56-7B26
FAA
Category C Brakes

BOEING
Flight Planning and Performance Manual

附 录 |

ENROUTE
All Engine

280/.78 Enroute Climb
ISA + 10 °C

PRESSURE ALTITUDE (FT)	UNITS MIN/LB NM/KTAS	BRAKE RELEASE WEIGHT (1000 LB)									
		180	170	160	150	140	130	120	110	100	90
41000	TIME/FUEL					29/4800	23/4000	20/3500	17/3000	15/2700	13/2400
	DIST/SPD					194/423	148/417	125/414	108/412	94/410	82/409
40000	TIME/FUEL					25/4300	21/3800	18/3300	16/2900	14/2600	12/2300
	DIST/SPD					158/416	133/413	115/411	100/409	88/407	77/406
39000	TIME/FUEL				26/4700	22/4100	20/3600	17/3200	15/2800	13/2500	12/2200
	DIST/SPD				169/415	141/411	121/409	106/407	93/406	82/405	72/403
38000	TIME/FUEL			28/5100	24/4400	21/3900	18/3500	16/3100	14/2800	13/2500	11/2200
	DIST/SPD			182/415	149/410	128/407	112/405	99/404	87/403	77/402	68/400
37000	TIME/FUEL		30/5500	25/4700	22/4200	19/3700	17/3300	15/3000	14/2700	12/2400	11/2100
	DIST/SPD		194/414	157/409	135/406	118/404	104/402	93/401	82/400	73/399	64/398
36000	TIME/FUEL	33/5900	27/5000	23/4400	21/4000	18/3600	16/3200	15/2900	13/2600	12/2300	10/2100
	DIST/SPD	209/413	165/407	142/404	124/402	110/400	98/399	87/398	77/397	69/396	61/395
35000	TIME/FUEL	28/5400	25/4700	22/4300	19/3800	17/3500	16/3100	14/2800	13/2500	11/2300	10/2000
	DIST/SPD	176/406	150/402	131/400	116/398	103/397	92/396	82/395	73/394	65/393	58/392
34000	TIME/FUEL	26/5100	23/4500	21/4100	19/3700	17/3400	15/3000	13/2700	12/2500	11/2200	10/2000
	DIST/SPD	159/401	138/398	122/396	108/395	97/393	87/392	78/391	69/390	62/389	55/389
33000	TIME/FUEL	24/4800	22/4400	20/4000	18/3600	16/3300	14/2900	13/2700	12/2400	10/2100	9/1900
	DIST/SPD	145/396	128/392	114/392	102/391	91/390	82/390	73/388	66/387	58/386	52/385
32000	TIME/FUEL	23/4600	20/4200	18/3800	17/3500	15/3100	14/2800	12/2600	11/2300	10/2100	9/1900
	DIST/SPD	133/390	118/389	105/387	95/386	85/385	76/384	69/383	61/383	55/382	49/381
31000	TIME/FUEL	21/4400	19/4000	17/3600	16/3300	14/3000	13/2700	12/2500	10/2200	9/2000	8/1800
	DIST/SPD	120/382	107/383	96/382	87/381	78/380	71/379	63/378	57/378	51/377	45/376
30000	TIME/FUEL	20/4100	18/3800	16/3500	15/3200	13/2900	12/2600	11/2400	10/2100	9/1900	8/1700
	DIST/SPD	109/379	98/378	88/377	80/376	72/375	65/374	59/374	53/373	47/372	42/371
29000	TIME/FUEL	18/3900	17/3600	15/3300	14/3000	13/2800	12/2500	10/2300	9/2100	8/1900	8/1700
	DIST/SPD	100/374	90/373	81/372	74/371	67/370	60/370	54/369	49/368	44/368	39/367
28000	TIME/FUEL	17/3800	16/3500	14/3200	13/2900	12/2700	11/2400	10/2200	9/2000	8/1800	7/1600
	DIST/SPD	91/369	83/368	75/367	68/367	62/366	56/365	50/365	45/364	41/363	36/363
27000	TIME/FUEL	16/3600	15/3300	14/3000	12/2800	11/2600	10/2300	9/2100	8/1900	8/1700	7/1500
	DIST/SPD	84/364	76/364	69/363	63/362	57/362	52/361	47/361	42/360	38/359	34/358
26000	TIME/FUEL	15/3400	14/3200	13/2900	12/2700	11/2500	10/2200	9/2000	8/1800	7/1700	6/1500
	DIST/SPD	77/360	70/359	64/359	58/358	53/358	48/357	43/357	39/356	35/355	31/354
25000	TIME/FUEL	14/3300	13/3000	12/2800	11/2600	10/2400	9/2200	8/2000	8/1800	7/1600	6/1400
	DIST/SPD	71/356	64/355	59/355	54/354	49/354	45/353	40/353	36/352	33/351	29/351
24000	TIME/FUEL	13/3100	12/2900	11/2700	11/2500	10/2300	9/2100	8/1900	7/1700	7/1500	6/1400
	DIST/SPD	65/352	59/351	54/351	50/350	45/350	41/349	37/349	34/348	30/348	27/347
23000	TIME/FUEL	13/3000	12/2800	11/2600	10/2400	9/2200	8/2000	8/1800	7/1600	6/1500	6/1300
	DIST/SPD	60/348	55/347	50/347	46/347	42/346	38/346	35/345	31/345	28/344	25/343
22000	TIME/FUEL	12/2900	11/2700	10/2500	9/2300	9/2100	8/1900	7/1700	7/1600	6/1400	5/1300
	DIST/SPD	55/344	51/344	46/343	42/343	39/343	35/342	32/342	29/341	26/340	23/340
21000	TIME/FUEL	11/2700	10/2500	10/2400	9/2200	8/2000	7/1800	7/1700	6/1500	6/1400	5/1200
	DIST/SPD	51/341	47/340	43/340	39/340	36/339	33/339	30/338	27/338	24/337	21/336
20000	TIME/FUEL	11/2600	10/2400	9/2300	8/2100	8/1900	7/1800	6/1600	6/1400	5/1300	5/1200
	DIST/SPD	47/337	43/337	39/337	36/336	33/336	30/335	27/335	25/334	22/334	20/333
18000	TIME/FUEL	9/2400	9/2200	8/2100	8/1900	7/1800	6/1600	6/1500	5/1300	5/1200	4/1100
	DIST/SPD	39/331	36/331	33/330	31/330	28/330	26/329	23/329	21/328	19/327	17/326
16000	TIME/FUEL	8/2100	8/2000	7/1900	7/1700	6/1600	6/1500	5/1300	5/1200	4/1100	4/1000
	DIST/SPD	33/325	30/325	28/324	26/324	23/324	21/323	19/323	18/322	16/321	14/320
14000	TIME/FUEL	7/1900	7/1800	6/1700	6/1600	6/1400	5/1300	5/1300	4/1100	4/1000	3/900
	DIST/SPD	27/319	25/319	23/319	21/318	19/318	18/318	16/317	15/316	13/316	12/315
10000	TIME/FUEL	6/1500	5/1400	5/1300	5/1200	4/1100	4/1000	4/1000	3/900	3/800	3/700
	DIST/SPD	17/309	15/308	14/308	13/308	12/308	11/307	10/307	9/306	8/305	7/304
1500	TIME/FUEL	2/600	2/600	2/600	2/600	2/500	2/500	2/400	1/400	1/400	1/300

FUEL ADJUSTMENT FOR HIGH ELEVATION AIRPORTS		AIRPORT ELEVATION	2000	4000	6000	8000	10000	12000
EFFECT ON TIME AND DISTANCE IS NEGLIGIBLE		FUEL ADJUSTMENT	-100	-300	-400	-600	-700	-900

Shaded area approximates optimum altitude at LRC.

附图 1-10　爬升性能图表(2)

ENROUTE
All Engine

BOEING
Flight Planning and Performance Manual

737-800/CFM56-7B26
FAA
Category C Brakes

280/.78 Enroute Climb
ISA + 15 °C

PRESSURE ALTITUDE (FT)	UNITS MIN/LB NM/KTAS	BRAKE RELEASE WEIGHT (1000 LB)									
		180	170	160	150	140	130	120	110	100	90
41000	TIME/FUEL						26/4300	22/3700	19/3200	17/2800	14/2500
	DIST/SPD						175/424	144/420	123/418	106/416	92/414
40000	TIME/FUEL					29/4700	24/4000	20/3500	18/3100	16/2700	14/2400
	DIST/SPD					190/423	154/419	131/416	113/414	99/413	86/411
39000	TIME/FUEL				31/5200	25/4400	22/3800	19/3400	17/3000	15/2700	13/2400
	DIST/SPD				206/423	164/418	139/415	121/413	105/411	92/410	81/408
38000	TIME/FUEL			34/5700	27/4700	23/4100	20/3700	18/3300	16/2900	14/2600	12/2300
	DIST/SPD			225/423	174/416	147/413	128/411	112/409	98/408	87/407	76/405
37000	TIME/FUEL			29/5100	25/4500	22/4000	19/3500	17/3100	15/2800	13/2500	12/2200
	DIST/SPD			185/415	156/411	135/409	118/407	104/406	92/405	81/403	72/402
36000	TIME/FUEL		31/5500	26/4800	23/4200	20/3800	18/3400	16/3000	14/2700	13/2400	11/2200
	DIST/SPD		196/414	164/410	142/407	125/405	110/404	98/402	87/401	77/400	68/399
35000	TIME/FUEL	33/5900	28/5100	24/4500	22/4100	19/3700	17/3300	15/3000	14/2600	12/2400	11/2100
	DIST/SPD	212/413	174/408	150/405	132/403	117/402	104/400	92/399	82/398	73/397	64/396
34000	TIME/FUEL	30/5500	26/4900	23/4400	21/3900	18/3500	16/3200	15/2900	13/2600	12/2300	10/2000
	DIST/SPD	187/407	159/404	139/401	123/400	109/398	98/397	87/396	77/395	69/394	61/393
33000	TIME/FUEL	28/5200	24/4700	22/4200	19/3800	18/3400	16/3100	14/2800	13/2500	11/2200	10/2000
	DIST/SPD	170/402	147/399	129/397	115/396	103/394	92/393	82/392	73/391	65/391	58/390
32000	TIME/FUEL	26/4900	23/4400	20/4000	18/3700	17/3300	15/3000	13/2700	12/2400	11/2200	10/1900
	DIST/SPD	153/396	134/394	119/392	107/391	96/390	86/389	77/388	68/387	61/386	54/385
31000	TIME/FUEL	23/4700	21/4200	19/3800	17/3500	16/3200	14/2900	13/2600	11/2300	10/2100	9/1900
	DIST/SPD	137/389	122/388	109/386	98/385	88/384	79/383	71/383	63/382	56/381	50/380
30000	TIME/FUEL	22/4400	20/4000	18/3700	16/3300	15/3000	13/2700	12/2500	11/2200	10/2000	9/1800
	DIST/SPD	124/384	111/382	100/381	90/380	81/379	73/379	65/378	59/377	52/376	46/376
29000	TIME/FUEL	20/4200	18/3800	17/3500	15/3200	14/2900	13/2600	11/2400	10/2100	9/1900	8/1700
	DIST/SPD	113/378	101/377	91/376	83/375	75/375	67/374	61/373	54/372	49/372	43/371
28000	TIME/FUEL	19/4000	17/3600	16/3300	14/3100	13/2800	12/2500	11/2300	10/2100	9/1900	8/1700
	DIST/SPD	103/373	93/372	84/371	76/371	69/370	62/369	56/369	50/368	45/367	40/367
27000	TIME/FUEL	18/3800	16/3500	15/3200	14/2900	12/2700	11/2400	10/2200	9/2000	8/1800	7/1600
	DIST/SPD	94/368	85/367	77/367	70/366	64/366	58/365	52/364	47/364	42/363	37/362
26000	TIME/FUEL	17/3600	15/3300	14/3100	13/2800	12/2600	11/2300	10/2100	9/1900	8/1700	7/1500
	DIST/SPD	86/364	78/363	71/362	65/362	59/361	53/361	48/360	43/360	39/359	35/358
25000	TIME/FUEL	16/3400	14/3200	13/2900	12/2700	11/2500	10/2200	9/2000	8/1800	7/1700	7/1500
	DIST/SPD	79/359	72/359	66/358	60/358	55/357	49/357	45/356	40/356	36/355	32/354
24000	TIME/FUEL	15/3300	13/3000	12/2800	11/2600	10/2400	9/2200	9/2000	8/1800	7/1600	6/1400
	DIST/SPD	73/355	67/355	61/354	55/354	50/353	46/353	41/352	37/352	33/351	30/351
23000	TIME/FUEL	14/3100	13/2900	12/2700	11/2500	10/2300	9/2100	8/1900	7/1700	7/1500	6/1400
	DIST/SPD	67/351	61/351	56/351	51/350	47/350	42/349	38/349	35/348	31/348	28/347
22000	TIME/FUEL	13/3000	12/2800	11/2600	10/2400	9/2200	9/2000	8/1800	7/1600	6/1500	6/1300
	DIST/SPD	62/348	56/347	52/347	47/346	43/346	39/346	36/345	32/345	29/344	26/343
21000	TIME/FUEL	12/2900	11/2700	10/2500	10/2300	9/2100	8/1900	7/1700	7/1600	6/1400	5/1300
	DIST/SPD	57/344	52/344	48/343	44/343	40/343	36/342	33/342	30/341	27/341	24/340
20000	TIME/FUEL	11/2700	11/2500	10/2300	9/2200	8/2000	8/1800	7/1700	6/1500	6/1400	5/1200
	DIST/SPD	52/341	48/340	44/340	40/340	37/339	33/339	30/338	27/338	25/337	22/336
18000	TIME/FUEL	10/2500	9/2300	9/2100	8/2000	7/1800	7/1700	6/1500	6/1400	5/1200	5/1100
	DIST/SPD	44/334	40/334	37/333	31/333	31/333	28/332	26/332	23/332	21/331	19/330
16000	TIME/FUEL	9/2200	8/2100	8/1900	7/1800	7/1700	6/1500	6/1400	5/1200	5/1100	4/1000
	DIST/SPD	36/328	34/328	31/327	28/327	26/327	24/326	22/326	19/326	18/325	16/324
14000	TIME/FUEL	8/2000	7/1900	7/1700	6/1600	6/1500	5/1400	5/1200	4/1100	4/1000	4/900
	DIST/SPD	30/322	28/322	25/322	23/321	21/321	20/321	18/320	16/320	14/319	13/318
10000	TIME/FUEL	6/1600	6/1500	5/1400	5/1300	4/1200	4/1100	4/1000	3/900	3/800	3/700
	DIST/SPD	19/312	17/311	16/311	15/311	13/311	12/310	11/310	10/309	9/308	8/307
1500	TIME/FUEL	2/600	2/600	2/600	2/500	2/500	2/400	1/400	1/400	1/300	

FUEL ADJUSTMENT FOR HIGH ELEVATION AIRPORTS			AIRPORT ELEVATION	2000	4000	6000	8000	10000	12000
EFFECT ON TIME AND DISTANCE IS NEGLIGIBLE			FUEL ADJUSTMENT	-200	-300	-500	-600	-800	-900

Shaded area approximates optimum altitude at LRC.

附图 1-10　爬升性能图表(3)

737-800/CFM56-7B26
FAA
Category C Brakes

BOEING

Flight Planning and Performance Manual

ENROUTE
All Engine

280/.78 Enroute Climb
ISA + 20 °C

PRESSURE ALTITUDE (FT)	UNITS MIN/LB NM/KTAS	BRAKE RELEASE WEIGHT (1000 LB)									
		180	170	160	150	140	130	120	110	100	90
41000	TIME/FUEL						32/4800	25/4000	22/3500	19/3000	16/2700
	DIST/SPD						215/431	170/426	142/423	122/421	105/420
40000	TIME/FUEL					35/5300	27/4400	23/3800	20/3300	18/2900	15/2600
	DIST/SPD					240/432	183/425	152/422	131/420	113/418	98/417
39000	TIME/FUEL					30/4800	25/4100	22/3600	19/3200	17/2800	14/2500
	DIST/SPD					197/424	163/421	139/418	121/416	105/415	92/414
38000	TIME/FUEL				32/5200	27/4500	23/3900	20/3500	18/3100	16/2700	14/2400
	DIST/SPD				211/423	173/419	148/416	129/415	113/413	99/412	86/411
37000	TIME/FUEL			34/5700	28/4900	25/4300	22/3800	19/3400	17/3000	15/2700	13/2300
	DIST/SPD			226/423	184/418	157/415	137/413	120/411	105/410	93/408	81/407
36000	TIME/FUEL		37/6200	30/5200	26/4600	23/4100	20/3600	18/3200	16/2900	14/2600	13/2300
	DIST/SPD		244/422	196/416	166/413	145/411	127/409	112/408	99/406	87/405	77/404
35000	TIME/FUEL	41/6800	33/5700	28/4900	25/4400	22/3900	19/3500	17/3100	15/2800	14/2500	12/2200
	DIST/SPD	270/422	211/415	177/412	154/409	135/407	119/406	105/404	93/403	82/402	73/401
34000	TIME/FUEL	35/6200	30/5300	26/4700	23/4200	21/3800	18/3400	16/3000	15/2700	13/2400	12/2200
	DIST/SPD	229/414	190/410	163/407	143/405	126/403	112/402	99/401	88/400	78/399	69/398
33000	TIME/FUEL	32/5800	28/5100	25/4500	22/4100	20/3700	18/3300	16/2900	14/2600	12/2400	11/2100
	DIST/SPD	204/408	173/405	151/403	133/401	118/399	105/398	93/397	83/396	74/395	65/394
32000	TIME/FUEL	29/5400	26/4800	23/4300	21/3900	18/3500	17/3200	15/2800	13/2500	12/2300	11/2000
	DIST/SPD	182/402	157/399	138/397	123/396	109/394	97/393	87/392	77/391	69/391	61/390
31000	TIME/FUEL	27/5100	24/4500	21/4100	19/3700	17/3400	16/3000	14/2700	13/2400	11/2200	10/1900
	DIST/SPD	161/395	141/393	125/391	112/390	100/389	90/388	80/387	71/386	64/386	56/385
30000	TIME/FUEL	25/4800	22/4300	20/3900	18/3500	16/3200	15/2900	13/2600	12/2300	11/2100	9/1900
	DIST/SPD	145/388	128/387	114/386	103/385	92/384	82/383	74/382	66/381	59/381	52/380
29000	TIME/FUEL	23/4500	21/4100	19/3700	17/3400	15/3100	14/2800	12/2500	11/2300	10/2000	9/1800
	DIST/SPD	131/383	117/381	105/380	94/380	85/379	76/378	68/377	61/377	55/376	48/375
28000	TIME/FUEL	21/4300	19/3900	17/3500	16/3200	14/2900	13/2700	12/2400	11/2200	9/1900	8/1700
	DIST/SPD	119/378	107/376	96/376	87/375	78/374	70/373	63/373	57/372	51/372	45/371
27000	TIME/FUEL	20/4100	18/3700	16/3400	15/3100	14/2800	12/2600	11/2300	10/2100	9/1900	8/1700
	DIST/SPD	109/371	98/371	88/371	80/370	72/370	65/369	59/368	53/368	47/367	42/367
26000	TIME/FUEL	19/3900	17/3500	15/3200	14/3000	13/2700	12/2500	11/2200	9/2000	8/1800	8/1600
	DIST/SPD	100/368	90/367	81/366	74/366	67/365	60/365	54/364	49/364	44/363	39/362
25000	TIME/FUEL	17/3700	16/3400	15/3100	13/2900	12/2600	11/2400	10/2100	9/1900	8/1700	7/1500
	DIST/SPD	91/363	82/363	75/362	68/362	62/361	56/361	50/360	45/360	40/359	36/358
24000	TIME/FUEL	16/3500	15/3200	14/3000	13/2700	11/2500	10/2300	9/2000	8/1800	8/1700	7/1500
	DIST/SPD	83/359	76/359	69/358	63/358	57/357	52/357	47/356	42/356	38/355	33/355
23000	TIME/FUEL	15/3300	14/3100	13/2800	12/2600	11/2400	10/2200	9/2000	8/1800	7/1600	6/1400
	DIST/SPD	77/355	70/355	63/354	58/354	53/353	48/353	43/352	39/352	35/351	31/351
22000	TIME/FUEL	14/3200	13/2900	12/2700	11/2500	10/2300	9/2100	8/1900	8/1700	7/1500	6/1400
	DIST/SPD	70/351	64/351	58/350	53/350	49/350	44/349	40/349	36/348	32/348	29/347
21000	TIME/FUEL	13/3000	12/2800	11/2600	11/2400	10/2200	9/2000	8/1800	7/1600	6/1500	6/1300
	DIST/SPD	64/348	59/347	54/347	49/346	45/346	41/346	37/345	33/345	30/344	26/344
20000	TIME/FUEL	13/2900	12/2700	11/2500	10/2300	9/2100	8/1900	8/1700	7/1600	6/1400	5/1300
	DIST/SPD	59/344	54/344	49/343	45/343	41/343	37/342	34/342	31/341	27/341	24/340
18000	TIME/FUEL	11/2600	10/2400	10/2200	9/2100	8/1900	7/1700	7/1600	6/1400	5/1300	5/1200
	DIST/SPD	50/337	45/337	42/337	38/337	35/336	32/336	29/335	26/335	23/334	21/334
16000	TIME/FUEL	10/2300	9/2200	8/2000	8/1900	7/1700	7/1600	6/1400	5/1300	5/1200	4/1000
	DIST/SPD	41/331	38/331	35/331	32/330	29/330	26/330	24/329	22/329	20/328	17/328
14000	TIME/FUEL	9/2100	8/1900	7/1800	7/1700	6/1500	6/1400	5/1300	5/1200	4/1100	4/900
	DIST/SPD	34/325	31/325	28/325	26/325	24/324	22/324	20/324	18/323	16/323	14/322
10000	TIME/FUEL	6/1600	6/1500	6/1400	5/1300	5/1200	4/1100	4/1000	4/900	3/800	3/700
	DIST/SPD	21/314	19/314	18/314	16/314	15/314	14/313	12/313	11/312	10/312	9/311
1500	TIME/FUEL	2/600	2/600	2/600	2/600	2/500	2/500	2/400	1/400	1/400	1/300

FUEL ADJUSTMENT FOR HIGH ELEVATION AIRPORTS		AIRPORT ELEVATION	2000	4000	6000	8000	10000	12000
EFFECT ON TIME AND DISTANCE IS NEGLIGIBLE		FUEL ADJUSTMENT	-200	-300	-500	-600	-800	-1000

Shaded area approximates optimum altitude at LRC.

附图 1-10　爬升性能图表(4)

Cruise tables are provided for Long Range Cruise and .79M speed schedules with both engines operating assuming normal engine bleed for air conditioning(2 bleeds/2 packs on). These tables are similar to the cruise control tables presented in the Performance Inflight chapter of the QRH and are expanded to include true airspeed and Max Cruise Thrust setting information.

To account for APU fuel burn, refer to the APU Operation During Flight table in the chapter 2 text.

ENROUTE	*BOEING*	737-800/CFM56-7B26
All Engine	**Flight Planning and Performance Manual**	FAA
		Category C Brakes

Long Range Cruise Table
41000 FT to 36000 FT

PRESS ALT (1000 FT) (STD TAT)		WEIGHT (1000 LB)									
		180	170	160	150	140	130	120	110	100	90
41 (-30)	%N1						93.7	90.2	87.8	86.0	84.0
	MAX TAT						-18	-9	-3		
	KIAS						234	235	232	228	222
	MACH						.793	.794	.786	.774	.754
	FF/ENG						2605	2344	2142	1998	1826
	KTAS						455	455	451	444	433
40 (-30)	%N1						91.1	88.7	86.7	85.0	82.8
	MAX TAT						-11	-5	1		
	KIAS						241	239	235	231	222
	MACH						.795	.790	.780	.767	.740
	FF/ENG						2532	2319	2134	1977	1814
	KTAS						456	453	447	440	425
39 (-31)	%N1					92.1	89.4	87.4	85.7	83.9	81.5
	MAX TAT					-13	-6	0			
	KIAS					246	246	243	239	233	222
	MACH					.795	.793	.785	.774	.757	.723
	FF/ENG					2725	2498	2304	2139	1972	1792
	KTAS					456	455	450	444	434	415
38 (-31)	%N1				93.2	90.1	88.0	86.3	84.8	82.8	80.2
	MAX TAT				-16	-8	-2				
	KIAS				251	252	250	246	242	234	220
	MACH				.793	.795	.788	.779	.767	.743	.704
	FF/ENG				2935	2678	2475	2299	2143	1961	1770
	KTAS				455	456	452	447	440	426	404
37 (-31)	%N1			93.9	90.6	88.5	86.8	85.3	83.7	81.5	79.0
	MAX TAT			-18	-9	-3	2				
	KIAS			257	258	256	254	250	244	233	219
	MACH			.792	.795	.791	.783	.773	.756	.726	.686
	FF/ENG			3146	2858	2648	2463	2304	2137	1939	1753
	KTAS			454	456	454	449	443	434	417	394
36 (-32)	%N1			91.1	89.0	87.3	85.8	84.4	82.6	80.3	77.7
	MAX TAT			-11	-5	0					
	KIAS			264	263	261	257	253	245	232	218
	MACH			.795	.793	.786	.777	.765	.743	.707	.669
	FF/ENG			3040	2823	2631	2463	2307	2126	1915	1736
	KTAS			456	455	451	446	439	426	406	384

Max TAT not shown where %N1 can be set in ISA + 30 °C conditions.
Increase/decrease %N1 required by 1% per 5 °C above/below standard TAT.
Increase/decrease fuel flow 3% per 10 °C above/below standard TAT.
Increase/decrease KTAS by 1 knot per 1 °C above/below standard TAT.
Shaded area approximates optimum altitude.

Max Cruise %N1

PRESS ALT (1000 FT)	TAT (°C)													
	-65	-60	-55	-50	-45	-40	-35	-30	-25	-20	-15	-10	-5	0
41	86.3	87.2	88.2	89.1	90.0	91.0	91.9	92.7	93.6	94.4	94.1	93.4	92.6	91.8
40	86.4	87.3	88.3	89.2	90.2	91.1	92.0	92.9	93.7	94.6	94.0	93.4	92.6	91.8
39	86.5	87.5	88.4	89.4	90.3	91.2	92.1	93.0	93.9	94.4	93.9	93.3	92.5	91.7
38	86.6	87.6	88.5	89.4	90.3	91.3	92.2	93.1	93.9	94.3	93.7	93.0	92.2	91.3
37	86.7	87.7	88.6	89.5	90.5	91.4	92.3	93.2	94.0	94.1	93.4	92.6	91.8	90.9
36	86.4	87.4	88.3	89.2	90.2	91.1	92.0	92.9	93.7	93.9	93.2	92.4	91.5	90.6

With engine anti-ice on, decrease limit %N1 by 0.9.
With engine and wing anti-ice on, decrease limit %N1 by 3.4.

附图 1-11　LRC 巡航性能图表(1)

737-800/CFM56-7B26
FAA
Category C Brakes

BOEING
Flight Planning and Performance Manual

ENROUTE
All Engine

Long Range Cruise Table
35000 FT to 30000 FT

PRESS ALT (1000 FT) (STD TAT)		WEIGHT (1000 LB)									
		180	170	160	150	140	130	120	110	100	90
35 (-30)	%N1		91.9	89.7	88.1	86.6	85.3	83.7	81.7	79.4	76.9
	MAX TAT		-10	-4	1						
	KIAS		270	270	268	264	261	255	244	231	219
	MACH		.795	.794	.788	.780	.771	.754	.726	.690	.655
	FF/ENG		3241	3013	2810	2630	2480	2314	2116	1907	1738
	KTAS		458	458	454	450	444	435	419	397	378
34 (-28)	%N1	92.7	90.4	88.7	87.3	86.0	84.6	82.9	80.8	78.5	76.3
	MAX TAT	-11	-5	0	5						
	KIAS	276	276	274	272	268	264	256	243	230	220
	MACH	.794	.795	.790	.783	.774	.762	.740	.707	.672	.644
	FF/ENG	3448	3207	2999	2811	2653	2497	2315	2102	1899	1746
	KTAS	460	460	458	453	448	441	429	409	389	373
33 (-26)	%N1	91.1	89.4	88.0	86.7	85.5	83.9	82.1	80.0	77.7	75.6
	MAX TAT	-5	0	5	9						
	KIAS	283	281	279	276	272	265	255	242	230	221
	MACH	.795	.791	.785	.777	.768	.751	.723	.690	.658	.634
	FF/ENG	3402	3190	2998	2831	2679	2507	2303	2095	1897	1758
	KTAS	462	460	457	452	446	437	421	401	383	369
32 (-24)	%N1	89.9	88.6	87.3	86.2	84.8	83.1	81.2	79.1	77.0	75.0
	MAX TAT	0	4	9							
	KIAS	288	286	283	279	274	265	253	241	231	222
	MACH	.792	.787	.779	.771	.758	.736	.705	.673	.647	.624
	FF/ENG	3381	3185	3010	2858	2692	2503	2290	2085	1902	1769
	KTAS	463	459	455	451	443	430	412	393	378	365
31 (-23)	%N1	89.1	87.9	86.8	85.5	84.1	82.3	80.4	78.3	76.4	74.4
	MAX TAT	4	8	12							
	KIAS	292	290	287	282	275	264	252	241	232	224
	MACH	.788	.781	.774	.763	.745	.718	.688	.659	.637	.615
	FF/ENG	3373	3193	3036	2877	2698	2489	2280	2080	1913	1784
	KTAS	462	458	454	448	437	421	404	387	374	361
30 (-21)	%N1	88.4	87.3	86.2	84.8	83.3	81.5	79.5	77.6	75.8	73.8
	MAX TAT	8	12								
	KIAS	297	294	290	284	275	263	251	242	234	225
	MACH	.782	.775	.767	.752	.729	.700	.671	.647	.627	.606
	FF/ENG	3379	3217	3061	2887	2687	2474	2267	2084	1922	1802
	KTAS	461	457	452	443	430	413	396	381	369	357

Max TAT not shown where %N1 can be set in ISA + 30 °C conditions.
Increase/decrease %N1 required by 1% per 5 °C above/below standard TAT.
Increase/decrease fuel flow 3% per 10 °C above/below standard TAT.
Increase/decrease KTAS by 1 knot per 1 °C above/below standard TAT.
Shaded area approximates optimum altitude.

Max Cruise %N1

PRESS ALT (1000 FT)	TAT (°C)													
	-60	-55	-50	-45	-40	-35	-30	-25	-20	-15	-10	-5	0	5
35	87.0	88.0	88.9	89.9	90.8	91.7	92.6	93.4	94.3	93.6	92.8	91.9	91.1	90.2
34	86.3	87.2	88.2	89.2	90.1	91.0	92.0	92.9	93.8	94.0	93.2	92.3	91.5	90.6
33	85.7	86.7	87.7	88.7	89.6	90.6	91.5	92.4	93.3	94.2	93.7	92.8	91.9	91.0
32	85.2	86.2	87.1	88.1	89.0	89.9	90.8	91.7	92.6	93.5	94.1	93.2	92.3	91.5
31	84.8	85.8	86.7	87.6	88.5	89.4	90.3	91.2	92.1	93.0	93.8	93.8	92.9	92.0
30	84.2	85.2	86.1	87.1	88.0	88.9	89.8	90.7	91.6	92.5	93.3	94.2	93.2	92.4

With engine anti-ice on, decrease limit %N1 by 1.1.
With engine and wing anti-ice on, decrease limit %N1 by 3.5.

附图 1-11　LRC 巡航性能图表(2)

飞行计划

ENROUTE
All Engine

BOEING

737-800/CFM56-7B26
FAA
Category C Brakes

Flight Planning and Performance Manual

Long Range Cruise Table
29000 FT to 24000 FT

PRESS ALT (1000 FT) (STD TAT)		WEIGHT (1000 LB)									
		180	170	160	150	140	130	120	110	100	90
29 (-20)	%N1	87.8	86.8	85.5	84.1	82.4	80.6	78.7	77.0	75.1	73.1
	MAX TAT	12	15								
	KIAS	301	298	292	284	273	262	251	243	235	226
	MACH	.777	.769	.757	.738	.711	.684	.658	.638	.618	.596
	FF/ENG	3399	3245	3072	2882	2667	2461	2263	2093	1933	1785
	KTAS	460	455	448	437	421	405	389	377	366	352
28 (-19)	%N1	87.3	86.1	84.8	83.3	81.6	79.8	78.0	76.4	74.5	72.4
	MAX TAT	15	19								
	KIAS	305	300	293	283	272	261	252	244	237	227
	MACH	.771	.760	.744	.721	.694	.668	.647	.628	.609	.584
	FF/ENG	3424	3255	3073	2864	2653	2445	2264	2101	1948	1794
	KTAS	458	452	443	429	413	397	384	373	362	347
27 (-17)	%N1	86.7	85.5	84.1	82.5	80.8	79.0	77.4	75.7	73.9	71.7
	MAX TAT	19									
	KIAS	308	302	293	282	271	261	254	246	238	227
	MACH	.763	.749	.729	.703	.678	.655	.637	.619	.599	.573
	FF/ENG	3438	3260	3057	2843	2637	2442	2272	2111	1963	1802
	KTAS	455	447	435	420	405	391	380	369	358	342
26 (-16)	%N1	86.0	84.7	83.2	81.7	79.9	78.4	76.8	75.1	73.2	71.0
	MAX TAT	22									
	KIAS	310	302	291	281	270	262	255	247	238	227
	MACH	.753	.735	.711	.687	.663	.645	.628	.610	.589	.561
	FF/ENG	3444	3248	3033	2827	2622	2444	2281	2128	1975	1808
	KTAS	451	441	426	412	398	386	376	366	353	336
25 (-14)	%N1	85.3	83.9	82.4	80.8	79.2	77.8	76.2	74.5	72.5	70.2
	MAX TAT										
	KIAS	310	301	290	280	271	264	256	249	239	227
	MACH	.740	.718	.695	.671	.652	.636	.619	.601	.577	.550
	FF/ENG	3439	3228	3018	2810	2622	2456	2295	2148	1986	1815
	KTAS	445	432	418	404	392	383	372	362	348	331
24 (-13)	%N1	84.5	83.1	81.6	80.0	78.6	77.1	75.6	73.8	71.8	69.4
	MAX TAT										
	KIAS	309	299	289	280	272	265	258	249	239	227
	MACH	.724	.701	.679	.658	.642	.626	.610	.591	.566	.539
	FF/ENG	3426	3213	3006	2807	2633	2471	2319	2165	1998	1825
	KTAS	437	424	410	398	388	378	369	357	342	326

Max TAT not shown where %N1 can be set in ISA + 30 °C conditions.
Increase/decrease %N1 required by 1% per 5 °C above/below standard TAT.
Increase/decrease fuel flow 3% per 10 °C above/below standard TAT.
Increase/decrease KTAS by 1 knot per 1 °C above/below standard TAT.

Max Cruise %N1

PRESS ALT (1000 FT)	TAT (°C)													
	-50	-45	-40	-35	-30	-25	-20	-15	-10	-5	0	5	10	15
29	85.6	86.5	87.5	88.4	89.3	90.2	91.1	92.0	92.8	93.7	93.5	92.6	91.8	90.8
28	85.1	86.1	87.0	87.9	88.8	89.7	90.6	91.5	92.3	93.2	93.6	92.7	91.8	90.8
27	84.7	85.7	86.6	87.5	88.4	89.3	90.2	91.0	91.9	92.8	93.6	92.8	92.0	91.0
26	84.4	85.3	86.2	87.1	88.0	88.9	89.8	90.6	91.5	92.4	93.2	93.0	92.1	91.2
25	84.0	85.0	85.9	86.8	87.7	88.5	89.4	90.3	91.1	92.0	92.9	93.2	92.4	91.4
24	83.8	84.7	85.6	86.5	87.4	88.3	89.2	90.1	90.9	91.8	92.6	93.2	92.4	91.5

With engine anti-ice on, decrease limit %N1 by 1.2.
With engine and wing anti-ice on, decrease limit %N1 by 3.1.

附图 1-11　LRC 巡航性能图表 (3)

737-800/CFM56-7B26
FAA
Category C Brakes

BOEING
Flight Planning and Performance Manual

ENROUTE
All Engine

Long Range Cruise Table
23000 FT to 18000 FT

PRESS ALT (1000 FT) (STD TAT)		WEIGHT (1000 LB)									
		180	170	160	150	140	130	120	110	100	90
23 (-12)	%N1	83.7	82.3	80.8	79.3	78.0	76.5	75.0	73.1	71.0	68.5
	MAX TAT										
	KIAS	308	298	288	281	274	267	260	250	239	227
	MACH	.706	.685	.664	.648	.633	.617	.601	.580	.555	.528
	FF/ENG	3412	3207	3000	2818	2654	2495	2348	2185	2012	1839
	KTAS	429	416	403	393	384	375	365	352	337	321
22 (-10)	%N1	82.9	81.4	80.0	78.7	77.3	75.9	74.3	72.4	70.2	67.7
	MAX TAT										
	KIAS	307	297	289	282	275	268	260	250	239	227
	MACH	.691	.670	.653	.638	.624	.609	.591	.568	.544	.518
	FF/ENG	3410	3199	3010	2840	2678	2528	2373	2205	2028	1852
	KTAS	421	408	398	389	380	371	360	346	331	316
21 (-9)	%N1	82.0	80.6	79.4	78.1	76.7	75.3	73.6	71.6	69.4	66.9
	MAX TAT										
	KIAS	305	297	290	284	277	270	260	250	239	227
	MACH	.675	.657	.643	.629	.615	.600	.580	.557	.533	.508
	FF/ENG	3404	3206	3029	2866	2709	2561	2397	2221	2042	1863
	KTAS	413	402	393	385	376	367	355	341	326	311
20 (-7)	%N1	81.2	79.9	78.7	77.4	76.1	74.6	72.8	70.8	68.5	66.1
	MAX TAT										
	KIAS	305	298	292	285	279	270	261	250	239	227
	MACH	.662	.647	.634	.620	.607	.590	.569	.547	.523	.498
	FF/ENG	3402	3221	3055	2894	2745	2587	2417	2236	2054	1874
	KTAS	406	397	390	381	373	362	350	336	321	306
19 (-6)	%N1	80.5	79.3	78.1	76.8	75.4	73.8	72.0	70.0	67.7	65.4
	MAX TAT										
	KIAS	306	299	293	287	280	271	260	250	239	227
	MACH	.651	.638	.625	.612	.598	.579	.558	.536	.513	.488
	FF/ENG	3413	3244	3080	2927	2776	2610	2430	2248	2064	1885
	KTAS	401	394	386	377	369	357	344	331	316	301
18 (-4)	%N1	79.8	78.7	77.5	76.2	74.7	73.1	71.3	69.2	67.0	64.7
	MAX TAT										
	KIAS	307	301	295	289	280	271	261	250	239	227
	MACH	.641	.629	.617	.604	.587	.568	.548	.526	.503	.479
	FF/ENG	3434	3270	3112	2966	2802	2627	2443	2258	2074	1898
	KTAS	397	390	382	374	364	352	339	326	312	297

Max TAT not shown where %N1 can be set in ISA + 30 °C conditions.
Increase/decrease %N1 required by 1% per 5 °C above/below standard TAT.
Increase/decrease fuel flow 3% per 10 °C above/below standard TAT.
Increase/decrease KTAS by 1 knot per 1 °C above/below standard TAT.

Max Cruise %N1

PRESS ALT (1000 FT)	TAT (°C)													
	-40	-35	-30	-25	-20	-15	-10	-5	0	5	10	15	20	25
23	85.5	86.4	87.2	88.1	89.0	89.9	90.7	91.6	92.4	93.2	92.4	91.6	90.6	89.6
22	85.3	86.2	87.1	87.9	88.8	89.7	90.5	91.4	92.2	93.0	92.4	91.6	90.7	89.7
21	85.0	85.9	86.8	87.7	88.6	89.4	90.3	91.1	91.9	92.8	92.4	91.6	90.7	89.7
20	84.8	85.7	86.6	87.4	88.3	89.2	90.0	90.8	91.7	92.5	92.3	91.5	90.6	89.7
19	84.5	85.4	86.3	87.2	88.0	88.9	89.7	90.6	91.4	92.2	92.4	91.7	90.8	89.9
18	84.2	85.1	86.0	86.8	87.7	88.5	89.4	90.2	91.0	91.9	92.6	91.7	90.9	90.0

With engine anti-ice on, decrease limit %N1 by 1.2.
With engine and wing anti-ice on, decrease limit %N1 by 2.9.

附图 1-11   LRC 巡航性能图表(4)

飞行计划

ENROUTE
All Engine

737-800/CFM56-7B26
FAA
Category C Brakes

Flight Planning and Performance Manual

.79M Cruise Table
41000 FT to 32000 FT

PRESS ALT (1000 FT)	KIAS STD TAT KTAS		WEIGHT (1000 LB)									
			180	170	160	150	140	130	120	110	100	90
41	233	%N1						93.5	90.0	88.0	86.6	85.4
	-29	MAX TAT						-18	-9	-3	1	
	453	FF/ENG						2583	2321	2162	2044	1973
40	239	%N1						90.8	88.6	87.1	85.9	84.9
	-29	MAX TAT						-11	-5	0		
	453	FF/ENG						2499	2318	2184	2079	1996
39	244	%N1					91.8	89.3	87.6	86.3	85.3	84.4
	-29	MAX TAT					-13		-1			
	453	FF/ENG					2697	2481	2331	2216	2121	2047
38	250	%N1				93.0	89.8	88.0	86.8	85.7	84.8	84.0
	-29	MAX TAT				-15	-8	-2	2			
	453	FF/ENG				2911	2649	2484	2357	2254	2170	2103
37	256	%N1			93.8	90.4	88.5	87.1	86.0	85.1	84.3	83.6
	-29	MAX TAT			-17	-9	-3	1				
	453	FF/ENG			3129	2822	2642	2504	2393	2299	2225	2161
36	262	%N1			90.8	88.9	87.5	86.3	85.4	84.6	83.8	83.2
	-29	MAX TAT			-10	-5	0					
	453	FF/ENG			3004	2804	2655	2536	2436	2353	2286	2223
35	268	%N1		91.7	89.6	88.1	87.0	86.0	85.2	84.5	83.8	83.2
	-27	MAX TAT		-10	-4	1	5					
	455	FF/ENG		3207	2982	2822	2694	2589	2498	2426	2362	2304
34	274	%N1	92.5	90.2	88.7	87.6	86.6	85.8	85.0	84.4	83.8	83.2
	-25	MAX TAT	-11	-4	0	4	8					
	457	FF/ENG	3416	3168	2998	2861	2750	2653	2573	2506	2444	2393
33	281	%N1	90.8	89.3	88.2	87.2	86.4	85.6	84.9	84.3	83.8	83.4
	-23	MAX TAT	-4	1	4	8						
	459	FF/ENG	3358	3179	3035	2917	2815	2727	2656	2592	2532	2489
32	287	%N1	89.8	88.7	87.7	86.9	86.1	85.5	84.9	84.3	83.8	83.5
	-20	MAX TAT	0	4	8	11						
	462	FF/ENG	3361	3212	3087	2982	2887	2810	2743	2680	2627	2593

Max TAT not shown where %N1 can be set in ISA + 30 °C conditions.
Increase/decrease %N1 required by 1% per 5 °C above/below standard TAT.
Increase/decrease fuel flow 3% per 10 °C above/below standard TAT.
Increase/decrease KTAS by 1 knot per 1 °C above/below standard TAT.
Shaded area approximates optimum altitude.

Max Cruise %N1

PRESS ALT (1000 FT)	TAT (°C)													
	-60	-55	-50	-45	-40	-35	-30	-25	-20	-15	-10	-5	0	5
41	87.4	88.3	89.3	90.2	91.1	92.0	92.9	93.7	94.6	94.4	93.5	92.8	92.1	91.3
40	87.5	88.4	89.4	90.3	91.2	92.1	93.0	93.8	94.7	94.5	93.6	92.9	92.2	91.5
39	87.6	88.5	89.5	90.4	91.3	92.2	93.1	94.0	94.8	94.6	93.7	93.0	92.4	91.8
38	87.7	88.6	89.6	90.5	91.4	92.3	93.2	94.1	94.9	94.6	93.8	93.1	92.5	91.9
37	87.8	88.7	89.7	90.6	91.5	92.4	93.3	94.2	95.0	94.7	93.9	93.2	92.6	92.0
36	87.5	88.5	89.4	90.3	91.2	92.1	93.0	93.9	94.7	94.6	93.8	93.1	92.5	91.9
35	87.2	88.2	89.1	90.0	90.9	91.8	92.7	93.6	94.4	95.0	94.4	93.6	92.9	92.3
34	86.4	87.4	88.3	89.3	90.2	91.1	92.1	93.0	93.9	94.7	94.5	93.8	93.1	92.6
33	85.8	86.8	87.8	88.7	89.7	90.6	91.6	92.5	93.4	94.3	94.7	94.0	93.3	92.8
32	85.3	86.3	87.2	88.2	89.1	90.0	90.9	91.8	92.7	93.6	94.5	94.1	93.3	92.8

With engine anti-ice on, decrease limit %N1 by 1.0.
With engine and wing anti-ice on, decrease limit %N1 by 3.4.

附图 1-12　M0.79 巡航性能图表(1)

737-800/CFM56-7B26
FAA
Category C Brakes

BOEING
Flight Planning and Performance Manual

.79M Cruise Table
31000 FT to 25000 FT

PRESS ALT (1000 FT)	KIAS / STD TAT / KTAS		WEIGHT (1000 LB)									
			180	170	160	150	140	130	120	110	100	90
31	293	%N1	89.2	88.2	87.4	86.7	86.0	85.4	84.8	84.3	83.9	83.7
	-18	MAX TAT	3	7	10	13						
	464	FF/ENG	3391	3262	3152	3052	2969	2899	2834	2774	2729	2703
30	300	%N1	88.7	87.9	87.1	86.5	85.9	85.3	84.8	84.4	84.1	83.9
	-16	MAX TAT	7	10	13	15						
	466	FF/ENG	3439	3325	3222	3133	3059	2992	2928	2875	2839	2819
29	306	%N1	88.3	87.6	86.9	86.3	85.8	85.3	84.8	84.5	84.2	84.1
	-14	MAX TAT	10	13	15	18						
	468	FF/ENG	3501	3395	3300	3222	3153	3088	3028	2983	2955	2939
28	313	%N1	88.1	87.4	86.8	86.3	85.8	85.3	84.9	84.6	84.4	84.3
	-11	MAX TAT	13	16	18	20						
	470	FF/ENG	3569	3470	3388	3317	3250	3187	3135	3098	3077	3064
27	320	%N1	87.8	87.2	86.7	86.2	85.8	85.3	85.0	84.8	84.7	84.6
	-9	MAX TAT	16	18	20	22	23					
	472	FF/ENG	3643	3557	3484	3416	3351	3293	3249	3221	3204	3195
26	326	%N1	87.6	87.1	86.7	86.2	85.8	85.4	85.1	85.0	84.9	84.8
	-7	MAX TAT	18	20	22	24	25					
	474	FF/ENG	3731	3655	3586	3520	3458	3408	3372	3351	3337	3331
25	333	%N1	87.5	87.1	86.6	86.2	85.8	85.5	85.3	85.2	85.1	85.1
	-5	MAX TAT	21	22	24	26	27	28				
	476	FF/ENG	3832	3762	3695	3630	3575	3533	3505	3488	3478	3474

Max TAT not shown where %N1 can be set in ISA + 30 °C conditions.
Increase/decrease %N1 required by 1% per 5 °C above/below standard TAT.
Increase/decrease fuel flow 3% per 10 °C above/below standard TAT.
Increase/decrease KTAS by 1 knot per 1 °C above/below standard TAT.
Shaded area approximates optimum altitude.

Max Cruise %N1

PRESS ALT (1000 FT)	TAT (°C)														
	-45	-40	-35	-30	-25	-20	-15	-10	-5	0	5	10	15	20
31	87.7	88.6	89.5	90.4	91.3	92.1	93.0	93.9	94.2	93.4	92.7	92.2	91.6	91.0
30	87.0	87.9	88.8	89.7	90.6	91.5	92.4	93.3	94.1	93.5	92.9	92.3	91.8	91.1
29	86.4	87.3	88.2	89.1	90.0	90.9	91.8	92.7	93.5	93.6	93.0	92.3	91.8	91.2
28	85.8	86.7	87.6	88.5	89.4	90.3	91.2	92.0	92.9	93.7	93.1	92.5	92.0	91.4
27	85.3	86.2	87.1	88.0	88.8	89.7	90.6	91.5	92.3	93.2	92.7	92.7	92.1	91.5
26	84.7	85.6	86.5	87.4	88.3	89.2	90.0	90.9	91.8	92.6	93.3	92.8	92.2	91.6
25	84.2	85.1	86.0	86.9	87.8	88.7	89.5	90.4	91.2	92.1	92.9	92.8	92.3	91.8

With engine anti-ice on, decrease limit %N1 by 1.2.
With engine and wing anti-ice on, decrease limit %N1 by 3.1.

附图 1-12　*M*0.79 巡航性能图表（2）

737-800/CFM56-7B26
FAA
Category C Brakes

BOEING

Flight Planning and Performance Manual

ENROUTE
All Engine

Descent
.78/280/250

PRESSURE ALTITUDE (FT)	TIME (MIN)	FUEL (LB)	DISTANCE (NM) LANDING WEIGHT (1000 LB)			
			90	110	130	150
41000	26	730	103	118	129	136
39000	25	720	98	112	123	131
37000	24	710	93	107	118	125
35000	23	700	89	102	112	120
33000	23	690	85	98	108	115
31000	22	680	81	93	102	109
29000	21	660	76	87	96	102
27000	20	650	71	81	89	95
25000	19	630	67	76	83	89
23000	18	610	62	71	77	82
21000	17	590	57	65	71	76
19000	16	570	53	60	65	69
17000	15	540	48	55	59	63
15000	14	520	44	49	54	56
10000	10	430	31	34	36	38
5000	7	330	18	19	20	21
1500	4	250	9	9	9	9

Allowances for a straight-in approach are included.

.78/250

PRESSURE ALTITUDE (FT)	TIME (MIN)	FUEL (LB)	DISTANCE (NM) LANDING WEIGHT (1000 LB)			
			90	110	130	150
41000	28	780	115	129	139	144
39000	27	770	110	124	133	139
37000	27	750	105	118	127	133
35000	25	740	99	111	120	125
33000	24	720	93	104	113	118
31000	23	710	87	98	106	111
29000	22	690	82	92	99	104
27000	21	670	76	85	92	97
25000	20	650	71	79	86	90
23000	19	630	65	73	79	83
21000	18	600	60	67	72	76
19000	16	580	54	61	66	69
17000	15	550	49	55	59	62
15000	14	520	44	49	52	55
10000	10	430	31	34	36	38
5000	7	330	18	19	20	21
1500	4	250	9	9	9	9

Allowances for a straight-in approach are included.

附图 1-13　下降性能图表

737-800/CFM56-7B26
FAA
Category C Brakes

BOEING
Flight Planning and Performance Manual

FLIGHT PLANNING
Simplified Flight Planning

Holding Planning
Flaps Up

PRESSURE ALTITUDE (FT)	TOTAL FUEL FLOW (LB/HR)									
	WEIGHT (1000 LB)									
	180	170	160	150	140	130	120	110	100	90
41000						5190	4580	4150	3720	3330
35000		6440	5900	5430	5000	4600	4220	3900	3540	3190
30000	6460	6050	5660	5280	4900	4530	4160	3860	3520	3220
25000	6290	5910	5540	5160	4800	4450	4130	3820	3580	3260
20000	6260	5900	5550	5220	4900	4570	4240	3910	3650	3330
15000	6350	6010	5670	5320	4980	4640	4310	3990	3730	3400
10000	6390	6050	5710	5380	5050	4720	4400	4070	3740	3490
5000	6440	6110	5780	5450	5130	4800	4480	4160	3840	3620
1500	6540	6210	5880	5560	5230	4910	4600	4280	3970	3750

This table includes 5% additional fuel for holding in a racetrack pattern.

ENROUTE
All Engine

BOEING
Flight Planning and Performance Manual

737-800/CFM56-7B26
FAA
Category C Brakes

Holding
Flaps Up

PRESSURE ALTITUDE (FT)		WEIGHT (1000 LB)									
		180	170	160	150	140	130	120	110	100	90
41000	%N1						93.4	89.4	86.8	84.2	81.6
	KIAS						217	212	203	192	182
	FF/ENG						2600	2290	2070	1860	1660
35000	%N1		91.5	89.1	87.2	85.3	83.4	81.5	79.3	77.0	74.4
	KIAS		250	244	236	227	218	209	199	190	180
	FF/ENG		3220	2950	2720	2500	2300	2110	1950	1770	1590
30000	%N1	87.3	85.8	84.3	82.7	81.0	79.1	77.1	74.9	72.4	69.9
	KIAS	257	249	241	233	224	216	207	198	188	179
	FF/ENG	3230	3030	2830	2640	2450	2270	2080	1930	1760	1610
25000	%N1	83.1	81.7	80.1	78.5	76.7	74.7	72.7	70.6	68.2	65.5
	KIAS	254	246	239	231	223	214	205	196	187	179
	FF/ENG	3150	2960	2770	2580	2400	2220	2060	1910	1790	1630
20000	%N1	78.8	77.3	75.7	74.0	72.3	70.5	68.4	66.2	64.0	61.6
	KIAS	252	245	237	229	221	213	204	195	186	179
	FF/ENG	3130	2950	2780	2610	2450	2290	2120	1950	1820	1660
15000	%N1	74.6	73.1	71.6	69.9	68.0	66.2	64.4	62.4	59.8	57.2
	KIAS	250	243	235	228	220	212	203	195	186	179
	FF/ENG	3180	3010	2830	2660	2490	2320	2160	1990	1860	1700
10000	%N1	70.5	69.0	67.5	66.0	64.4	62.5	60.3	58.1	55.9	53.5
	KIAS	249	242	234	227	219	211	202	194	186	179
	FF/ENG	3200	3020	2860	2690	2520	2360	2200	2030	1870	1750
5000	%N1	66.7	65.4	63.8	62.0	60.1	58.3	56.5	54.5	52.3	49.9
	KIAS	248	241	234	226	218	210	202	193	186	179
	FF/ENG	3220	3050	2890	2730	2560	2400	2240	2080	1920	1810
1500	%N1	64.0	62.4	60.8	59.2	57.5	55.8	54.0	52.0	49.8	47.5
	KIAS	247	240	233	225	218	209	201	192	186	179
	FF/ENG	3270	3110	2940	2780	2620	2460	2300	2140	1990	1870

This table includes 5% additional fuel for holding in a racetrack pattern.

附图 1-14　等待性能图表

737-800/CFM56-7B26
FAA
Category C Brakes

BOEING
Flight Planning and Performance Manual

ENROUTE
All Engine

Long Range Cruise Maximum Operating Altitude
Max Cruise Thrust
ISA + 10 °C and Below

WEIGHT (1000 LB)	OPTIMUM ALT (FT)	TAT (°C)	MARGIN TO INITIAL BUFFET 'G' (BANK ANGLE)				
			1.20 (33°)	1.25 (36°)	1.30 (39°)	1.40 (44°)	1.50 (48°)
180	31200	-7	33900*	33900*	33900*	33000	31600
170	32400	-10	35300*	35300*	35300*	34200	32800
160	33700	-13	36600*	36600*	36600*	35500	34100
150	35100	-16	37900*	37900*	37900*	36800	35500
140	36500	-18	39200*	39200*	39200*	38300	36900
130	38000	-18	40600*	40600*	40600*	39800	38500
120	39700	-18	41000	41000	41000	41000	40100
110	41000	-18	41000	41000	41000	41000	41000
100	41000	-18	41000	41000	41000	41000	41000
90	41000	-18	41000	41000	41000	41000	41000

ISA + 15 °C

WEIGHT (1000 LB)	OPTIMUM ALT (FT)	TAT (°C)	MARGIN TO INITIAL BUFFET 'G' (BANK ANGLE)				
			1.20 (33°)	1.25 (36°)	1.30 (39°)	1.40 (44°)	1.50 (48°)
180	31200	-2	32100*	32100*	32100*	32100*	31600
170	32400	-4	34000*	34000*	34000*	34000*	32800
160	33700	-7	35600*	35600*	35600*	35500	34100
150	35100	-10	37000*	37000*	37000*	36800	35500
140	36500	-13	38200*	38200*	38200*	38200*	36900
130	38000	-13	39600*	39600*	39600*	39600*	38500
120	39700	-13	41000	41000	41000	41000	40100
110	41000	-13	41000	41000	41000	41000	41000
100	41000	-13	41000	41000	41000	41000	41000
90	41000	-13	41000	41000	41000	41000	41000

ISA + 20 °C

WEIGHT (1000 LB)	OPTIMUM ALT (FT)	TAT (°C)	MARGIN TO INITIAL BUFFET 'G' (BANK ANGLE)				
			1.20 (33°)	1.25 (36°)	1.30 (39°)	1.40 (44°)	1.50 (48°)
180	31200	4	29100*	29100*	29100*	29100*	29100*
170	32400	1	31600*	31600*	31600*	31600*	31600*
160	33700	-2	33800*	33800*	33800*	33800*	33800*
150	35100	-5	35600*	35600*	35600*	35600*	35500
140	36500	-7	37000*	37000*	37000*	37000*	36900
130	38000	-7	38300*	38300*	38300*	38300*	38300*
120	39700	-7	39700*	39700*	39700*	39700*	39700*
110	41000	-7	41000	41000	41000	41000	41000
100	41000	-7	41000	41000	41000	41000	41000
90	41000	-7	41000	41000	41000	41000	41000

*Denotes altitude thrust limited in level flight, 100 fpm residual rate of climb

附图 1-15　高度能力/机动能力图表

ENROUTE
Engine Inoperative

BOEING
Flight Planning and Performance Manual

737-800/CFM56-7B26
FAA
Category C Brakes

ENGINE INOP
MAX CONTINUOUS THRUST

Long Range Cruise Diversion Fuel and Time
Based on .78/280/250 descent and includes APU fuel burn.

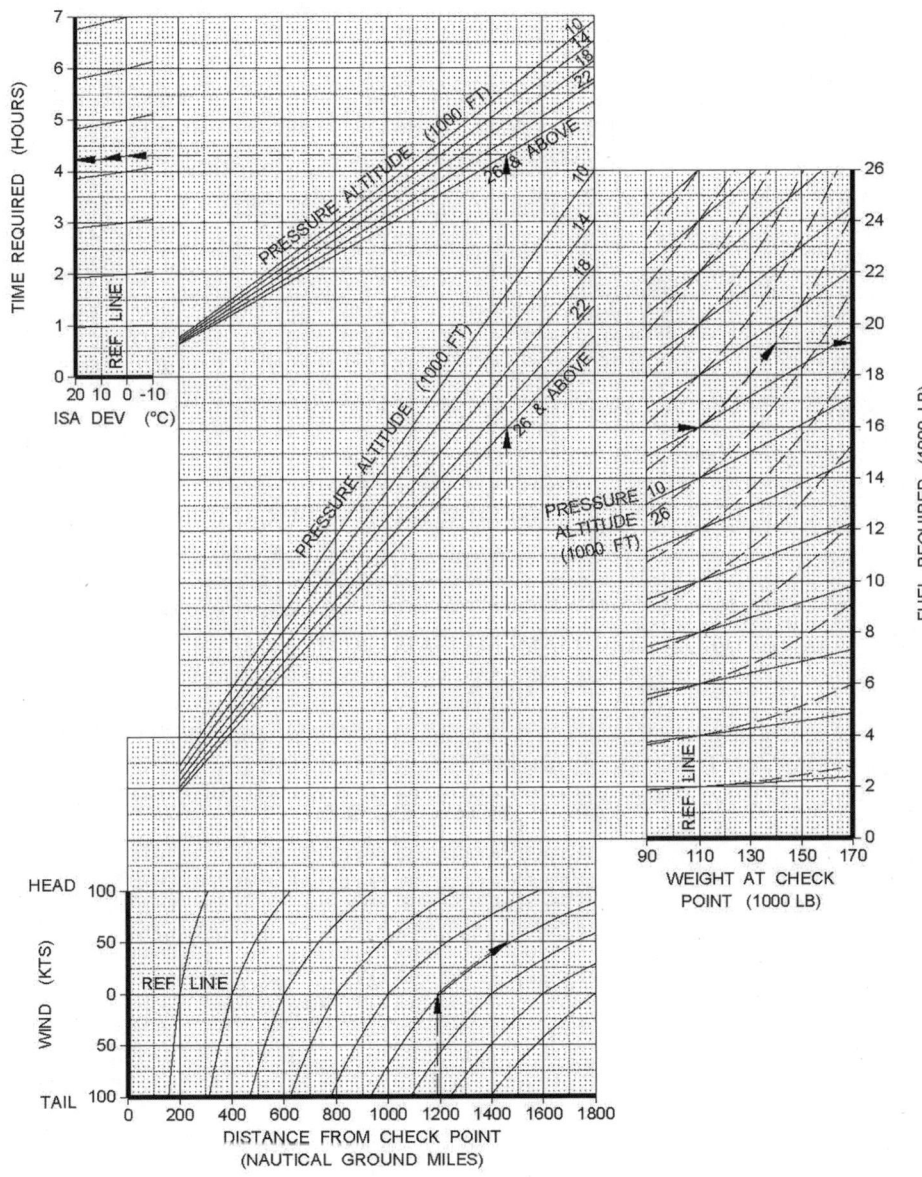

附图 1-16　空中改航图表

附录 2   公务机飞行计划相关图表

豪客 800XP 飞行手册图表

发动机防冰关断

高度(英尺)		1500	5000	10000	15000	20000	25000	30000	35000
重量(磅)	指示空速节(KIAS)	总燃油流量(磅/小时)							
16000	165	1038	1010	971	938	899	867	850	849
17000	170	1086	1057	1017	981	940	909	891	879
18000	175	1135	1105	1060	1023	982	951	934	922
19000	180	1185	1154	1107	1066	1025	995	978	967
20000	185	1236	1204	1153	1111	1068	1038	1022	1011
21000	190	1288	1250	1197	154	1111	1080	1065	1054
22000	194	1339	1297	1243	1197	1156	1125	1109	1098
23000	198	1387	1343	1287	1240	1199	1169	1150	1139
24000	202	1433	1355	1332	1283	1242	1213	1193	1181
25000	206	1481	1434	1378	1328	1283	1258	1236	1222
26000	210	1527	1481	1425	1375	1333	1303	1279	1264

8XC098

注释:温度每高于国际标准大气温度 10 ℃,总燃油流量增加约 50 磅/小时

温度每低于国际标准大气温度 10 ℃,总燃油流量减小约 50 磅/小时

附图 2-1   等待燃油流量图表

高速程序
国际标准大气
高高度

着陆重量 17000 磅

37000 英尺		39000 英尺		41000 英尺		43000 英尺		高度
飞行阶段		飞行阶段		飞行阶段		飞行阶段		
时间	燃油	时间	燃油	时间	燃油	时间	燃油	空中距离
分钟	磅	分钟	磅	分钟	磅	分钟	磅	海里
								100
34	953	34	910	35	882			200
48	1365	48	1288	48	1224	49	1176	300
61	1777	61	1666	62	1567	63	1487	400
74	2189	75	2044	76	1909	76	1799	500
88	2601	89	2422	89	2251	90	2110	600
101	3014	102	2800	103	2593	104	2422	700
115	3426	116	3178	116	2935	118	2734	800
128	3838	129	3557	130	3277	132	3050	900
141	4252	143	3935	144	3621	146	3366	1000
155	4669	156	4314	158	3968	160	3682	1100
168	5086	170	4696	171	4315	174	3998	1200
182	5503	184	5078	185	4662	188	4313	1300
196	5920	197	5460	199	5009	203	4631	1400
209	6337	211	5842	213	5356	217	4954	1500
223	6754	225	6223	227	5703	232	5276	1600
236	7171	238	6605	241	6050	246	5599	1700
250	7589	252	6987	255	6403	260	5921	1800
263	8011	266	7369	269	6759	275	6244	1900
277	8433	280	7755	284	7116			2000
291	8855	294	8146	298	7472			2100
305	9278	308	8537	312	7828			2200
318	9700	322	8928	327	8184			2300
		336	9319	341	8540			2400
		351	9710					2500
								2600
								2700

8XC041-2

附图 2-2　飞行阶段燃油和时间图表（1）

着陆重量 20000 磅

37000 英尺		39000 英尺		41000 英尺		43000 英尺		高度
飞行阶段		飞行阶段		飞行阶段		飞行阶段		
时间	燃油	时间	燃油	时间	燃油	时间	燃油	空中距离
分钟	磅	分钟	磅	分钟	磅	分钟	磅	海里
								100
35	997	36	961	36	933			200
49	1413	49	1342	50	1290	50	1261	300
62	1829	63	1722	64	1637	64	1579	400
76	2245	76	2103	77	1984	79	1897	500
89	2661	90	2483	91	2331	93	2220	600
103	3076	104	2864	105	2679	108	2542	700
116	3495	117	3244	119	3031	122	2866	800
130	3914	131	3630	133	3382	137	3191	900
144	4334	145	4015	147	3734	151	3500	1000
157	4753	159	4400	162	4086	164	3790	1100
171	5173	173	4786	176	4441	179	4117	1200
184	5592	187	5171	190	4799	194	4452	1300
198	6015	201	5557	205	5158	210	4789	1400
212	6437	215	5948	219	5517	224	5113	1500
226	6860	229	6341	234	5875	239	5435	1600
240	7282	243	6734	248	6198	253	5711	1700
253	7704	258	7127	262	6503	267	5911	1800
267	8127	272	7520	277	6814	282	6067	1900
281	8549	286	7914	291	7156			2000
295	8971	300	8307	306	7532			2100
309	9394	315	8700	321	7926			2200
322	9816	329	9093	337	8327			2300
		343	9486	352	8715			2400
		357	9879					2500
								2600
								2700

8XC042−2

注释:阴影部分用于插值计算。

附图 2-2　飞行阶段燃油和时间图表(2)

着陆重量 17000 磅

21000 英尺		23000 英尺		25000 英尺		27000 英尺		高度
改航		改航		改航		改航		
时间	燃油	时间	燃油	时间	燃油	时间	燃油	空中距离
分钟	磅	分钟	磅	分钟	磅	分钟	磅	海里
								20
								40
14	377							60
18	460	18	461	18	462	18	466	80
22	543	22	540	21	537	21	539	100
26	626	25	620	25	612	25	611	120
30	709	29	699	29	688	28	684	140
34	792	33	778	32	763	32	756	160
38	875	37	857	36	838	35	828	180
41	958	40	936	39	914	39	901	200
45	1041	44	1015	43	989	42	973	220
49	1124	48	194	47	1065	46	1045	240
53	1207	52	1173	50	1140	49	1118	260
57	1290	55	1252	54	1215	53	1190	280
61	1373	59	1332	58	1291	56	1262	300
65	1456	63	1411	61	1366	60	1335	320
69	1539	67	1490	65	1441	63	1407	340
72	1622	70	1569	69	1517	67	1479	360
76	1705	74	1648	72	1592	70	1552	380
80	1788	78	1727	76	1667	74	1624	400
84	1871	82	1806	79	1743	77	1696	420
88	1954	86	1885	83	1818	81	1769	440
92	2037	89	1965	87	1893	84	1841	460
96	2120	93	2044	90	1969	88	1913	480
100	2202	97	2123	94	2044	91	1986	500
103	2285	101	2202	98	2120	95	2058	520
107	2368	104	2281	101	2195	98	2130	540
111	2451	108	2360	105	2270	102	2203	560
115	2534	112	2439	108	2346	105	2275	580
119	2617	116	2518	112	2421	109	2347	600

注释:忽略温度变化对时间和燃油的影响。

附图 2-3 改航阶段燃油和时间图表

不同高度和温度时的最大起飞重量

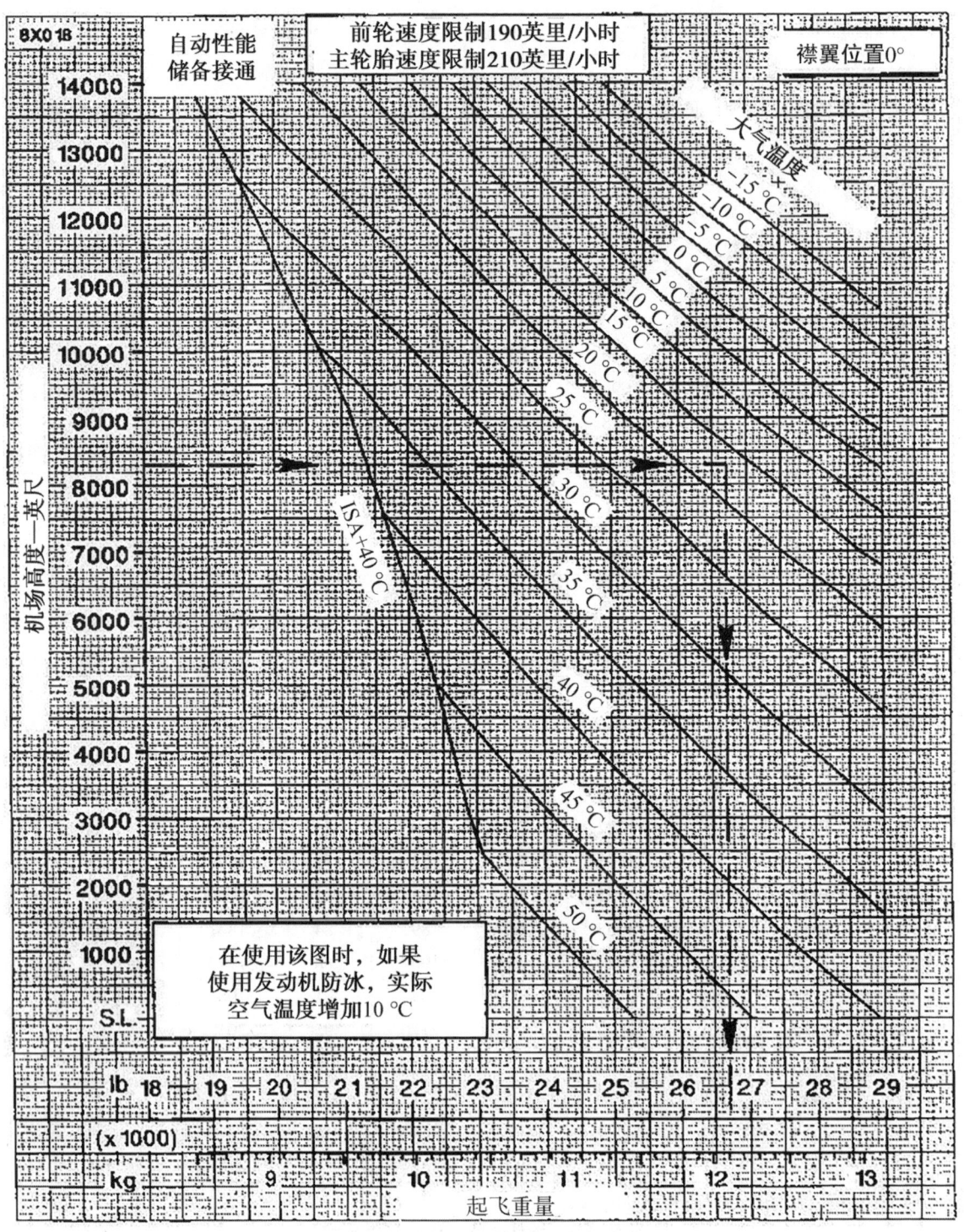

附图 2-4　起飞重量图表

D=10000 ft AM （平均海面之上）时的起飞重量

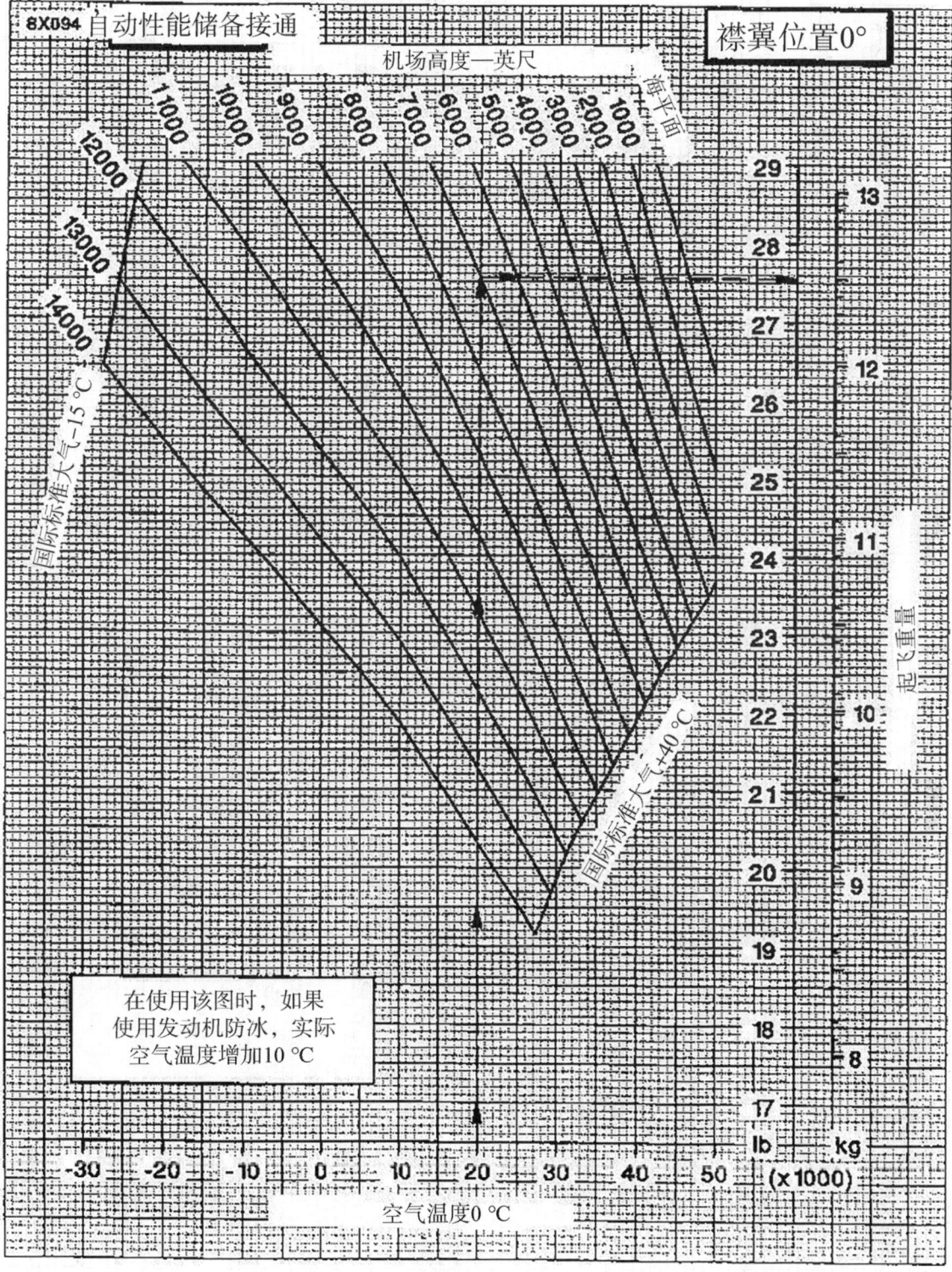

附图 2-5 *D* = 10000 ft AM 时的起飞重量

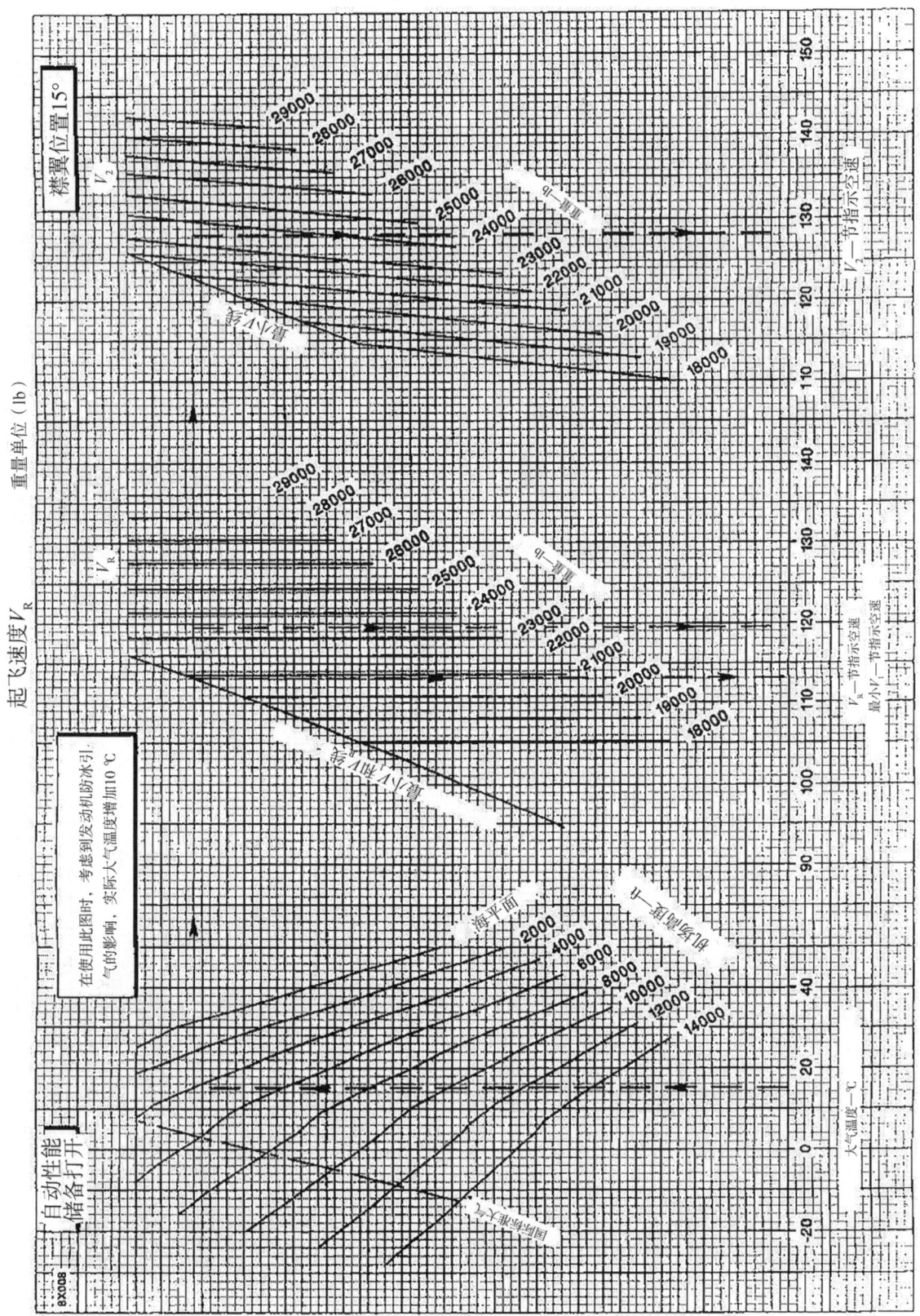

附图 2-6 起飞速度

不同高度和温度时的最大着陆重量

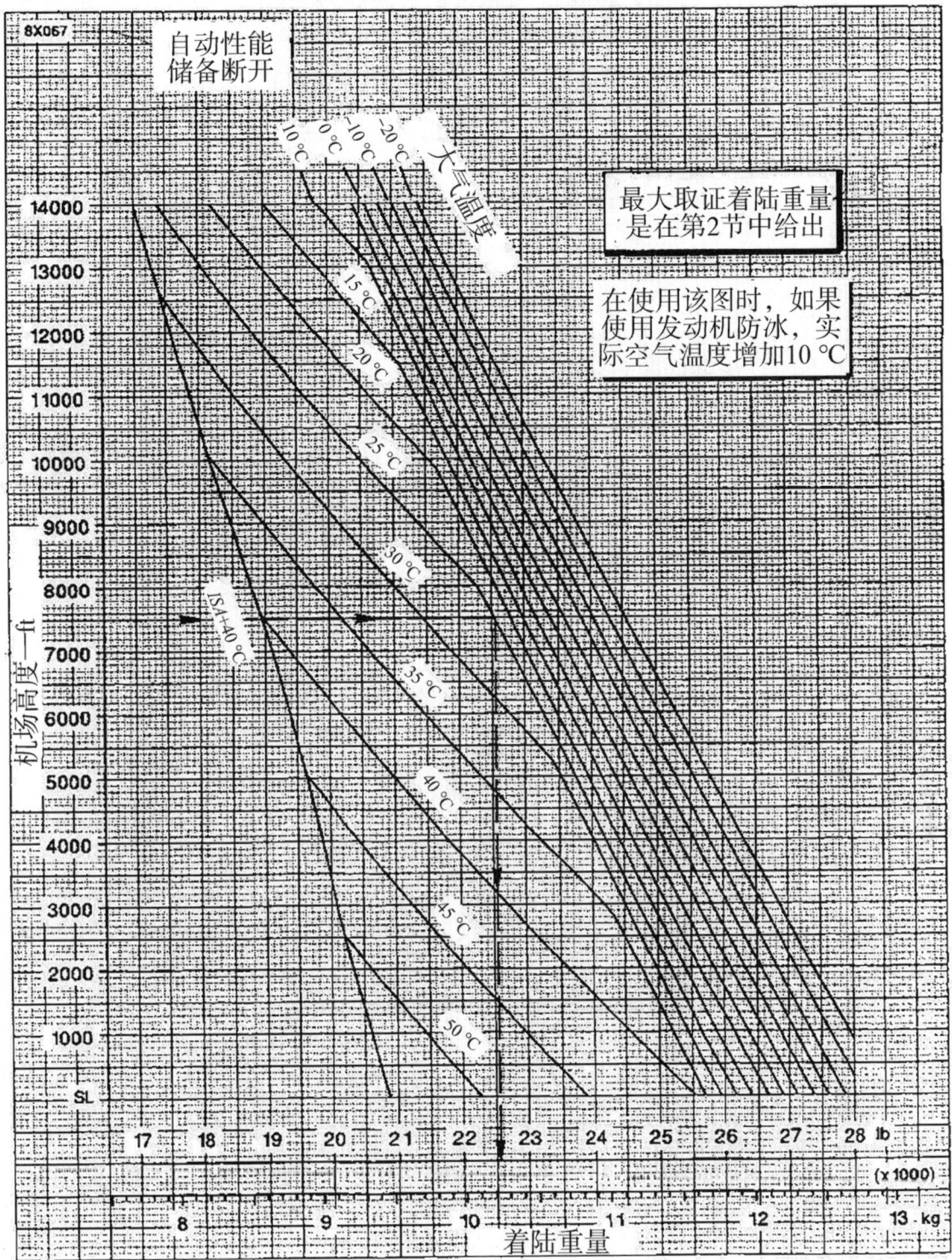

附图 2-7　着陆重量限制

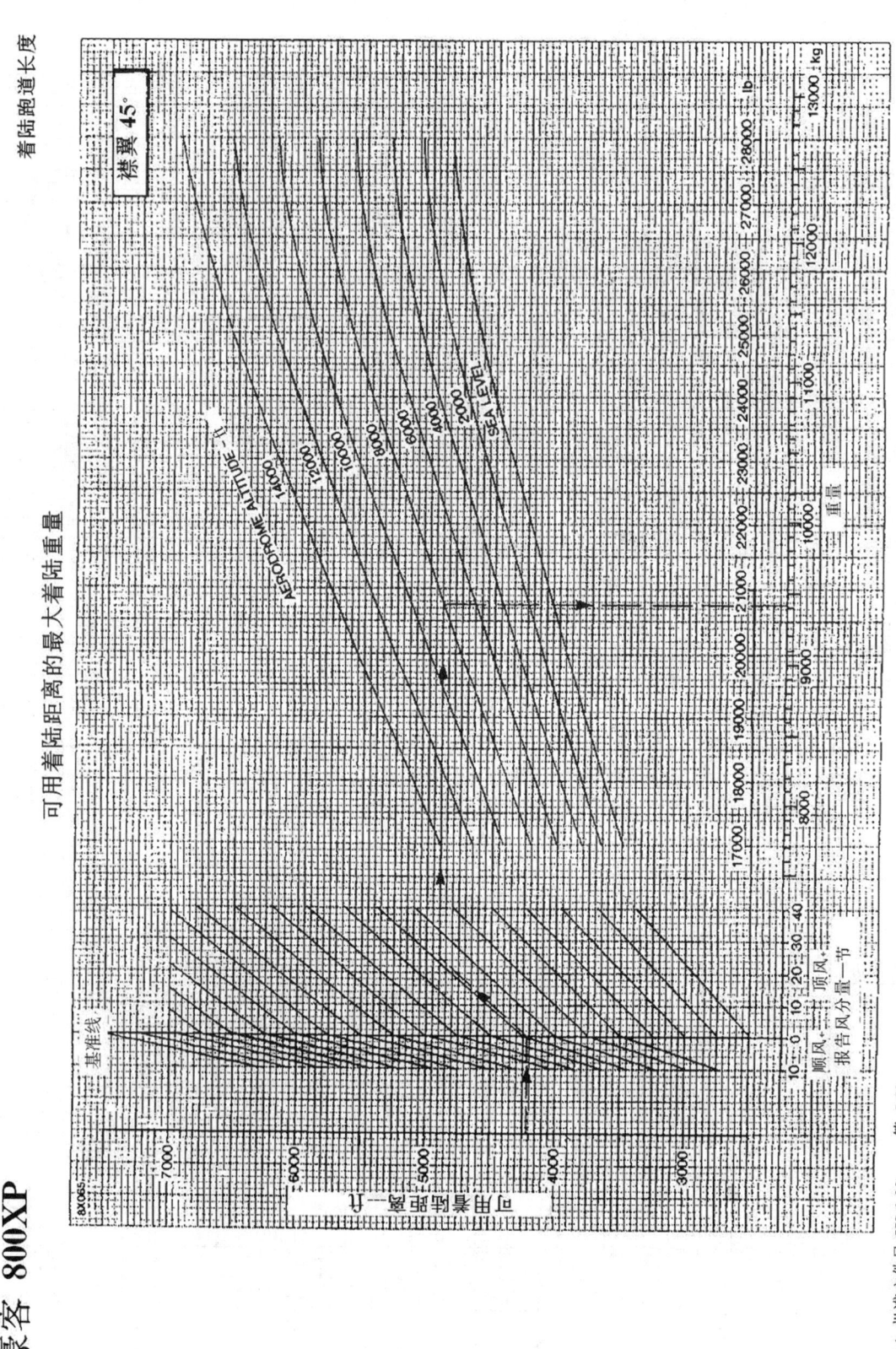

豪客 **800XP**

着陆跑道长度

可用着陆距离的最大着陆重量

附图 2-8　着陆跑道长度

附录 3　ETOPS 飞行计划相关图表

Regulations require that flights conducted over a route that contains a point further than one hour's time at "normal one engine inoperative speed" from an adequate diversion airport comply with rules set up specifically for extended range operation with two engine airplanes. If the airplane meets these requirements, the approved Airplane Flight Manual (AFM) will contain a statement to this effect.

This section provides the planning and inflight performance information necessary to conduct operations under these regulations. Information is presented for several different driftdown/cruise speed schedules. The information is based on drifting down using Max Continuous Thrust at the Mach number shown until reaching the corresponding airspeed and maintaining that airspeed throughout the remainder of the driftdown and level cruise. Based on route requirements, operators may select any one of these speeds for their operation, or may elect to use an entirely different speed which is not contained in this section.

Area of Operation(Diversion Distance)

The maximum diversion distance used to establish the Area of Operation may be obtained from this chart. The Area of Operation is defined as the region within which the operator is authorized to conduct extended range operation. The distance to the diversion airport from any point along the intended route must be covered within the approved time using the single engine cruise speed(assuming still air and ISA conditions) selected by the operator. Enter the chart for the appropriate speed with the weight at the point of diversion using the speed schedule and time selected and read the maximum diversion distance.

Critical Fuel Reserves

Regulations require reserve planning to include a "Critical Fuel Scenario" calculation. The information shown is the fuel required to satisfy the flight profile as described on the charts. This information is shown for two engine operation at Long Range Cruise, as well as single engine operation at LRC, 280, 290, 310 and 330 KIAS. The fuel required is the greater of the two engine fuel and the single engine fuel for the speed schedule selected. This fuel is compared to the amount of fuel normally onboard the airplane at that point in the route. If the fuel required by the critical fuel reserves exceeds the amount of fuel normally expected, the fuel load must be adjusted accordingly.

To determine the fuel required, enter the appropriate chart with the desired diversion distance, correct for the anticipated wind, proceed vertically to the weight at critical point and read the fuel required to the right. Adjustments for non-standard conditions are shown at the bottom of the page.

The data does not include an allowance for performance deterioration. Regulations require a 5% allowance for performance deterioration unless a value has been established by the operator for inservice deterioration. However, as noted on the chart, an additional 5% allowance is included in the chart to account for wind errors.

Net Level off Weight

These charts are used to determine the maximum weight for terrain clearance based on net engine inoperative performance for the speed schedules shown and is for planning purposes only. Regulations require terrain clearance flight planning based on net performance which is the gross(or actual) gradient performance degraded by 1.1%.

Net level off weight is determined by entering the appropriate chart at the left with the desired level off pressure altitude, adjusting for anti-ice as shown, proceeding across to the correct ISA deviation line and reading the allowable weight at the bottom. The net level off altitude may be obtained by entering the chart with weight and proceeding up to the correct ISA deviation line and reading across to the altitude on the left.

Driftdown/Cruise Range Capability

These charts show range capability from the start of driftdown/cruise when the driftdown/cruise procedure outlined in this section is followed. Charts are shown for each of the speed schedules discussed previously.

To determine the fuel required, enter the appropriate chart with the desired ground distance, correct for the anticipated wind, and proceed vertically to the start of driftdown weight and read the fuel required to the right. The time required is read to the left from the single weight line on top. The chart may be used in reverse to determine the range capability knowing the fuel remaining from the start of driftdown.

ETOPS

Area of Operation

Long Range Cruise Critical Fuel Reserves

280 KIAS Cruise Critical Fuel Reserves

290 KIAS Cruise Critical Fuel Reserves

310 KIAS Cruise Critical Fuel Reserves

330 KIAS Cruise Critical Fuel Reserves

Long Range Cruise Net Level off Weight

280 KIAS Net Level off Weight

290 KIAS Net Level off Weight

310 KIAS Net Level off Weight

330 KIAS Net Level off Weight

Long Range Cruise Driftdown/Cruise Range Capability

M.76/280 KIAS Driftdown/Cruise Range Capability

M.79/290 KIAS Driftdown/Cruise Range Capability

M.79/310 KIAS Driftdown/Cruise Range Capability

M.79/330 KIAS Driftdown/Cruise Range Capability

737-800/CFM56-7B26
FAA
Category C Brakes

BOEING
Flight Planning and Performance Manual

ENGINE INOP
MAX CONTINUOUS THRUST

Area of Operation
Based on standard day and driftdown starting at or near optimum altitude.

SPEED (M/KIAS)	WEIGHT AT DIVERSION (1000 LB)	DIVERSION DISTANCE (NM) TIME (MINUTES)														
		60	70	80	90	100	110	120	130	140	150	160	170	180	190	200
.76/280	180	386	448	510	572	634	696	758	820	882	944	1006	1068	1130	1192	1254
	170	391	454	517	580	643	705	768	831	894	957	1019	1082	1145	1208	1271
	160	395	458	522	585	649	712	776	839	903	966	1030	1093	1157	1220	1283
	150	399	463	527	592	656	720	784	848	912	977	1041	1105	1169	1233	1297
	140	404	469	533	598	663	728	793	858	922	987	1052	1117	1182	1246	1311
	130	408	473	539	604	670	735	801	866	931	997	1062	1128	1193	1259	1324
	120	411	477	543	609	675	741	807	873	939	1005	1071	1137	1203	1269	1335
	110	414	480	547	613	680	747	813	880	946	1013	1079	1146	1212	1279	1345
	100	416	482	549	616	683	750	817	884	951	1018	1085	1152	1218	1285	1352
	90	417	484	551	618	685	753	820	887	954	1021	1089	1156	1223	1290	1357
.79/290	180	395	458	522	585	648	712	775	838	901	965	1028	1091	1155	1218	1281
	170	401	465	529	593	657	721	786	850	914	978	1042	1106	1171	1235	1299
	160	404	468	533	598	663	728	793	858	922	987	1052	1117	1182	1247	1312
	150	408	473	538	604	669	735	800	866	931	997	1062	1128	1193	1259	1324
	140	411	477	543	610	676	742	808	874	940	1006	1072	1138	1204	1270	1336
	130	416	482	549	616	682	749	815	882	949	1015	1082	1149	1215	1282	1349
	120	417	484	552	619	686	753	820	887	954	1021	1089	1156	1223	1290	1357
	110	420	487	555	622	690	757	825	893	960	1028	1095	1163	1230	1298	1365
	100	421	489	556	624	692	760	828	895	963	1031	1099	1167	1234	1302	1370
	90	422	490	558	626	694	762	830	898	966	1034	1102	1170	1238	1306	1374
.79/310	180	410	475	540	606	671	737	802	867	933	998	1063	1129	1194	1260	1325
	170	414	480	546	612	678	744	810	876	943	1009	1075	1141	1207	1273	1339
	160	417	483	550	617	684	751	818	885	951	1018	1085	1152	1219	1286	1352
	150	420	488	555	623	690	758	825	893	960	1028	1095	1163	1230	1298	1366
	140	424	492	560	629	697	765	833	901	969	1038	1106	1174	1242	1310	1379
	130	427	495	564	633	702	770	839	908	976	1045	1114	1183	1251	1320	1389
	120	428	497	566	635	704	773	842	911	980	1049	1118	1187	1256	1325	1394
	110	429	499	568	637	706	776	845	914	983	1053	1122	1191	1260	1330	1399
	100	429	499	568	638	707	777	846	916	985	1054	1124	1193	1263	1332	1402
	90	429	499	568	638	707	777	846	916	985	1055	1124	1194	1264	1333	1403
.79/330	180	421	488	554	621	688	755	821	888	955	1022	1089	1155	1222	1289	1356
	170	424	491	559	626	694	761	829	896	963	1031	1098	1166	1233	1301	1368
	160	426	494	562	630	698	766	835	903	971	1039	1107	1175	1243	1311	1379
	150	429	497	566	634	703	771	839	908	976	1045	1114	1182	1250	1319	1387
	140	431	500	569	638	707	776	844	913	982	1051	1120	1189	1258	1327	1395
	130	433	502	571	641	710	779	848	917	987	1056	1125	1194	1263	1333	1402
	120	433	503	572	642	711	781	850	920	989	1059	1128	1198	1267	1337	1406
	110	435	504	574	644	713	783	853	922	992	1062	1131	1201	1270	1340	1410
	100	434	504	574	643	713	783	853	922	992	1062	1131	1201	1271	1341	1410
	90	433	503	573	642	712	782	852	921	991	1061	1130	1200	1269	1339	1408
LRC	180	395	458	521	583	646	708	771	833	895	957	1019	1081	1142	1204	1265
	170	393	456	519	581	644	706	768	830	892	954	1016	1077	1139	1200	1261
	160	390	452	514	576	638	700	762	824	885	946	1008	1069	1130	1191	1251
	150	387	449	511	573	634	696	757	818	879	940	1001	1061	1122	1182	1243
	140	386	447	509	570	631	692	752	813	874	934	994	1055	1115	1175	1234
	130	385	446	507	568	628	689	749	810	870	930	990	1050	1110	1169	1229
	120	382	442	503	564	625	685	745	805	865	925	985	1044	1104	1163	1223
	110	379	440	500	560	621	681	740	800	860	919	979	1038	1097	1156	1215
	100	373	433	493	553	613	672	732	791	850	909	968	1026	1085	1143	1202
	90	368	427	486	545	604	663	722	780	838	897	955	1013	1071	1129	1186

附图 3-1　改航半径图表

BOEING
Flight Planning and Performance Manual

ALL ENGINES

Long Range Cruise Critical Fuel Reserves

Based on:
Emergency descent to 10000 ft
Level cruise at 10000 ft
250 KIAS descent to 1500 ft
15 minute hold at 1500 ft
One missed approach, approach and land
5% allowance for wind errors
Includes APU fuel burn

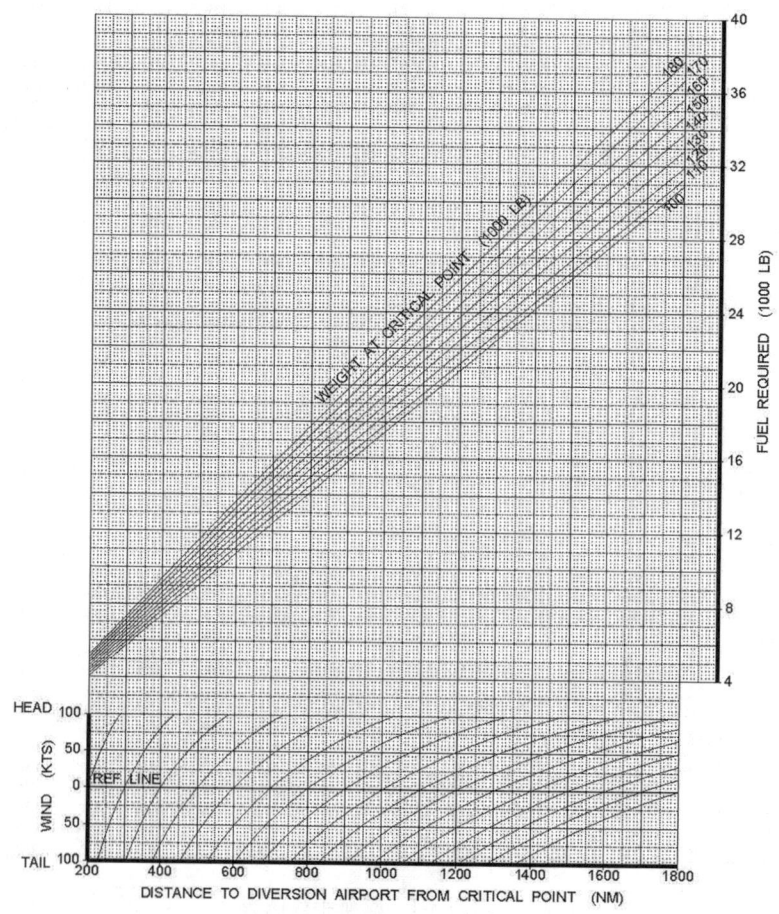

Increase fuel required by 0.5% for each 10 °C hotter than ISA conditions.
If icing conditions exist, increase fuel required by 15% to account for engine and wing anti-ice on (7%) and ice accumulation on unheated surfaces (8%).
Allowance for performance deterioration not included.
Compare the fuel required for all engine and engine inoperative critical fuel reserves and use the higher of the two.

附图 3-2　全发座舱失压临界油量

　　预报有结冰时增加由图表查出的临界油量的 15%，其中 7% 是因防冰消耗的油量，另外 8% 是因结冰阻力大消耗的油量。

BOEING

Flight Planning and Performance Manual

ENGINE INOP
MAX CONTINUOUS THRUST

Long Range Cruise Critical Fuel Reserves

Based on:
Emergency descent to 10000 ft
Level cruise at 10000 ft
250 KIAS descent to 1500 ft
15 minute hold at 1500 ft
One missed approach, approach and land
5% allowance for wind errors
Includes APU fuel burn

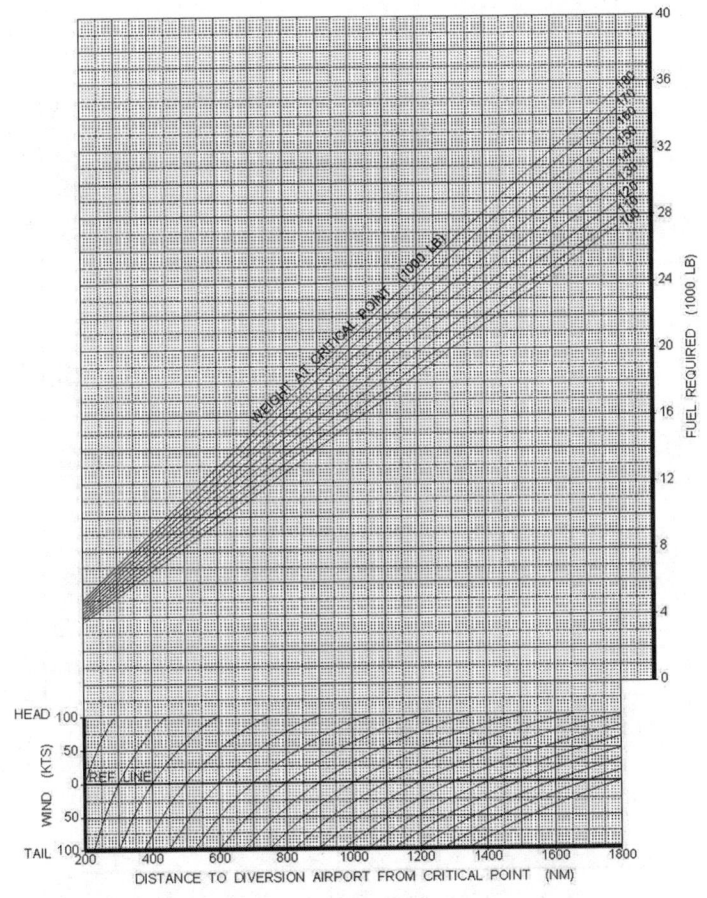

Increase fuel required by 0.5% for each 10 ℃ hotter than ISA conditions.
If icing conditions exist, increase fuel required by 15% to account for engine and wing anti-ice on (6%) and ice accumulation on unheated surfaces (9%).
Allowance for performance deterioration not included.
Compare the fuel required for all engine and engine inoperative critical fuel reserves and use the higher of the two.

附图 3-3　一发失效及座舱失压临界油量（LRC 巡航）

　　预报有结冰时增加由图表查出的临界油量的 15%，其中 6% 是使用机翼和发动机防冰消耗的油量，另外 9% 是由于未加温表面有结冰多消耗的油量。

✈ BOEING

Flight Planning and Performance Manual

ENGINE INOP
MAX CONTINUOUS THRUST

280 KIAS Cruise Critical Fuel Reserves
Based on:
Emergency descent to 10000 ft
Level cruise at 10000 ft
250 KIAS descent to 1500 ft
15 minute hold at 1500 ft
One missed approach, approach and land
5% allowance for wind errors
Includes APU fuel burn

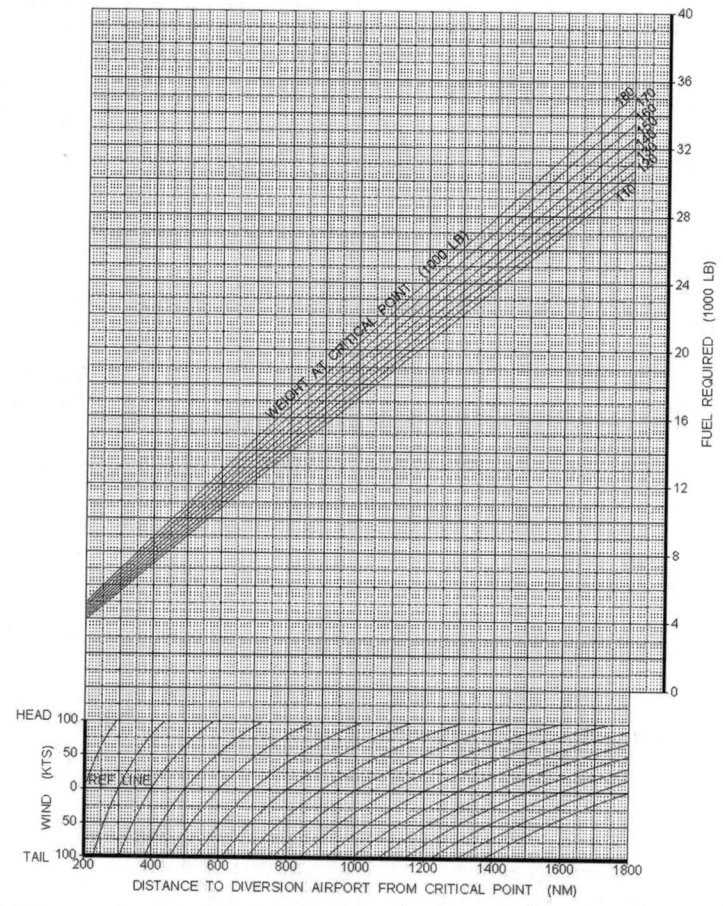

Increase fuel required by 0.5% for each 10 ℃ hotter than ISA conditions.
If icing conditions exist, increase fuel required by 16% to account for engine and wing anti-ice on (6%) and ice accumulation on unheated surfaces (10%).
Allowance for performance deterioration not included.
Compare the fuel required for all engine and engine inoperative critical fuel reserves and use the higher of the two.

附图 3-4　一发失效及座舱失压临界油量（280KIAS 巡航）

737-800/CFM56-7B26
FAA
Category C Brakes

BOEING
Flight Planning and Performance Manual

ENGINE INOP
MAX CONTINUOUS THRUST

290 KIAS Cruise Critical Fuel Reserves
Based on:
Emergency descent to 10000 ft
Level cruise at 10000 ft
250 KIAS descent to 1500 ft
15 minute hold at 1500 ft
One missed approach, approach and land
5% allowance for wind errors
Includes APU fuel burn

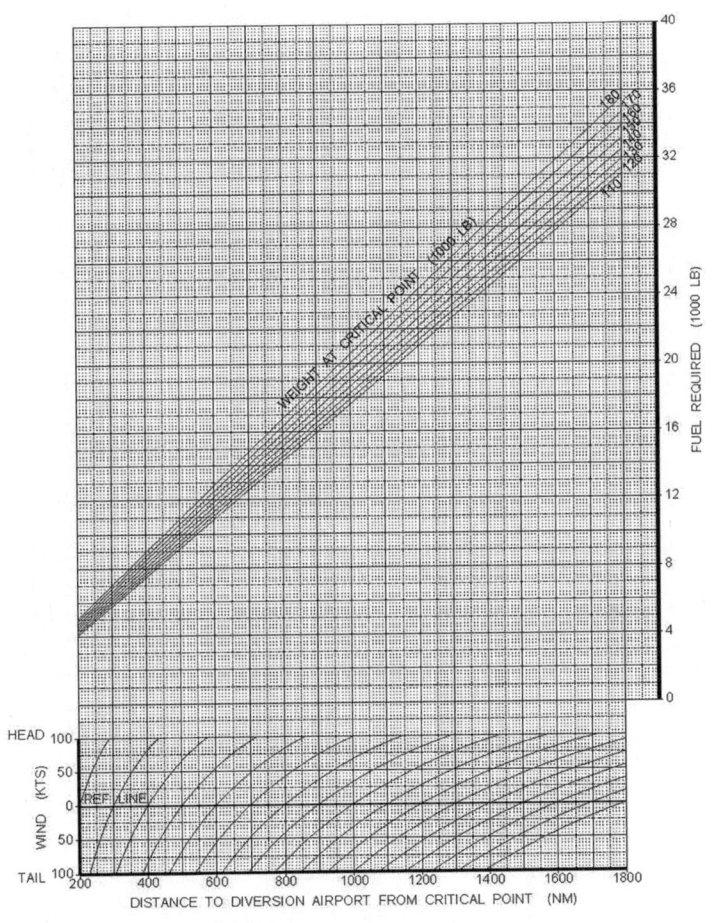

Increase fuel required by 0.5% for each 10 °C hotter than ISA conditions.
If icing conditions exist, increase fuel required by 16% to account for engine and wing anti-ice on (6%) and ice accumulation on unheated surfaces (10%).
Allowance for performance deterioration not included.
Compare the fuel required for all engine and engine inoperative critical fuel reserves and use the higher of the two.

附图 3-5 一发失效及座舱失压临界油量(290KIAS 巡航)

BOEING

Flight Planning and Performance Manual

737-800/CFM56-7B26
FAA
Category C Brakes

ENGINE INOP
MAX CONTINUOUS THRUST

310 KIAS Cruise Critical Fuel Reserves
Based on:
Emergency descent to 10000 ft
Level cruise at 10000 ft
250 KIAS descent to 1500 ft
15 minute hold at 1500 ft
One missed approach, approach and land
5% allowance for wind errors
Includes APU fuel burn

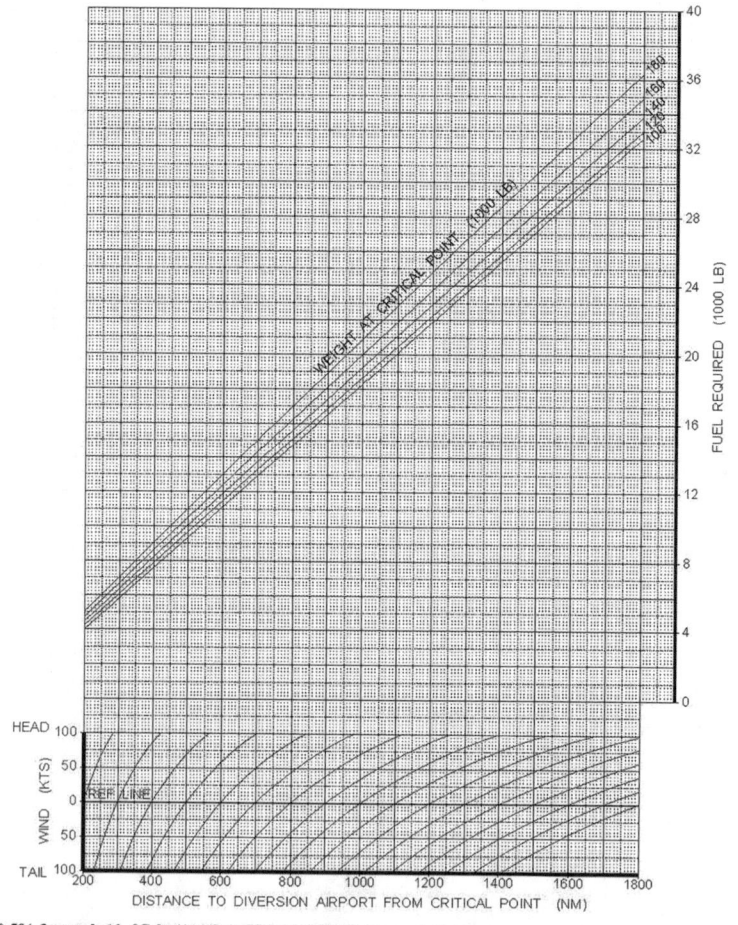

Increase fuel required by 0.5% for each 10 °C hotter than ISA conditions.
If icing conditions exist, increase fuel required by 17% to account for engine and wing anti-ice on (5%) and ice accumulation on unheated surfaces (12%).
Allowance for performance deterioration not included.
Compare the fuel required for all engine and engine inoperative critical fuel reserves and use the higher of the two.

附图 3-6　一发失效及座舱失压临界油量(310KIAS 巡航)

328

737-800/CFM56-7B26
FAA
Category C Brakes

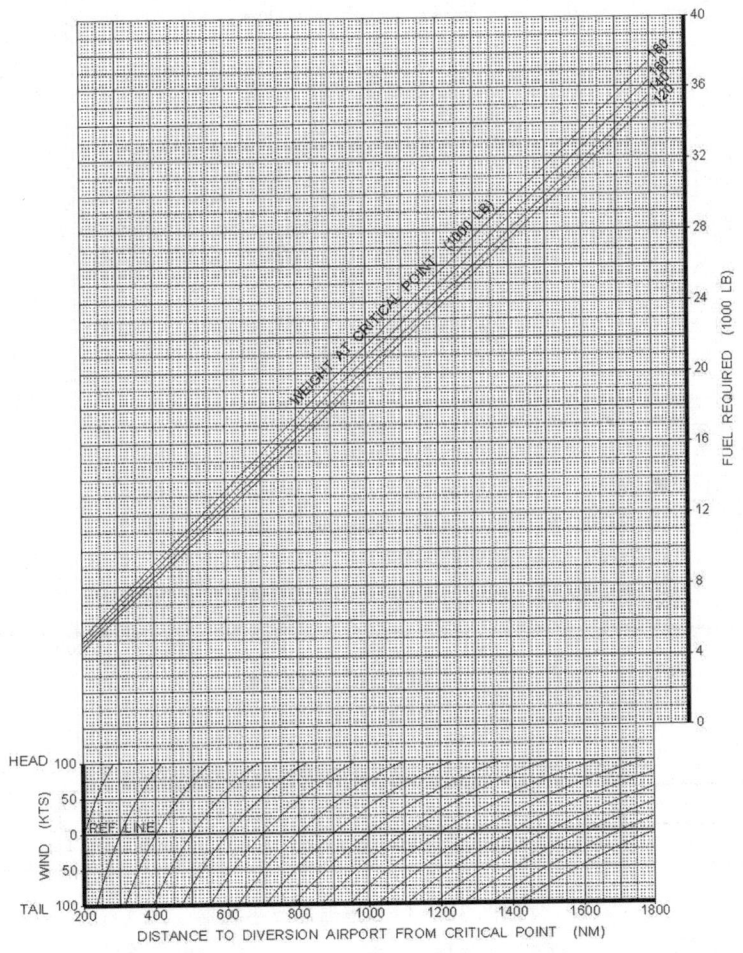

✈ BOEING
Flight Planning and Performance Manual

ENGINE INOP
MAX CONTINUOUS THRUST

330 KIAS Cruise Critical Fuel Reserves
Based on:
Emergency descent to 10000 ft
Level cruise at 10000 ft
250 KIAS descent to 1500 ft
15 minute hold at 1500 ft
One missed approach, approach and land
5% allowance for wind errors
Includes APU fuel burn

Increase fuel required by 0.5% for each 10 °C hotter than ISA conditions.
If icing conditions exist, increase fuel required by 18% to account for engine and wing anti-ice on (5%) and ice accumulation on unheated surfaces (13%).
Allowance for performance deterioration not included.
Compare the fuel required for all engine and engine inoperative critical fuel reserves and use the higher of the two.

附图 3-7 一发失效及座舱失压临界油量(330KIAS 巡航)

BOEING

Flight Planning and Performance Manual

ENGINE INOP
MAX CONTINUOUS THRUST

Long Range Cruise Net Level Off Weight

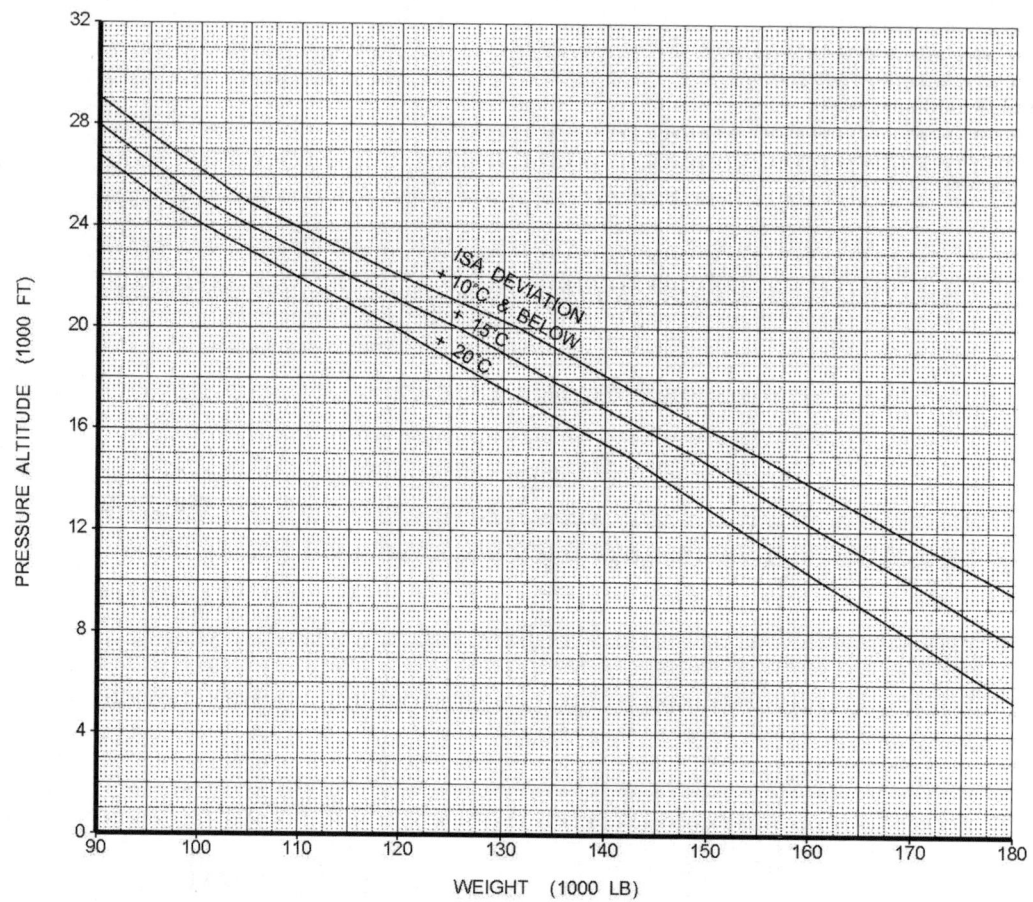

Anti-Ice Adjustments

ANTI-ICE CONFIGURATION	LEVEL OFF WEIGHT ADJUSTMENT (1000 LB)										
	PRESSURE ALTITUDE (1000 FT)										
	8	10	12	14	16	18	20	22	24	26	28
ENGINE ONLY	-5.5	-5.3	-5.1	-4.9	-4.6	-4.2	-3.7	-3.5	-3.3	-2.9	-2.6
ENGINE AND WING	-19.6	-19.2	-18.1	-16.8	-15.9	-15.2	-14.6	-13.2	-11.9	-11.2	-10.6

附图 3-8　LRC 巡航净改平重量

注意:图中的重量是在改平时的重量。

B737-800/CFM56-7B26 的飘降巡航航程能力图表如下所示:

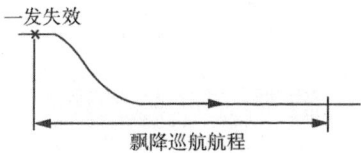

737-800/CFM56-7B26
FAA
Category C Brakes

Flight Planning and Performance Manual

FLIGHT PLANNING
ETOPS

ENGINE INOP
MAX CONTINUOUS THRUST

Long Range Cruise Driftdown/Cruise Range Capability
Includes APU fuel burn

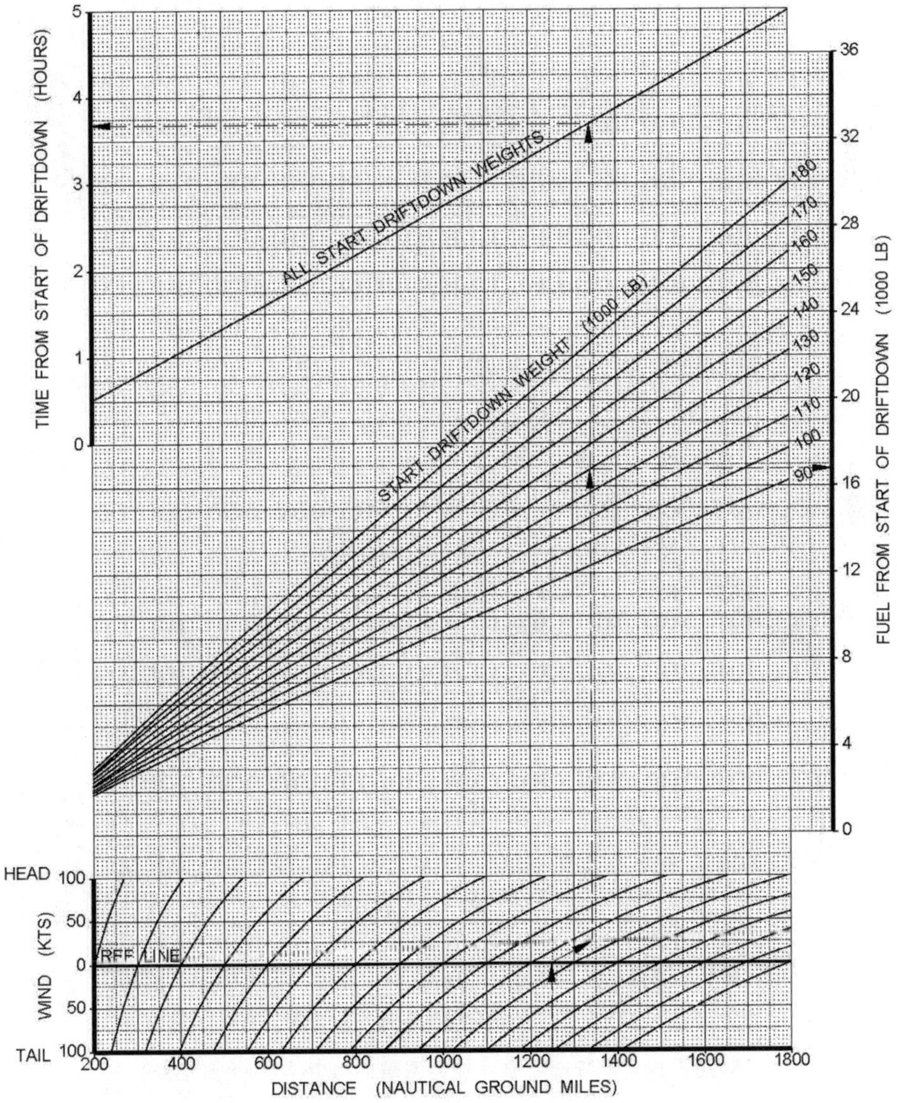

附图 3-9　飘降巡航航程能力图表(LRC 巡航)

飞行计划

FLIGHT PLANNING
ETOPS

737-800/CFM56-7B26
FAA
Category C Brakes

BOEING

Flight Planning and Performance Manual

ENGINE INOP
MAX CONTINUOUS THRUST

.76M/280 KIAS Driftdown/Cruise Range Capability
Includes APU fuel burn

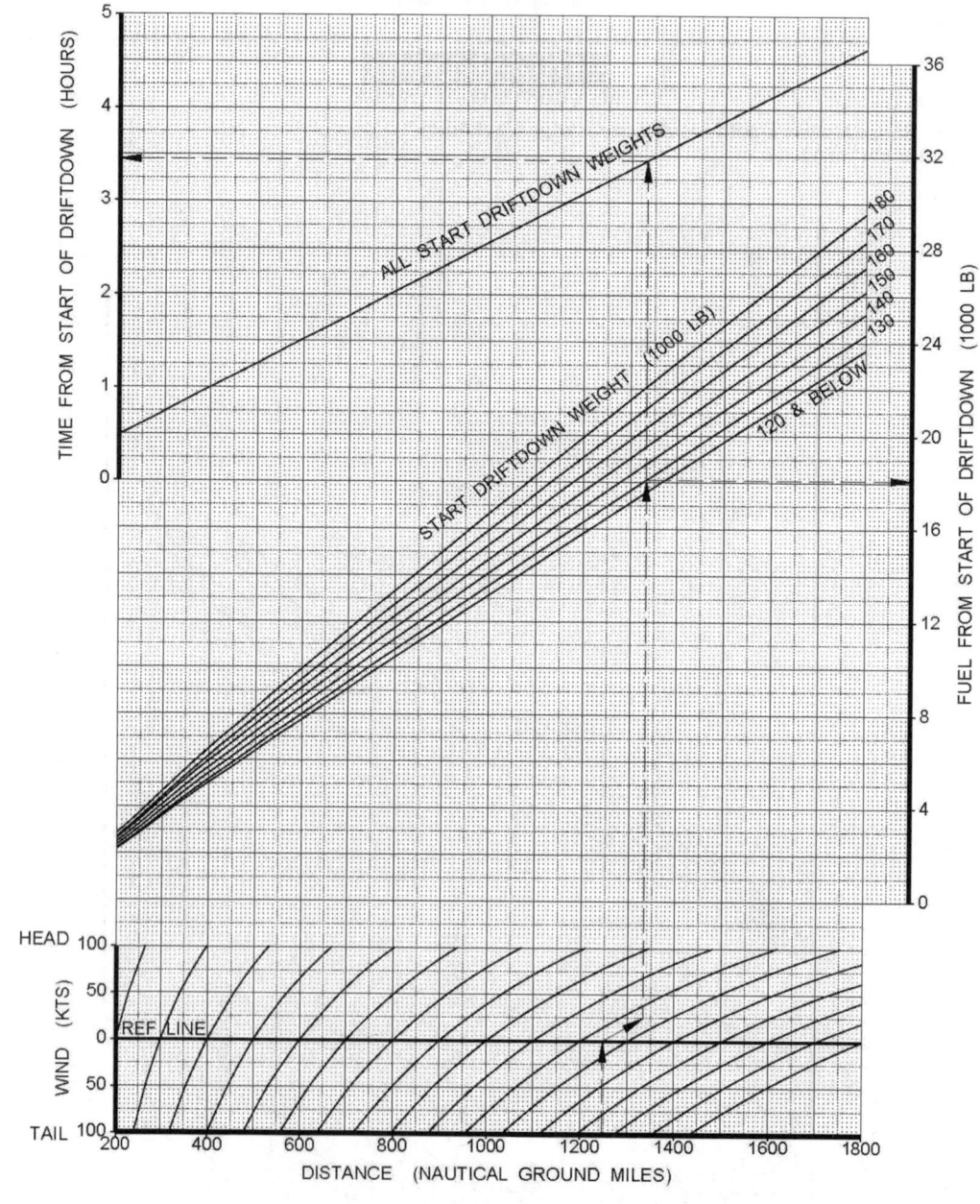

附图 3-10　飘降巡航航程能力图表（*M*0.76/280KIAS 巡航）

737-800/CFM56-7B26
FAA
Category C Brakes

BOEING
Flight Planning and Performance Manual

FLIGHT PLANNING
ETOPS

ENGINE INOP
MAX CONTINUOUS THRUST

.79M/330 KIAS Driftdown/Cruise Range Capability
Includes APU fuel burn

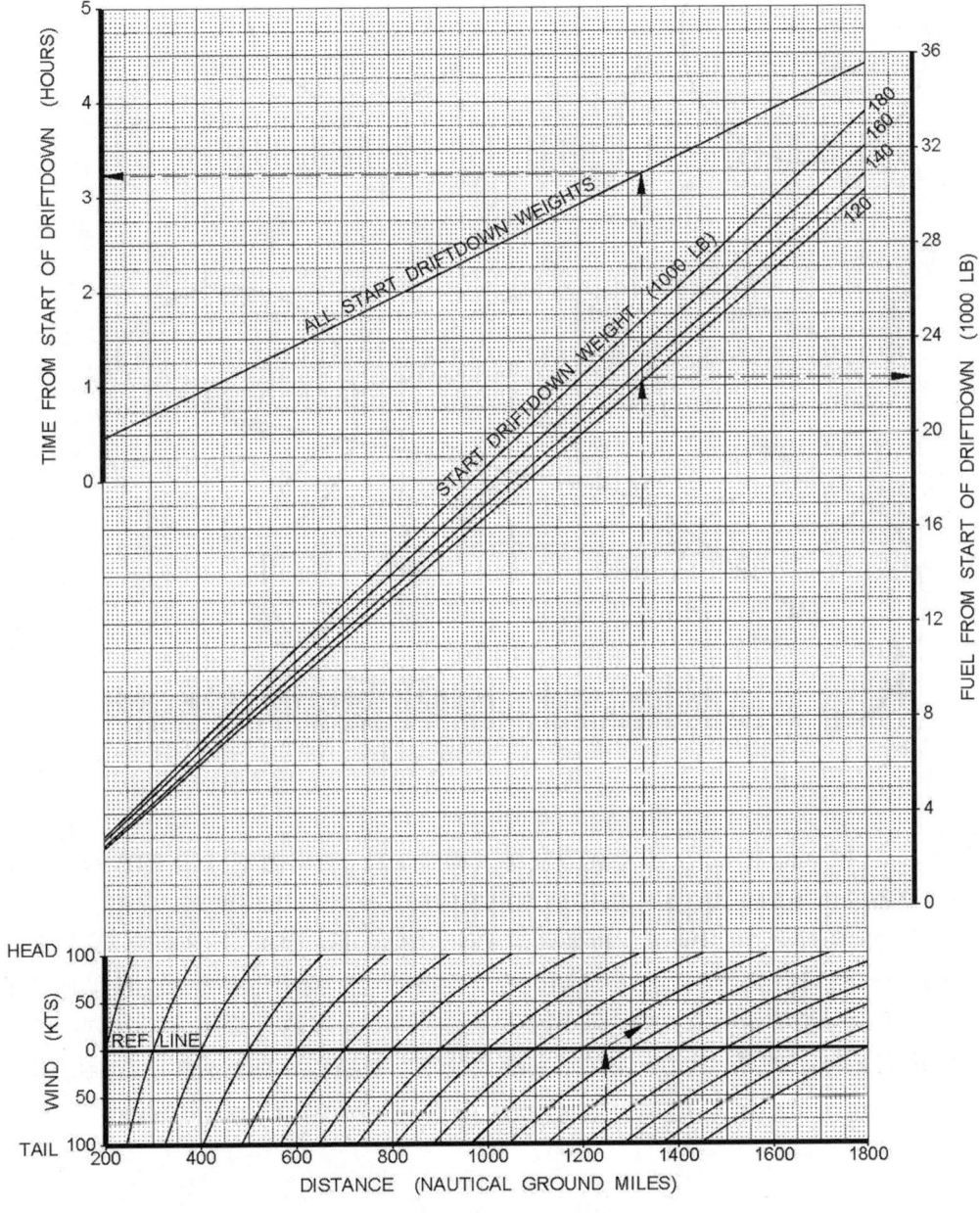

附图 3-11　飘降巡航航程能力图表（*M*0.79/330KIAS 巡航）

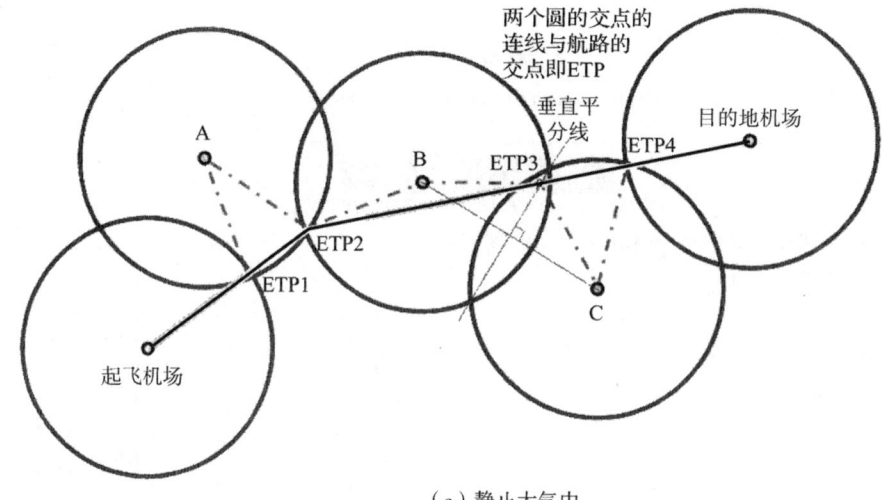

（a）静止大气中

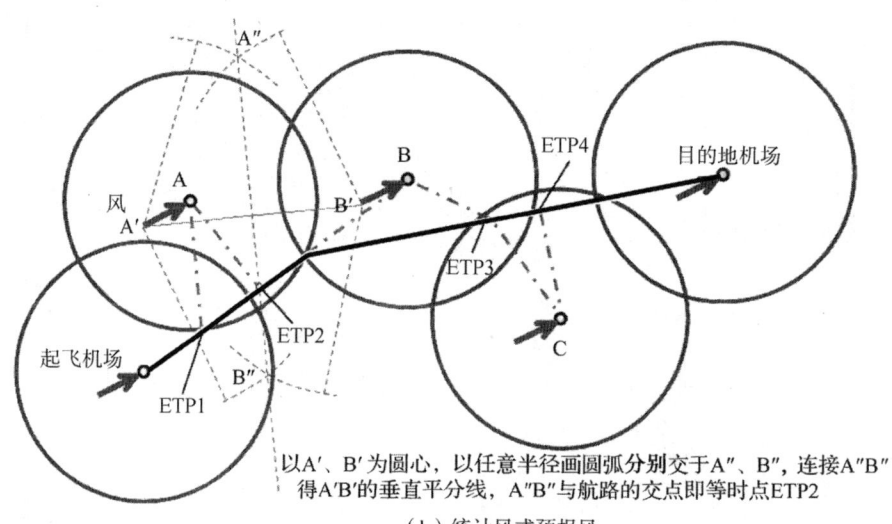

以A′、B′为圆心，以任意半径画圆弧分别交于A″、B″，连接A″B″
得A′B′的垂直平分线，A″B″与航路的交点即等时点ETP2

（b）统计风或预报风

附图 3-12 作图法确定 ETP 的方法

在 Oceanic Plotting Chart 上按上述方法画线确定 ETP,然后由图读出 ETP 的经纬度并量出由 ETP 到备降机场的距离(也可由 ETP 和机场的经纬度计算出距离)。

在静止大气条件下,可以通过分析来确定 ETP。然而,静止大气下的 ETP 只是理论层面,因为在 ETOPS 飞行计划过程中,ETP 和相关改航距离的实际使用需要考虑沿改航航迹的预测风分量。由于没有简单的闭合式方程可用于执行此计算,大多数计算机飞行计划的制作通过迭代方法确定 ETP 上的风,用此方法求解到两个适宜的航路备降机场的相同改航时间问题。

附录4　关于不可预期燃油的政策优化规定

随着科技的发展,飞机可靠性增加,气象预报准确性和及时性提高,机场、空管的保障能力增强,航空公司运行控制能力不断提升,在保证安全的前提下,合理减少飞机携带的燃油,特别是不可预期燃油,有利于促进低碳绿色飞行,降低燃油消耗,增加业载,提高运行效率,从而提升航空公司竞争力。

不可预期燃油主要是为补偿不可预见因素导致的额外燃油消耗。不可预见因素包含可能对飞往目的地机场的燃油消耗产生影响的情况,比如飞机偏离预定燃油消耗数据、偏离预报的气象条件、偏离计划航路和/或巡航高度层等,但不包括飞行计划阶段已预知且已考虑过的影响正常计划航路和高度剖面的因素,比如最低设备清单(MEL)/构型偏离清单(CDL)、跑道关闭、台风、火山灰、空域限制等。

不可预期燃油最低标准是在制订运行飞行计划和签派放行阶段,为补偿随机出现的不可预见因素所需的燃油量占计划航程油量百分比的最小值,但在任何情况下不得低于以等待速度在目的地机场上空450 m(1500 ft)高度上在标准条件下飞行15 min所需的燃油量。

121.663条(c)款中提出,"尽管有本规则第657条和第659条的规定,若安全风险评估结果表明合格证持有人能够保持同等的安全水平,局方仍可以颁发运行规范批准合格证持有人使用不同的燃油政策"。2019年10月22日下发的《航空承运人不可预期燃油政策优化与实施指南》(AC-121-FS-136)为航空承运人实施燃油政策优化提供了政策支持。

6.实施不可预期燃油最低标准5%的要求

6.1　飞机性能监控

6.1.1　航空承运人应建立飞机性能监控能力,针对每一特定机身/发动机组合对飞机性能的理论值与实际数据的差异进行持续监控、分析和比较。

6.1.2　航空承运人应及时修正飞行管理计算机中的性能修正系数,以确保飞行管理计算机使用的性能修正系数与运行飞行计划的数据保持一致。

6.1.3　航空承运人实施了对飞机燃油消耗有影响的维护(特别是飞机换发或飞机构型发生变化)时,应及时修正运行飞行计划系统中燃油消耗的相关数据。

6.2　签派放行

6.2.1　航空承运人的运行飞行计划应基于准确的气象数据,在整条飞行计划航路上应当使用精确度等同或高于1.25度(全球1.25纬度乘1.25经度网格)网格化模型的高空风信息。

6.2.2　航空承运人应制定政策和程序,确保机长和飞行签派员在确定起飞油量时,能够满足不可预期燃油最低标准。起飞总油量应当满足121.659条所要求释压、飘降油量、延程运行临界燃油的适用条款。

6.2.3　对于每一个实施优化的不可预期燃油政策(如5%)的运行,航空承运人应当在签派单或飞行放行文件中进行标注。

6.3　飞行中监控

6.3.1　航空承运人应按照《航空承运人运行监控实施指南》(AC-121-FS-2019-133)咨询通告的要求,建立运行监控系统和程序,用于监控飞机的位置、剖面、异常下降、返航备降等,以及实现飞机的实际燃油与飞行计划中的计划燃油比较的功能。系统在达到所设定告警条件时应立即告警,航空承运人应采取相应的措施。

6.3.2 航空承运人应当建立相应的应急程序,以应对航路上发生驾驶舱燃油指示表故障、燃油消耗异常、地空通信失效等情况,应急程序应包括但不限于:机组报告程序、应急处置预案等。

7.实施不可预期燃油最低标准3%的附加要求

航空承运人在实施不可预期燃油最低标准3%时,除了满足"实施不可预期燃油最低标准5%的要求"外,还应当制定政策和程序满足本节附加要求。

7.1 航空承运人在实施不可预期燃油最低标准3%运行时,应当基于每一个飞机制造商和型号/城市对/目的地机场到达时间窗口组合,按照本通告第8节中的要求,确定运行飞行计划的可用 $PBCF$ 值(基于性能的不可预期燃油)。

7.2 如果 $PBCF$ 值大于3%,航空承运人不得使用3%的不可预期燃油政策。

7.3 在实施不可预期燃油最低标准3%的运行的签派放行时,所有驾驶舱燃油量指示必须正常工作。

7.4 航空承运人在实施不可预期燃油最低标准3%运行的签派放行时,应当指定一个3%航路备降机场,并在签派单或飞行放行单上注明。3%航路备降机场选择参见本通告附录一。在签派放行后及运行实施中,航空承运人应加强对3%航路备降机场天气的监控。

7.5 航空承运人应制定政策和程序,确保机长和飞行签派员在放行时能够确定起飞油量满足3%航路备降机场要求。

7.6 实施不可预期燃油最低标准3%的运行,不得与二次放行同时使用。

7.7 航空承运人应当建立程序,在起飞延误可能导致飞机预达时间超出初始计划的到达时间窗口时,对造成延误的原因进行分析,并对后续运行的天气情况、空域拥挤程度等影响因素进行评估,决定是否能够继续使用该航班预定的 $PBCF$ 值。

7.8 对于用于运行飞行计划的每一个 $PBCF$ 值,航空承运人应当至少每季度进行一次重新计算。生成/复制 $PBCF$ 值的所有数据必须留存供局方检查。

7.9 当航空承运人所用飞机制造商和型号的任何在机体结构、发动机状况、构型上的变更可能影响燃油消耗时,都应当对 $PBCF$ 值进行重新评估。

附录 5　缩略语

注:不同的公司以及不同的手册、图表对同一事物可能使用不同的缩略语,为便于查询,把在飞行计划和配载平衡图表中遇到的缩略语一并列出。

06Z／12Z	世界(协调)时 06 点、12 点
A	乘务员座位
ACN	飞机分类号、飞机等级号
A/C REG	飞机注册号
A/C	空调系统,空调引气
ADDRESS	收电人地址
ADLT、ADT、AD	成年人
ADP	冲压空气驱动的泵
AFR	实际剩余油量
A/F	成年女性旅客
A/I	防冰,防冰引气
ALLOWED TRAFFIC LOAD	允许业载
ALTN ×××	改航×××(备降×××)
APS	已准备好、可以使用的飞机
APU	辅助动力装置
A/P MMO	飞机性能限制的最大允许使用马赫数
A/P VMO	飞机性能限制的最大允许使用表速
A/P	飞机性能
ARVL FUEL	到目的地机场时剩余油量
A4A	美国航空运输协会
ATA	实际到达时间
ATD	实际离场时间
AVG	平均
AVTAS	平均真空速
AW	航路;航路代号
AWY	航路;航路代号
B	行李;公务舱
BA	平衡力臂
BAG	行李
BASIC APS WT	基本使用重量
BASIC OPTG WT	基本使用重量
BOW	基本使用重量
BRW	松刹车重量
BRWT	松刹车重量
BTU	英国热量单位
BULK	散货舱

BURN	消耗的油量
Clb	爬升
C	货物
CA	中国民航航班号的头两个字母
CAB BAG	装在客舱中的行李
CAPT	机长
CARGO HOLD	货舱
CAS	修正空速
CCAR	中国民用航空条例
CG	重心
CH	小孩(单独占座位)
CHD	小孩(单独占座位)
CI	成本指数
CL/CPT	客舱等级/货舱位置
CMP	构型、维修、程序(文件),由飞机生产厂家制定
COF	公司备份油量
COMP	计算
CONF	布局,指头等舱和普通舱的座数,如写:F18/Y21
CONT	航线应急油量
CONTAINER	集装箱
CORR	修正值
CP	临界点、关键点
CPM	集装箱、货板分布信息
CREW	机组/乘务组人数,如写:3/7
CRIT	临界的,关键的
CRZ	巡航
DCT	(由某点)直飞(某点)
DEG	恶化、衰退
DES	下降
DESFL	下降油量
DEST MNVR	目的地机场机动(进近)
DEST	目的地机场
DIR	(由某点)直飞(某点)
DIST	累计距离
DLI	Dead Load Index,干使用指数与各货舱中货物、行李、邮件指数之和(即不能移动的载荷的指数)
DOC	直接使用成本
DST	航段距离
DTGO	剩余距离,待飞距离
E	经济舱,即普通舱
ECCF	经济巡航成本函数
EEP	延程段的进入点
EFR	预计剩余油量
ELDW	预计着陆重量

ELE	机场标高
ENG	发动机
EOA	地址结束
EOM	报文(信息)结束
EPLD	预计业载
EPR	发动机压力比
ETA	预计到达时间
ETD	预计离场时间
ETE	预计航路(飞行)时间
ETO	预计飞越时间
ETOPS	双发延程飞行
ETOW	预计起飞重量
ETP	等时点;预计穿越时间
EZFW	预计无油重量
F/C	头等舱
F	头等舱
FAA	美国联邦航空局
FAR	美国联邦航空条例
FF	一发燃油流量
FL NO	航班号
FLT	航班号(FLIGHT)
FM	燃油里程
FMC MMO	飞行管理计算机控制的最大允许使用马赫数
FMC VMO	飞行管理计算机控制的最大允许使用表速
FMC	飞行管理计算机
FMS	飞行管理系统
FOA	飞机上的油量
FOC	飞行运行控制中心
FOD	到目的地机场时剩余油量
FREQY	频率
FT COMPT(AFT HOLD)	后舱
FUEL REM	剩余油量
FWD COMPT	前舱
FWD HOLD	前舱
FWD LINE	代表重心前限的线
G/S	地速
G	厨房
GND	地面距离,航程
GS	地速
HARM	部件所在位置到 0 站位的距离、水平力臂
HMG	液压驱动发电机
I	不占座位的婴儿
IAS	表速
IATA	国际航空运输协会

ICAO	国际民用航空组织
INF	不占座位的婴儿
ISA DEV	与 ISA 的温度偏差
ISA	国际标准大气
KIAS	以节给出的表速
L	厕所
LAT	纬度
LDGWT	着陆重量
LDM	装载信息
LGW	着陆重量
LHV	燃油低热值
LMC	最后一分钟变化
LONG	经度
LRC	远航速度
LWA	目的地机场的着陆重量
LWD	备降机场的着陆重量
LWT	着陆重量
M12	负 12
M	邮件；男性旅客
MAC	马赫数；平均空气动力弦
MAGTRK	磁航迹
MBRW	最大松刹车重量
MC	磁航迹，即地速与磁北夹角
MID COMPT	中舱
MLWA	备降机场最大允许着陆重量
MLWD	目的地机场最大允许着陆重量
MMO	最大允许使用马赫数
MNO	马赫数
MORA	最低偏航高度（单位 100 ft）
MPL	最大允许业载
MTK	磁航迹
MTOW	最大允许起飞重量
MZFW	最大无油重量
NAM	空中海里
NGM	地面海里
NM	海里
OAT	巡航高度上的实际气温
ORIGINATOR	发电人地址
OWE	使用空机重、营运空重
P/L	业载
P.A.	气压高度
P12	正 12
PAD	需要减载时，可优先减下去的旅客，如本公司职工、持免票或优惠票的旅客
PALLET	货板，货盘

PANTRY	与正常配餐标准相比,增加或减少的配餐重量
PASS	旅客
PAX	旅客
PC	百分号,百分之……
PCN	跑道分类号、等级号
PL	业载、商载
PLN FUEL	计划油量
POSN	位置,航路点(名称)
PSGR	旅客
RECHARGE	指向谁收取加油费,该项可填客户的代号
RECLR PT	二次放行点
RECLR	再次放行,二次放行
REFILE	二次放行
REGN	注册号
REM	剩余
REQD	所需的
REV	修改,修正,修订
RF	二次放行
RSRV	航线应急油量
S	贮藏室
SEATING CONDITION	就座状态,不满员时才填,如:F14/Y80
SI	补充信息、补充说明
SMC	几何平均弦
SOC	系统运行控制中心
SR	每上升 1000 ft 风速增量(节)
STAB SET	安定面(配平)调定值
STD DAY	标准日即尚可再增加的业载
STD	标准的
T/C	旅游舱(经济舱)
T/C	真航迹,地速与真北夹角
T/H	真航向,真空速与真北夹角
T	总计
TAIL STAND	尾撑杆,尾支架
TAS	真空速
TAT	总温
TAXW	滑行重量,即停机坪重量
TDV	与 ISA 的温度偏差
TIME	SITA 飞行计划中的累计时间
TMP	温度,与 ISA 的温度偏差
TOC	爬升顶点
TOCG	起飞重心
TOD	下降顶点
TOF	起飞油量
TOGW	起飞重量

TOW	起飞重量
TR	过站(指机上在经停站中转的客人数,货物、邮件件数、重量)
TRK	真航迹(角)
TRP	对流层顶
TTME	累计时间
UD	上层舱(UP DECK)
ULD	Unit Load Device,集装箱(Container)或货盘(Pallet),在载重平衡图中,ULD 表示要填写它们的重量,如它们已计入使用空机重,则无此项
UNDERLOAD	空余吨位
VERSION	客舱布局,同 CONF
VIP	要客
VMO	最大允许使用表速
W/C	风分量
WCP	风分量
WT	重量
WTOC	飞机在爬升顶点重量
WTOD	飞机在下降顶点重量
Y	普通舱
ZFCG	无油重心
ZFW	无油重量
ZFWT	无油重量
ZND	航段距离
ZNT	航段时间
ZTM	航段时间

参考文献

[1] 中国民用航空局.大型飞机公共航空运输承运人运行合格审定规则(CCAR-121-R8)[Z].2024.

[2] 中国民用航空局.航空承运人不可预期燃油政策优化与实施指南(AC-121-FS-136)[Z].2019.

[3] 中国民用航空局.延程运行和极地运行(AC-121-FS-2019-009R2)[Z].2019.

[4] 中国民用航空局.航空器重量与平衡控制规定(AC-121-FS-135)[Z].2019.

[5] 中国民用航空局.一般运行和飞行规则(CCAR-91-R4)[Z].2022.

[6] 中国民用航空局.高原机场运行(AC-121-FS-2015-21R1)[Z].2015.

[7] ICAO. Extended Diversion Time Operations(EDTO)Manual(DOC.10085)[Z].2017.

[8] ICAO. Flight Planning and Fuel Management Manual(Doc.9976)[Z].2015.

[9] 中国民用航空局飞行标准司.飞机航线运营应进行的飞机性能分析[Z].2001.

[10] 傅职忠,谢春生,王玉.飞行计划[M].北京:中国民航出版社,2012.

[11] 陈治怀,谷润平,刘俊杰.飞机性能工程[M].北京:兵器工业出版社,2006.

[12] Boeing. B737-800 Flight Planning and Performance Manual[Z].2008.

[13] Boeing. B737-800 Operations Manual[Z].2008.

[14] Boeing. B737-800 Mast Minimum Equipment List[Z].2008.

[15] Boeing. B737-800 Load Sheet & Balance Chart[Z].2008.

[16] Airbus. A320 Weight & Balance Manual[Z].2004.

[17] Airbus. Getting To Grips with Aircraft Performance[Z].2002.

[18] Airbus. Getting To Grips with Weight and Balance[Z].2002.

[19] Airbus. Getting To Grips with Fuel Economy[Z].2004.

[20] Airbus EOSID Workshop. Flight Operation Support and Line Assistance[Z].2003.

[21] Airbus Caroline STEMART. High Altitude Operations[Z].2003.

[22] SITA.SITA Flight Planning Service Dispatchers Handbook[Z].1990.